AF760113

Progress in Slow Sand and Alternative Biofiltration Processes

Progress in Slow Sand and Alternative Biofiltration Processes

Further Developments and Applications

Edited By

Nobutada Nakamoto, Nigel Graham, M. Robin Collins and Rolf Gimbel

IWA PUBLISHING

Published by **IWA Publishing**
Alliance House
12 Caxton Street
London SW1H 0QS, UK
Telephone: +44 (0)20 7654 5500
Fax: +44 (0)20 7654 5555
Email: publications@iwap.co.uk
Web: www.iwapublishing.com

First published 2014
© 2014 IWA Publishing

Apart from any fair dealing for the purposes of research or private study, or criticism or review, as permitted under the UK Copyright, Designs and Patents Act (1998), no part of this publication may be reproduced, stored or transmitted in any form or by any means, without the prior permission in writing of the publisher, or, in the case of photographic reproduction, in accordance with the terms of licenses issued by the Copyright Licensing Agency in the UK, or in accordance with the terms of licenses issued by the appropriate reproduction rights organization outside the UK. Enquiries concerning reproduction outside the terms stated here should be sent to IWA Publishing at the address printed above.

The publisher makes no representation, express or implied, with regard to the accuracy of the information contained in this book and cannot accept any legal responsibility or liability for errors or omissions that may be made.

Disclaimer
The information provided and the opinions given in this publication are not necessarily those of IWA and should not be acted upon without independent consideration and professional advice. IWA and the Author will not accept responsibility for any loss or damage suffered by any person acting or refraining from acting upon any material contained in this publication.

British Library Cataloguing in Publication Data
A CIP catalogue record for this book is available from the British Library

ISBN 9781780406374 (Paperback)
ISBN 9781780406381 (eBook)

Contents

Preface

This book is the latest in an international series, beginning in 1988, with updated additions in 1994, 1996 and 2006, that is devoted to the general topic of biological filtration processes. Initially the focus was exclusively on slow sand filtration but the scope has broadened in the intervening period to include gravel roughing filters, biological activated carbon beds, riverbank filtration, soil-aquifer treatment and household biosand filters. Such processes are applied in a wide range of applications and at various scales as an alternative to, or in combination with, conventional treatment processes. While differing in their physical configuration, structure and design, and mode of operation, they are similar in that they depend on natural biological communities and activities that interact with the associated physico-chemical separation processes. For this reason, it is advantageous to incorporate information about recent advances in the various methods of bio-filtration in a single reference book such as this. Thus, it is hoped that awareness and knowledge of developments in one aspect of bio-filtration will stimulate research and developments in other aspects.

It is clear that worldwide interest in biofiltration processes is growing since their dependence on 'natural' treatment phenomena make them inherently more sustainable than conventional, higher-rate technologies. This is particularly the case for small-community water systems and household point-of-use devices. Developments in these applications are summarised in a number of chapters within the book with valuable examples of 'ecological purification systems' in the Asia-Pacific region, and international experiences with household biosand filters.

The book is compiled from contributions presented at the 5th International Slow Sand and Alternative Biological Filtration Conference, held in Nagoya, Japan in June 2014. The timing of the conference was significant since it coincided with the centenary of the use of slow sand filtration in Nagoya, and the continued importance and appropriateness of slow sand filtration is evidenced by the recent modernization of the filters at the Nabeyaueno treatment plant by the City of Nagoya. The main themes of the conference encompassed the following: overviews of biofiltration systems; slow sand filtration process behaviour, treatment performance and process developments; biosand filters; alternative biofiltration process behaviour, treatment performance and process developments; river bank filtration and groundwater recharge. Thus, the book provides a state-of-the-art summary of developments in biofiltration technologies and systems from studies conducted around the world, authored by academicians, practitioners. system users, designers, and manufacturers.

In providing the basis for this book the editors would like to give special recognition and thanks to the organisers, sponsors and supporters of the 5th International Conference, including the IWA, the City of Nagoya Waterworks & Sewerage Bureau (Japan), the Ministry of Health, Labor and Welfare (Government of Japan), the Japan Water Works Association, the Japan Society on Water Environment, the Federation of Japan Water Industries Incorporated, the Japan Water Research Center, the Japan International Cooperative Agency, the Japan Small Scale Water Works Association, the Japan Water Forum, the Life of Water Chubu Forum, the Nagoya Research Institute of WA-Links for Future Development, the Institute of Water and Environment Business (Japan), the Community Water Supply Support Center of Japan, Nagoya Water and Sewage General Service Co Ltd, Hazen and Sawyer (USA), Imperial College London (UK), the Water Treatment Technology Assistance Center at the University of New Hampshire (USA) and the University of Duisburg-Essen (Germany). The editors also extend appreciation to Sombo Yamamura for his invaluable coordination of the organisations in Japan and to the distinguished members of the Conference Programme Advisory Panel (listed separately) for their assistance in the selection of the conference papers. Finally, the editors wish to gratefully acknowledge Tom Bond, Vaso Partinoudi, and especially Takashi Naito for their valuable administrative assistance. Both the conference and this book represent a continuing international collaboration which seeks to make a worthwhile contribution to the development and application of bio-filtration technology in the field of drinking water supply and wastewater management worldwide.

N. Nakamoto
Shinshu University and Community Water Supply Support Center, Japan

N.J.D. Graham
Imperial College London, UK

M.R. Collins
University of New Hampshire, USA

R. Gimbel
University of Duisburg-Essen, Germany
June 2014

Conference programme advisory panel

Dr René van der Aa: Delft University of Technology/Waternet, The Netherlands

Prof. Dr Avner Adin: Hebrew University of Jerusalem, Israel

Dr Michiriro Akiba: National Institute of Public Health, Japan

Prof. Dr Luiz Di Bernardo: University of Sao Paulo, Brazil

Prof. Dr Ed Bouwer: Johns Hopkins University, USA

Dr Luiza Campos: University College London, UK

Dr Michael Chipps: Thames Water Utilities, UK

Prof. Dr Hans-Curt Flemming: University of Duisburg-Essen/IWW, Germany

Dr Kebreab Ghebremichael: University of South Florida, USA

Prof. Dr Johannes Haarhoff: University of Johannesburg, South Africa

Prof. Dr Peter Huck: University of Waterloo, Canada

Dr Yunho Lee: Gwangju Institute of Science and Technology-GIST, Korea

Prof. Dr Fusheng Li: Gifu University, Japan

Prof. Dr Barry Lloyd: University of Surrey, UK

Prof. Dr Jun Ma: Harbin Institute of Technology, China

Dr Hans-Joachim Mälzer: IWW, Germany

Prof. Dr Victor Muhandiki: Nagoya University, Japan

Prof. Dr C Ojha: IIT Roorkee, India

Dr Declan Page: CSIRO, Australia

Prof. Dr Jack Schijven: University of Utrecht, The Netherlands

Prof. Dr Kazunari Sei: Kitasato University

Prof. Dr Scott Summers: University of Colorado, Boulder, USA

Prof. Dr Wolfgang Uhl: TU Dresden, Germany

Mrs Bénédicte Welte: Eau de Paris, France

Part I
General Overview

Chapter 1
Slow sand filtration: recent research and application perspectives

N. J. D. Graham and M. R. Collins***
**Department of Civil & Environmental Engineering, Imperial College London, South Kensington, London SW7 2AZ, UK*
***Department of Civil & Environmental Engineering, University of New Hampshire, Durham, NH, USA*

ABSTRACT
Slow sand filtration (SSF) has been widely used in the field of water treatment for over 100 years and while important features of the process have been studied in detail, many aspects of the process remain poorly understood. In addition, modifications to the process and pre-treatment methods are required to enhance SSF performance to meet more demanding treated water quality objectives and reduced costs. This paper highlights some of these aspects and refers to recent research that has investigated the nature and role of the biomass in the process, the benefits of applying surface fabric layers, pre-ozonation and GAC sub-layers, and the development of a deterministic-type mathematical model to simulate the SSF process. In all cases, the research has provided useful insights but much futher work is required to consolidate the findings and bring benefits to the application of SSF in practice.

Keywords Slow sand filtration; biofiltration; fabric layers; pre-oxidation; GAC amendments; process modelling

INTRODUCTION

The process of slow sand filtration (SSF) has been widely used in the field of water treatment for over 100 years but while important features of the process have been studied in detail, many aspects of the process remain poorly understood. This is partly explained by the greater attention that has given to the more widely applied rapid filtration process, but also because of the inherent complexity of SSF. Much is known of the ecology and fundamental dynamics of the filter process in qualitative terms (eg. common biological species and their interactions), but a comprehensive *quantitative* description of the process remains to be established.

In the 1990's slow sand filtration was the subject of a series of international conferences (Graham, 1988; Collins and Graham, 1994; Graham and Collins, 1996), professional guidance manuals (AWWA, 1991; ASCE, 1991), and literature reviews (eg. Lambert and Graham, 1995), which provided an extensive body of reference material and details of ongoing research studies. Since then, there has continued to be a steady flow of publications in the scientific literature concerned with various aspects of SSF research and plant operation (e.g. Gimbel *et al*., 2006). Broadly, these have been in four areas of SSF application, namely: drinking water treatment where the SSF is the principal treatment step (eg. for small community water supplies after rudimentary pre-treatment); drinking water treatment where the SSF is a secondary process following extensive prior treatment (Rachwal *et al*., 1996); the tertiary treatment of wastewaters (eg. Muhammad and Morris Hooke, 2003), particularly for effluent re-use; horticultural applications, such as the disinfection of recirculating nutrient solutions (Garibaldi *et al*., 2003). Only developments in drinking water applications will be described in this paper.

The key areas of general interest in SSF include pre-treatment, process mechanisms, treatment performance, modelling, and process enhancement. Associated with these are specific limitations or gaps in our current knowledge, such as the following:

i) a comprehensive, quantitative description of the fundamental process mechanisms;
ii) the relationship between the influent water quality and the nature of the SSF schmutzdecke;
iii) the removal of natural and synthetic organic substances;
iv) predicting filter run time;
v) methods of increasing filtration rates and filter run time;
vi) enhanced cleaning technologies.

Aspects of SSF performance that continue to be the subject of study are the nature of the biomass (in the schmutzdecke and filter bed), the mechanisms of treatment, the impact and role of animals, the removal of pathogens and specific substances (e.g. pesticides, pharmaceutical and endocrine disrupting compounds), and the benefits of covering and media amendments. In view of the broad range of the above topics, particular aspects of previous, and potentially future, research will be referred to here. These are: the nature and development of the biomass in SSF; the application of fabric layers to support the schmutzdecke; the effect of ozone as a pre-treatment for SSF; the incorporation of GAC sub-layers for enhanced organics removal; the development of a deterministic process model.

BIOMASS

The high degree of water treatment achieved by SSFs is partly explained by the slow filtration rate (0.1 - 0.3 m/h) and fine effective size of the sand (0.1 - 0.3 mm), but is also attributed to biological processes in the layer of material that accumulates above the sand surface (schmutzdecke) and within the upper layers of the sand bed. The schmutzdecke is also believed to be primarily responsible for the progressive increase of head loss observed during filter operation. Despite its current and historical importance as a water treatment process, the fundamental biological composition and mechanisms affecting water purification and head loss development during SSF runs remain poorly defined (Haig *et al*., 2011). Microbial biomass in the schmutzdecke and filter sand bed has been quantified using a range of microbiological methods (e.g. Duncan, 1988; Yordanov *et al*., 1996), but the different approaches used to measure biomass concentrations and the inconsistent units and sampling intervals adopted have confounded inter-study comparisons of biomass development and behaviour during SSF. Most reports are of single measurements of the net biomass production at the end of a filter run, prior to cleaning, and these emphasise the significant variability apparent in schmutzdecke and sand biomass accumulation in operational slow sand filters. The collection of representative samples of schmutzdecke and sand material during filter operation is difficult in practice, and the lack of a simple routine method for measuring microbial biomass are probable reasons for the limited amount of field-scale investigation of the biological mechanisms of SSF. Detailed analyses of biomass growth in the schmutzdecke and within the sand bed during filter operation would improve understanding of the complex and fundamental interactions between the biological and physico-chemical processes operating in the filter and enable the development of mechanistic models for SSF operational management (e.g. prediction of head loss rate, run time, frequency of sand cleaning and renewal).

One recent study has considered the development of microbial biomass in the sand and schmutzdecke layer in pilot and full scale slow sand filters, operated both with and without light shading (Campos *et al.*, 2002). Through the use of random sampling and specially developed analytical methods, it has been observed that the interstitial microbial biomass in an uncovered sand bed increases with time (Fig. 1.1) and decreases with sampling depth. In contrast, there was only a small accumulation of sand biomass with time in a parallel covered filter, but no relationship was apparent between biomass concentration and depth.

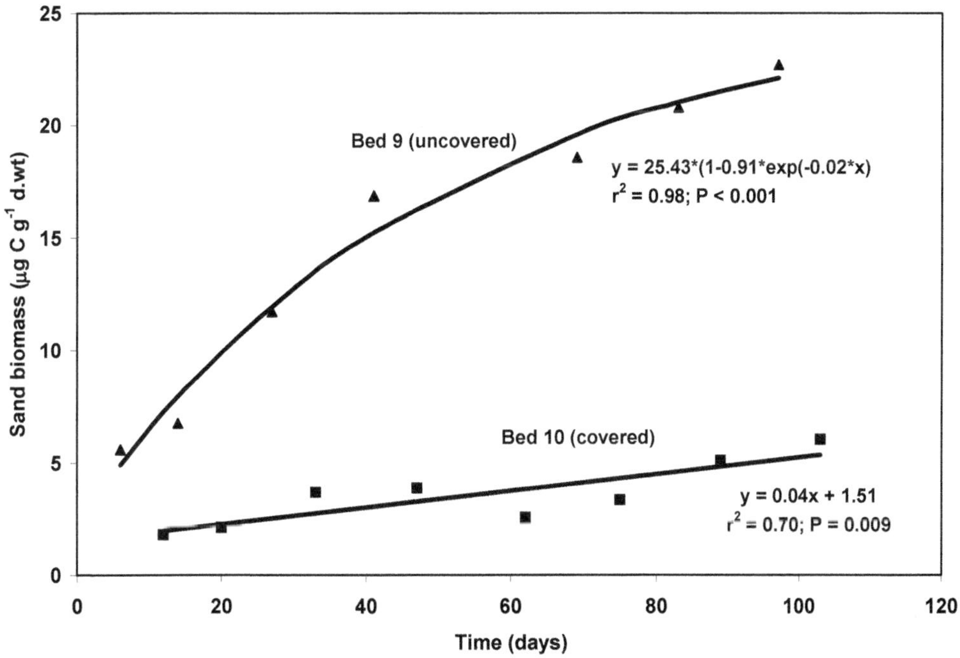

Figure 1.1 Example of biomass growth in the 0-10 cm sand layer of an uncovered and covered SSF with time (Campos *et al.*, 2002).

Biomass accumulation and thickness of the schmutzdecke layer from the uncovered filter bed were highly variable and showed no consistent patterns of spatial or temporal development (Fig. 1.2). The substantial spatial variability of the schmutzdecke was in contrast to the relatively uniform patterns of biomass growth observed in the sand. It was speculated that microbial biomass in the sand of uncovered filters is largely related to carbon inputs from photosynthetic activity in the schmutzdecke and involves mechanisms that spatially distribute carbon substrate from the schmutzdecke to the sand. Since TOC and DOC removals were similar in both covered and uncovered filters (~20%) it is believed that relatively small biomass populations are sufficient to remove residual labile carbon in the influent water. As expected, there was no discernible development of a schmutzdecke layer in the covered filter.

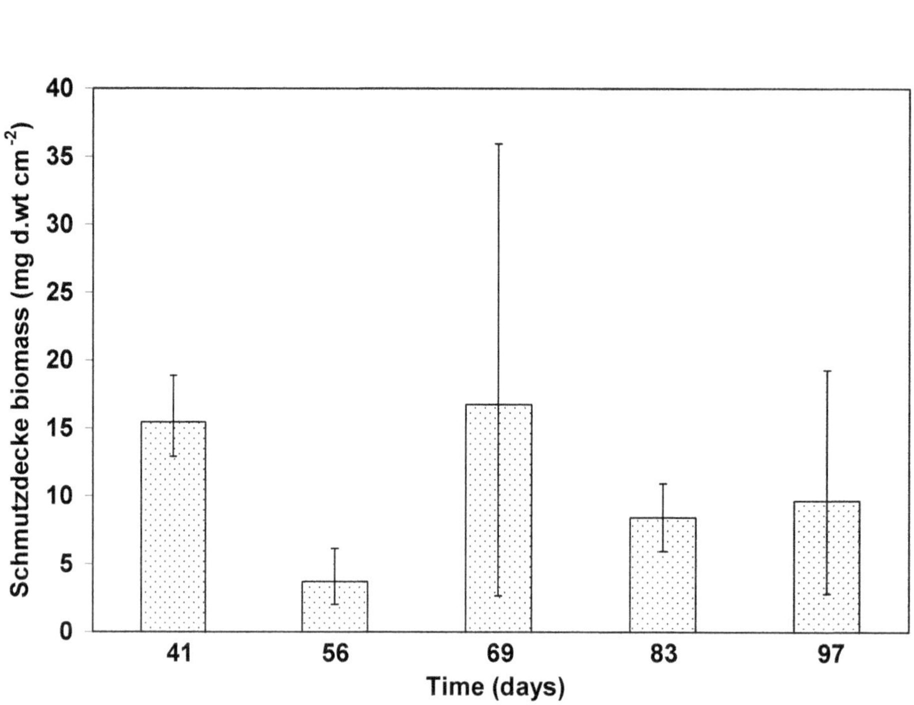

Figure 1.2 Example of variation in mean schmutzdecke biomass of an uncovered SSF with run time (vertical bars – min and max values) (Campos *et al.*, 2002).

FABRIC LAYERS

Previous research has shown that SSFs protected by surface layers of synthetic fabrics are capable of operating at significantly longer run times compared to un-protected SSFs (up to 8 times as long; Graham and Mbwette, 1991). Although a typical SSF bed is up to 1 m in depth, the majority of the filtration processes are concentrated in the upper 2-3 cm of the sand bed. With the use of fabric-protected slow sand filtration, and through careful selection of fabric type and specification, the filtration processes can be concentrated in the fabric layers of 2-3 cm thickness. Hence, solids penetration through to the sand can be prevented. As a consequence of the porosity of fabrics being significantly greater than that of sand, for example 90% compared to 45-50%, the rate of fabric blocking is much lower than the rate of sand blocking with unprotected slow sand filters. Therefore the rate of head loss development during the filter run period is reduced which leads to an extension of the filter run time in between filter cleaning operations. As well as extended run times, fabric-protected slow sand filtration can also eliminate the need for sand removal, and off-line cleaning, through the prevention of solids penetration by the fabrics. Therefore, an extended filter run time and reduced maintenance effort with fabrics can result in a substantial reduction in filter operational costs. Previous research at Imperial College (Mbwette, 1989; Graham and Mbwette, 1991) has demonstrated that by using six layers (3 cm) of a moderate-to-high density polypropylene fabric placed on the surface of slow sand filters, the filter run times can be extended by up to a factor of eight, and regularly by a factor of four, compared to a conventional slow sand filter. After a filter run, the fabrics can be removed and in the case of synthetic fabrics, can be cleaned and reused. A limitation of the use of synthetic fabrics is their cost, which can considerably increase the capital costs of a slow sand filtration project, especially if the importation of the fabrics is necessary. In view of the widespread non-availability of synthetic non-woven fabrics in developing countries, and the above-mentioned high costs of purchase and importation, there is a need to evaluate the potential use of *natural* fabric materials that may be locally available and produced.

A preliminary evaluation of the use of filtration fabrics made from *natural* materials, as a method for protecting slow sand filters in developing countries using indigenous materials and locally available textiles technology, was carried out some time ago (Luxton and Graham, 1998). The properties of the materials are shown in Table 1.1, with the natural materials (jute, abaca, sisal and flax) compared to a synthetic non-woven fabric (FiberTex). Coir was also included owing to its high resistance to biodegradation. The results of pilot scale tests showed that a combination of four layers of jute below a single layer of coir (3 cm total thickness) could achieve a doubling of the SSF run time. In addition, solids penetration through the fabrics into the sand was prevented and the fabrics displayed no significant deterioration. These reults were very encouraging but the development of natural fabrics as a SSF enhancement still requires further investigation and field demonstration in order to confirm their potential.

Table 1.1 Specifications and properties of selected fabrics.

Property	FiberTex (Polypropylene)	Jute 1	Jute 2	Abaca	Sisal	Flax
Fabric thickness (mm)	4.8	5.5	5.5	7	6	5
Fabric density (kg/m^3)	100	142	147	100	150	170
Fibre density (kg/m^3)	910	1440	1440	1450	1450	1540
Mean fibre diameter (μm)	35	50	50	280	200	40
Fabric porosity (%) (calculated)	89.0	90.1	89.8	93.1	89.7	89.0
Specific surface area (m^2/m^3) (calculated)	12559	7889	8167	985	2069	11039

OZONE PRE-TREATMENT

As principally a filtration process, SSF is able to remove only a minor proportion of the total influent organic matter, commonly expressed as colour and DOC, by sorption and microbial degradation. It has been reported that DOC removal may range between 5 and 40%, but typically less (9-15%) for humic-type DOC (Lambert and Graham, 1995). Similarly, the removal of trace natural and synthetic compounds is generally poor (e.g. pharmaceuticals; Kuhlmann *et al.*, 2006). However, in some cases the extent of treatment can be substantial depending on the presence of an active schmutzdecke and the properties of the specific compound, such as polar aromatic sulphonates (Eichhorn *et al.*, 2002) and microcystins (Grutzmacher *et al.* (2002).

When ozonation is employed prior to SSF, the beneficial impact on the overall removal of natural and synthetic organic substances can be considerable, mainly through the combined effects of direct oxidation and increased biodegradability. Ozone causes substantial structural changes to humic substances which include: a strong and rapid decrease in colour and UV-absorbance due to a loss of aromaticity and depolymerisation; a small reduction in TOC (eg. 10%

at 1 mgO_3/mgC); a slight decrease in the large apparent molecular weight fractions, and a slight increase in the smaller fractions; a significant increase of the carboxylic functions; and the formation of ozonation byproducts (Graham, 1999). The ozonation by-products have been reported to be mainly aldehydes (formaldehyde, acetaldehyde, glyoxal, methylglyoxal) and carboxylic acids (formic, acetic, glyoxylic, pyruvic and ketomalonic acids) (Camel and Bermond, 1998); in addition, glyoxalic acid and hydrogen peroxide have been identified as fulvic acid by-products. Ozone can also disrupt algal cells leading to an increase in assimilable organic carbon (AOC) (Muller *et al.*, 2003).

Whilst it is generally believed that the SSF can remove the additional biodegradable organic material and by-products generated by the ozone, only a small number of studies have confirmed this. A summary of the removal of DOC by pre-O_3 and SSF has been presented previously, and is given in Table 1.2; it can be seen that the removal values are significantly greater compared to SSF without pre-O_3 (i.e. 16-18%). Ozone by-products have been identified in a few studies and comprise low molecular weight aldehydes whose formation increased with ozone dose, but measured concentrations (total <100 μg/l) were reduced to below detection limits by subsequent SSF (Graham, 1999).

Table 1.2 DOC removals typically achieved by pre-O_3 and SSF (Graham, 1999).

Water source	DOC (mg/l)	Ozone dose	Removal (%)
River Dee, UK	2.9-6.1*	<5 mgO_3/l	18-55
Seagahan, UK	na†	3.1-4.8 mgO_3/l	25
Lake Vyrnwy, UK	2.4-4.8*	1.1-2.5 mgO_3/l	26.5
Norsborg, Sweden	na	0.2-1.0 $mgO_3/mgTOC$	20-30 (>8°C)
River Dee, UK	3.0-7.9	0.5 $mgO_3/mgTOC$	28
Portsmouth, USA	7.6-8.2	0.5, 1 $mgO_3/mgDOC$	20-35

*TOC (mg/l); †not available

The precise effect of ozone on the ecology of SSF is neither well understood nor well researched. A few studies of this aspect have been carried out (e.g. Yordanov *et al.*, 1996) in which various techniques have been used to characterise the schmutzdecke biomass. These have provided some evidence that pre-O_3 has an effect on the *surface* microbiology of SSFs. The indication of an increased bacterial population on the filter surface seems to be consistent with the higher levels of BDOC in the influent water after ozonation. Further evidence, indirectly, of the impact of pre-ozonation on the fundamental behaviour of the schmutzdecke is the effect the ozone has on filter headloss development and thus, filter run time. A summary of some experiences with pre-O_3 on SSF run time is shown in Table 1.3. It is clear that the influent water quality and seasonal effects are important and these may relate to the level of nutrients, temperature and algal activity.

Table 1.3 Effect of pre-O_3 on the run time of SSF (Graham, 1999).

Water source	Ozone dose	Effect on headloss rate / run time
River Dee, UK	<5 mgO_3/l	Increased headloss rate in winter / lower rate in summer
Seagahan, UK	3.1-4.8 mgO_3/l	Reduced run length
Lake Vyrnwy, UK	1.1-2.5 mgO_3/l	Increased run length in winter /shorter run length in summer
Lake Vyrnwy, UK	2.5-3.5 mgO_3/l	Increased run length (by 34%) in winter and spring
River Dee, UK	0.5 mgO_3/mgTOC	Increased headloss rate (shorter run length) in winter
Portsmouth, USA	0.5, 1 mgO_3/mgDOC	Increased headloss rate

The ozone pre-treated SSFs in Portsmouth (NH, USA) were observed to remove 40-70% of the UV absorbance (254 nm) and the THM formation potential (THMFP) compared to the 10-15% removals for the conventional slow sand filters (Eighmy *et al*., 1993). As can be seen in Figures 1.3 and 1.4, a significant fraction of these elevated UV and THMFP removals (up to half of the overall removals) were observed in the water column above the sand media surface along with significant increases of water column bacterial populations compared to the non-ozonated SSFs.

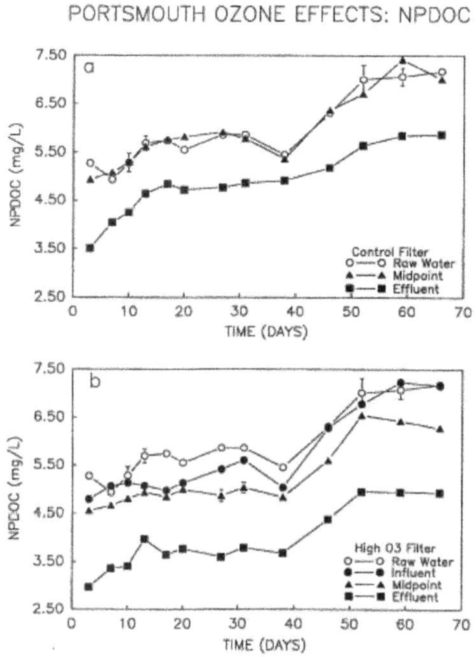

Figure 1.3 Effect of ozone pre-treatment on NPDOC (non-purgable dissolved organic carbon) at Portsmouth filters.

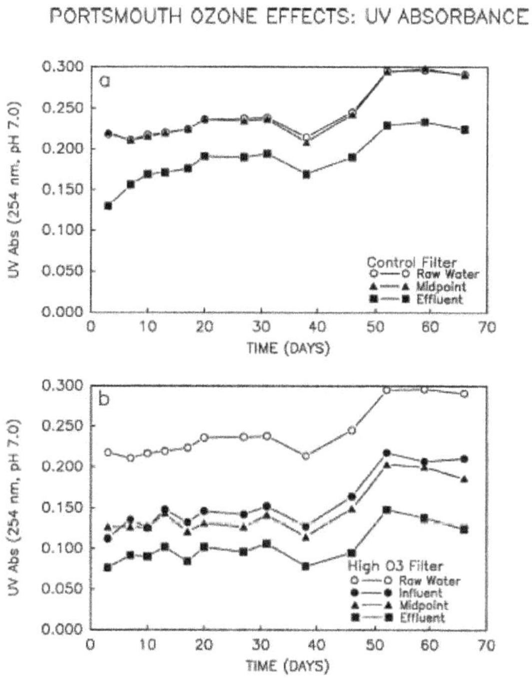

Figure 1.4 Effect of ozone pre-treatment on UV absorbance at Portsmouth filters.

Ozone pre-treatment were also observed to cause dramatic increases in filter headloss development causing filter run times to be less than ½ to ¼ of the control SSFs. These elevated headloss levels occurred in the schmutzdecke due to accelerated accumulations of biomass. In separate pilot filter runs, the rate of headloss development in pre-ozonated SSFs was significantly reduced by using larger sized sand media (up to 0.70 mm effective size) without a measurable reduction in treatment performance.

The primary ozonation by-product (OBP) observed in the pilot filter runs on two different water sources were formaldehyde. Production of the formaldehyde appeared to be a function of the ozone: DOC ratio. Formaldehyde was readily removed by the biological slow sand filtration process, as were acetaldehyde and glyoxal. In virtually all cases, SSF effluent OBPs measured in this study were below detection limits.

GAC AMENDMENTS

GAC placed as a sub-layer in the sand media of a slow sand filter can significantly enhance the removal of organic precursors as quantified by dissolved organic carbon (DOC). Research on GAC amended SSFs began at the University of New Hampshire back in 1987. A more mechanistic understanding of DOC removal by the enhanced SSF began in the mid 1990s (Page, 1997). In a comparative pilot study between a control SSF and two GAC amended SSFs (7.5 cm and 15 cm GAC sub-layers), DOC removals reached pseudo steady-state removals of 12%, 28% and 46%, respectively, after 200-300 days of continuous operation as shown in Figures 1.5 and 1.6. Removals of the biodegradable fraction of DOC, i.e. BDOC, were roughly the same for all three SSFs suggesting, as expected, that the elevated removals of DOC by the GAC amended SSFs were due to adsorption. Thusly, the pseudo steady-state removals by adsorption were 16% and 34% for the 7.5 cm and 15cm GAC amended SSFs, respectively.

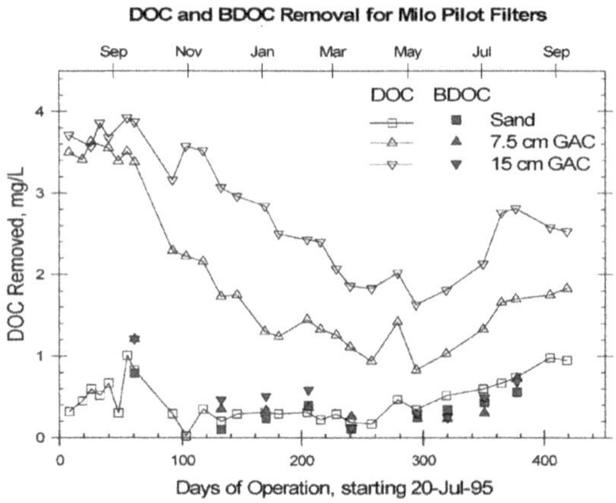

Figure 1.5 DOC and BDOC removal for Milo pilot filters.

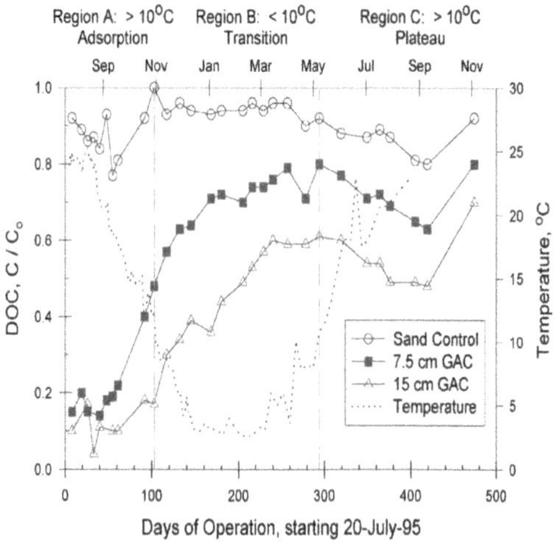

Figure 1.6 Influence of temperature on DOC removal trends.

A summary depiction of DOC removals by both biodegradation and adsorption from the 7.5 cm GAC amended SSF is shown in Figure 1.7. As expected, adsorption is the dominant DOC removal mechanism for the first 10000-15000 GAC bed volumes (BVs). Once steady-state removals by adsorption were observed, biodegradation started to achieve proportional removals of DOC, especially as the temperature rises above 10° C.

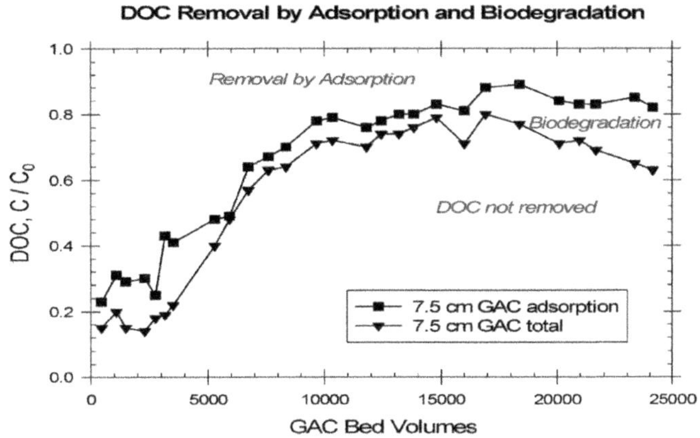

Figure 1.7 DOC removal by adsorption and biodegration by 7.5 cm GAC amended SSF.

So, are elevated steady-state DOC removals observed for GAC amended SSFs due to enhanced biodegradation, slow adsorption by the GAC, or from bioregeneration of GAC adsorption sites? Since biomass levels (data not shown) and BDOC removals (see Figure 1.5) were similar in sand and GAC sub-layers, there is little evidence in this study to support an enhanced biodegradation removal mechanism.

Since DOC removals by adsorption continued at a constant rate even after 400+ days (12000-23000 GAC BVs), there is evidence to support slow adsorption or bioregeneration of the GAC sublayer. The slow adsorption mechanism is suggested by the steady-state adsorption rates being proportional to the GAC empty-bed-contact-time (EBCT). The bioregeneration mechanism is suggested by the data showing adsorption rates increasing with rising temperature. Additional long-term characterizations will be required to distinguish between these two possibilities.

SSF MODELLING

To date, very little progress has been made in the development of a comprehensive SSF process model. Such a model would provide a valuable, quantitative decription of the interactions between the physico-chemical and biological processes with time and location within the filter, and the dependence of filter performance on influent water quality. In addressing this deficiency, the author (Graham) and colleagues have previously worked on developing a deterministic simulation model of the SSF process which incorporates the fundamental physico-chemical and biological dynamics within a classical filtration framework. The model was based on a temporal and spatial finite difference method and was subsequently calibrated and pseudo-verified using operational data from pilot-scale SSF units.

The model developed by Campos (2002) considered two different forms of SSF, where the filter bed is covered or not. The uncovered filter was conceptualised as a reactor receiving full light interception. Three different zones are identified: the supernatant water, the schmutzdecke layer, and the sand bed. The schmutzdecke is defined as a uniform layer on the top of the sand formed by filamentous algae. The SSF model assumes a linear function to simulate the growth of the schmutzdecke with time, and that a new layer is produced each day. By contrast, the covered filter was conceptualised as a closed reactor without light interception. Therefore, supernatant water quality is unchanged and no schmutzdecke development occurs in the covered system. Nevertheless, particles are retained and microbial biomass develops within the sand bed, contributing to progressive headloss.

The SSF model employed classical filtration theory in which changes in particle removal and headloss are related to the bulk specific deposit in the schmutzdecke and filter media. The total bulk specific deposit is assumed to consist of biomass, inert material and non-living particulate organic carbon (POC). The biomass was formed by the removal of algae and removal/growth of bacteria and protozoa within the pores of the schmutzdecke and sand bed. In the covered filter, algae and algal effects were ignored (Campos, 2002). The dynamic interactions between the principal groups of microorganisms including: algae, bacteria and protozoa, are modelled using Monod-type kinetic equations. The filtration performance of the filter media is defined in the model by the removal of particulate material from water and is represented by a combination of headloss and filtration coefficient functions. The computer program was developed in Visual Basic with Application to Excel for the model simulation and the data required for simulation consist of influent water quality concentrations (inert material, chlorophyll-a, bacteria, protozoa, non-living POC, dissolved organic carbon (DOC), phosphorus, nitrate-nitrogen, ammonium-nitrogen), temperature, flow rate, filter surface area, supernatant water level, filter bed depth, duration of filtration runs, and several other input parameters (Campos, 2002). A full description of the model has been given elsewhere (Campos *et al.*, 2006a, 2006b).

Results from the testing of the model show that, while still in a preliminary form, it was able to reproduce many of the observed features of SSF, including the pattern of headloss development (Fig. 1.8).

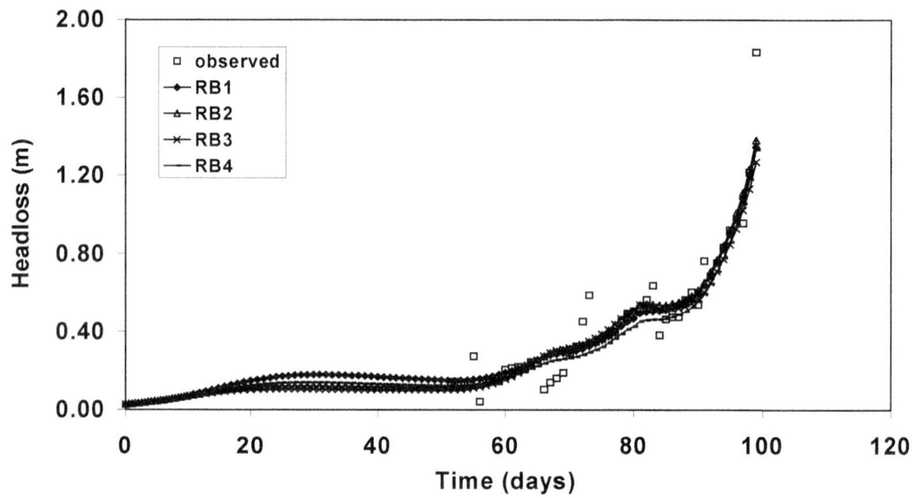

Figure 1.8 Example of model simulation of SSF headloss with time (Campos *et al.*, 2006a).

Typically, the major contribution to headloss in uncovered SSF occurs in the schmutzdecke and in the immediate underlying sand (top 2 to 4 cm), but it is difficult experimentally to differentiate the contributions between these. In this regard, the model has the advantage of being is able to do this and the results shown in Fig. 1.9 provide an example for one particular SSF run. In this case, the schmutzdecke contributed very little headloss during the first 50 days of the run, but became increasing important with further time and was the major component of overall headloss at the end of the run. The model provides valuable insight into the fundamental behavior of the filter process. For example, it has shown that the growth of the schmutzdecke in uncovered filters is beneficial in preventing rapid clogging of the sand bed, and significantly reducing the headloss that would otherwise occur.

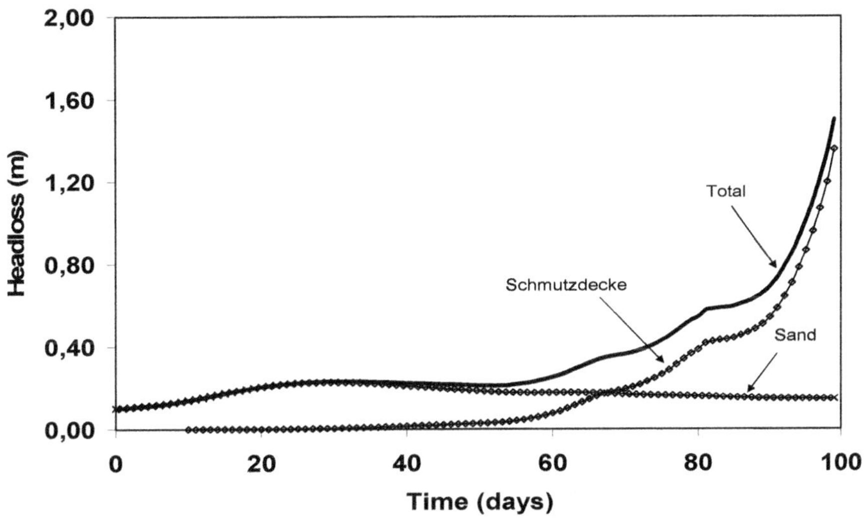

Figure 1.9 Example of model prediction of sand and schmutzdecke headloss development with time (Campos *et al.*, 2006b).

In the model the importance and role of the schmutzdecke are demonstrated based on changes in the 'lag phase', and in the variation of the thickness and biomass of the filamentous algae. The 'lag phase' represents the time period that elapses from the start of the filtration run for the growth of the first schmutzdecke layer, and consequently, when filtration across the schmutzdecke is simulated. The total absolute specific deposit (i.e. sum of inert material, algae, bacteria, protozoa, and non-living POC) in the top 10 cm of sand beneath the schmutzdecke increased as the assumed 'lag phase' is increased, as more impurities reach the sand bed when the schmutzdecke is not developed, emphasising the role of the schmutzdecke in removing particles in uncovered filters. The model predicted that the consequence of the increase in total absolute deposit of material in the top 10 cm of sand, as the 'lag phase' was increased, would be a more rapid headloss development, and thus a shorter run time, as illustrated in Fig. 1.10. This behaviour is consistent with obervations of full-scale plants reported by Nakamoto *et al.* (1996) and with earlier studies, such as Hirschi and Sims (1991) who reported that increasing schmutzdecke growth improves particle removal performance. In addition, Huisman and Wood (1974) noted in their seminal text on SSF that less suspended matter reaches the filter medium when the schmutzdecke is well established, and this may lengthen the interval between successive filter cleaning of uncovered filters.

The model has also provided insight into the interactions in the filter bed between micro-organisms, and shown how the presence of deposits accumulated from previous runs, are important to the overall process performance (Campos *et al.*, 2006b). Overall, while the model has significant limitations owing to its complexity, need for calibration, and over-simplification of biological processes, it has the potential to provide a comprehensive representation of the fundamental nature of SSF and could form the basis of an operational design and control tool.

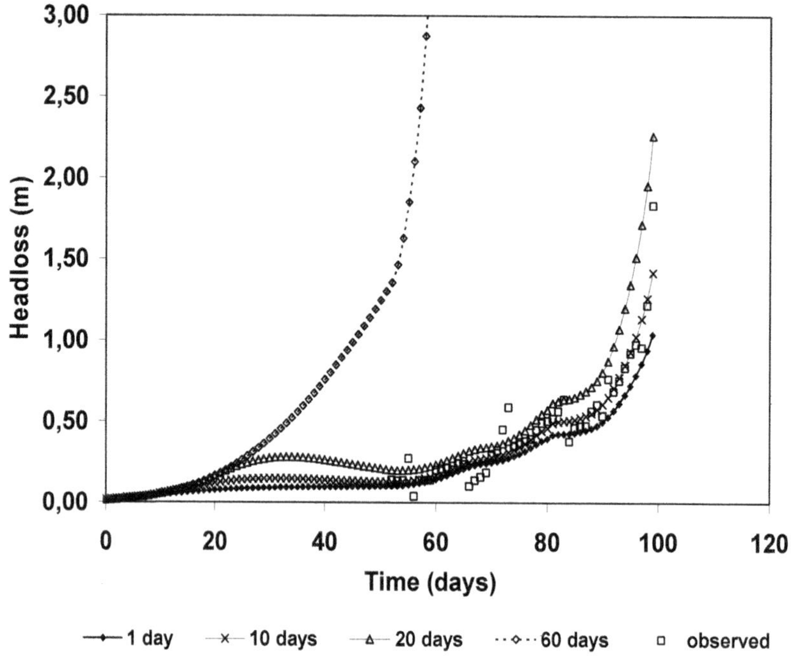

Figure 1.10 Influence of'lag phase' of the schmutzdecke growth on total headloss.

CONCLUSIONS

The main conclusions of this paper are the following:

(1) In the last 25 years SSF technology has advanced considerably in terms of process performance and developments, but a comprehensive description of the fundamental process behavior is still lacking.

(2) Microbial biomass in the schmutzdecke and filter bed has been studied and quantified, and a reasonable understanding of the spatial and temporal patterns of biomass development has been obtained. However, many aspects of the fundamental nature, dynamics and inter-relationships of the microbial community in SSFs remain uncertain and require further research.

(3) A significant improvement in SSF performance and operation can be obtained by the application of non-woven fabric layers to support the schmutzdecke/active treatment zone and the optimal properties required have been established. Preliminary studies of natural fibrous materials have been encouraging in terms of their performance and further studies are needed to demonstrate the potential benefits, particularly for small community systems in developing countries.

(4) The performance of SSFs for removing/treating organic substances, both natural and anthropogenic contaminants, is generally poor. However, both ozone pre-treated SSF and the GAC amended SSF have been shown to give elevated removals of DOC and organic precursors that appear to be sustainable. However, ozone pre-treatment can cause dramatic increases in SSF headloss development causing filter run times to be significantly shorter than conventional SSFs. Using larger

sized sand media, however, may mitigate shorten filter runs without significant reductions in overall treatment performance.

(5) Both biodegradation and adsorption are important DOC removal mechanisms in GAC amended SSFs but their relative importance will be a function of temperature. The ability of the GAC to extend its useful life in an amended SSF may be due to slow adsorption and/or bioregeneration. Additional research is needed to distinguish between these two mechanisms;

(6) The modelling of SSF is still in its infancy reflecting the complex nature of the process. A recently developed model was able to simulate particular features of the basic process, based on classical filtration theory and simple representations of the process biology. Further development of a deterministic model is recommended in order to understand more fully the process behaviour in a quantitative manner, and to explore the impacts/benefits of process modifications, such as pre-treatment, covering beds and alternative filter media.

ACKNOWLEDGEMENTS

The authors acknowledge the contributions of the following colleagues to aspects of the research described in the paper: Mr Paul Luxton, Dr Luiza Campos, Professor Stephen Smith, Ms Michelle Su, Dr Steven Lambert, Dr Taylor Eighmy, Dr Jim Malley Jr, Jim Fenstermacher and Tom Page.

REFERENCES

ASCE (1991) *Slow Sand Filtration*, G.S. Logsdon (Ed.), American Society of Civil Engineers, New York, USA.

AWWA (1991) *Manual of Design for Slow Sand Filtration*, D. Hendricks (Ed.), American Water Works Association Research Foundation, Denver, Co, USA.

Camel, V. and Bermond, A. (1998) The use of ozone and associated oxidation processes in drinking water treatment. *Water Res.*, **32**, 3208-3222.

Campos, L.C. (2002) Modelling and simulation of the biological and physical processes of slow sand filtration. Ph.D. Thesis, University of London, UK.

Campos, L.C., Su, M.F.J., Graham, N.J.D., and Smith, S.R. (2002) Biomass development in slow sand filters. *Water Res.*, 36(18), 4543-4551.

Campos, L.C., Smith, S.R. and Graham, N.J.D. (2006a) Deterministic-based model of slow sand filtration – I: Model development. *J. Environ. Eng. (ASCE)*, 132, 872-886.

Campos, L.C., Smith, S.R. and Graham, N.J.D. (2006b) Deterministic-based model of slow sand filtration – II: Model application. *J. Environ. Eng. (ASCE)*, 132, 887-894.

Collins, M.R. and Graham, N.J.D. (1994) (Eds.) *Slow Sand Filtration: An International Compilation of Recent Scientific and Operational Developments*, American Water Works Association, Denver, USA.

Duncan, A. (1988) The ecology of slow sand filters. In *Slow Sand Filtration: Recent Development in Water Treatment Technology*, N.J.D. Graham (ed.), Ellis Horwood, pp.163-180.

Eichhorn, P., Knepper, T.P., Ventura, F. and Diaz, A. (2002) The behaviour of polar aromatic sulfonates during drinking water production: a case study on sulfophenyl carboxylates in two European waterworks. *Water Res.*, **36**, 2179-2186.

Eighmy, T.T., Collins, M.R., Malley, J.P., Royce, J. and Morgan, D. (1993) *Biologically Enhanced Slow Sand Filtration for Removal of Natural Organic Matter.* AWWA Research Foundation Final Report, Denver, CO.

Garibaldi, A., Minuto, A., Grasso, V. and Gullino, M.L. (2003) Application of selected antagonistic strains against Phytophthora cryptogea on gerbera in closed soilless systems with disinfection by slow sand filtration. *Crop Protection*, **22**(8), 1053-1061.

Gimbel, R., Graham, N.J.D. and Collins, M.R. (2006) (Eds.) *Recent Progress in Slow Sand and Alternative Biofiltration Processes*, IWA Publishing, London, U.K.

Graham, N.J.D. (1988) (Ed.) *Slow Sand Filtration - Recent Developments in Water Treatment Technology*, Ellis Horwood Ltd., Chichester, U.K.

Graham, N.J.D. (1999) Removal of humic substances by oxidation/biofiltration processes – a review. *Wat. Sci. Tech.*, **40**(9), 141-148.

Graham, N.J.D. and Collins, M.R. (1996) (Eds.) *Advances in Slow Sand and Alternative Biological Filtration*, John Wiley & Sons, Ltd., Chichester, U.K.

Graham, N.J.D. and Mbwette, T.S.A. (1991) Protected slow sand filtration: specification of non-woven synthetic fabric layers. *Water Supply*, **9**, 157-164.

Grutzmacher, G., Bottcher, G., Chorus, I. and Bartel, H. (2002) Removal of microcystins by slow sand filtration. *Env. Toxicol.*, **17**(4), 386-394.

Haig, S.J., Collins, G., Davies, R.L., Dorea, C.C. and Quince, C. (2011) Biological aspects of slow sand filtration: past, present and future. *Wat. Sci. Tech.: Wat Supply*, **11**(4), 468-472.

Hirschi, S.D. and Sims, R.C. (1991) Particles and microorganisms in slow rate sand filtration. *International Slow Sand Filtration Workshop*, 27-30, October, Durham, USA.

Huisman, L. and Wood, W.E. (1974) *Slow Sand Filtration*, World Health Organisation, Geneva.

Kuhlmann, B., Zullei-Seibert, N., Nolte, J. and Grote, M. (2006) Behaviour of selected drugs during slow sand filtration. In *Recent Progress in Slow Sand and Alternative Biofiltration Processes*, R. Gimbel, N.J.D. Graham and M.R. Collins (Eds.), IWA Publishing, London, U.K., 188-193.

Lambert, S.D. and Graham, N.J.D. (1995) A comparative evaluation of the effectiveness of potable water filtration processes. *J. Water SRT – Aqua,* **44**, 38-51.

Luxton P.J. and Graham, N.J.D. (1998) Preliminary evaluation of natural fabrics in slow sand filtration. *J Water SRT - Aqua*, **47**(3), 95-106.

Mbwette, T.S.A. (1989) *The performance of fabric protected slow sand filters*. PhD Thesis, Imperial College London.

Muhammad, N. and Morris Hooke, A. (2003) Biomass characterization of slow sand filtration schmutzdecke and its effects on filter performance. *Environ Tech.*, 24, 43-50.

Muller, K.C., Forster, R., Gammeter, S. and Hambsh, B. (2003) Influence of ozonated cyanobacteria on bacterial growth in rapid sand filters. *J. Water SRT – Aqua,* 52(5), 333-340.

Nakamoto, N., Yamamoto, M., Sakai, M., Nozaki, K., Iwase, N., Yasuda, M., and Kitada, T. (1996) Role of filamentous diatom as an automatic purifier in a slow sand filter. In *Advances in Slow Sand and Alternative Biological Filtration.* N.J.D. Graham and R. Collins (Eds.), John Wiley & Sons, U.K., 139-148.

Page, T.G. (1997) GAC sandwich modification to slow sand filtration for enhanced removal of natural organic matter. Masters Thesis, University of New Hampshire, Durham, NH.

Rachwal, A.J., Bauer, M.J., Chipps, M.J., Colbourne, J.S. and Foster, D.M. (1996) Comparisons between slow sand and high rate biofiltration. In *Advances in Slow Sand and Alternative Biological Filtration.* N.J.D. Graham and R. Collins (Eds.), John Wiley & Sons, UK, 3-10.

Yordanov, R.V., Lamb, A.J., Melvin, M.A.L. and Littlejohn, J. (1996) Biomass characteristics of slow sand filters receiving ozonated water. In *Advances in Slow Sand and Alternative Biological Filtration.* N.J.D. Graham and R. Collins (Eds.), John Wiley & Sons, UK, 107-118.

Chapter 2

A Review of Biologically-Based Drinking Water Treatment Processes for Organic Micropollutant Removal

R. S. Summers, K. Shimabuku*, T. L. Zearley***

**Department of Civil, Environmental and Architectural Engineering, University of Colorado at Boulder, Boulder CO 80309 -0428*

***AAAS Science and Technology Fellow*

ABSTRACT

Organic micropollutants are increasingly being detected in the world's drinking water sources. The need exists for sustainable drinking water treatment options to remove these pollutants. The research studying biotechnologies, such as biological filtration and managed aquifer recharge is growing. This review compiles removal efficiencies observed in biological drinking water systems to reveal the extent of such technologies' effectiveness for a range of organic micropollutants.

Keywords Biological filtration; managed aquifer recharge

INTRODUCTION

Increasing contamination of freshwater systems with thousands of agricultural, industrial, and wastewater derived compounds is a key environmental problem (Schwarzenbach et al., 2006) Subsequently the list of organic micropollutants identified in drinking water requiring attention continues to grow and cost-effective and adequate treatment of drinking water must be developed and refined (Schwarzenbach et al., 2006). Several organic pollutants persist during conventional drinking water treatment techniques such as coagulation, sedimentation and chlorination. Advanced treatment barriers, such as ozone, advanced oxidation processes, and activated carbon, can be effective at removing micropollutants, but can be cost-prohibitive as natural organic matter reduces their efficiency. Biodegradation is a central mechanism utilized in managed aquifer recharge (MAR) systems such as riverbank filtration, lake bank filtration, and artificial recharge and in in-plant biological filtration (BF) systems, and have the potential for removing micropollutants. These treatment options are sustainable and require minimal energy and material inputs which has motivated more recent research quantifying micropollutant degradation in batch, laboratory, and in-situ experiments. This article provides a table comparing the removal efficiencies in research projects that investigated micropollutant treatment options for drinking water where biodegradation is the main or a contributing mechanism.

METHODOLOGY

Research articles were compiled from studies examining biofiltration employing inert or GAC media, and MAR including riverbank filtration, lake bank filtration, and artificial recharge. Removal attributed to these treatment techniques are both biodegradation and, in many cases, initially adsorption process. MAR removals also have additional mechanisms such as filtration, dispersion, and mixing (Worch et al., 2002; Snyder et al., 2007). Although biodegradation has been identified as the most important removal mechanism when adsorption sites are exhausted and pollutants desorb in MAR systems (Maeng et al., 2011), its contribution to the reported removals are uncertain. However, batch experiment removal efficiencies reported here are largely dominated by biodegradation. Batch experiments usually monitored degradation rates in controlled laboratory settings with inoculated sand in ground or surface water. Most of the removals reported are for biotransformation, not mineralization, but mineralization was monitored in several of the batch experiments. Some MAR research also tracked the formation and degradation of the parent compound's metabolites, which are reported as well. Most of the removals shown were directly found in a table or graph, while others provided data from which removals were calculated. In Table 2.1 the first column organizes contaminants into either nonpoint or wastewater derived categories as well as the appropriate class such as pesticides or hormones. These classifications are not absolute as the same compound could be introduced to the environment as a point source from an industrial wastewater plant or from a nonpoint source like gasoline burning engines. The three technologies reviewed were biological filtration, managed aquifer recharge (including riverbank filtration, lake bank filtration, and artificial recharge) and batch experiments (not included in Table 2.1). When available, the studies' influent concentrations, removals, residence times and literary source are listed. Table 2.2 lists pertinent attributes to the experimental designs and explanations of the superscripts found in Table 2.1. For more details the reader should refer to the original articles sited.

Table 2.1 Organic micropollutant removal efficiencies in biological filters and managed aquifer recharge.

Type	Compound	*Biological filter*				*Managed Aquifer Recharge*			
		Inf. Conc. (ng/L)	*Removal (%)*	*RT (min)*	*S*	*Inf. Conc. (ng/L)*	*Removal (%) A.W., P.W.*	*RT (d)*	*S*
Nonpoint Sources									
Gasoline Additive									
	Benzene					28	<0-61, 66	10.5-28	n
	Ethylbenzene					18	11-79, 79	10.5-28	n
	Hemellitene					34	58-85, 87	10.5-28	n
	MTBE					178	25-50,	15-70	a
	Naphthalene					8	<0-61, 75	10.5-28	n
	o-Xylene					12	<0-62, 65	10.5-28	n
	Pseudocumene					100	76-93, 94	10.5-28	n
	Toluene					117	4.3-86.3, 87	10.5-28	n
	Benzo[a]pyrene	94	89[S], 89[A]	2	u				
Naturally Occurring									
	MIB	5-50	90	10	c				
		100-200	46	7	d				
		50	50	2-7 d	l				
		600	>90	60	m				
		84	93[a], 99[b]	7.9[a], 15.8[b]	z				
Pesticides									
2,4-Dichlorophenoxyacetic acid (2,4-D)		171	68[a], 77[b]	7.9[a], 15.8[b]	z				
	Acetochlor	266	8[a], 17[b]	7.9[a], 15.8[b]	z	1460-4350	72-98	6	w
	Aldicarb	31	49[a], 72[b]	7.9[a], 15.8[b]	z				

(Continued)

Table 2.1 Organic micropollutant removal efficiencies in biological filters and managed aquifer recharge. (*Continued*)

Atrazine	10000	50	217	f	7800-12500	7.2-92	6	w
	142	15^{S},83^{A}	2	u				
	9	0^{a}, 3^{b}	7.9^{a}, 15.8^{b}	z				
Bentazone					25	<0-20	<0	j
					10	n/a, <0-100	100	h
Bromacil					54	n/a, 100		h
					140	n/a 48		h
Carbaryl	96	3^{a}, 17^{b}	7.9^{a}, 15.8^{b}	z				
Cyanazine					1350-3490	15-95	6	w
DDT	77	·87^{S}, 85^{A}	2	u				
o,p'-DDA (DDT metabolite)					5	0- <0,>0	<60	k
					<10	0-100,<0		h
p,p'-DDA (DDT metabolite)					20	<0-50,0-50	<60	k
					15	<0-100, <0		h
DEET	147	>8^{S},80^{A}	2	u				
Diazinon	4	12^{a}, 40^{b}	7.9^{a}, 15.8^{b}	z				
Dimethenamid					560	9-36	6	w
Dimethoate	38	75^{a}, 81^{b}	7.9^{a}, 15.8^{b}	z				
Diuron	92	0^{a}, 8^{b}	7.9^{a}, 15.8^{b}	z				
Flufenacet					190	58-63	6	w
Lindane	150	7^{S}, 91^{A}	2	u				
Malaoxon	132	16^{a}, 49^{b}	7.9^{a}, 15.8^{b}	z				
Methomyl	176	12^{a}, 5^{b}	7.9^{a}, 15.8^{b}	z				
Mecoprop					15	<0-33,0-33		k
					20	n/a ,25-100	100	h
Metolachlor	130	4^{S}, 79^{A}	2	u	48-340	n/a <0-87		h
	284	7^{a}, 9^{b}	7.9^{a}, 15.8^{b}	z	3600-4720	50->93.6	6	w
Molinate	205	85^{a}, 95^{b}	7.9^{a}, 15.8^{b}	z				
Pentachlorophenol					80	0-100	hrs-wks	s
Prometon	132	3^{a}, 0^{b}	7.9^{a}, 15.8^{b}	z	29	28		k
Propazine					120	<0-8	6	k
Simazine	52	7^{a}, 8^{b}	7.9^{a}, 15.8^{b}	z	50	20-40	6	k
Sunscreen compound								
Oxybenzone	66	14^{S}, 98^{A}	2	u				
Analgesic								
Acetaminophen	39	6^{S}, 95^{A}	2	u				
	306	59^{a}, 79^{b}	7.9^{a}, 15.8^{b}	z				
Hydrocodone	74	14^{S}, 92^{A}	2	u				
Antibiotics								
Clarithromycin					8.9	100^{an} 100°, 100	30	i
Clindamycin					31	26^{an} 93°, 100	30	i
Erythromycin	111	31^{S}, 78^{A}	2	u				
	104	15^{a}, 22^{b}	7.9^{a}, 15.8^{b}	z				
Roxithromycin					11	100^{an} 100°, 100	30	i
Sulfamethoxazole	77	5^{S}, 63^{A}	2	u	151	99^{an} 52°, 99	30	i
	230	2^{a}, 4^{b}	7.9^{a}, 15.8^{b}	z	485	9-100, 95	<30-140	g
					463^{AR}	35-100, 74	4-50	g
					7	n/a, 71		h
Trimethoprim	102	20^{S}, 94^{A}	2	u	12	100^{an} 100°, 100	30	i
Anticonvulsant								
Carbamazepine	112	5^{S}, 90^{A}	2	u	330	<0-35,79-96	<60	k
	85	0^{a}, 2^{b}	7.9^{a}, 15.8^{b}	z	470	<0-55,55	<60	j
					235	<0-100, 92		h
					480	0, n/a	<3	p

(Continued)

Table 2.1 Organic micropollutant removal efficiencies in biological filters and managed aquifer recharge. (*Continued*)

Dilantin	93	19^S, 80^A	2	u				
Primidone					600	<0-17, 33-83	<60	k
					105	<0-100, 86		h
Antidepressant					135	<0-41, 26	<60	j
Fluoxetine	79	97^S,>99^A	2	u				
Antiseptic								
Triclosan	85	37^S, 97^A	2	u				
	190	90	7.9	z				
Dry Cleaning Product								
Tetrachloroethylene					600	<0-67, n/a	hrs-wks	s
Flame Retardant								
TCEP	85	10^S, 80^A	2	u	315	<0-19,33-78	<60	k
					250	25-100, 58		h
TCIPP					2100	14-46, 81-91	<60	k
					510	25-100, 65		h
Fragrance Compounds								
1,8-Cineole					n.g	97, n/a	18	n
4-Terpineol					n.g	99, n/a	18	n
α-Terpineol (Degradation product of 4-terpineol)					85	99,n/a	18	n
5-Methylheptan-3-one					n.g	99, n/a	18	n
Galaxolide	48	19^S, 74^A	2	u				
Limonene					14	83, n/a	18	n
Menthol					139	94, n/a	18	n
Musk Ketone	170	10^S, 83^A	2	u				
Hemorrheologic agent								
Pentoxifylline	98	13^S, 90^A	2	u				
	268	39^a, 38^b	7.9^a, 15.8^b	z				
Hormones								
Androstenedione	157	41^S, 97^A	2	u				
Cholesterol					680	n/a, <0		h
					510	n/a, 100		h
Estradiol	147	1^S, 94^A	2	u				
Estriol	143	<1^S, 92^A	2	u				
Estrone	150	<1^S, 95^A	2	u				
Ethinylestradiol	145	1^S, 91^A	2	u				
	316	12^a, 22^b	7.9^a, 15.8^b	z				
Progesterone	142	52^S, 99^A	2	u				
β-sitosterol					710	n/a, <0		h
					1100	n/a, 100		h
Testosterone	144	35^S, 96^A	2	u				
Lipid regulator								
Bezafibrate					60	83-100,100	<60	k
					170	88-100, 100		h
					30	100,100	<60	j
Clofibric acid	263	35^a, 52^b	7.9^a, 15.8^b	z	60	58-92,<0	<60	k
					50	<0-100,<0		h
					20	50-75,75	<60	j
Gemfibrozil	135	13^S, 74^A	2	u				
	228	70^a, 94^b	7.9^a, 15.8^b	z				
Manufacturing additives								
1,3-dimethylbenzene					230	78-100, n/a	hrs-wks	s
1,5-naphthalenesulfonic acid					368	2-49, 48	<30-140	g
					360^{AR}	49-58, 53	4-50	g
2,4,5-Trichloroaniline	n.g.	50	2080	x				
2,4,6-Trichloroaniline	n.g.	50	1188	x				
2,7-naphthalenesulfonic acid					102	<0-55, 39	<30-140	g
					102^{AR}	30-66, 6	4-50	g
2-Chloro-5-nitroaniline	n.g.	50	594	x				
2-Chloroaniline	n.g.	50	176	x				

(*Continued*)

Table 2.1 Organic micropollutant removal efficiencies in biological filters and managed aquifer recharge. (*Continued*)

2-methylaniline	n.g.	50	46	x				
2-Nitroaniline	n.g.	50	1808	x				
3,4-Dichloroaniline	n.g.	50	457	x				
3-Chloro-4-methylaniline	n.g.	50	89	x				
3-Chloroaniline	n.g.	50	59	x				
4-Bromoaniline	n.g.	50	62	x				
4-Chloroaniline	n.g.	50	59	x				
4-Nitroaniline	n.g.	50	1664	x				
4-n-nonylphenol								
4-tert-octylphenol								
Aniline	n.g.	50	62	x				
benzenesulfonamide	50	0	288	r				
Bisphenol A	311	64^{a}	7.9^{a}	z				
Butyl Acetate					n.g.	84	18	n
Fluorene	167	28^{S}, $>94^{A}$	2	u				
Isophorone					n.g.	95	18	n
o-toluenesulfonamide	140	0	288	r				
p-toluenesulfonamide	1300	90	288	r				
Tributyl phosphate	149	16^{a}, 24^{b}	7.9^{a}, 15.8^{b}	z				
NPS					20	<0-25,<0	<60	k
						<0-34,<0	<60	j
Diclofenac	108	13^{S}, 75^{A}	2	u	25	<0-40, 60->80	<60	k
	252	21^{a}, 28^{b}	7.9^{a}, 15.8^{b}	z	50	20-100, 100		h
					135	67->96, 93	<60	j
Ibuprofen	118	30^{S}, 83^{A}	2	u				
	276	95	7.9	z				
Indomethacine					15	>66.7-100, 100	<60	k
					20	>50-100, 100	<60	j
Naproxen	50	3^{S}, 82^{A}	2	u				
	170	72^{a}, 86^{b}	7.9^{a}, 15.8^{b}	z				
Phenazone					150	10^{an} 66°, n/a	<15	q
					340	91, n/a	<3	p
DP (Phenazone metabolite)					30	$<0^{an}$ $<0^{o}$, n/a	<15	q
					280	5, n/a	<3	p
					170	<0-32.4, <0	<60	k
AAA (Phenazone metabolite)					530	45^{an} 91°, n/a	<15	q
					650	96, n/a	<3	p
FAA (Phenazone metabolite)					340	36^{an} 72°, n/a	<15	q
					320	89, n/a	<3	p
Propyphenazone (Phenazone metabolite)					145	31-62, <0-79	<60	k
					120	75-92, 67	<60	j
					80	100, n/a	<3	p
					280	<0-39, 82		h
Smoking Degradate								
Cotinine	92	23^{a}, 39^{b}	7.9^{a}, 15.8^{b}	z				
Stimulant								
Caffeine	102	8^{S}, 93^{A}	2	u				
	118	67^{a}, 80^{b}	7.9^{a}, 15.8^{b}	z				
Tranquilizer								
Diazepam	110	15^{S},84^{A}	2	u				
Meprobamate	133	4^{S}, 71^{A}	2	u				
X-ray contrast medium								
Iopromide	120	11^{S}, 42^{A}	2	u	841	82-100, 99	<30-140	g
	556	13^{a}, 3^{b}	7.9,15.8^{b}	z	737^{AR}	89-100, 99	4-50	g

RT: Residence time, S: source (see Table 2.2), A.W.: Removals found at both monitoring and production wells
P.W.: Removals found at production wells , n.g.: not given, n/a: not applicable

Table 2.2 Research summarized in Table 2.1 including source identification and superscript definitions.

Source	Reference	System description with superscript definitions
a	Achten et al. (2002)	Riverbank filtration dominated by oxic conditions
b	Bradley et al. (2007)	Microcosm with oxic river sediment
c	Brown and Lauderdale (2006)	Bench scale experiment employing biologically acclimated GAC filters
d	Chae et al. (2006)	Experimental sand filters with two year acclimated sand
e	Druzina and Stegu (2007)	Batch experiment using groundwater
f	Goux et al. (2000)	Laboratory scale sand filter inoculated bacteria
g	Grunheid et al. (2005)	Riverbank filtration wells under oxic/anoxic condition and artificial recharge (**AR**) wells dominated by aerobic conditions
h	Heberer (2011)	River and lake bank filtration
i	Heberer et al. (2008)	Lake bank filtration under anoxic (**an**) and oxic/anoxic (**o**) conditions
j	Heberer and Adam (2004)	Artificial groundwater recharge pond (same site as Massmann et al. 2006)
k	Heberer et al. (2004)	River/ lake bank filtration
l	Ho et al. (2007)	Bench scale sand filters inoculated with taste and odor degrading bacteria
m	Ishida and Miyaji (1992)	Filters using aluminum oxide beads inoculated with MIB degrading bacteria and fed modified river water.
n	Juttner (1999)	Riverbank filtration under oxic/suboxic conditions with dilution identified as important removal mechanism
p	Massmann et al. (2006)	Average removals in the infiltrate from artificial groundwater recharge pond with varying redox conditions. (same site as Heberer and Adam 2004)
q	Massmann et al. (2008)	Lake bank filtration under anoxic (**an**) and oxic/anoxic (**o**) conditions
r	Richter et al. (2008)	Experimental rapid sand filter
s	Schwarzenbach et al. (1983)	Riverbank filtration with varying redox conditions
u	Snyder et al. (2007)	Pilot filters employing biological activated carbon (**A**) and biological sand (**S**)
w	Verstraeten et al. (2002)	Riverbank filtration with well oxygenated river water transitioning to anaerobic in the subsurface
x	Worch et al. (2002)	Biological test filter employing a non-adsorbing pumice stone
z	Zearley and Summers (2012)	Laboratory scale sand filters top 7.9 min (**a**) and total 15.8 min (**b**) of empty bed contact time

RESULTS AND DISCUSSION

The articles reviewed are not exhaustive. However, relatively few studies have investigated drinking water treatment systems where biodegradation is an important removal mechanism for organic micropollutant removal. Zearley and Summers (2012) provide the most extensive data as 34 compounds were evaluated with acclimated inert media, such that all of the removal cn be attributed to biological mechanism. Significant research has explored organic pollutant biodegradation removal efficiencies appropriate to wastewater treatment (Onesios et al., 2009). These removals were not included here because of the relatively high concentrations of target compounds and organics under wastewater relevant conditions. Conclusions on the microbial metabolism of micropollutants at a certain concentration in one environment may not be applicable at other concentrations or in other environments (Wang et al., 1985). Massmann et al. (2006) showed at an MAR site that under aerobic conditions phenazone was completely removed, but it was only partially removed under anaerobic conditions. In a rapid sand filter Chae et al. (2006) found marked impacts of temperature and acclimation on MIB removal. Acclimation is a very important aspect and many studies have not properly addressed the impact. Introducing new compounds to a biomass system will often lead to an under estimation of the removal potential, as there has not been sufficient time for acclimation. It may take months for biomass to acclimate to many micropollutants as they do not serve as the primary substrate.

Studies, such as Worch et al. (2002), have related compound structure and physiochemical properties to predict biodegradability of micropollutants. Such research is needed for practitioners to get a sense of what performance can be expected from drinking water treatment technologies which rely on biodegradation. Biodegradation pathways are also necessary to identify degradation products, their toxicity, and subsequently a risk assessment, since complete mineralization in these systems is rare (Ying et al., 2007). Further research is also needed to understand other environmental factors influencing biological degradation such as organic matter sources and synergistic negative effects of micropollutants on microorganism activity (Maeng et al., 2011).

Results from the batch experiments are most difficult to directly utilize in practical systems and as such results are not included in Table 2.1. In particular acclimation with batch systems is critical, as unacclimated biomass will yield an underestimation of removal efficiencies. In batch systems Ternes et al. (2002) evaluated nonadapted microorganisms in sand with anoxic groundwater and oxic surface water, while Snyder et al. (2004) used biological sand acclimated at a minimum of one month in river water Comparing the removals of carbamazepine between these two studies reveals 54% removal and no removal of the compound with acclimated aerobic microorganisms and nonadapted anoxic microorganisms, respectively. This discrepancy in removals shows that for the same compound it is difficult to predict biodegradability. Ying et al. (2003) used batch experiments with sediments in anaerobic and aerobic groundwater.

CONCLUSIONS

There has been some research conducted on biological removals of micropollutants in drinking water systems, but this research lags compared to the wastewater field. Generally, micropollutant removals cannot be predicted based on compound class, e.g., pesticide, PhAC, PCP, but dependent on background source water characteristics, influent concentration, operating conditions, and micropollutant physiochemical properties.

ACKNOWLEDGEMENTS

The authors wish to acknowledge National Water Research Institute Fellowship, the Malcolm Pirnie, Inc. Fellowship, and an EPA Science To Achieve Results (STAR) Fellowship (#91730001) for partial funding.

REFERENCES

Achten, C., Kolb, A. & Puttmann, W., 2002. Occurrence of methyl tert-butyl ether (MTBE) in riverbank filtered water and drinking water produced by riverbank filtration 2. *Environ. Sci. & Technol.*, (*36*)17, 3662-3670.

Bradley, P.M., Barber, L.B., Kolpin, D.W., McMahon, P.B. & Chapelle, F.H., 2007. Biotransformation of caffeine, cotinine, and nicotine in stream sediments: Implications for use as wastewater indicators. *Environmental Toxicology and Chemistry*, (*26*)6, 1116-1121.

Brown, J.C. & Lauderdale, C.V., 2006. Efficient, Simultaneous Destruction of Multiple Drinking Water Contaminants Using Biological Filtration. *FLORIDA WATER RESOURCES JOURNAL.*

Chae, S.H., Summers, R.S., Kim, S.M. & Ahn, H.W., 2006. Biofiltration of MIB and Geosmin under Varying Influent Conditions. *AWWA-Water Quality and Technology Conference.*

Druzina, B. & Stegu, M., 2007. Degradation study of selected organophosphorus insecticides in natural waters. *International Journal of Environmental Analytical Chemistry*, (*87*)15, 1079-1093.

Goux, S.J. et al., 2000. Biodegradation of atrazine in sand sediments and in a sand-filter. *App. Micro. and Biotechnol.*, (*54*)4, 589-596.

Grunheid, S., Amy, G. & Jekel, M., 2005. Removal of bulk dissolved organic carbon (DOC) and trace organic compounds by bank filtration and artificial recharge. *Water Research*, (*39*)14, 3219-3228.

Heberer, T., 2011. Occurrence and fate of pharmaceuticals during bank filtration–preliminary results from investigations in Germany and the United States.

Heberer, T. & Adam, M., 2004. Transport and Attenuation of Pharmaceutical Residues During Artificial Groundwater Replenishment. *Environmental Chemistry*, (*1*)1, 22-25.

Heberer, T., Massmann, G., Fanck, B., Taute, T. & Dunnbier, U., 2008. Behaviour and redox sensitivity of antimicrobial residues during bank filtration. *Chemosphere*, (*73*)4, 451-460.

Heberer, T. et al., 2004. Field studies on the fate and transport of pharmaceutical residues in bank filtration. *Ground Water Monitoring and Remediation*, (*24*)2, 70-77.

Ho, L., Hoefel, D., Bock, F., Saint, C.P. & Newcombe, G., 2007. Biodegradation rates of 2-methylisoborneol (MIB) and geosmin through sand filters and in bioreactors. *Chemosphere*, (*66*)11, 2210-2218.

Ishida, H. & Miyaji, Y., 1992. Biodegradation of 2-Methylisoborneol by Oligotrophic Bacterium Isolated from a Eutrophied Lake. *Water Science and Technology*, (*25*)2, 269-276.

Juttner, F., 1999. Efficacy of bank filtration for the removal of fragrance compounds and aromatic hydrocarbons. *Water Science and Technology*, (*40*)6, 123-128.

Maeng, S.K., Sharma, S.K., Lekkerkerker-Teunissen, K. & Amy, G.L., 2011. Occurrence and fate of bulk organic matter and pharmaceutically active compounds in managed aquifer recharge: A review. *Water Research*, (*45*)10, 3015-3033.

Massmann, G., Dunnboer, U., Heberer, T. & Taute, T., 2008. Behaviour and redox sensitivity of pharmaceutical residues during bank filtration - Investigation of residues of phenazone-type analgesics. *Chemosphere*, (*71*)8, 1476-1485.

Massmann, G. et al., 2006. The impact of variable temperatures on the redox conditions and the behaviour of pharmaceutical residues during artificial recharge. *Journal of Hydrology*, (*328*)1-2, 141-156.

Onesios, K.M., Yu, J.T. & Bouwer, E.J., 2009. Biodegradation and removal of pharmaceuticals and personal care products in treatment systems: a review. *Biodegradation*, (*20*)4, 441-466.

Richter, D., Massmann, G. & Dunnbier, U., 2008. Behaviour and biodegradation of sulfonamides (p-TSA, o-TSA, BSA) during drinking water treatment. *Chemosphere*, (*71*)8, 1574-1581.

Schwarzenbach, R.P. et al., 2006. The challenge of micropollutants in aquatic systems. *Science*, (*313*)5790, 1072-1077.

Schwarzenbach, R.P., Giger, W., Hoehn, E. & Schneider, J.K., 1983. Behavior of Organic-Compounds during Infiltration of River Water to Groundwater - Field Studies. *Environ. Sci. & Technol.*, (*17*)8, 472-479.

Snyder et al., S.A.S., E.C. Wert, H. Lei, P. Westerhoff and Y. Yoon, 2007. Removal of EDCs and pharmaceuticals in drinking and reuse treatment processes.Warter Research Foundation Report, Denver CO.

Snyder, S.A. et al., 2004. Biological and physical attenuation of endocrine pisruptors and pharmaceuticals: Implications for water reuse. *Ground Water Monitoring and Remediation*, (*24*)2, 108-118.

Ternes, T.A. et al., 2002. Removal of pharmaceuticals during drinking water treatment. *Environ. Sci. & Technol.*, (*36*)17, 3855-3863.

Verstraeten, I.M., Thurman, E.M., Lindsey, M.E., Lee, E.C. & Smith, R.D., 2002. Changes in concentrations of triazine and acetamide herbicides by bank filtration, ozonation, and chlorination in a public water supply. *Journal of Hydrology*, (*266*)3-4, 190-208.

Wang, Y.S., Madsen, E.L. & Alexander, M., 1985. Microbial-Degradation by Mineralization or Co-Metabolism Determined by Chemical Concentration and Environment. *J. of Agricul. and Food Chem*, (*33*)3, 495-499.

Worch, E., Grischek, T., Bornick, H. & Eppinger, P., 2002. Laboratory tests for simulating attenuation processes of aromatic amines in riverbank filtration. *Journal of Hydrology*, (*266*)3-4, 259-268.

Ying, G.G., Kookana, R.S. & Dillon, P., 2003. Sorption and degradation of selected five endocrine disrupting chemicals in aquifer material. *Water Research*, (*37*)15, 3785-3791.

Ying, G.G., Yu, X.Y. & Kookana, R.S., 2007. Biological degradation of triclocarban and triclosan in a soil under aerobic and anaerobic conditions and comparison with environmental fate modelling. *Environmental Pollution*, (*150*)3, 300-305.

Zearley, T., Summers, R. S., 2012. Removal of Trace Organic Micropollutants by Drinking Water Biological Filters *Environ. Sci. Technol.*, 46(17):9412-9.

Chapter 3

Overview on the current condition of slow sand filtration and its challenges in Japan

*S. Yamamura, Dr.Eng.**

**W&E Research Institute, Japan*

ABSTRACT

The author gives the overview on the present condition of slow sand filtration (herein after referred to SSF) in Japan. SSF was introduced to Japan as the modern water supply technology and played significant roles in the development of Japan's water works before the World War II. However, rapid sand filtration (herein after referred to RSF) became more popular in new plans after the 1960's, reflecting limited space availability in urban areas. SSF received pressures by urban development and source water degradation, and some SSF plants were forced to close. As the result, the amount of purified water by SSF was reduced to 40% of the peak time in the past. Having said that, the number of SSF still amounts to 2,400 in Japan, of which small scale SSFs are dominant. They require technical assistance and renovation of technologies.

Keywords Slow sand filtration; SSF, RSF, renovation, operation and maintenance

INTRODUCTION

The 5th International Slow Sand and Alternative Biological Filtration Conference provide experts to share information and discuss future measures and to improve the recognition of SSF and biological filtration. With those in mind, the author highlights the current condition of SSF in Japan, aiming at contributing to sharing information.

This report reviews the literatures and statistical data of the water service retroactively to the past about the use situation of SSF in Japan. The change by times was observed in order to investigate the factor of the change. Furthermore, some SSF facilities in Japan were visited to grasp their problems by interviews.

HISTORICAL OBSERVATION OF JAPAN'S SSF

The situation from the foundation time to prewar days

The SSF, first founded in the U.K. in 1829, spread in Japan as an essential method to purify river water, as it shows remarkable functions on removals of suspended matter and bacteria. The first plant of SSF was completed in 1887 at Yokohama-Nogeyama, followed by many other cities including Tokyo-Yodobashi (1898).

Some of plants completed in this time are still in use after 100 years, which include Hakodate-Akagawa-lower block (1888), Okayama-Mino (1904), Shimonoseki-Takao (1905), and Sasebo- Yamanota (1906). Other plants whose origins back to this time include Aomori-Yokouchi (1909), Nagoya-Nabeyaueno (1914), Osaka-Kunijima (1914), Matsue-Inbe (1917) and Ueda-Someya (1923).

In the 1930's, some researches were conducted in order to extend filtration continuation days and to increase filtration speed. The experimental studies on "two stages' filtration", begun in Osaka from 1929 and in Kyoto from 1925, were addressed this objective and to reduce frequency of sand-layer scratching. It was confirmed experimentally that expected results were achieved by pre-filtration, which lead to further study for practical use. Moreover, there were a lot of endeavors to increase the filtration speed, which had been almost fixed to 3m/day conventionally. Tokyo-Sakai water purification plant demonstrated in 1931 that filtration speed can be increased to 9m/day and proposed a remodeling plan of the filtration pond structure. The demand to rationalize the operation of SSF required the proof by basic studies; many works were conducted from 1935 to 1940, with a careful attention on filtration.

The SSF plants completed in this period include Kanazawa-Sue (1932), Morioka-Yonai (1934), Kakuda-Oda (1934) and Kanagawa-Tanigahara (1942).

The situation to date after the World War II

After the war, reinforcement of the chlorine sterilization was ordered under the occupation policy of the Operation Army, and chlorine dose were increased to several times, even much, higher than prewar time. Accordingly the burden of a filter basin, especially those for bacterial removal was eased remarkably. On the other hand, progress of the chemical coagulation basin in the postwar period achieved remarkable effects in removal of suspended matter. At the time of high turbidity of raw water, chemical coagulation was also applied at regular type sedimentation basin for SSF, and the burden on the suspended matter removal by SSF basin was also reduced.

In this way, the burden to a filtration function was eased not only from bacterial removal but also from suspended matter removal, as a result, RSF spread quickly and widely. Moreover, large water purification plants, which faced with shortage of lands for new installation and extension, chose RSF which can process a lot of water at a narrower site. On the other hand, the SSF, which needed good quality of raw water, faced with the problem on degradation of raw water quality by contamination of source environment. This also led to reduction on the operating ratio of SSF.

The SSF plants completed in this period include Shimane-Imazu (1969), Hiroshima-Noda (1977), and Shimane-Gozu (1985). The most recent completions were reported in Shimane-Mishiro (2011), and Nagoya-Nabeyaueno renovation (to be completed in 2014).

The present conditions and change of the plants ability

As shown in Table 3.1, there are about 2,400 purification plants using SSF totally in Japan now, including small scale waterworks. Except plants founded before the war, there are many small-scales plants in local cities. As shown in Figure 3.1, their purified water in a year covers 4 % of total amount. On the other hand, small scale waterworks shows a higher percentage of 24% on SSF than 22% on RSF; this is due to the reason that the design criteria on governmental subsidy strongly advised the use of SSF as far as possible.

Table 3.1 Details of the distribution of SSF in Japan

Water works 2011 (with population more than 5001)			
Capacity	Annual purification amount 1000m3	Daily amount of purified water m3/day	Number
X>30000m3/day	31,888	304,060	6
30,000> X > 10,000m3/day	206,482	565,704	34
10,000> X > 5,000 m3/day	85,038	232,981	32
5,000> X > 1,000 m3/day	107,542	294,636	125
1,000 m3/day >X	10,694	29,299	79
	520,738	1,426,679	276
	average	5,169m3/day	

small scale water supply	2011
Block of prefectures	Number
Hokkaido	136
Tohoku	153
Kanto	70
Chubu	331
Kansai	337
Chugoku	275
Shikoku	361
KyushuOkinawa	499
All Japan	2162

The data for 40 years from 1968 to 2008 shows that the amount of purified water of the SSF of water works (with population 5001 or more) decreased to 40%, as shown in Figure 3.2. However, SSF of bulk water supply showed a slight increase in this period. The decrease in this period caused mainly by the change in large cities such as Tokyo-Yodobashi-plant (closed in 1965 for relocation)and Osaka-Kunijima-plant (closed in 1974 due to aggravation of raw water quality). However, in Japan, about 70 % of the capacity has been preserved after 50 years even after the operating ratio decrease, as shown in Figure 3.3.

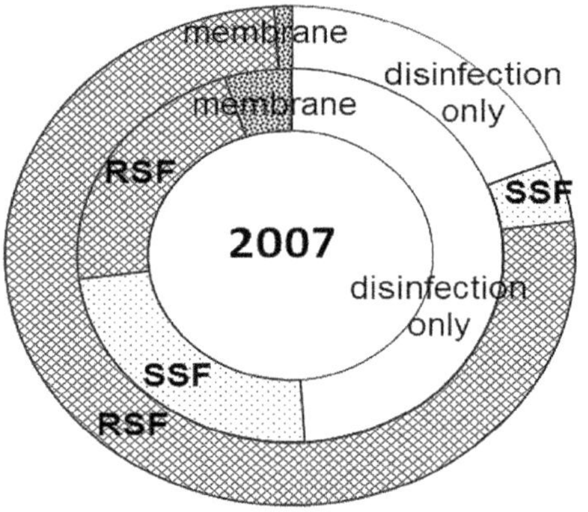

	disinfect ion only	SSF	RSF	membrene
small scale <5000	49%	24%	22%	5%
water supply >5001	19%	4%	77%	1%

inner circle: small scale < 5000
outer circle: water supply > 5001

Figure 3.1 Share of amount of water by purification methods

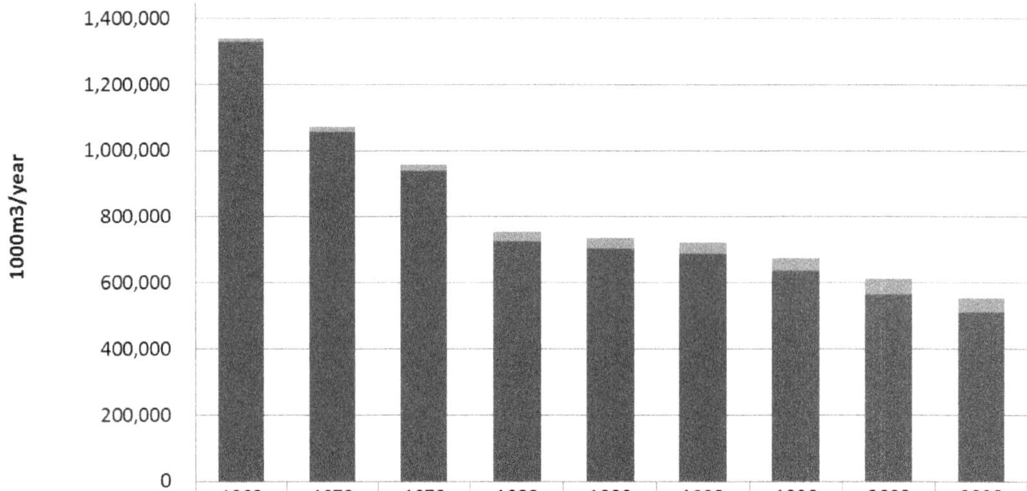

	1968	1973	1978	1983	1989	1993	1998	2003	2008
Bulk water supply	11,503	17,836	19,955	27,244	32,785	34,434	36,841	47,437	44,177
Waterworks	1,329,66	1,055,41	938,423	726,107	703,278	688,671	636,585	564,139	510,179

Figure 3.2 Purification amount by SSF (1968-2008) in Japan

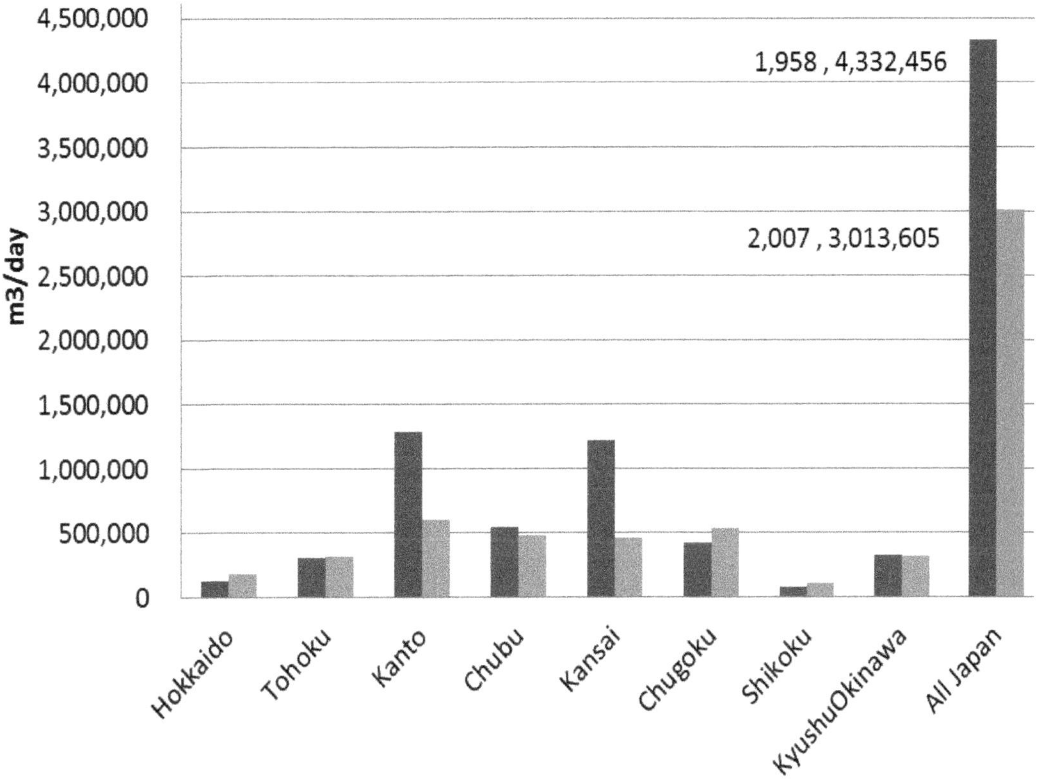

Figure 3.3 Comparison of SSF capacities by block of prefectures between 1958 and 2007

In Figure 3.4, the transition of the number of installations of the SSF in small-scale water supply is shown. Although about 2,700 installations existed in 1965 and the number exceeded 3,000 installations for about 25 years after the 1970s to, it decreased and has become about 2,200 installations by now. However, SSF is still playing a role of main water-purifying facilities in small water supplies. RSF in small-scale water supply showed the peak of about 1700 installation at around year 2000, but it decreased after that and is now about 1500. The height of the bar chart of each year shows the number of small-scale water supply, and, the latest number became below the half in 1965. Although SSF was about 20% of the number of the whole in 1965, as there were many water projects that did not have water-purifying facilities other than chlorination, the recent figure of SSF shows 33% of total.

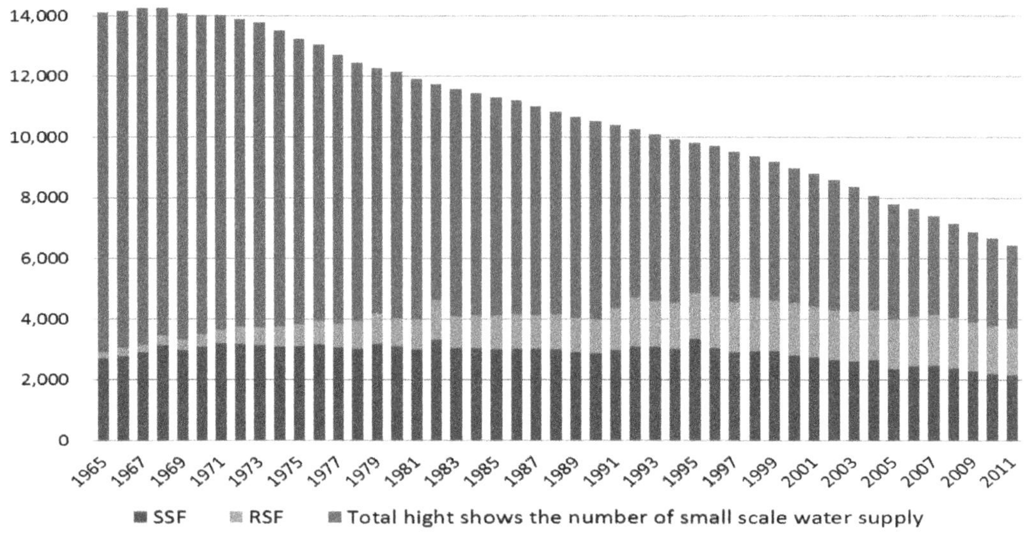

Figure 3.4 SSF in small-scale waterworks in Japan (1965-2011)

MAIN FACTORS ON SSF'S DECREASE IN JAPAN

Pressure from urban development

Since SSF needs larger land space compared with RSF, development around water purification plants or environmental condition change may cause concerns about more effective use of land, and thus continuation of SSF plant may become difficult. Abolition of Tokyo-Yodobashi water purification plant, constructed in 1898 and had SSF of a 240,000-m3/day, is the representative case being subject to development pressure.

Shinjuku area, which used to be a suburbs agricultural area, changed into the area of importance for traffic and commercial purpose in 1920's, and then a large land of the water purification plant drew the attention of relocation. By late in the 1950's, Capital Region Development Commission planned to build the Shinjuku high-rise complex at 960,000 m^2, including the site of Yodobashi water purification plant of 340,000 m^2. The Yodobashi water purification plant transferred the function to the Higashi-Murayama water purification plant, and was abolished in 1965.

Water source quality and environmental changes

The SSF is the effective water-purifying method, when source water quality is good, and it may become difficult to use, if aggravation of source water quality progresses. Abolition of the Osaka Kunijima's SSF is the representative case which was subject to the influence of change of source environment. The Osaka-Kunijima water purification plant completed on a scale of 151,800 m3/day in 1914, and had SSF of 601,000 m3/day at the time of the maximum. However, the declining water quality of raw water ran around from 1960, and the chemical coagulation-sedimentation system was introduced to cope with the change. In addition to shortening of filtration sustaining days, the outflow of manganese resulting from lack of dissolved oxygen and the problem of water-coloring by a post-chlorination arose. Moreover, the frequency of scratching sand layer increased and actual filtration capability declined to 130,000 m3 at the time of the worst, and, it was estimated as 400,000m3 at the best condition. As a result, SSF in Osaka-Kunijima was obliged to change to RSF, and came to stop operation in 1974.

Increase of the problems related to maintenances cost

Scratching of bio-filter has been most typical and important work on operation and maintenance of SSF, which requires mastery of skills and physical strength. With decline in the operation efficiency by expert's shortage, improvement in the operation efficiency by mechanization is needed. For this reason, bicycle trailers and conveyor belts came to be used in the SSF pond cleaning work, and also the run type-scratching machine was invented. Such efforts to mechanize work were paid. However, increases of maintenance cost by the rise of personnel expenses, and difficulty on recruiting talented-people make continuation of SSF more and more difficult.

Other factors

In recent years, the water works that takes raw water from sources with fear of contamination by Cryptosporidium came to be asked for strengthening of the measure. Although SSF is the water-purifying processing that can remove Cryptosporidium, water suppliers are requested not only to meet the standard turbidity (2 degree or less), but also to maintain the turbidity of water of filter basin exit at 0.1 degree or less. Fine operation management, such as using the fully adjusted turbid meter in that case, is required. Some cases were reported recently to choose membrane process when renovation of SSF was planned.

THE PROBLEM AND CHALLENGES OF SSF IN JAPAN

In Japan, some of SSFs have problems on operation and maintenance with the change of source river quality, and a rise in management cost of the facilities including personnel expenses. On the other hand, a lot of facilities face with aging problems and their concerns include degradation of the function and earthquake resistance of facilities. Although positive aspects on SSF has widely been recognized, such as good water quality, easy operation and maintenance and less expensive operation cost, some negative aspects are also raised, such as necessary large land, difficulty to cope with aggravation of raw water quality, not fitting for rationalization of maintenance management, such as automatic control. Moreover, in spite of general understanding that O&M is easy, the reality is different and efforts to deal with subjects are reported, such as increased efforts to cope with aggravation of raw water quality and difficulties to form good biological slime. In a middle and small waterworks, there is a tendency that renovation plans have not been progressed easily due to financial difficulties: development of renovation technique without spending big expense (or prolongation-of-lifetime) is desired.

From the consumers' viewpoint, SSF gives an effective message that purification is carried out with natural biological process, which do not require artificial additives, in order to improve consumers' safe image on tap water. Practical use of SSF is effective also for a posture of water works as a business unit. It would be possible to solve negative aspects of SSF, if firm researches are conducted to activate R&D on SSF not only by waterworks but by private companies, as there are possibilities of new technical development which harnesses the feature of SSF. In overseas, there are some examples of SSF combined with ozonization.

The following aspects are raised as examples of future examination subjects but not limited to: (1) pre-treatment system to cope with aggravation of raw water quality, e.g. roughing filtration, ozonization; (2) the intake managerial operation system interlocked with raw water turbidity measurement; (3) water-purifying turbidity management and an alarm control system; (4) consulting/construction of prolongation-of-lifetime of a superannuated systems; (5) practical use of SSF combined with ozonization and BAC treatment; (6) development of the future type slow sand filtration system

In addition, formation of the network of waterworks concerned with SSF would be necessary to exchange information and experience. This requires the center function that accumulates the technology and experience about SSF. Moreover, a network relevant to international cooperation is also expected.

CONCLUSIONS

The main conclusions of this work are:

1) The data for 40 years from 1968 to 2008 shows that the amount of purified water of the SSF of water works (with population 5001 or more) decreased to 40%, however, about 70 % of the capacity has been preserved after 50 years even after the operating ratio decrease.
2) The data of existing 2,400 SSF plants is a proof of necessity on technical development for improvement and renovation of plants for the future.
3) Formation of the network of waterworks concerned with SSF would be necessary to exchange information and experience.

ACKNOWLEDGEMENTS

The author wishes to acknowledge the important contribution of Prof..Masao Yamada, Prof.Nobutada Nakamoto and Mr.Tomoaki Isaji for providing comments with certain aspects of this work.

REFERENCES

JWWA (1967) Japan's History of Water Works.

Tokyo Metropolitan Water Works Bureau (1966), History of Yodobashi Water Purification Plant.

Tokyo Metropolitan Water Works Bureau (1999), A Hundred Years of Modern Water Supply in Tokyo, Sectional View.

Osaka City Water Works Bureau (1982), Eighty Years of Water Supply of Osaka City.

JWWA, Japan Water Supply Statistics (yearly publication)

The website of Nagoya City Waterworks & Sewerage Bureau, http://www.water.city.nagoya.jp/english/

Chapter 4

The Centenary Slow Sand Filtration in Nagoya City – A 100 Year Trend of Water Quality

K. Mori
Nagoya City Waterworks & Sewerage Bureau

ABSTRACT

A slow sand filtration system is one of water purification methods. When raw water flows through a sand layer, impurities are removed by a biofilm which is grown on the surface of the sand layer. The data of the water treated by slow and rapid sand filtration systems, which had been accumulated for nearly 100 years, were analyzed to watch the long-term trend. As 14 indexes including evaporation residue, hardness and standard plate count, have been measured since our establishment, the past and present water qualities can be compared. During past 100 years, raw water was severely polluted twice, around 1960 and during 1980s. It is inferred that the causes of the pollution are industrial wastewater for the former and domestic wastewater for the latter. Both slow and rapid sand filtration systems have capacity to remove the pollutant in domestic wastewater. But, a slow sand filtration system could not enough remove the pollutant by industrial wastewater. The cause may be the interference in the activity of the biofilm by industrial pollutant. Generally speaking, colour is removed well by a rapid sand filtration system. But, it is found that, without the interference by industrial pollutant, colour is also removed well by slow sand filtration system. On the other hand, a slow sand filtration system is known to remove ammonium nitrogen (NH_4^+-N) well, which was realized by comparing the chlorine injection rates in slow and rapid sand filtration systems. This shows that slow sand filtration system has an advantage in inhibiting the chloramine forming reaction which causes chlorinous odor.

Keywords Slow sand filtration; removal of ammonium nitrogen; chloramine; chlorinous odor

INTRODUCTION

A slow sand filtration system is one of water purification methods. When raw water flows through a sand layer, impurities are removed by a biofilm which is grown on the surface of the sand layer. Compared with a rapid sand filtration system, a slow sand filtration system can be operated with simple facilities and maintenance, but it needs a large site and advantageous raw water quality. So it is a minor method in Japan today.

Nagoya City Waterworks & Sewerage Bureau has operated the slow sand filtration system since 1914. And this year, 2014, is the 100th anniversary. Water environment of this area have changed for the times since our establishment. The data of the water quality, which had been accumulated for nearly 100 years, were analyzed to watch the long-term trend.

HISTORY OF THE SLOW SAND FILTRATION

Supplying water started in 1914 in the Bureau. At that time, eight slow sand filters were built in Nabeyaueno Water Purification Plant. All the water to supply was treated by slow sand filtration. More six slow sand filters were established in 1928.

(a) Setting of Supporting Layer
(Porous Brick and Gravel Layer)

(b) Setting of Filter Sand

(c) Building of the Service Reservoir (in 1913)

(d) After Starting to Supply Water

Figure 4.1 Photographs of the slow sand filtration facility in the pioneer days

In 1936, the rapid sand filters were established to increase the capacity of Nabeyaueno Water Purification Plant. Then, two plants, Oharu and Kasugai Water Purification Plants were also established in 1946 and in 1969, respectively. At these two plants, all the amount of the water is treated by rapid sand filtration.

As a result, judging from the quantity of water, our main system is rapid sand filtration (Table 4.1). A slow sand filtration system is rare among large cities in Japan because it needs the advantageous raw water quality. The system has been our symbol of "Nagoya, the City with Delicious Water".

Table 4.1 Plants and the capacities

Filtering System / Plan	SLOW Sand Filtration	RAPID Sand Filtration
Kasugai	–	590,000 m^3/day
Nabeyaueno	140,000 m^3/day (Under Reconstruction)	150,000 m^3/day
Oharu	–	544,000 m^3/day
Total	140,000 m^3/day	1,284,000 m^3/day

The overage slow sand filtration facilities are being rebuilt to celebrate our 100th anniversary. The rebuilding is now in the final phase.

OUTLINE OF EXISTING DATA

Table 4.2 shows the existing data now in the Bureau.

Table 4.2 Existing data about the slow sand filtration

Name	Contents	Target Indexes	Period of Statistics
Annual Statistics	Maximum, minimum and average values of the analysis results done several times a year	**Water Temperature** **Turbidity** **Colour** **Taste** **Odor** **Alkalinity** **Chloride** **Sulfate** **Nitrate Nitrogen** **Nitrite Nitrogen**	1915～1929
Monthly Statistics	Maximum, minimum and average values of the analysis results done several times a month		Jan. 1925～ Mar. 1949

(Continued)

Table 4.2 Existing data about the slow sand filtration (*Continued*)

		Potassium Permanganate ($KMnO_4$) Consumption **Total Solids** **Hardness** **Standard Plate Count** (14 indexes in total)	
Monthly Reports	Values of the analysis results done once a month	In addition to the above 14 indexes, **Free Carbon Dioxide** **Acidity** **pH** **Iron** **Residual Chlorine** **Total Coliforms** (20 indexes in total in the year 1951) After that, number of indexes analyzed is increased with the development of analyzing instruments and technology (heavy metals, VOCs and pesticides). More than 300 indexes are analyzed at the present time.	Jan. 1951～ Present
Monthly Series of Daily Reports	Extracts from the "Daily Report", measured to keep water treatment well managed	**Water Temperature** **Turbidity** **Colour** **pH** **Alkalinity** **Potassium Permanganate ($KMnO_4$) Consumption** **Electric Conductivity** **Residual Chlorine** (filtered water only) **Operational Conditions** Air Temperature Treating Water Quantity Injection Rates of Coagulants and Disinfectants	Dec. 1960～ Present

TRENDS OF THE RAW WATER QUALITY

To analyze the 100-year trend of the raw water quality, the Annual Statistics, the Monthly Statistics and the Monthly Reports were studied. In this study, the data in the Monthly Statistics and the Monthly Reports are converted to the annual average for statistical reason.

Several indexes were found to show the characteristic trends. Three trends of them are as follows.

INDEXES INCREASING SINCE THE ESTABLISHMENT

Hardness shows the trend increasing since the establishment. Though the statistic periods are short, Electric Conductivity and Sulfate also show this trend (Fig. 4.2). Hardness is calculated from the concentration of Calcium and Magnesium. Electric Conductivity reflects the concentration of dissolved anions and cations mainly from inorganic salts.

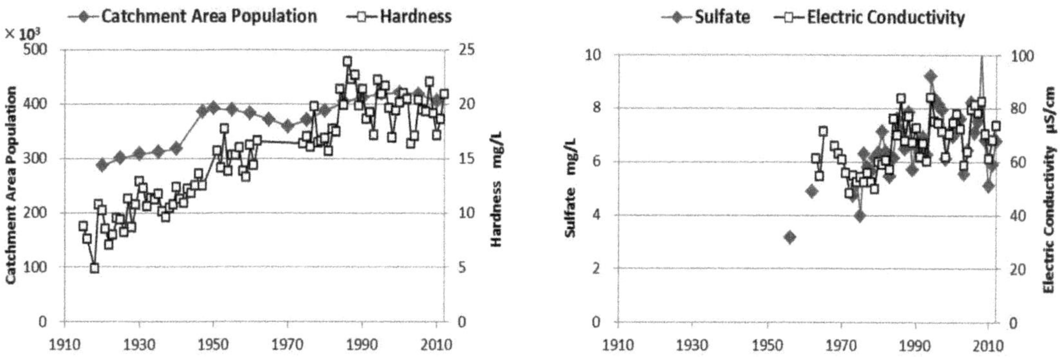

Figure 4.2 Long-term trends of the Population of the catchment area, and Hardness, Sulfate and Electric Conductivity of the raw water

These indexes are caused by inorganic pollution and hardly removed by a normal sewage treatment. The population of the catchment area shows the similar trend to them. These trends seem to be caused by domestic wastewater from the areas along the Kiso River.

INDEXES HAVING PEAKS AROUND 1960

Turbidity, Potassium Permanganate ($KMnO_4$) Consumption and Colour show the trend having peaks around 1960 (Fig. 4.3 and 4.4). Industrial wastewater from the manufacturing plants along the Kiso River may be the main cause of the trend. These results have led to more severe legal regulations and the campaign to keep the water resource clean, then the water quality of Kiso River has improved[1).]

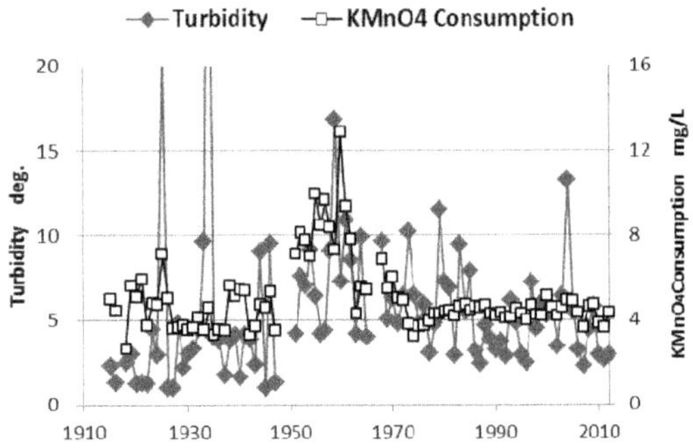

Figure 4.3 Long-term trends of Turbidity and $KMnO_4$ Consumption of the raw water

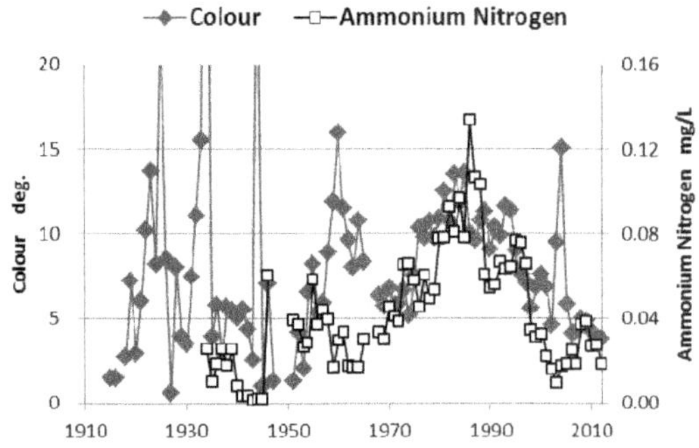

Figure 4.4 Long-term trends of Colour and Ammonium Nitrogen of the raw water

INDEXES HAVING PEAKS DURING 1980s

Colour and Ammonium Nitrogen (NH_4^+-N) show this trend (Fig. 4.4). Turbidity and $KMnO_4$ Consumption don't have peaks on 1980s. From this decade, sewage treatment became popular along the Kiso River. So these pollutants are probably caused by domestic wastewater specially night soil[2].

TRENDS OF THE QUALITY OF THE SLOW SAND FILTER EFFLUENT

Similar to the raw water, the quality of the slow sand filter effluent was also studied.

Hardness, Electric Conductivity and Sulfate keep increasing (Fig. 4.5). These trends are same to the ones of the raw water above described. Generally, these indexes are not removed by filtration so this trend is expected result.

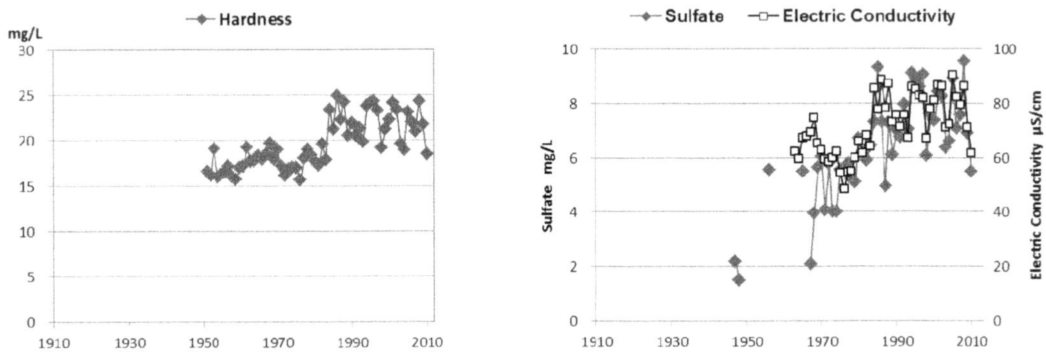

Figure 4.5 Long-term trends of Hardness, Sulfate and Electric Conductivity of the filtered water

Then the four indexes which have peaks in the raw water quality, $KMnO_4$ Consumption, Turbidity, NH_4^+-N and Colour, were investigated especially during the peak period for both slow and rapid sand filter effluents (Fig. 4.6).

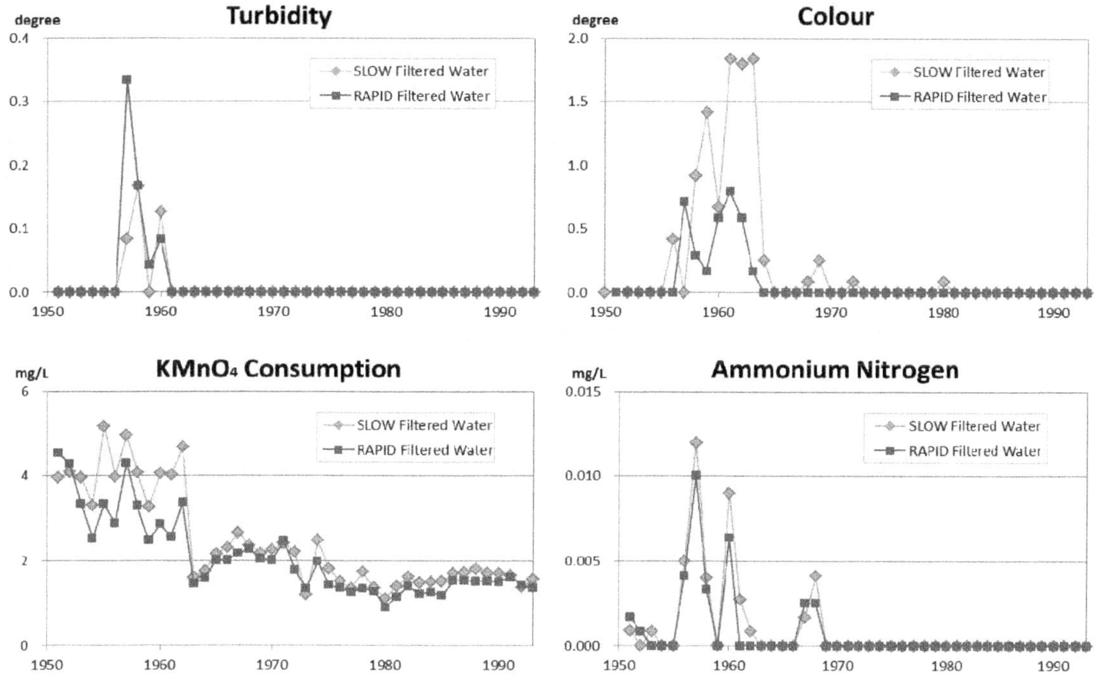

Figure 4.6 Comparison of four indexes between slow and rapid filtered water

Turbidity and Colour can be removed by both slow and rapid sand filtrations. In almost all cases, these two indexes are below the detection limits. But in some cases around 1960, the period that Turbidity and Colour showed the peak in the raw water, Turbidity and Colour were not removed completely. The cases which Turbidity and/or Colour were detected in the filter effluents were counted and shown in Table 4.3. Generally, a rapid sand filtration is better on removing Colour than a slow sand filtration. Table 4.3 shows the same result.

Table 4.3 Number of cases which turbidity and/or colour were detected and the maximum concentration (units in degree)

Index	SLOW Sand Filter Effluent cases/max conc.	RAPID Sand Filter Effluent cases/max conc.	Water Quality Standard
Turbidity	4 times/1.5	8 times/1.5	2
Colour	57 times/5.0	34 times/3.0	5

As mentioned above, there are two peaks of raw water Colour, one around 1960 and the other in the 1980s (Fig. 4.4). On the other hand, Fig. 4.6 shows only one peak of treated water Colour around the year 1960. This shows that the causes of two peaks of Colour in raw water are not the same. The former peak around 1960 may be caused by industrial wastewater. The Kiso River was heavily polluted at that time by the surrounding manufacturing plants. Coagulation was inhibited in the rapid sand filtration and the activity of microorganism lowered in the slow sand filtration. In contrast, in 1980s, both rapid and slow sand filtrations were able to remove Colour caused by domestic wastewater enough.

Both rapid and slow sand filtrations were able to remove $KMnO_4$ Consumption. $KMnO_4$ Consumption in the filtered water was lower than the one in the raw water. But there were peaks on both the slow and rapid sand filtered water. No clear different tendency to remove $KMnO_4$ Consumption was found between two filtrations.

For NH_4^+-N, strange trends were found. Like Turbidity and Colour, NH_4^+-N was removed well by both the slow and rapid sand filtrations in most cases, but detected in the filtered water sometimes around 1960.

This is strange. Generally, a slow sand filtration removes NH_4^+-N well and a rapid sand filtration hardly removes it in the filtration process (Fig. 4.7). Data used in the discussion are about the slow and rapid sand filtered water drawn after disinfection process. Even if NH_4^+-N was not removed completely in the filtration process, NH_4^+-N would react with chlorine in the following disinfection process. Normally, there are no possibilities that NH_4^+-N is detected in the chlorine-disinfected water. According to the data of Residual Chlorine of the same water samples, disinfectant was injected enough and the water was surely disinfected.

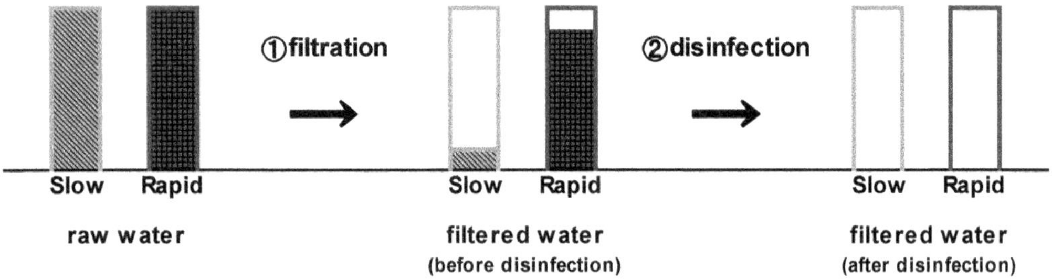

① Ammonium Nitrogen in raw water is mostly removed in slow sand filtration by its biological activity, but hardly removed in rapid sand filtration.
② In disinfection process, Ammonium Nitrogen is completely consumed by disinfectant to form chloramines.

Figure 4.7 Two Removing Processes of Ammonium Nitrogen

There are not enough data and information to discuss this strange phenomenon. The pollutants like amine type compounds in industrial wastewater may cause "false positive" in Nessler's method[3)].

MERIT OF SLOW SAND FILTRATION

As mentioned above, slow sand filtration removes NH_4^+-N better than rapid sand filtration because it removes NH_4^+-N by biological activity. To examine this advantage, It is necessary to compare NH_4^+-N concentrations of the slow sand filter effluent with the one of the rapid sand filter effluent before the chlorine-disinfection. Unfortunately, such data are not available.

Then, instead of NH_4^+-N before the disinfection process, the data of the Chlorine Injection Rates from "Monthly Series of Daily Reports" were noticed. NH_4^+-N reacts with Chlorine with the ratio of about 1:10 and consumes it. Chloramines, which causes chlorinous odor, are formed during this reaction. So, to disinfect the water containing NH_4^+-N, excess chlorine is needed according to NH_4^+-N concentration.

Comparing the Chlorine Injection Rates, in most cases, the rate in the slow sand filtration is lower than the one in the rapid sand filtration (Fig. 4.8). This tendency is more remarkable in 1980s, the period that the peak of NH_4^+-N was found in the raw water. That is to say, the slow sand filter effluent contains less NH_4^+-N than the rapid sand filter effluent because slow sand filtration removes NH_4^+-N better than rapid sand filtration.

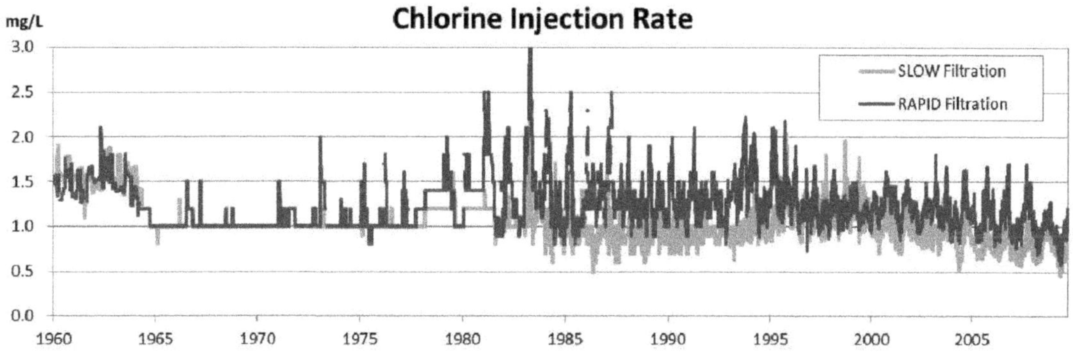

Figure 4.8 Comparison of Chlorine Injection Rates between slow and rapid filtration systems

In case that the raw water contains high concentration of NH_4^+-N, slow sand filtration has the merit to avoid the chlorinous odor in the treated water.

Nowadays, sewage treatment system is popular along the Kiso River, so there is not so much risk that the raw water is contaminated by high concentration of NH_4^+-N like in 1980s. But the author and the colleagues, members to aim "the Most Delicious Water in Japan" should regard this function as important.

CONCLUSIONS

As a result of this study to investigate the accumulated data, main conclusions are as follows.

(1) During past 100 years, raw water was severely polluted twice, around 1960 and during 1980s. It is inferred that the causes of the pollution are industrial wastewater for the former and domestic wastewater for the latter;

(2) Both slow and rapid sand filtration systems have capacity to remove the pollutants in domestic wastewater;

(3) With pollution by industrial wastewater, neither slow nor rapid sand filtration systems could not remove the pollutants enough;

(4) Generally speaking, Colour is removed well by rapid sand filtration system. But, it is found that, without the interference by industrial pollutants, Colour is also removed well by slow sand filtration system;

(5) Slow sand filtration system is known to remove Ammonium Nitrogen (NH_4^+-N) well. Comparing the Chlorine Injection Rates in slow and rapid sand filtration systems, NH_4^+-N is better removed by slow sand filtration system. This shows that slow sand filtration system has an advantage in inhibiting the chloramine forming reaction, which causes chlorinous odor.

REFERENCES

1) Nagoya City Waterworks Bureau (1964) Fifty-year History of the Waterworks, City of Nagoya.
2) Isaji, C. (2008) Estimate of the ammonium nitrogen load and concentration in water resource area region. *Kosui*, **8**, 100-104.
3) Japan Water Works Association (1978) Analyzing methods for waterworks.

Chapter 5

The centenary slow sand filtration in Nagoya City - the reconstruction of the slow sand filters

*A. Okuyama**

**Nagoya City Waterworks and Sewerage Bureau, 1-1, Sannomaru 3-Chome, Naka-ku, Nagoya, JAPAN*

ABSTRACT

The waterworks of Nagoya City reaches the centenary in 2014. The slow sand filters in Nabeyaueno Water Purification Plant (WPP) had worked since the waterworks started in 1914. Because those facilities became too old for further use, they have been totally under reconstruction since 2010. The slow sand filtration process requires better quality raw water. Nagoya City Waterworks and Sewerage Bureau is committed to maintaining raw water quality through water resource conservation measures and continuing the slow sand filtration system that makes the best use of the high quality raw water.

Keywords Slow sand filter; reconstruction

INTRODUCTION

Start of waterworks in Nagoya City dates back to 1914. The slow sand filters in Nabeyaueno WPP, the oldest one of three plants of Nagoya waterworks, are one of the largest of its kind in Japan. Construction of first eight slow sand filters were started in May 1910 and completed in March 1914. Since water supply was started in September 1914, they have supplied reliable, safe and tasty water continuously. After six filters were newly added, the filtration capacity of fourteen filters in total became 140,000m^3/day in 1928.

Since the facilities had operated nearly 100 years, they were too old for further use and had to be renewed and raised seismic performance. Therefore all filters have been under reconstruction since 2010 to restart water supply on the centenary of Nagoya waterworks in 2014. The new slow sand filters total twelve and the filtration capacity is 140,000m^3/day.

This paper describes the feature of Nagoya's slow sand filtration system and the reconstruction scheme of the filters.

GEOMORPHOLOGY AND WATERWORKS IN JAPAN

Japan is an archipelago, consists of four large islands and more than 6,000 small islands. Japan is mountainous country with two thirds of its land covered in forests. Therefore rivers in Japan have the characteristics of running short in distance and having rapid streams. Kiso Rever, the water source of Nagoya waterworks, is the seventh longest river in Japan, running 229km long (Figure 5.1). The average precipitation in Japan is 1,668 mm/year, which is about twice as much as that of the world average 815mm/year. These geographical features provide water utilities abundant water sources.

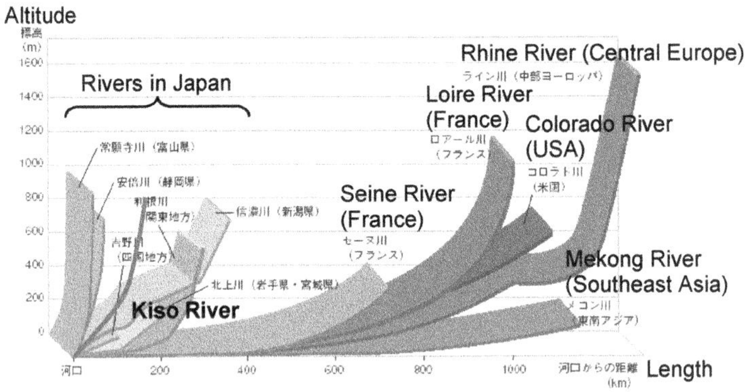

Figure 5.1 Rivers around the world Ministry of Land, Infrastructure, Transport and Tourism. Rivers around the world. Online.

Japanese waterworks have reached about 98% penetration rate at the present day, since the first modern waterworks of Japan has been introduced in Yokohama City in 1887. And Japan still remains as one of some countries, you can drink tap water.

River surface water including dams comprises 72.6% of the water sources (Figure 5.2). 77.7% of water treatment system is the rapid sand filtration, while the slow sand filtration is 3.4% (Figure 5.3). As the slow sand filtration system requires better quality water resource and larger land than others, they are now used less and less in large cities.

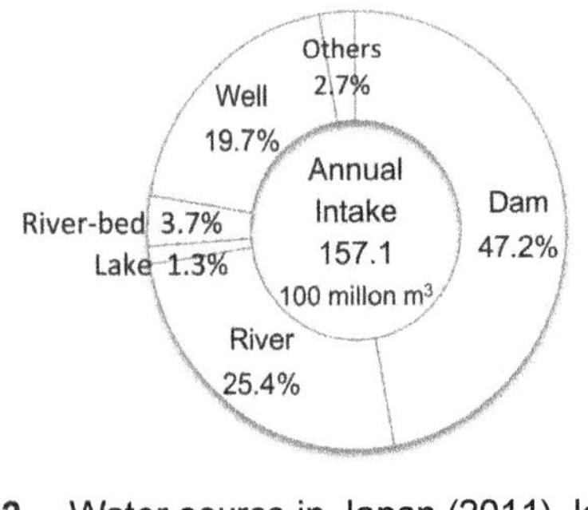

Figure 5.2 Water source in Japan (2011) Japan Water Works Association

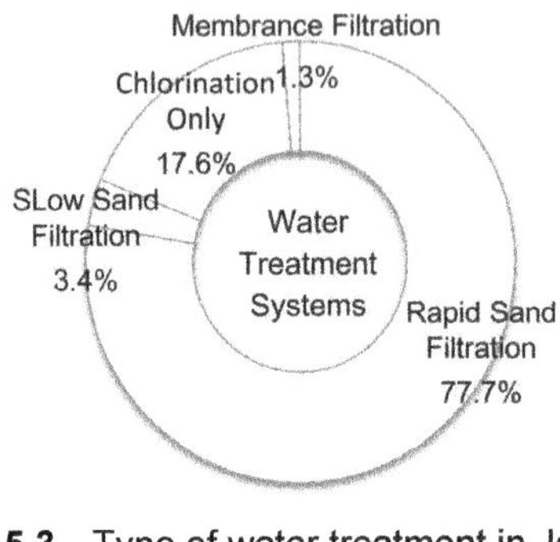

Figure 5.3 Type of water treatment in Japan (2011) Japan Water Works Association

WATERWORKS AND SLOW SAND FILTRATION IN NAGOYA

Surface water of Kiso River is taken in at two intake facilities, and sent to three WPPs in Nagoya waterworks. Then purified water is distributed to customers directly or through service reservoirs (Figure 5.4). Most of the water treatment system of Nagoya is rapid sand filtration. There is also slow sand filtration in Nabeyaueno WPP, the oldest plant of Nagoya waterworks (Table 5.1).

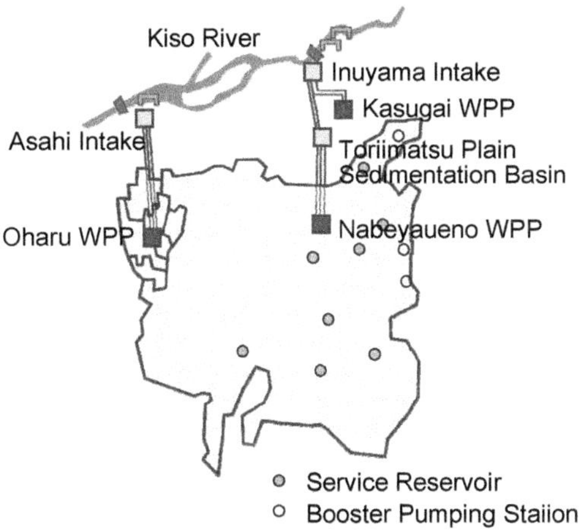

Figure 5.4 Waterworks facilities of Nagoya City

Table 5.1 Water supply capacity of Nagoya City

Intake	Water purification plant	
	Name	Water supply capacity (m^3/day)
Inuyama	Kasugai WPP	Rapid sand filtration 590,000
	Nabeyaueno WPP	Slow sand filtration 140,000
		Rapid sand filtration 150,000
Asahi	Oharu WPP	Rapid sand filtration 544,000
Total		1,424,000

Figure 5.5 shows the slow sand filtration system of Nagoya City. Raw water is taken at Inuyama Intake facility and sent to Nabeyaueno WPP, through the plain sedimentation basin. Filtered water at the slow sand filters is disinfected by sodium hypochlorite, stored in treated water reservoir and pumped to the service reservoirs for distribution.

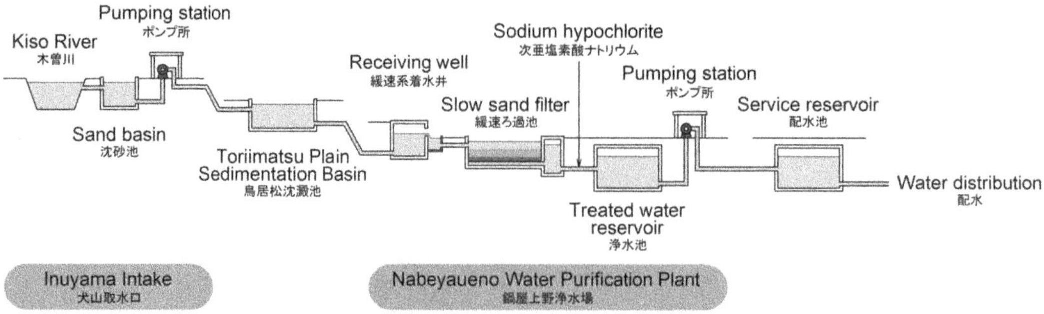

Figure 5.5 Slow sand filtration system of Nagoya City

BETTER DRINKING WATER

Tap water of Nagoya was ranked as the second tastiest water among main twelve cities in Japan by the Ministry of Welfare (the present name the Ministry of Health, Labour and Welfare) in 1984. Nagoya's good quality tap water has been also mentioned as a great advantage of living in Nagoya City by public opinion poll. One of the factors that good quality water has been maintained is its good water resource, the natural water quality of Kiso River. Total Dissolved Solids (TDS), including minerals causing a bitter taste is 63 mg/l of raw water, relatively low among main cities in Japan. In addition, the raw water is soft water, the hardness of 18mg/l, which is familiar with Japanese people.

Nagoya City Waterworks and Sewerage Bureau has made constant effort to protect the water environment throughout the Kiso River basin, such as planting trees for water conservation forest, promoting exchanges of waste water treatment technology between local governments along Kiso River, and supporting markets of a special product of the basin area to make the grass roots partnership and so on.

RECONSTRUCTION AND FEATURE

The slow sand filters of Nabeyaueno WPP are the oldest plant in Nagoya waterworks and the filtration process is only possible with good quality raw water. Thus the filters themselves are widely known among the general public. Nagoya waterworks is continuing to use the slow sand filtration system as a symbol of reliable, safe and tasty drinking water of Nagoya.

STRUCTURE OF THE SLOW SAND FILTER

Former filters were too old and didn't have enough earthquake-resistance. The body of new filters are earthquake-resistance structure with reinforce concrete and soil improvement was completed under the bottom of filters against liquefaction of ground by earthquake. Facilities are rearranged based on the previous operation experience to improve efficiency of purification as well as operation and management.

Installation of air bricks. Air bricks with many perforations of centimeters squared is used for the drainage system instead of standard bricks. The adoption of air bricks will promote smooth filtration across the all filter, reducing eccentric flow and head loss.

ENVIRONMENT-FRIENDLINESS

Recently there has been an increased attention to the slow sand filtration as an environment-friendly technique that requires no chemicals and relatively little mechanical equipment to treat water. Not only the process of slow sand filtration was selected again as an environmental technology, but also a large quantity of the existing filter sands and bricks were reused.

Reuse of filter sands. Approximately 14,000m^3 of sands used for the former sandbed (totally 24,000m^3) were reused for a new sandbed. The reused sands were washed additionally to meet the requirement for the filter media before laid in the filters.

Reuse of bricks. 1.1 million bricks in good condition of 2.2 million bricks used for the underdrainage system underneath the sandbed will be reused in various ways such as pavement, because they have been in the filters for the hundred years and symbolize the history of Nagoya waterworks.

CONCLUSIONS

Nagoya waterworks reaches the centenary in 2014. The slow sand filters worked since the start of the waterworks are now under reconstruction to newly start water supply in 2014. Nagoya City Waterworks and Sewerage Bureau determined to maintain source water in good condition cooperatively with people in the Kiso River basin, to continue use of slow sand filtration based on the good quality raw water, and thereby positively to supply better drinking water for the next 100 years.

Chapter 6

Value of Adenosine Tri-Phosphate and Total Cell Count for the assessment of general microbial water quality after sand filtration

Aleksandra Magic-Knezev, Luc Zandvliet , Willem A. Oorthuizen**, Ed J.van der Mark***

**Het Waterlaboratorium, Lucasweg 2, 2031 BE Haarlem, The Netherlands*

***Dune water company Dunea, Zoetermeer, The Netherlands*

ABSTRACT

In the present study we evaluated the suitability of two culture independent methods for the assessment of general microbial water quality after three sand filtration systems: dune recharge (DR), rapid sand filtration (RSF) and slow sand filtration (SSF). General microbial water quality was furthermore compared with the occurrence of faecal indicators in the DR. The concentration of total adenosine tri-phosphate (ATP) and total cell count (TCC) in water were measured during the period of 16 months in effluents of DR, RSF and SSF at three production locations of the Dune water company Dunea. Dune filtrates exhibited relatively stable ATP values that were not affected by temperatures between 8-18 °C. TCC values increased at increased temperature on account of LNA cells while HNA cell numbers remained stable. RSF reduced the ATP concentration and TCC values by 65-89% and 16-24%, respectively, but occasionally bacteria were also released from RSF. SSF reduced the ATP concentration and TCC values by 80-95% and 45-52%, respectively. SSF filters with the shorter lifetime exhibited better reduction of ATP and TCC values. ATP and TCC are valuable parameters for the routine monitoring of general microbial water quality and can be used to assess the efficiency of processes applied during drinking water production.

Keywords Sand filtration; ATP; total cell count; flowcytometry

INTRODUCTION

The infiltration of pre-treated surface water and abstraction from the dunes (Dune Recharge: DR) and slow sand filtration (SSF) are the two major disinfection steps applied by Dune water company Dunea, the water supply company of the Hague area of the Netherlands. During the DR and SSF pathogen microorganisms are removed (8 log10 and 0.4-6 log10, respectively, Schijven et al 1998, Dullemont et al. 2006) what ensure the distribution of drinking water without disinfectant at the very low level of infection risk ($<10^{-4}$ person per year). Although both filtration systems are robust, incidental presence of faecal indicators in large sample volumes (>20 L) in the effluent of DR system indicates shortcomings in (a part of) the abstraction system. The sampling of large volumes of water for the detection of faecal indicators is time consuming and too expensive to be used with high frequency for the routine monitoring. An alternative, sensitive, simple and rapid tool is therefore required for the routine monitoring of microbial water quality after major disinfection steps.

The hygienic aspect of water quality is routinely assessed by the analysis of faecal contaminants, but general microbial parameters, such as heterotrophic plate count (HPC), can also give valuable information about microbial water quality. HPC is used for the routine monitoring of drinking-water treatment process and abnormal changes in HPC levels indicate problems in the treatment (Bartam et al. 2003). Although HPC is still widely applied for the assessment of drinking water quality, insight in low culturability of aquatic bacteria (Staley and Konopka 1985) stimulated application of culture independent methods, such as ATP and total cell count, for the assessment of general microbial water quality (van der Wielen and van der Kooij 2010, Vital et al. 2010). Development of sensitive and rapid culture independent methods for general microbial water quality, such as adenosine triphosphate (ATP) concentration (Holm-Hansen and Booth 1966) and total cell count (TCC) (Hammes et al. 2008) opens new possibilities for routine monitoring of general microbial water quality in process of drinking water production and distribution.

In the present study we evaluated the suitability of two rapid, culture independent methods for the assessment of general bacteriological water quality. For this purpose, the concentration of total ATP and TCC were analysed during the period of 16 months in effluents of DR, rapid sand filtration (RSF) and SSF at three production locations of Dunea. Fluctuation in parameters for general microbial water quality in DR was compared with occurrence of faecal indicators that was obtained from routine monitoring program. Furthermore, the effect of RSF and SSF on the ATP concentration and on number, integrity and size of bacteria in water was evaluated and obtained results were used to define water quality guidelines for sand filtration at investigated locations.

METHODOLOGY

Dunea takes source water from the Meuse River. Pre-treated source water is transported to three dune areas with natural lakes, where it recharges the groundwater: Berkheide, Solleveld and Meijendel. After an average residence time of 2 months, the water is abstracted from the dunes (DR) and post-treated by softening, powdered activated carbon, aeration, rapid sand filtration (RSF), and slow sand filtration (SSF) before being pumped into the distribution system.

Water samples were taken each 14 days during a period of 16 months in effluents of DR, RSF and SSF. The concentration of total ATP in water was measured as light emission in the luciferine/luciferase assay (Holm-Hansen and Booth 1966) with the luminometer (Advance, Celsis, Netherlands) using reagent kit for bacterial ATP (Celsis). The ATP concentration was calculated from the relative light units (RLU) using the coefficient of the linear regression analysis of the calibration curve for the free ATP (Celsis) in autoclaved tap water in the range 1 – 250 ng L-1. TCC was measured as previously described by Hammes et al. (2008). Briefly, water samples were stained at 37 °C for 15 min with SYBR Green I and propidium Iodide (PI) (1:100 dilution in DMSO; Sigma). Samples were then diluted to final cell concentrations below 2 10^5 cells mL^{-1} in bottled mineral water (Evian, France) filtered trough 0.22 μm filters (Millipore). TCC was measured by C6 flow cytometer (BD Accuri cytometers, Belgium) equipped with a 50 mW laser emitting light of 488 nm. Fluorescence (Fl1 = 533± 30 nm, Fl3 > 670 nm) and sideward and forward scattered light were collected at flow rate of 35 μL min^{-1} during 2 minutes. Electronic gating was used to separate green (SYBR) fluorescence from intact cells (FL1; 533 nm), and red (PI) fluorescence from membrane compromised cells (FL3; >670 nm) and instrument noise or sample background. Cells occurred in two clusters as low (LNA) and high (HNA) nucleic acid content bacteria.

Faecal indicators were measured according to standard methods described by Dutch legislation (NEN). The difference between data sets was tested using Wilcoxon signed-rank test for the 95% confidence interval.

RESULTS AND DISCUSSION

Dune filtration

The ATP concentration in dune filtrate at the Meijendel location was structurally higher than in Solleveld and Berkheide and exhibited more variation in the examined parameters (Figure 6.1a,b Table 6.1). Although the total cell numbers were lowest at the location Meijendel, the number of HNA cells was similar for the three locations. The fraction of HNA cells was however highest in dune filtrate at location Meijendel. Interpretation of the existence of two clusters with low (LNA) and high nucleic acid content (HNA) in the flowcytometric analysis is still a matter of debate (Bouvier et al. 2007). The ATP concentration did not change with the temperature (8–18 °C) (Figure 6.1a), while the number of cells increased at higher temperatures (Figure 6.1b). Increase in cell numbers was caused by increase in number of LNA cells. Therefore, stable ATP concentration and stable number of HNA cells can be considered as indicators of stable production process at different temperatures. Abnormalities in production process can be detected as elevated cell numbers, especially HNA, and elevated ATP concentration.

The ATP concentration in the dune filtrate was in most of the samples < 100 ng l^{-1}. However, at the location Meijendel part of the extraction system allocated as P10 frequently exhibited increased ATP values (Figure 6.2a). TCC values were elevated more often in different parts of the extraction system than the ATP concentration. Increased ATP or TCC values are indicating functional deterioration in these parts of the system already recognized by the operators at Dunea as frequent deviations in various water quality parameters (Figure 6.2 a, b).

Table 6.1 Average values with standard error of ATP concentration, Total cell count (TCC), number of HNA cells and percentage of HNA cells in dune filtrates of Meijenedel, Solleveld and Berkheide. N = number of measurements.

Location	N	ATP	TCC	HNA	HNA
		ng l^{-1}	N ml^{-1}		%
Meijendel	32	91.7 ± 110.1	3.8 10^5 ± 1.1 10^5	1.0 10^5 ± 7.7 10^4	27 ± 6
Solleveld	32	47.3 ± 25.8	5.8 10^5 ± 9.9 10^4	1.2 10^5 ± 3.0 10^4	20 ± 4
Berkheide	32	31.8 ± 18.2	5.0 10^5 ± 1.1 10^5	1.0 10^5 ± 4.6 10^4	19 ± 3

Coincidence of faecal indicators with increased values of the ATP concentration and TCC values was regularly observed (Figure 6.3). The low frequency of sampling for faecal indicators and measurement at the level of the detection limit decreases chance of detection of faecal indicators. It was therefor not possible to establish clear correlation between occurrence of faecal indicators and parameters for general microbial water quality. However, elevated ATP and TCC levels were observed regularly at the location where incidental faecal contaminants are detected. At other two locations, where the ATP and TCC levels showed less fluctuation presence of faecal indicators in dune recharge is rarely observed.

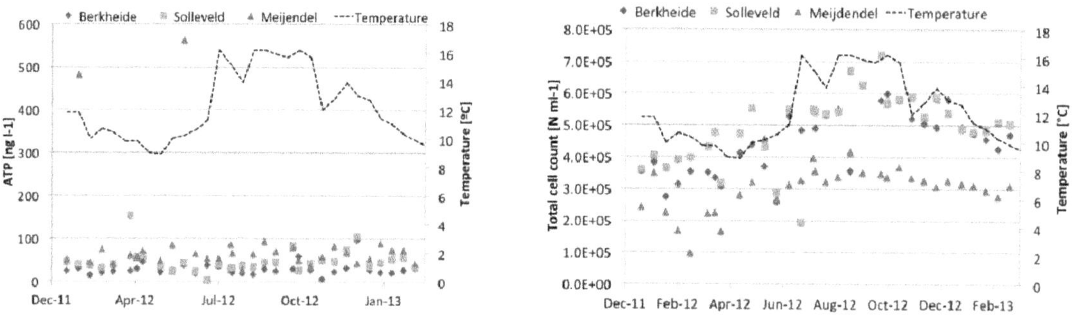

Figure 6.1 The ATP (a) concentration and TCC (b) in the dune filtrate at locations Meijendel, Berkheide and Solleveld and temperature of extracted water.

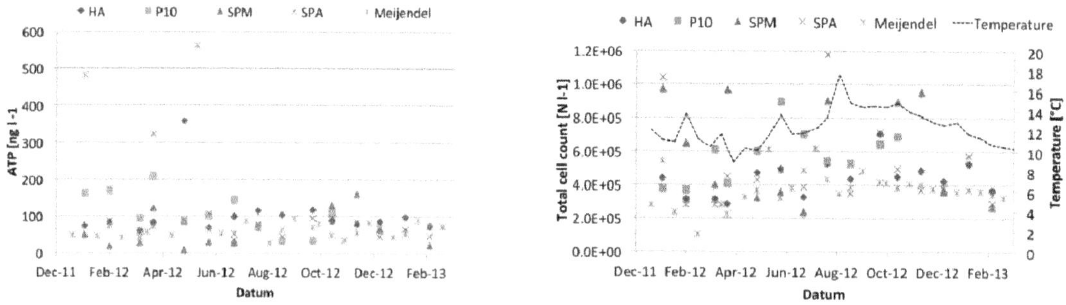

Figure 6.2 ATP concentration (a) and TCC (b) in different parts (HA, P10, SPM, SPA) of the extraction system at the location Meijendel and temperature of extracted water.

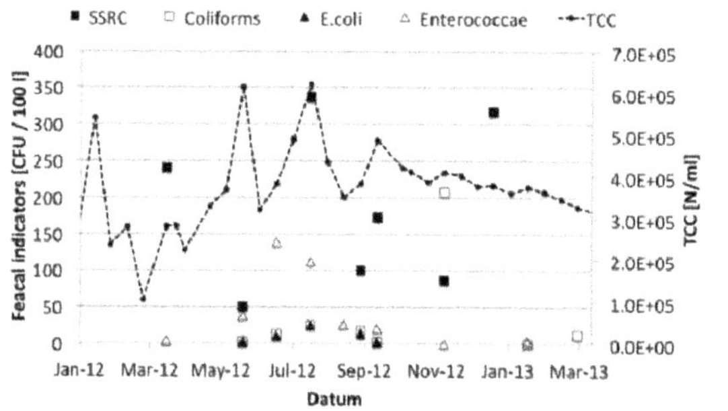

Figure 6.3 Concentration of faecal indicators, SSRC – spores of sulphite reducing clostridia, coliforms, E.coli and faecal enterococci [CFU 100 l^{-1}] and TCC [N ml^{-1}] in the dune filtrate of Meijendel.

Rapid sand filtration

After the RSF the ATP and TCC were reduced 65-89% and 16-24%, respectively (Figure 6.4 a, b). HNA cells were reduced better (32 – 48%) than the LNA cells (10-20%) (Figure 6.5 a, b). The range of ATP concentration after RSF was very similar for 3 locations. RSF at the location Berkheide were able to reduce less ATP and TCC and showed more fluctuation than RSF's at other two locations.

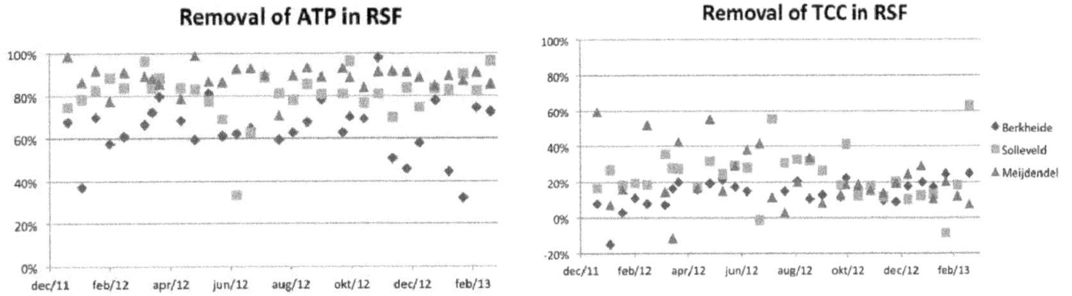

Figure 6.4 Removal of ATP (a) and TCC (b) in the RSF filters at locations Meijendel, Berkheide and Solleveld.

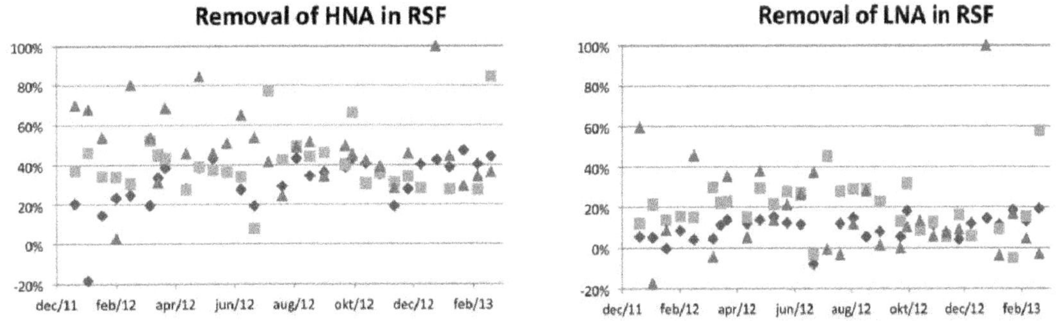

Figure 6.5 Removal of HNA (a) and LNA (b) cells in the RSF filters at locations Meijendel, Berkheide and Solleveld.

The reduction of the ATP concentration and TCC values after the RSF is a result of simultaneous processes in these filters. Bacteria from the influent are removed in a filter by physical filtration, attachment and inactivation. However, indigenous microbial community is also developed in a sand filter, and therefore, bacteria can be released in the effluent. Reduction of ATP concentration and TCC values after RSF indicate that the removal and inactivation of bacteria from the influent are more dominant than the release of bacteria from the filter. Better reduction of HNA cells than the LNA cells also indicate difference in composition of bacterial communities in the influent and the effluent. However, incidentally we observed that in some (rare) occasions bacterial numbers were higher in the effluent of RSF filter. Increased particle load after backwashing of RSF (ref) suggests that backwashing of the RSF may also cause release of bacteria into the effluent. For the range of temperatures from this study (8-18 °C), reduction of ATP and TCC values in RSF is temperature independent.

Slow sand filtration

The SSF further improve general microbial water quality by reducing the ATP by 80-95% and TCC by 45-52% (Figure 6.6 a, b). HNA cells were removed in average by 65-78% while LNA removal ranged between 39-49% (Figure 6.7 a, b). Increase in HNA or LNA cell numbers was not observed after the SSF. Indicating that SSF always improve general microbial water quality, although it can be expected that the release of indigenous microorganisms also occurs in SSF. However, better filtration, attachment and inactivation of microorganisms from the influent, result in higher reduction of ATP and TCC values by SSF then by RSF. The removal of ATP and TCC is most efficient and most stable at the recently renovated location Solleveld. The age of the SSF in Solleveld is 7 years, while SSF at other two locations are in average > >15 years old.

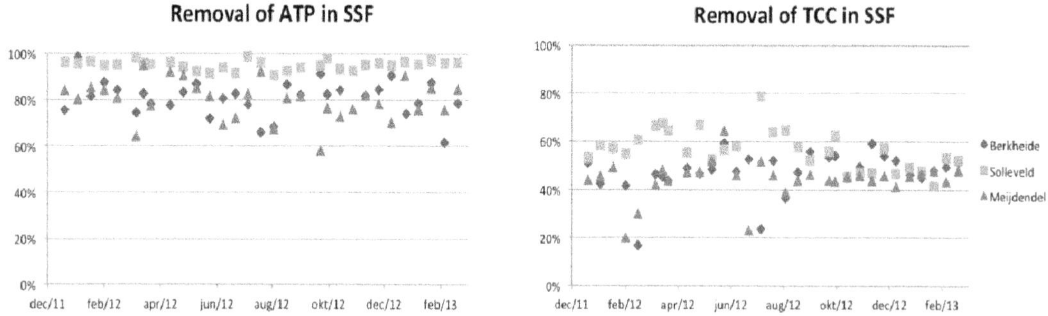

Figure 6.6 Removal of ATP (a) and TCC (b) in the SSF filters at locations Meijendel, Berkheide and Solleveld.

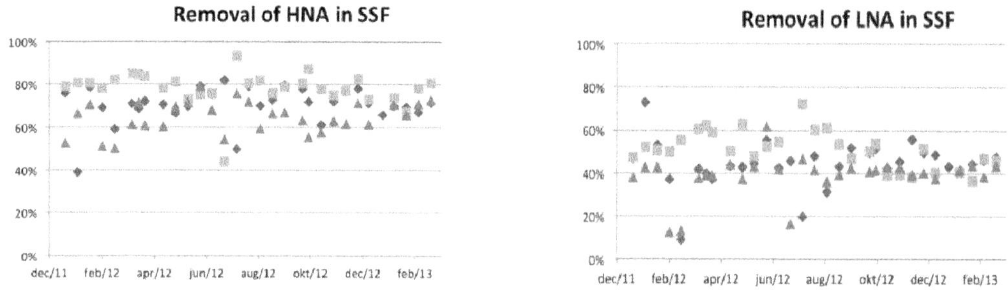

Figure 6.7 Removal of HNA (a) and LNA (b) cells in the SSF filters at locations Meijendel, Berkheide and Solleveld.

SSF are applied for the removal of pathogenic microorganisms and are reported to achieve much higher removal efficiency (2-6 log units) for specific microorganisms than observed in this research (Dullemont et al. 2006). This may be explained by the higher inactivation rate of specific microorganisms, such as *E.coli* or *Campylobacter,* than the inactivation rate of aquatic bacteria. Furthermore, aquatic bacteria are active or can even grow in SSF and may be actively released from the SSF, but pathogenic microorganisms and indicators of faecal origin are not able to grow under conditions in SSF.

CONCLUSIONS

Each treatment step exhibit clear, location specific general microbial water quality. The ATP and TCC values provide information about different aspects of microbial water quality and should optimally be used in combination. The TCC and number of LNA cells are directly related to water temperature while the ATP concentration seems not to correlate with the temperature for this data set. Both filtration steps improve general microbial water quality, but as expected, the SSF are more powerful in reduction of bacteria. The age of SSF may have an important impact on bacterial activity in the effluent. The combination of ATP and TCC has good potential to provide rapid and sensitive information about efficiency and stability of a treatment step. Additional data are necessary to examine the coincidence of the occurrence of faecal indicators and elevated levels of ATP and TCC.

REFERENCES

Bartram J., Cotruvo J., Exner M., Fricker C., and Glasmacher A. (2003) Heterotrophic plate counts and drinking-water safety: The significance of HPCs for water quality and the human health. IWA Publishing, London, UK.

Bouvier T., del Giorgio P.A., and Gasol J. M.. (2007) A comparative study of the cytometric characteristics of High and Low nucleic-acid bacterioplankton cells from different aquatic ecosystems. *Environmental Microbiology* **9**(8), 2050–2066

Dullemont Y.J., Schijven J.F., Hijnen W.A.M., Colin M., Magic-Knezev A., and Oorthuizen W.A. (2006) Removal of micro organisms by slow sand filtration. Proceedings of the IWA specialized Biological and alternative filtration conference, Mulheim May 2-6 2006, Germany. IWA Publishing, London, UK

Hammes, F., Berney M., Wang Y., Vital M., Köster O.and Egli T. (2008). "Flow-cytometric total bacterial cell counts as a descriptive microbiological parameter for drinking water treatment processes." *Water Res.*, **42**(1-2), 269-277.

Hammes F., Goldschmidt F., Vital M., Wang Y., and Egli T. (2010) Measurement and interpretation of microbial adenosine tri-phosphate (ATP) in aquatic environments. *Water Res.* **44**(13),3915-3923.

Schijven J.F., Hoogenboezem W., Nobel P.J., Medema G.J., and Stakelbeek A., (1998). Reduction of FRNA-bacteriophages and faecal indicator bacteria by dune infiltration and estimation of sticking efficiencies. *Water Science and Technology* **38**: 127-131.

Staley J.T., and Konopka A. (1985) Measurement of in Situ Activities of Non photosynthetic Microorganisms in Aquatic and Terrestrial Habitats. Annual Review of Microbiology Vol. **39**, 321-346

Van der Wielen P.W.J.J., and van der Kooij D. (2010). Effect of water composition, distance and season on the adenosine triphosphate concentration in unchlorinated drinking water in the Netherlands. *Water Res.*, **44**,4860–4867.

Vital M., Dignum M., Magic-Knezev A., Ross P., Rietveld L., and Hammes F. (2012) Flow cytometry and adenosine tri-phosphate analysis: Alternative possibilities to evaluate major bacteriological changes in drinking water treatment and distribution systems. *Water Res.*, **46**, 4665-4676.

Chapter 7

Evaluation of dissolved organic matter fractions removal due to biodegradation

E. Vasyukova, R. Proft, W. Uhl

Technische Universität Dresden, Institute of Urban Water Management, Chair of Water Supply Engineering, 01062 Dresden, Germany

ABSTRACT

The objective of this study was to evaluate the biodegradability of dissolved organic carbon and organic nitrogen fractions of different composition and origin, as well as to understand how these NOM fractions are removed during full-scale drinking water treatment. Samples across three drinking water plants treating surface waters of different origin were investigated. The observed microbial catabolism of natural organic matter of the whole molecular weight spectrum points out the need for control of all fractions by treatment processes in order to minimize the risk of microbial proliferation in drinking water distribution systems. The results furthermore indicate the importance of an optimised coagulation procedure as the main process to remove to a large extent the high molecular weight DOC and DON including their biodegradable part.

Keywords Biodegradation; dissolved organic carbon and nitrogen; drinking water; LC-OCD

INTRODUCTION

The removal of natural organic matter (NOM) from surface water is an important issue for the drinking water industry. NOM in drinking water is considered not only a precursor of potentially carcinogenic disinfection by-products during chlorination, but can also serve as energy source for bacterial (re)growth and may, thus, favour formation of biofilms in drinking water distribution systems (Reckhow et al., 1990; Volk et al., 2001). Bacterial proliferation in drinking water systems, in turn, may pose some risks including violation of quality standards, deterioration of aesthetic (e.g. increased turbidity, colour, taste and odour) and hygienic water quality as well as bio-corrosion and biofouling problems on membranes and pipes (Servais et al., 1995; Ndiongue et al., 2005).

One of the factors that influence the development of biofilms in drinking water distribution systems is the concentration of biodegradable organic matter (Servais et al., 1995). This matter represents a NOM fraction, which is bioavailable and may serve as energy and carbon source for heterotrophic bacterial growth (van der Kooij et al., 1982). Molecular weight is often mentioned as an important factor influencing the microbial utilization of NOM. It has been shown that the bioreactivity or utilization of organic matter by bacteria decreases along a continuum size (from large to small) and diagenic state (from fresh to old), while high molecular weight DOM (>1kDa) is more bioreactive and less diagenically altered than the low molecular weight DOM (<1kDa) (Amon and Benner, 1996).

Since chlorination does not completely prevent biological regrowth in the distribution systems, the biodegradable fraction of NOM should be removed as far as possible during drinking water treatment in order to provide a biologically stable water. NOM removal in various drinking water treatment processes depends not only on operational parameters but also on the NOM composition and properties, such as molecular weight distribution and physical-chemical character (Collins et al., 1986 Matilainen et al., 2002; Baghoth et al., 2011). The objective of this study was, therefore, to evaluate the biodegradability of organic carbon and organic nitrogen NOM fractions of different composition and origin, as well as to understand how these NOM fractions are removed during full-scale drinking water treatment.

METHODOLOGY

Three waterworks in the region of Dresden (Germany), which treat surface water with different organic matter concentrations and origin (reservoir, river, river bank filtrate), were investigated in the period from August to October 2011 (Table 7.1). Their source waters are characterised by low to high organic matter content (3-8 mg/L DOC on average) and low turbidity (2-5 NTU). Samples of raw water and after each treatment step were collected during four sampling campaigns (n=4) in pre-cleaned DOC-free vials and stored untreated at 4°C until analysis, which followed within a week.

Table 7.1 The studied waterworks and their treatment steps.

Treatment step	ALTB	COS	HOST
Raw water	Reservoir DOC ~ 8 mg/L	Reservoir DOC ~ 8 mg/L	River DOC ~ 8 mg/L
Coagulation	$FeCl_3$	$Al_2(SO_4)_3$	$Al_2(SO_4)_3$
Sedimentation	yes	yes	yes
Filtration	Anthracite coal/Sand	Sand	Expanded clay/Sand
Ozonation	yes	no	no
Subsurface infiltration	no	no	yes
Activated carbon filtration	Biological AC	no	Granular AC
Disinfection	Cl_2/ClO_2	ClO_2	Cl_2

The characterisation of NOM-fractions was performed using a liquid chromatography system with organic carbon and organic nitrogen detectors (LC-OCD-OND, Model 8; DOC Labor, Germany) based on the Gräntzel thin-film reactor (Huber et al., 2011). The method is used to fractionate DOC and DON into five fractions according to their apparent molecular weight: biopolymers, humic substances, building blocks, low molecular weight (LMW) acids and LMW neutrals. Due to interferences with nitrate, only dissolved organic nitrogen of the biopolymer and humic fractions could be quantified. Total DON concentration was thus calculated as a sum of these two fractions. An accompanying software (ChromCALC) was used for interpretation of the chromatograms on calculation of the respective group's C- and N-concentrations.

The DOC-concentration of proteinic matter in the biopolymer fraction was calculated using the following presumptions:

1) All organic nitrogen in the biopolymer fraction originates from proteins;
2) A typical DOC-to-DON ratio for proteins is 3:1 (Westerhof and Mash, 2002; Huber et al., 2011), and
3) DOC of the biopolymers consists primarily of polysaccharides and proteins, i. e. $DOC_{biopolymers} = DOC_{polysaccharides} + DOC_{proteins}$

Biodegradable dissolved organic matter assays were performed for two sampling campaigns (n=2). According to the technique of Joret et al. (1998), the inoculation was performed with the natural assemblage of sessile bacteria that previously grew in water of the same origin during 3 weeks incubation. 1 L of the sample was incubated in a glass bottle containing 100 mL quartz sand during 14 days in the dark at temperature of 22±1°C. After the incubation, samples were taken for DOC and DON analysis. The biodegradable dissolved organic carbon and dissolved organic nitrogen concentrations (BDOC and BDON) were determined by subtracting the DOC and DON concentrations after the incubation time from the initial corresponding concentrations of a sample.

RESULTS AND DISCUSSION

The considered raw waters exhibited low to high organic matter content with mean concentrations ranging from 3 to 8 mgC/L DOC and from 0.15 to 0.3 mgN/L (Fig. 7.1). High organic matter concentrations are typical for waters draining the upland regions of the Ore Mountains (ALTB) that are characterised by upland moors and podzol soils. The biodegradable fraction of organic matter was in the range of 16-26 % DOC and 8-15 % DON and was the highest for the organic-rich reservoir water. Humic substances were the prevailing fraction in all considered waters contributing to 52-62 % to the total DOC (not shown). High biopolymers content was observed in organic-rich water due to algae presence (up to 0.8 mgC/L and 0.2 mgN/L). Organic-poor waters (COS and HOST) exhibited lower biopolymer concentrations with <0.2 mgC/L and <0.03 mgN/L on average. The DOC/DON ratio of the source waters varied between 25 mgC/mgN (ALTB and HOST) and 38 mgC/mgN (COS) on average and was relatively high for natural waters indicating on a contribution of plants and soils (allochthonous material) to a greater amount of NOM (Westerhoff and Mash, 2002; Lee et al., 2006). Proteins constituted 63 % of biopolymer-bound DOC in organic-rich reservoir water, whereas an inverse distribution was observed for other source waters. LMW acids were not detected in all samples.

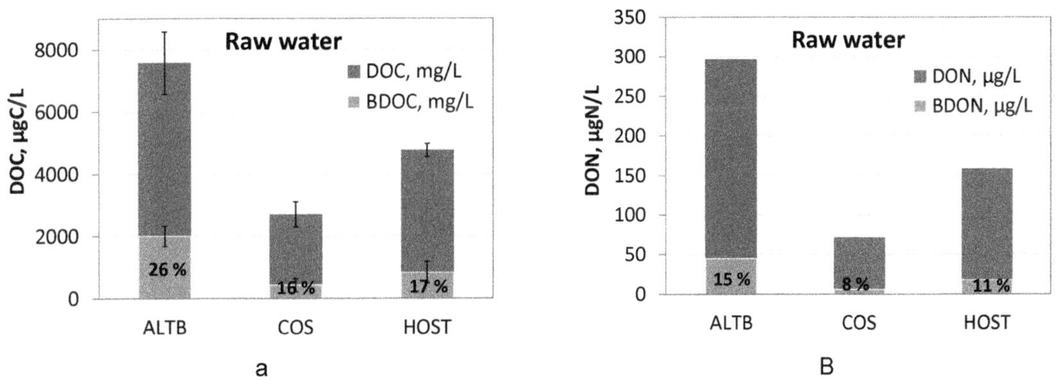

Figure 7.1 Total and biodegradable DOC (a) and DON (b) in the considered raw waters. Error bars represent 95 % confidence interval.

The biodegradability of the NOM fractions was found to be independent of their origin and initial concentrations, which varied within three orders of magnitude. Biopolymers were the most readily biodegradable fraction demonstrating 40 to 100 % removal as both dissolved organic carbon and nitrogen in bioassays (Fig. 7.2a,b). Furthermore, polysaccharide-like DOC was removed by biodegradation slightly better than proteinaceous-like DOC (not shown). The biodegradability of low molecular weight fractions, such as building blocks and LMW neutrals, was lower and in a few samples their production was observed (Fig. 7.2c). Finally, an insignificant biodegradation (<20 %) or an increase of humic-bound dissolved organic carbon and nitrogen was observed in bioassays (Fig. 7.2a,b). An increase of humic substances after incubation may suggest an abiotic polymerisation of lower MW compounds to humic-like structures or a microbial production of those. In contract, an increase of LMW fractions, i. e. building blocks and neutrals, may be a result of breakdown of high MW substances or microbial transformation processes by extracellular enzyme activity.

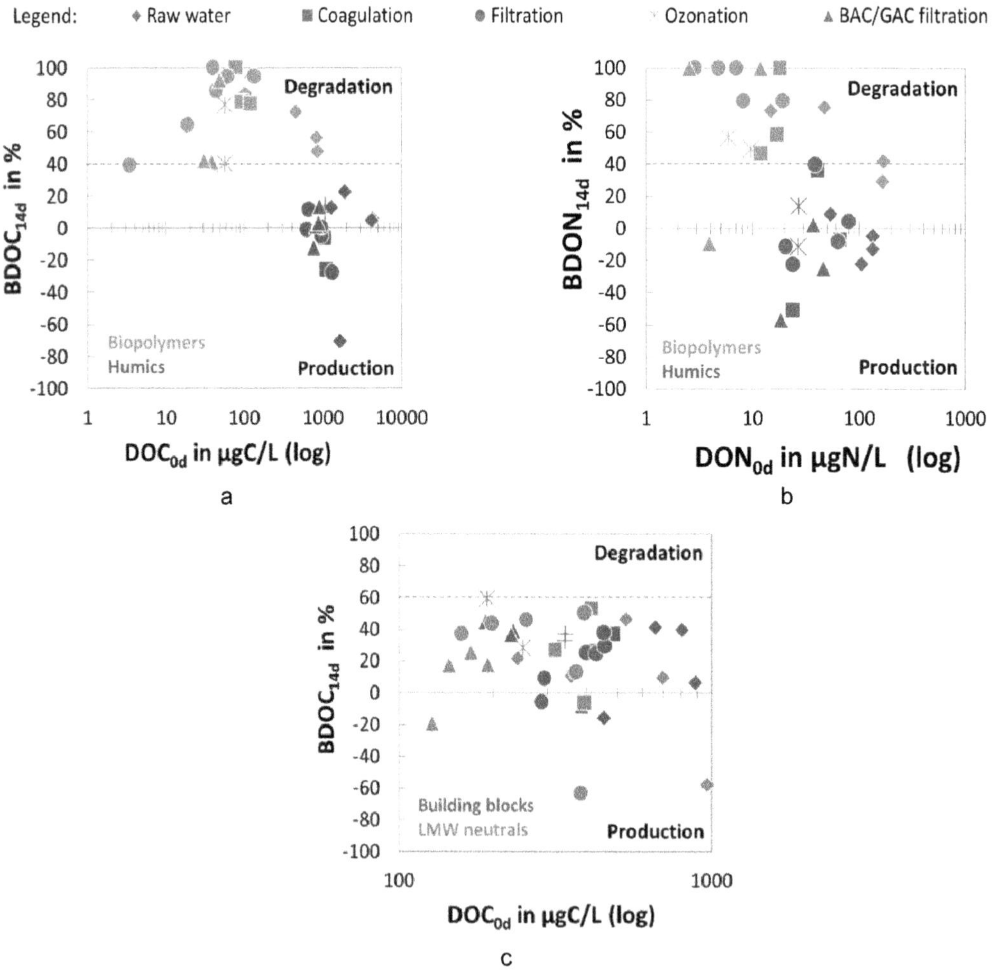

Figure 7.2 Relationship between the biodegradable (BDOC) and total DOC and DON of the respective NOM fractions in waters sampled after each treatment step: (a) and (b) biopolymers and humic substances and (c) building blocks and low molecular weight (LMW) neutrals.

The kinetics of microbial degradation of different DOC fractions is exemplarily illustrated in Fig. 7.3 for one sampling campaign at the waterworks ALTB. About 60 % of biopolymers in raw water were labile and subject of relatively rapid biodegradation within the first four days of incubation, when more than 0.17 mgC/L were removed (Fig. 7.3a). Thr biopolymer degradation rates during the first four days of incubation were lower in the effluent waters of the different treatment stages than in raw water (e. g. only 0.1 mgC/L removed within four days of incubation in coagulated water). This suggests that coagulation was able to remove labile biopolymers. The biodegradation kinetics for the specific biopolymer fractions including proteins and polysaccharides as well as DON is presented below in Fig. 7.4.

A substantial amount of humic substances was degraded within 14 days of incubation in the raw water sample (about 0.6 mgC/L or 13 %; Fig. 7.3b). The results demonstrate that the enzymatic potential of bacterial population responded fairly rapidly to a proportion of humics during the first seven days of incubation, when about 0.5 mgC/L were removed. Previous studies showed that humic substances can be utilised by bacteria by up to 30 % assuming that this fraction contains a proportion of biodegradable biomolecules, such as peptides and carbohydrates, which may serve as microbial substrate and which are weakly bound to humics (Steinberg 2001). However, no similar trend of biodegradation was noticed for effluets of different treatment steps across the waterworks.

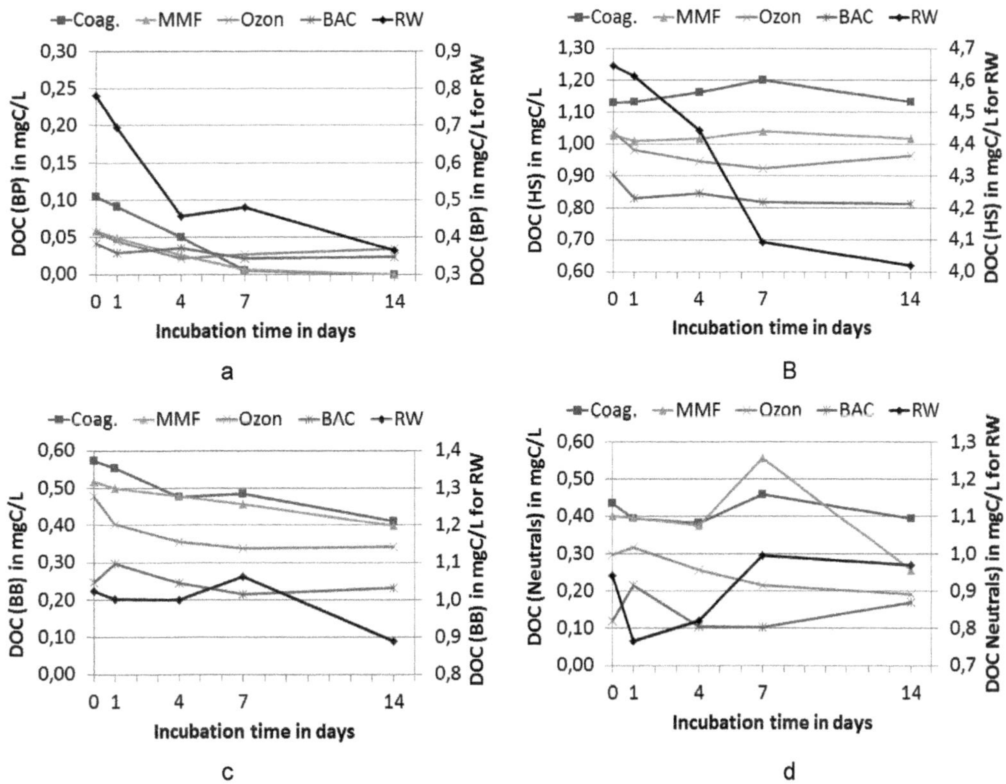

Figure 7.3 Kinetics of microbial degradation of DOC fractions from different treatment processes at waterworks ALTB during 14 days of incubation illustrated for a) biopolymers, b) humic substances, c) building blocks and d) LMW neutrals. Coag. stands for coagulation, MMF for multi-media filtration, Ozon. for ozonation, BAC for biological activated carbon filtration, RW for raw water. Note the different scale ranges.

Building blocks, that are considered to be humic substances hydrolysates, appeared to be less reactive to biodegradation as only 0.1 to 0.2 mgC/L were removed within 14 days of incubation in all samples (Fig. 7.3c). The results of LMW neutrals were found to be variable for different waters. No considerable removal by biodegradation could be noticed for raw and coagulated water, while about 0.15 mgC/L were degraded in the multi-media filter effluent and 0.1 mgC/L in ozone-treated water (Fig. 7.3d). A similar trend was also observed for BDOC tests of another sampling campaign in ALTB. Since LMW neutrals are present in all natural waters, including groundwaters, this fraction is presumably refractory. The increase of LMW neutrals in some BDOC samples might indicate breakdown products of higher molecular weight DOC fractions.

The biodegradation kinetics of DON as well as proteins and polysaccharides of the biopolymer fraction is shown in Fig. 7.4 for two bioassays using ALTB raw water. Polysaccharides exhibited significantly higher degradation rates than proteins within the first four days of incubation (0.3mgC/L vs. 0.05mgC/L). However, while proteins appeared to be utilised continuously over the entire incubation time, polysaccharides were degraded almost completely during the first four days of incubation. These results demonstrates the need to control the polysaccharides to reduce microbial activity in the distribution systems.

No clear trend in the removal of biodegradable organic matter with respect to DOC and DON fractions was observed by treatment processes across the considered waterworks (not shown). However, between 30 and 90 % of BDOC of the biopolymer fraction could be removed by coagulation and was closely associated (proportional in percentage) with the removal of biopolymer-bound BDON in this treatment step. The biodegradation behaviour of the remaining small parts of BDOC and BDON during river bank and biologically activated carbon filtration of the considered waterworks was found to be similar to that observed in bioassays for different waters (not shown). These two processes removed biopolymers to the greatest extent, followed by low molecular weight compounds and humic substances.

Overall, all considered treatment plants could achieve very low final DOC and DON concentrations at the end of treatment (~1.3 mgC/L and ~0.025 mgN/L), whatever their initial levels. The entire treatment trains removed between 45 and 76 % of the bulk DOM on average, while BDOM levels were reduced by a higher percentage (67 to 90%) demonstrating preferential removal of biodegradable organic matter over non-biodegradable. Biodegradable DOC in treated waters ranged between 0.1 and 0.2 mgC/L, and BDON was removed completely (<0.01 mgN/L) during treatment. The remaining biopolymer fraction in treated waters (<0.05mgC/L) was mainly constituted of polysaccharides, as proteins were removed almost completely from all waters due to the complete removal of organic nitrogen from this fraction. Humic-bound DON prevailed in treated waters and ranged between 0.02 and 0.03 mgN/L.

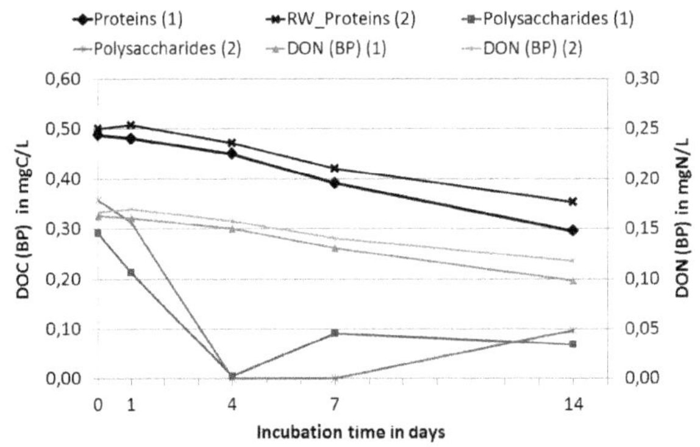

Figure 7.4 Microbial degradation of proteins and polysaccharides as well as DON of the biopolymer fraction during 14 days of incubation in two bioassays using ALTB raw water

CONCLUSIONS

This study provided important insights into the biodegradation of the natural organic matter fractions as characterised by the LC-OCD method. The biodegradability (both BDOC and BDON) of single fractions was found to be independent of their initial concentrations. Biopolymers were the most readily and quickly biodegradable fraction (i.e. removed in bioassays) compared to others, followed by low molecular weight neutrals and building blocks. An insignificant biodegradation or an increase of humic-bound BDOC and BDON was observed suggesting a possible production of transformation products due to biopolymer degradation or production of soluble microbial products. Most of the BDOC and BDON were removed by coagulation. This result indicates the importance of an optimised coagulation procedure as the main process to remove as far as possible the high molecular weight DOC and DON including their biodegradable part. The observed microbial catabolism of NOM of the whole molecular weight spectrum points out the need for control of all NOM fractions by treatment processes in order to minimize the risk of microbial proliferation in drinking water distribution systems.

ACKNOWLEDGEMENTS

This work was performed as part of the REGKLAM and IWAS projects, funded by the German Federal Ministry for Education and Research (BMBF) under grants number FKZ 01LR0802B and 02WM1166, respectively. The authors greatly appreciate this financial support. We are also grateful to the staff of the involved waterworks for providing data and samples, and Jana Brückner for technical assistance.

REFERENCES

Amon, R.M.W.; Benner, R. (1996). Bacterial utilization of different size classes of dissolved organic matter. *Limnology and Oceanography* **41**(1):41-51.

Baghoth, S.; Sharma, S.; Guitard, M.; Heim, V.; Croué, J.-P.; Amy, G. (2011). Removal of NOM-constituents as characterized by LC-OCD. *Journal of Water Supply: Research and Technology*, **60**(7):412-423.

Collins, M.; Amy, G.; Steelink, C. (1986). Molecular weight distribution, carboxylicacidity and humic substances content of aquatic organic matter: implications for removal during water treatment. *Environmental Science and Technology*, **20**:1028-32.

Huber, S. A.; Balz, A.; Abert, M.; Pronk, W. (2011). Characterisation of aquatic humic and non-humic matter with size-exclusion chromatography-organic carbon detection-organic nitrogen detection (LC-OCD-OND). *Water Research*, **45**: 879-885.

Joret, J.S.; Levi, Y.; Dupin, T.; Gilbert, M. (1998). In Proc. of AWWA Annual Water Qualty Conference 1998. Denver, Colo.: AWWA.

van der Kooij, D.; Visser, A.; Hijnen, W.A.M. (1982) Determining the concentration of easily assimilable organic carbon in drinking water. *J. Am. Water Works Assoc.* **74**, 540-545.

Lee, W.; Westerhoff, P.; Esparza-Soto, M. (2006). Occurrence and removal of dissolved organic nitrogen in US water treatment plants. J. *Am. Water Works Assoc.* **98**(10): 102-110.

Matilainen, A.; Lindqvist, N.; Korhonen, S.; Tuhkanen, T. (2002). Removal of NOM in the different stages of the water treatment process. *Environment International*, **28**: 457– 465.

Ndiongue, S.; Huck, P.M.; Slawson, R.M. (2005). Effects of temperature and biodegradable organic matter on control of biofilms by free chlorine in a model drinking water distribution system. *Water Research* **39**(6): 953-964.

Reckhow, D. A.; Singer, P. C.; Malcom, R. L. (1990). Chlorination of humic materials: by-product formation and chemical interpretation. *Environmental Science and Technologies*, **11**:1655-1664.

Servais, P.; Laurent, P.; Randon, G. (1995). Comparison of the bacterial dynamics in various french distribution systems. *J. Water SRT Aqua*, **44** (1):10-17.

Steinberg, C. E. (2001). Ecology of Humic Substances in Freshwaters. Berlin Heidelberg: Springer.

Volk, C. J.; Robinson, B.; Zhu, H. W.; Kaplan, L. A. (2001). Total organic carbon monitoring and drinking water treatment plant performance. Themes in potable water. **41**:854-858. Chicago: Devision of Environmental Chemistry, American Chemical Society.

Westerhoff, P.; Mash, H. (2002). Dissolved organic nitrogen in drinking water supplies: a review. *J. Water Supply: Res. Technol.-AQUA*, **51**(8):415-448.

Chapter 8

Comparison between different filter systems as a post treatment after tertiary ozonation

M. Stapf, U. Miehe*, B. Wiedemann*, M. Jekel***

**Berlin Centre of Competence for Water (KWB), Cicerostraße 24, 10709 Berlin, Germany*

***Technical University Berlin, Dept. of Water Quality Control, Straße des 17. Juni 135, 10623 Berlin, Germany*

ABSTRACT

To cope with occurring traces of organic contaminants in the effluent of waste water treatment plants, ozonation is a suitable technical treatment method. However, there is an ongoing discussion about the necessity of a post-treatment of ozonation effluents to remove possible toxic ozonation by-products. This study compares a dual media filter (DMF) and a biological activated carbon filter (BAC), which were used for ozonation post-treatment, and were also designed as coagulation filters for tertiary phosphor removal. The results of this study demonstrate that both rapid filters performed similarly in respect to DOC reduction and oxygen demand, and could also be used for tertiary phosphorus removal without any impairments. A comparison of a serial mode of the DMF and the BAC with a slow sand filter, which was used as a surrogate for an infiltration pond, showed that this two-stage process could increase the degradation of the DOC, but was not able to remove the entire biodegradable DOC.

Keywords: Ozonation, dissolved organic carbon, deep bed filtration, tertiary treatment, trace organic contaminants (TrOCs)

INTRODUCTION

The occurrence of trace organic contaminants (TrOCs) like pharmaceuticals and personal care products in the urban water cycle and the options for their removal is a subject of vivid discussion. One possible technology to reduce TrOCs is ozonation of secondary waste water treatment plant (WWTP) effluent. As ozonation increases the biodegradability of dissolved organic carbon (DOC) and also might possibly produce toxic by-products, a post-treatment via biologically active filtration is discussed or in some cases even recommended (Abegglen et al. 2009). Not many investigations have been carried out yet, focusing on the kind and mode of operation of post-treatment during tertiary filtration. While Abegglen et al. (2009) used a mono-layer sand filter, biological activated carbon is used at several sites in Australia

(Reungoat et al. 2011) and was said to be more effective for biological DOC removal compared to a sand filter.

Therefore, different biological filter systems for ozonation post-treatment are compared in this study regarding their total DOC reduction and the additional effects of coagulation if the filters are also used for tertiary phosphorus removal.

METHODOLOGY

The pilot plant in this study was located at the effluent of the WWTP Ruhleben in Berlin (Germany), which is designed for the treatment of 1.6 million population equivalents. The pilot plant (Figure 8.1) was in operation between August 2012 and September 2013. The ozonation unit was designed for a maximum flow rate of up to 10 m^3/h. The reactor consisted of three columns to provide a hydraulic retention time (HRT) between 12 to 63 minutes. The produced ozone was injected via a ceramic membrane at the bottom of the first column. The ozonation was programmed to maintain a specific ozone dose of 0.5 or 0.7 mg O_3/mg DOC using a constant ozone dose during the first trial phase. During the second trial phase an online UV-VIS probe (Spectro::lyser, S::can, Austria) located at the influent of the ozonation was used to regulate the ozone dose in proportion to the influent UV absorption. Occurring nitrite in the effluent of the WWTP, which reduces the efficiency of the ozonation due to additional ozone consumption (3.43 mg O_3/mg-N), could not be compensated by the control system of the ozonation. As a post-treatment of the ozonation three subsequent filters were installed: a dual media filter (DMF), a biological activated carbon filter (BAC) and a slow sand filter (SSF).

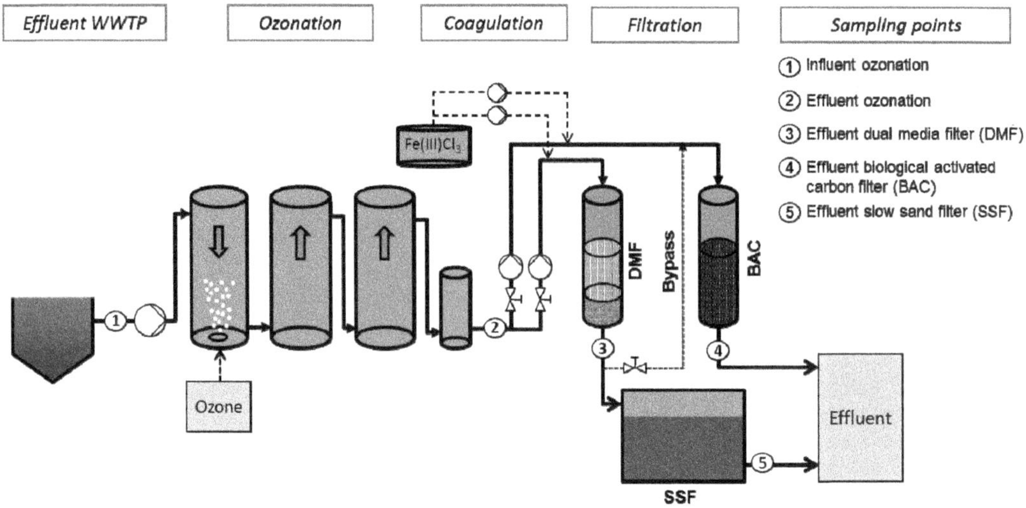

Figure 8.1 Scheme of the Pilot plant at WWTP Ruhleben

The DMF and the BAC were designed as coagulation filters to remove phosphorus by injecting 3 mg/L Fe into each feed and were operated at a filter velocity of 6 to 10 m/h. The SSF, which served as a surrogate for an infiltration pond, was operated at a filter velocity of 1 to 3 m/d. Because the SSF was subsequent to the DMF the feed was almost free of suspended solids. The granular activated carbon in the BAC should not only reduce occurring TrOCs due to adsorption, but also support the growth of microorganisms as a carrier media. To prevent clogging both rapid filters were flushed daily (air/water) during regular operation. The filter designs and materials can be found in Table 8.1.

Table 8.1 Filter design and filter material

	Design	Filter material
DMF	Column: Ø,i: 0.3 m Water level: 3.6 m	Anthracite: 120 cm (1.4 – 2.5 mm) Sand: 85 cm (0.7 – 1.25 mm)
BAC	Column: Ø,i: 0.3 m Water level: 3.6 m	Granular activated carbon: 185 cm AquaSorb® CS, Jacobi, 6 x 12 mesh
SSF	Tank (L x B x H): 1.5 m x 0.73m x 1.62 m Water level: 1.4 m	Sand: 100 cm (0.7 – 1.25 mm) Gravel: 20 cm (3.15 – 5.6 mm)

Sampling was performed twice a week: Mixed samples were taken at all effluents, except the effluent of the SSF, at which grab samples were taken, during the times when the DMF and the BAC were in operation. The samples were analysed by the laboratories of the Berliner Wasserbetriebe. The characteristics of the effluent of the WWTP can be found in Table 8.2.

Table 8.2 Characteristics of the WWTP effluent. Mean values during project time (mixed samples).

Conventional parameters	n		
UV absorbance (SAK_{254})	92	30 ± 3	[1/m]
Dissolved organic carbon (DOC)	91	12.5 ± 1.7	[mg/L]
Chemical oxygen demand (COD)	92	41 ± 6	[mg/L]
Filtered chemical oxygen demand ($COD_{filtered}$)	46	34 ± 5	[mg/L]
Total suspended solids (TSS)	83	6.9 ± 3	[mg/L]
Nitrite ($N\text{-}NO_2$)	55	0.3 ± 0.2	[mg/L]
pH	51	7.5 ± 0.2	[-]

During the first trial phase, both filters were operated for seven months to establish biological activity in the rapid filters and exhaust the adsorption capacity of the granular activated carbon in the BAC. The used specific ozone doses were 0, 0.5 or 0.7 mg O_3/mg DOC. During the second trial phase both rapid filters were operated for 10 weeks in parallel mode at a filter velocity of 8 m/h, of which five weeks included coagulation with 3 mg/L Fe. This was done to evaluate the impact of the coagulation and to compare both filters regarding their DOC removal.

Furthermore, to estimate the possible benefit of a two-stage process and to achieve a higher biodegradation in the BAC, it was operated subsequently to the DMF (serial mode) using the bypass (Figure 8.1) during the following 10 weeks, of which also five weeks included coagulation. Therefore the flushing interval of the BAC was reduced from daily to once in a week, because the coagulant was only dosed into the feed of the DMF and most of the suspended solids were removed before entering the BAC. Additionally, the filter velocity in the BAC was reduced compared to the DMF (4 m/h vs. 8 m/h) to increase the contact time in the BAC. During the second trial phase ozonation was constant at 0.7 mg O_3/mg DOC. Because of the serial mode it was also possible to compare the additional short-term biodegradation in the BAC with the medium-term biodegradation in the SSF.

RESULTS AND DISCUSSION

This chapter sums up the results of the parallel mode within the entire project to compare the performance of the DMF and the BAC, which is followed by the results of the parallel and serial mode during the second trial phase. Finally the different reactions in the filters regarding the TrOCs are illustrated.

The total reduction of the DOC at the effluents of the DMF and the BAC in respect to the WWTP effluent in the period of the parallel mode during the entire project is shown in Figure 8.2. During the start-up phase (~ 3,600 bed volumina) without ozonation but with coagulation, more DOC was removed in the BAC (17 %) compared to the DMF (10 %), due to adsorption at the granular activated carbon in the BAC. After the start-up phase the adsorption capacity of the granular activated carbon in the BAC for DOC was exhausted, and ozonation started with a specific ozone dose of 0.5 mg O_3/mg DOC. With ozonation a similar level of DOC reduction occurred in both filters (20 % – 22 %), while an increase of the specific ozone dose to 0.7 mg O_3/mg DOC also increased the level of DOC reduction in both filters (27 % – 28 %). The improved DOC reduction at higher ozone doses was assumed to be due to a better biodegradability of the DOC because no mineralisation of the DOC was observed in the ozonation.

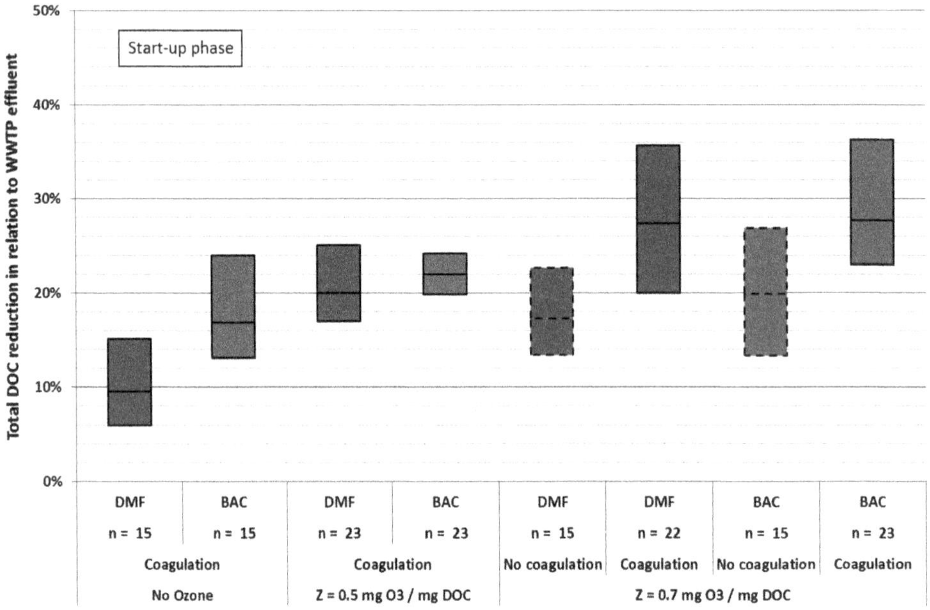

Figure 8.2 DOC reduction due to ozonation and post-treatment with DMF or BAC. Results of the parallel mode within the entire project shown as box plots (25 %, 50 % and 75 % quantiles).

At the high specific ozone dose but without coagulation, the DOC reduction drops in both filters to a lower level (17 % - 20 %). Therefore, it could be estimated that coagulation itself was responsible for about 10 % of the total DOC reduction. A major advantage of the BAC in comparison to the DMF, as it is stated in Reungoat et al. (2011), was not observed.

The results of the comparison of the parallel and serial mode during the second trial phase are shown in Figure 8.3. A comparison of the results of the second trial phase with the overall performance during the entire project (Figure 8.2) illustrates an improved DOC reduction during parallel mode with coagulation (36 % vs. 28 %). Because the second trial phase took place during summer (Mai to September), higher water temperatures are a possible explanation for this increase. The better performance of the BAC during times without coagulation was within the variability of the DOC reduction and, therefore, it should not be overestimated by stating an advantage of the BAC over the DMF. The results of the serial mode are used to estimate the difference between the short- and medium-term degradation.

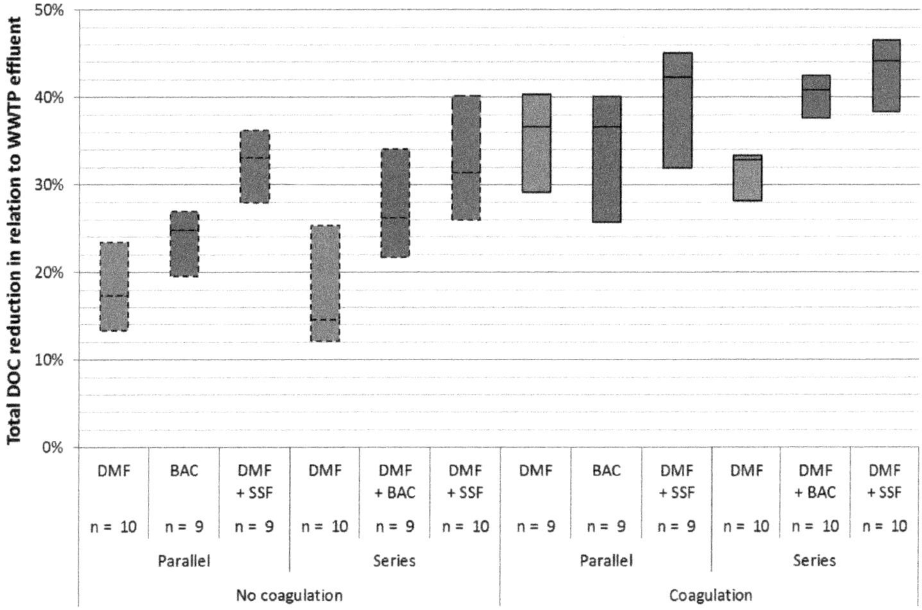

Figure 8.3 DOC reduction due to ozonation and post-treatment. Results shown as box plots (25 %, 50 % and 75 % quantiles) during the second trial phase at a specific ozone dose of 0.7 mg O_3/mg DOC.

Without coagulation, the additional medium-term degradation in the SSF with an empty bed contact time of about 9 hours could be estimated to about 17 %, whereas with coagulation the additional degradation was lower (6 – 11 %). In contrast, the short-term DOC degradation in the BAC with a contact time of about 27 minutes resulted in an additional reduction of 8 % with coagulation and in 12 % without it. As a result it can be stated that the two-stage process (DMF + BAC) can increase the total DOC reduction, but it was not sufficient to remove the entire biodegradable DOC.

The additional phosphorus removal by coagulation increased the total level of DOC removal but did not impair the biological activity, even though the concentration of readily available phosphorus (as ortho-phosphate) was even below the LOQ of 10 µg/L. Also the oxygen consumption as an indicator for biological activity in the filters was not affected. During parallel mode both rapid filters performed similarly in respect to DOC reduction and oxygen demand, while the total oxygen consumption almost doubled indicating a similar biological activity in the BAC during the serial mode (Figure 8.4). Because the feed of the rapid filters was always supersaturated with dissolved oxygen (> 30 mg/L) due to the ozonation, the filters were fully aerobic. This was also the case while ammonia was measured in the feed of the filters resulting in higher oxygen demands due to nitrification processes.

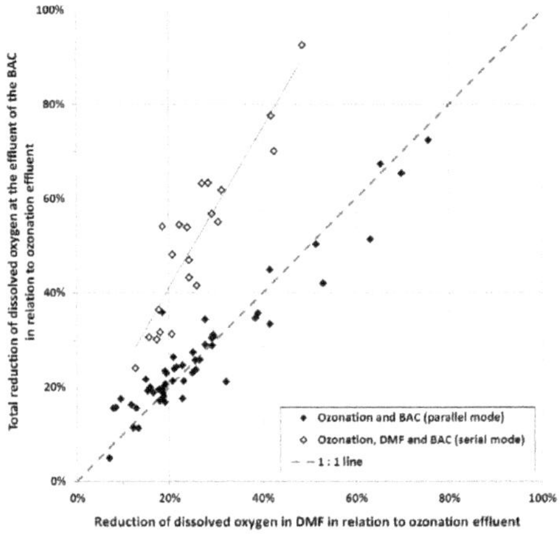

Figure 8.4 Reduction of dissolved oxygen during parallel and serial mode in the rapid filters in relation to the ozonation effluent

Finally, regarding a further reduction of TrOCs the results depend on the specific substances that are taken into account. While using a specific ozone dose of 0.7 mg O_3/mg DOC, substances that are known to be oxidised very well by ozone like carbamazepine (anti-epileptic), diclofenac (anti-inflammatory) or sulfamethoxazole (antibiotic) are already reduced below LOQ at the effluent of the ozonation. Thus, no changes in concentration were observed in the effluents of the filters (data not shown) and a comparison was impossible. But other substances, which were not easily oxidised like benzotriazole (corrosion inhibitor), acesulfame (artificial sweetener) or iopromide (x-ray contrast media), were measurable at the effluent of the ozonation. These results agree with other sources (Bahr et al. 2007, Abegglen et al. 2009).

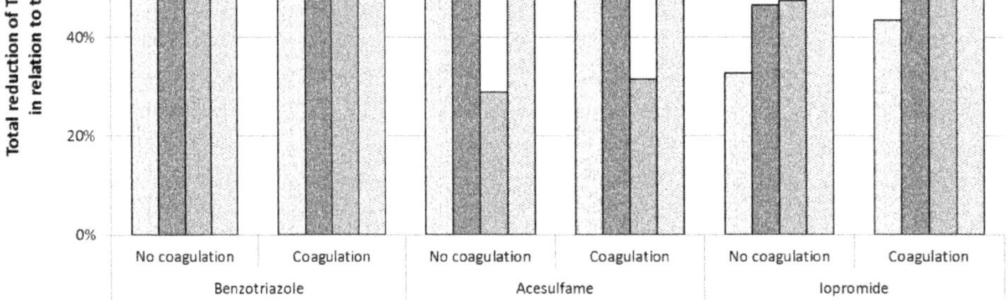

Figure 8.5 Total reduction of TrOCs at different stages during serial mode at a specific ozone dose of 0.7 mg O3 /mg DOC

The total reductions of the latter three TrOCs at the different stages are shown in Figure 8.5. The results illustrate that the post-treatment of the ozonation effluent with the DMF and the SSF rarely affected the reduction of benzotriazole and acesulfame. The reduction of acesulfame in the SSF during coagulation was not confirmed by other measurements during the parallel mode (data not shown). After more than 20,000 bed volumia adsorption processes were still observed in the BAC, resulting in an increase of the total reduction of benzotriazole to more than 95% in the effluent of the BAC. However, the drop of the total acesulfame reduction at the BAC compared with the total reduction at the DMF indicated a release of acesulfame. A possible explanation for that is that strong absorbents like benzotriazole or other ones can displace weak absorbents such as acesulfame. Therefore higher concentrations of such weak absorbents can be seen at the effluent of granular activated carbon filters (Worch, 2012). As for iopromide the results showed that both sand filters (DMF and SSF) were able to reduce the concentration of iopromide, whereas concentrations in the BAC were unchanged. Measurements during the parallel mode confirmed these results (data not shown), but up to now no explanation for the different results for DMF/SSF and BAC was found. In general, coagulation itself did not influence the reduction of the TrOCs.

CONCLUSIONS

This study showed that a dual media filter (DMF) and a biological activated carbon filter (BAC), which were used as a post-treatment after ozonation, performed similarly regarding DOC removal and oxygen demand. As the filters were used for tertiary phosphorus removal, no impairment of DOC removal or reduction of oxygen consumption due to coagulation was observed. Thus phosphorus limitation seems unlikely. Moreover, coagulation with 3 mg/L Fe itself increased the total DOC reduction by about 10%. Higher ozone doses improved the biodegradability of the DOC and therefore increased the total DOC reduction. In combination with coagulation, a total DOC reduction of about 30 % in relation to the effluent of the WWTP was possible with each rapid filter. Results of the serial mode of the DMF and the BAC in comparison to a serial mode of the DMF and a slow sand filter (SSF) showed that the additional short-term biodegradation in the BAC led to a higher DOC removal but was not sufficient to remove the entire biodegradable DOC.

Regarding the further reduction of TrOCs in the subsequent filters of the ozonation, different reactions were observed: After more than 20,000 bed volumina, a further adsorption of benzotriazole in the BAC came along with a release of acesulfame out of the granular activated carbon probably due to displacement. The further decrease of iopromide in the DMF and the SSF could not be observed in the BAC.

As a final conclusion it can be stated that both rapid filters performed similarly in respect to DOC reduction and oxygen demand, and could also be used for tertiary phosphorus removal without any impairments.

REFERENCES

Abegglen, C., Escher, B., Hollender, J., Koepke, S., Ort, C., Peter, A., Siegrist, H., von Gunten, U., Zimmermann, S. et al. (2009) *Ozonung von gereinigtem Abwasser. Schlussbericht Pilotversuch Regensdorf*, EAWAG Aquatic Research, 80 pages

Bahr, C., M. Ernst & M. Jekel (2007) *Pilotuntersuchungen zur kombinierten oxidativ-biologischen Behandlung von Klärwerksabläufen für die Entfernung von organischen Spuren- und Wirkstoffen und zur Desinfektion.* Schriftenreihe Kompetenzzentrum Wasser Berlin.

Reungoat, J., Escher, B.I., Macova, M., Keller J. (2011) *Biofiltration of wastewater treatment plant effluent: Effective removal of pharmaceuticals and personal care products and reduction of toxicity*, Water Research 45, pages 2751 – 2762

Worch, E. (2012) *Adsorption technology in water treatment. Fundamentals, processes and modeling.* De Gruyter, Berlin, Boston.

Part II

Slow Sand Filtration – Process Behaviour

Chapter 9

Food chain is the key in ecological purification system: new concept and new name of slow sand filter

N. Nakamoto
Prof. Emeritus of Shinshu University, Ueda, Japan
Community Water Supply Support Center of Japan, 1218-5, Suwagata, Ueda, Nagano, 386-0032, Japan (e-mail: cwscnkmt@yahoo.co.jp)

ABSTRACT
Slow sand filter is based on the biological community. High filter rate of 40cm/h had been adopted as standard flow rate by Thames Water Utilities. Impurity of dissolved and particulate substance is eliminated during several minutes passing through the active sand layer. Food chain is the key for the purification process. Unwilling action for small organisms is not good. Gentle for small organisms is the key for Ecological Purification System which is a new concept and a new name of slow sand filter.

Keywords food chain, ecological purification system, filter rate

INTRODUCTION

The present vertical type of slow sand filter was devised by James Simpson in 1829 to provide a safe drinking water in London. This water was free of pathogens. At that time, they believed the clear water was made by the mechanical filtration with fine sand and under slow flow rate (Figure 9.1). This system was called an English filter which was widely spread to the world.

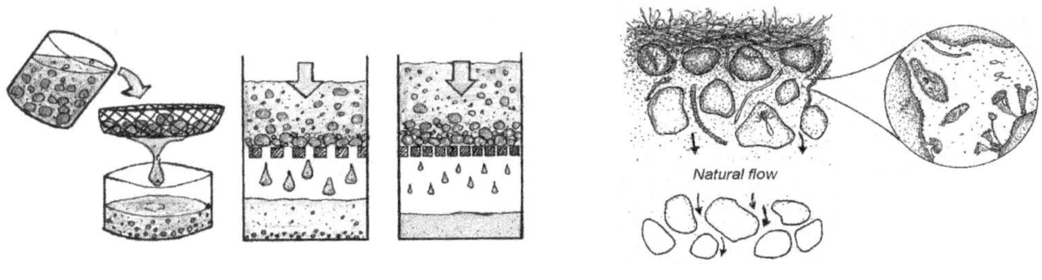

Figure 9.1 Mechanical filter image of slow sand filter with fine sand under slow flow rate. Real purification system is based on the role of small organisms under natural downward current.

They noticed there was biological function in the purification process and they called it as biological filter. They believed the main function of purification process was mechanical filtration. However, the real action of purification is based on the food chain among the small organisms which live in the top of sand layer. People misunderstood the role of algae in this system (Sakai and Nakamoto 1994).

The passing time through the top of the biological active layer is only several minutes. In the case of slow sand filter, there is a constant downward current. Sand on the filter is not moved even the sudden change of the flow rate. This is gentle for small organisms. However too much load of suspended solid to the filter leads the accumulation the mud on the sand surface and it blocks the filter. In order to reduce the load of suspended solid from a river, sedimentation pond and roughing filter are installed.

The SSF image easily misleads the misunderstanding of the main purification action. The author proposes the new name of Ecological Purification System to the traditional name of Slow Sand Filter.

UPGRADE FILTER RATE

The slow rate of 10cm/h was adopted by James Simpson in 1827 (Baker 1949). Then, the standard filter rate of 20cm/h (4.8m/d) was used as an English filter. Thames Water Utilities Ltd reported that iron bacteria and some algae were sometimes observed in filtrate when the filter was being operating too slowly for too long (Toms and Bayley 1988). Living organisms were sensitive to anaerobic environment. Small animals escaped from this environment and rushed to better water. Then, they tried to shift to upgrade rate of 40cm/h (9.6m/d) at Ashford common WTP at least in 1985 (Duncan 1988, Woodward and Ta 1988) and operated up to 53cm/h (12.7m/d) (Rachwal et al. 1988). When the author studies on algal development pattern at the Thames filters during 1994 to 1996, Thames Water Utilities Ltd still continued the experiment on the upgrade trials at the practical filters of Ashford Common WTP.

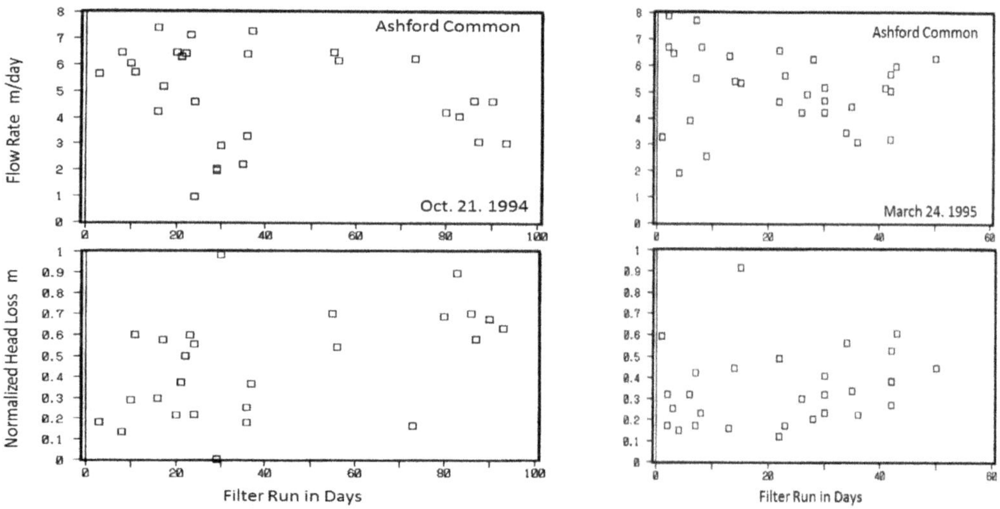

Figure 9.2 Filter rate and filter resistance as normalized head loss of Ashford Common WTP of Thames Water Utilities Ltd in October 21, 1994.

Filter resistance as Normalized Head Loss (calculated at 4.8 m/d) did not related with the filter rate and with the length of filter run. Activity of small organisms among sand layer does not affected under this upgrade rate. They confirmed upgrade rate made a better quality of filtrate than normal standard rate of 20cm/h (4.8m/d). Then, all of Thames filters shifted to the upgrade rate of 40cm/h (9.6m/d). When small organisms grow well, the filter resistant may not rapidly increase during long filter run except severe cold season.

ALGAL GROWTH IS NOT SERIOUS

Algae grow well in a shallow filter and produce oxygen by photosynthesis in day time. Algae consume the oxygen during the night time. When the algal biomass becomes large, the diurnal fluctuation in dissolved oxygen (DO) becomes large in filtrate (Nakamoto et al 1996). In case of slow filter rate, DO becomes severe low concentration in night. Small animals near the surface sand layer escape into the deep sand layer. Then animals leak in the filtrate. They believe that algal growth leads filter clog and algae produce odor substance. But these phenomena are caused by too slow filtering rate.

People misunderstood the role of algae. Algae produce oxygen by photosynthesis and food for heterotrophic animals. Growth of algae means reduction of nutrient in water. Algal biomass increases on the sand surface. When a filtration rate shifts to faster rate, the DO variation becomes small in filtrate water. Small fluctuation of environmental parameters is gentle for small organisms in sand layer. Algal mat easily lifts up to the surface by oxygen bubbles in shallow depth (Nakamoto and Kato 2006). Floating algal mat easily escapes through the scum outlet. The continuous culture system of algae is important to keep aerobic condition. Effective removal of floating scum is good to keep favorite DO concentration during night. Algae produce oxygen when they grow. When they die, they consume much oxygen. During the night, oxygen consumption becomes high. Sometimes, low oxygen makes unpalatable effluent water under low DO concentration.

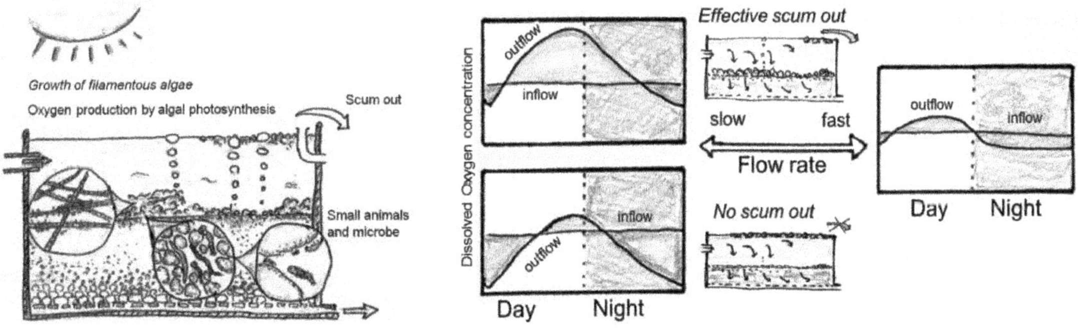

Figure 9.3 Algae promote the biological activity. Aerobic environment is essential for this system.

FOOD CHAIN IS THE KEY FOR PURIFICATION

The clear safe water can be made by the role of small organisms at the top of sand layer. A slow sand filter is an ecosystem where various organisms work under aerobic condition. The small animals can move freely among the sand particles under the vertical flow where sand particles do not move even any change of flow rate. Small animals like Protozoa catch and trap any particulate matter including germ cell. Small microscopic animals are essential to reduce germ bacteria. There is a food chain from small protozoa to small animals such as insect larvae, *Oligocheata*, etc. They live near the surface of the sand layer. Algae grow well on the surface of the slow sand filter. Turbid matters do not enter in deep sand layer. Various small animals collect them near the surface of the sand layer. Depth of Small animals works within several cm from the top. Passing time of the blacken sand layer is only several minutes

(Figure 9.4). Almost all the suspended matters are collected. Animals are also a fecal pellet maker. Real decomposition occurs inside the fecal pellet. Aerobic condition is essential for small animals to live. This system is similar structure of soil profile in natural forest. The filtrate is clear and delicious water during long filter run.

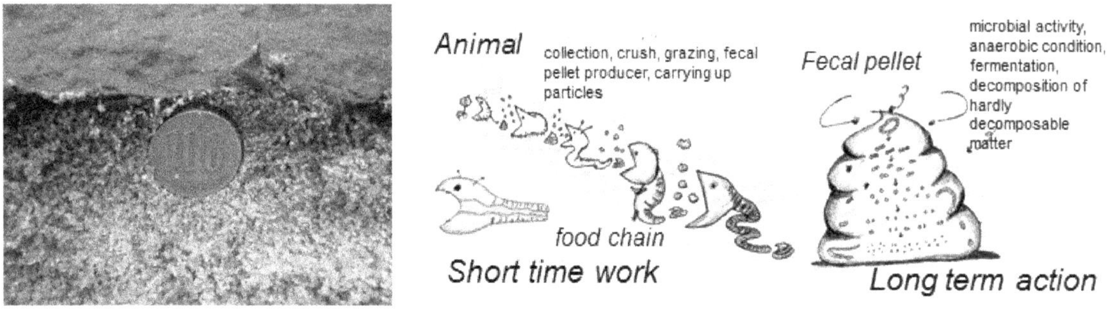

Figure 9.4 Top profile of sand filter. Images of food chain and long term action in fecal pellet.

SCRAPE THE SAND SURFACE

When the filter resistance increases, the surface of the sand layer is scraped. A sedimentation pond is common pre-treatment to reduce the suspended matter from a river in Japan. And almost all the filters do not block in warm period. The severe increase of the filter resistance is not observed so frequently. However algal bloom is observed in a source reservoir and in a filter pond. They do not care for the filter resistance. Then, periodical scrape of the sand surface is performed in almost all the filter plants in Japan.

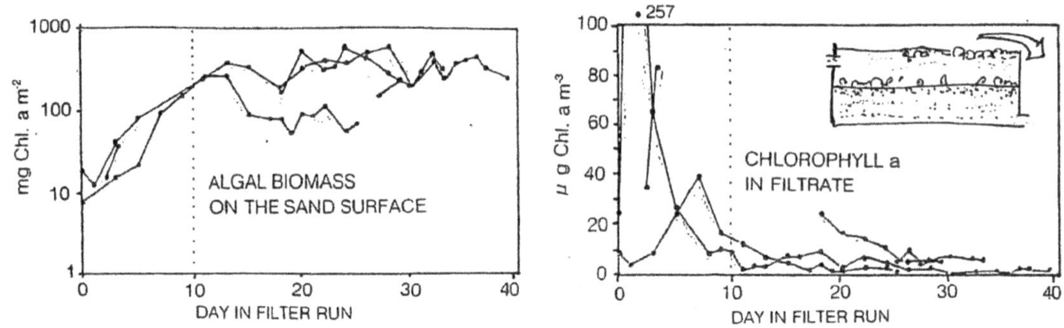

Figure 9.5 Development of algal mat and filtrate during the filter run. Filtrate water becomes clear water within 10 days. (Nakamoto et al 1997).

They believe that slower filter rate makes better quality of filtrate. Then, the filter rate was reduced to slow rate. Frequent scrape the surface mud is not good. Small organisms in the dirty sand layer are important to purify the raw water (Duncan 1988, Yamamoto and Nakamoto 1996, 1997, Yamamoto et al. 2000). Nakamoto et al 1997 checked the development of algal mat and the filtrate during the filter run. Continuous culture system was observed at 10 days after scrapping in Ueda, Japan. Then the filtrate became clear. Scrape the surface dirty layer was an elimination of the active biological layer.

SUCCESSION OF ALGAE AND GRAZING ACTIVITY

I checked the developmental pattern of algal mat at Thames filters, in London during 1994 to 1996. Filamentous green alga of *Cladophora* was dominated in summer. They called blanket weed for mat of *Cladophora*. However filamentous diatom of *Melosira* dominated in winter. My Schmutzdeck sampler is suitable to take diatom mat, quantitatively (Nakamoto 1993). But filamentous green algae are large and long in comparison with filamentous diatom. Their cell wall is hard. I could not take algal mat in summer, completely. However, I could check the developmental pattern of algal mat on the Thames filters (Figure 9.6 and 9.7).

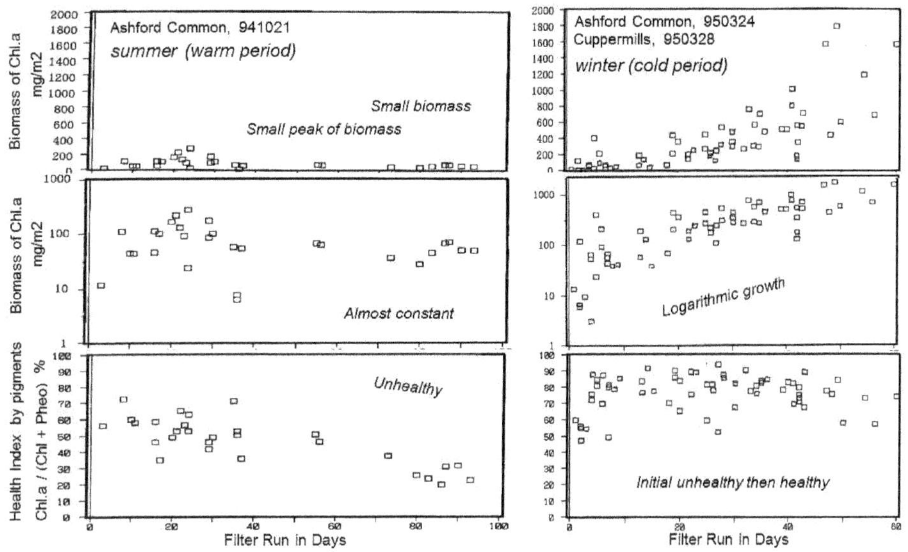

Figure 9.6 Development of algal mat in cold period (March, 1995) and in warm period (October, 1994) in Thames filters (Ashford Common and Cuppermills).

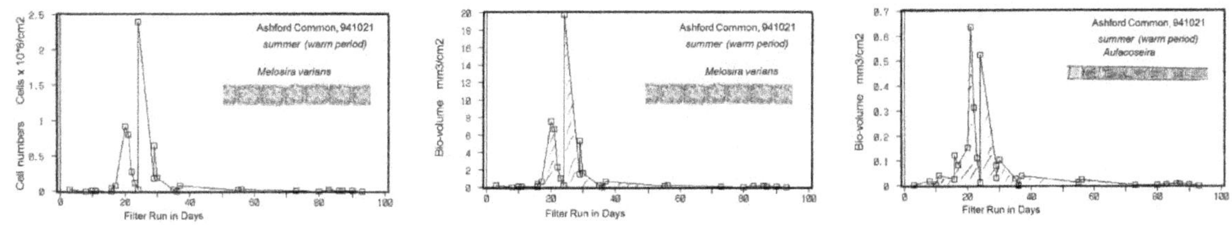

Figure 9.7 Biomass changes of filamentous diatom during filter run in warm period at Ashford Common WTP (October, 1994).

I found there was a peak at around 3 weeks of filter run in summer. But the diatom peaks were not remarkable. The algal biomass as chlorophyll pigment in summer was smaller than the biomass of filamentous diatom of *Melosira* in winter. Health index of algal mat was checked using pigment analysis (Lorenzen 1967). In winter, algal community was unhealthy at the initial. And then algae became healthy and this condition continued during long filter run. However, algal community in summer became gradually toward unhealthy condition. I checked the cell number and bio-volume of filamentous diatom (*Melosira* and *Aulacoseira*). The peak at around 3 weeks was remarkable. There are many papers on the ecology of midge (Chironomidae) in Thames filters (Duncan 1988, Hurley and Wotton 2006, Wotton

2002, Wotton and Hirabayashi 1999). Filamentous diatom was the best food for midge. But midge could not graze the green alga of *Cladophora*. Grazing effect on diatom was remarkable in warm period but it was weak in cold winter. Succession from diatom to green alga was caused by grazing animal.

Filamentous diatom of *Melosira* always dominated from April to December at Someya WTP in Ueda, Japan (Nakamoto et al. 1966). The length of filter run was short. They scraped periodically the surface mud. We could not see real succession. I confirmed that same succession of algae was observed in Japan in case of longer filter run (Nakamoto and Kato 206). The green algae are always remarkable in the filter pond even in initial days in the warm region in Japan. Grazing activity by animals is very strong in warm region.

ECOLOGICAL PURIFICATION SYSTEM

The author studied on the role of algae in slow sand filter from 1984 in Ueda, Japan. I pointed out the importance of a continuous culture system of filamentous algae in slow sand filter system. Vertical profile of top of a sand layer is similar to the cross section of a soil profile.

Biological activity is high in the boundary between different environments. Suitable organisms grow in different environment. In the case of slow sand filter, supernatant water on the sand layer is like a pond but there is a downward current to the sand layer. If the filter rate is slow, planktonic algae or phytoflagellate grow in the supernatant water. If the flow rate is high, filamentous algae grow on the sand surface of the sand layer. And algal photosynthesis is active on the shallower depth. Algal growth is related with concentration of nutrient and strength of radiation. Animal growth is related with amount of food. Animal biomass is large near the surface. Its biomass decrease suddenly with the depth (Duncan 1988). Animals usually move around for food near the surface. Animals in deep layer are hungry condition for food.

Heterotrophic organisms such as microbe and small animals are poikirothermal organisms. The growth rate of these heterotrophic organisms is slow under low temperature condition. Then the grazing activity is also weak in cold condition (Figure 9.8). The biological activity of algae and animals is high in warm region and the oxygen consumption is large during the night. When the biomass is abundant, a large amount of dissolved oxygen is required by these organisms. In order to keep aerobic condition during the night, the high filter rate is essential to supply enough amount of dissolved oxygen.

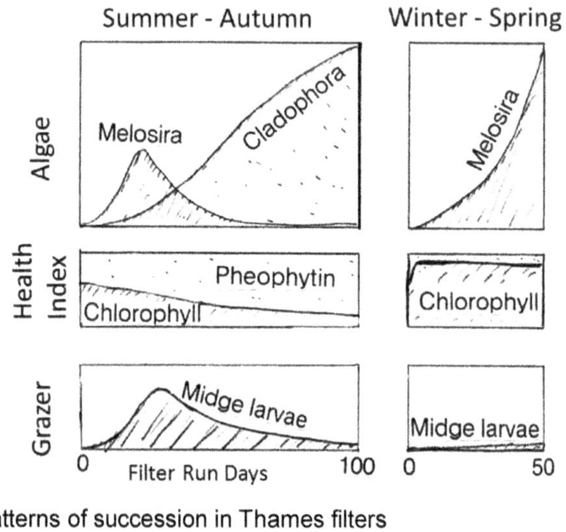

Figure 9.8 Schematic patterns of succession in Thames filters

There is frequent heavy rain fall in warm region then river water becomes easily muddy. The filter is easily blocked with the load of much amount of suspended matter. They thought the slow filter rate was better than a faster flow rate. They scraped the surface sand within short filter run before the biological community grew up. If they use enough residence time of sedimentation basin and roughing filters, suspended matters are reduced. Then slow sand filter does not clog. And algae and animals grow well in this sytem.

In case of warm region, or in warm season, we can see algal mat is grazed up on the surface of sand filter by the activity of grazing animals. In this time, the filtrate is good quality. Animals under hungry condition live in the top of sand layer. These animals trap any particle which enters the sand layer. In this case, the filter does not clog.

Covered filters are common in severe cold region. Algae cannot grow without sunshine. Impurity reduction function through the filter is rapidly developed in case of open filter. In case of covered filter, food supply by algae is nothing for animals. Therefore, it takes the time to develop the biological community in cold region and without sunshine. Heterotrophic organisms are always hungry. This condition is very important to trap impurity. Therefore, the water quality of filtrate by covered filters is same by open filters. Algal growth in filter is not always essential to develop a food chain as an ecological purification system. Aerobic condition for small organisms is essential in this system.

CONCLUSIONS

Slow sand filter was invented in cool climate region of UK. The water temperature is high in tropical region where the biological activity is always high. Small animals in the slow sand filter work in aerobic condition. The fundamental purification process was based on the food chain among the biological communities. In warm region, saturation concentration of dissolved oxygen is small. It is necessary to shift to higher flow rate to provide the enough amount of dissolved oxygen for small animals.

Slow sand filter has been believed that it was a mechanical filter with fine sand under slow current. However, the major contribution of the purification of the impurities is the food chain in this system. Unpleasant change for small organisms is not good such as sudden change of flow rate and water quality. We have to think acceptable moderate fluctuation for small organisms such as daily fluctuation in nature. The word of "**slow**" was "**gentle for organisms**". The original name of slow sand filter is not fair. The English filter of "**Slow Sand Filter**" is "**Ecological Purification System**". New concept of Ecological Purification System creates a new idea of a system design and a new maintenance technique. This new idea to make safe drinking water is now spread slowly but steady to Indonesia, Bangladesh, Sri Lanka, China, Pacific islands, Bolivia, Brazil and other countries in the world.

The main conclusions in this paper are:

(1) Slow sand filter is not mechanical filter. It is an ecological purification system.

(2) Large size of sand (around 0.5mm to 1mm) is better to reduce filter resistance. Sand is just habitat for small organisms.

(3) Faster flow rate (around 10m/d) is better to keep aerobic condition.

(4) Shallow water depth (around 50 cm) is better for photosynthetic algae.

(5) Remove the floating scum and algae to keep aerobic condition.

(6) Frequent scrape is not good. Longer filter run is better to growth of animals.

(7) Requirement for verification of Ecological Purification System by yourself.

ACKNOWLEDGEMENTS

The author wishes to acknowledge Thames Water Utilities for my co-operative research during 1994 to 1996. The author would like to thank Prof. Nigel Graham for the preparation of the research proposals. This research was carried out with the financial support of the British Council and the Great Britain Sasakawa Foundation.

REFERENCES

Baker, M.N. (1949) The quest for pure water. AWWA.

Duncan, A. (1988) The ecology of slow sand filters. In *Slow Sand Filtration: Recent Developments in Water Treatment Technology*, N.J.D. Graham (Ed.), Ellis Horwood Ltd., UK, 281-304.

Huisman, L. and Wood, W.E. (1974) *Slow Sand Filtration*, World Health Organization, Geneva.

Hurley, S.P. and Wotton, R.S. (2006) Chironomid midges: the forgotten water industry engineers? In *Recent Progress in Slow Sand and Alternative Biofiltration Processes.* Gimbel, R., N.J.D. Graham and R. Collins (Eds.), IWA, 51-59.

Lorenzen, C.J. (1967) Determination of chlorophyll and phaeopigments: spectrophotometric equations, *Limnol. Oceanogr., **12**,:343-346.*

Nakamoto, N. (1993) Schmutzdecke sampler reduces filter bed damage. *Opflow*, AWWA, **19**(7), 1-4.

Nakamoto, N., Yamamoto, M., Sakai, M., Nozaki, K., Iwase, N., Yasuda, M., Kitada, T., and Hirabayashi, K. (1996) Role of filamentous diatom as an automatic purifier in a slow sand filter. In *Advances in Slow Sand and Alternative Biological Filtration.* N.J.D. Graham and R. Collins (Eds.), John Wiley & Sons, UK, 139-148.

Nakamoto, N., Iwase, N., Nozaki, K. and Sakai, M. (1997) Inconsiderable amount of phytoplankton leakage by algal mat peeling from slow sand filter. *Jap. J. Wat. Treat. Biol.* ,**33**(4), 179-186.

Nakamoto, N., Yasuda, M., Sakai, M. and Nozaki, K. (2000) Oxygen production and bubble formation by algal mat on the bed in a slow sand filter bed. *Verh. Internat. Verein. Limnol.* **27**,3092-3096.

Nakamoto, N. and Kato, H. (2006) Development pattern of filamentous diatom and its condition related with midge larvae in slow sand filter. In *Recent Progress in Slow Sand and Alternative Biofiltration Processes.* Gimbel, R., N.J.D. Graham and R. Collins (Eds.), IWA, 68-73.

Rachwal, A.J., Bauer, M.J. and West, J.T. (1988) Advanced techniques for upgrading larger scale slow sand filters. In *Slow Sand Filtration: Recent Developments in Water Treatment Technology*, N.J.D. Graham (Ed.), Ellis Horwood Ltd., UK, 331-347.

Sakai, M. and Nakamoto, N. (1994) Role of filamentous algae in a slow sand filter. *Verh. Internat. Verein. Limnol.* **25**,1993-1995.

Toms, I.P. and Bayley, R.G. (1988) Slow sand filtration: an aproache to practical issues. In *Slow Sand Filtration: Recent Developments in Water Treatment Technology*, N.J.D. Graham (Ed.), Ellis Horwood Ltd., UK, 11-28.

Woodward, C.A. and Ta, C.T. (1988) Developments in modelling slow sand filtration. In *Slow Sand Filtration: Recent Developments in Water Treatment Technology*, N.J.D. Graham (Ed.), Ellis Horwood Ltd., UK, 349-366.

Wotton, R.S. (2002) Water purification using sand. *Hydrobiologia,* **469**, 193-201.

Wotton, R.S. and Hirabyashi, K. (1999) Midge larvae (Diptera, Chironomidae) as engineers in slow sand filter beds. Water Res., **33**(6),1509-1515.

Yamamoto, M. and Nakamoto, N. (1996) Depth distribution of suspended matter in the sand layer of slow sand filters that take the surface river water. *Jap. J. Wat. Treat. Biol.* , **32**(2), 123-129.

Yamamoto, M. and Nakamoto, N. (1997) Depth distribution of suspended matter in the sand layer of slow sand filters at Ashford common water works. *Jap. J. Wat. Treat. Biol.* , **33**(2), 91-96.

Yamamoto, M., Yamashita, M. and Nakamoto, N. (2000) Depth distribution of suspended matter in the sand layer of slow sand filter model treated by chlorination. *Jap. J. Wat. Treat. Biol.* , **36**(3), 137-142.

Chapter 10

A study of protistan movement speed and filtration rate in slow sand filter as an ecological purification system

*N. Iwase**, *N. Nakamoto***
**NJS Consultants Co., Ltd. 6-8, Tomihisa-cho, Shinjuku-ku, Tokyo, 162-0067, Japan (e-mail: njs-iwase@mbr.nifty.com)*
***Community Water Supply Support Center of Japan*

ABSTRACT

The filter rate of slow sand filter as an ecological purification system should be referred to the protistan movement speed. The fastest movement speed of protist was 1,208 μm/sec. The standard flow rate among the sand grains is 139 μm/sec calculated from the porosity 40 %. The first movement speed of protist was 8.7 times faster than the rate of water flow. The faster filtration rate of slow sand filter may be adopted on the movement speed of protists.

Key Words Slow sand filter, ecological purification system, protist, movement speed, filter rate

INTRODUCTION

The space size of a water treatment plant (WTP) is closely related to the filtration rate. And a construction cost is also influenced by the size of the civil engineering structure. If we can accept faster filtration rate for SSF, we can reduce the WTP area and the construction cost.

The standard filtration rate of 4.8 m/day of slow sand filter is well known as an English filter in the world. Recently, the double faster rate of 9.6 m/day has been adopted already by Thames Water Utilities Limited (Duncan 1988).

SSF is based on Ecological Purification System. The filtration rate can be adjustable to suitable rate depended on the different environment. Duncan (1988) carried out a survey of vertical profile of the top of sand layer on an ecological aspect at the rate of 9.6 m/d in Ashford Common WTP in London. In Japan, most of filters are operated less than 4 m/d. Nakamoto (2005) suggested that high filtration rate is better to keep aerobic condition in slow sand filter pond especially in a warm region.

Unger and Collins (2006) showed that the E. coli removal of the slow sand filter had a strong relationship with biological activity. In addition, they pointed that flagellates feed bacteria continuously. Protists such as ciliates in this system play an important role to collect colloidal particulate matter including bacteria and virus. A suitable filtration rate should be devised on the standpoint of free swimming protists. The protists are not washed away from the habitat in the top of the sand layer.

In this paper, we discuss the suitable filtration rate which can keep the protists activity in the sand layer in comparison with the filtration rate in the aperture of sand grains and their movement speed.

METHODS

Study Site

The study sites were Someya WTP of Ueda-city, Nagano prefecture (treatment capacity: 56,000 m^3/d, elevation: 500 m, raw water temperature: 0.0 - 22.5 °C, avg. 10.3 °C and raw water turbidity: 8 - 160 degree, avg. 11.0 degree) and Umagajo WTP of Seto city, Aichi-prefecture (treatment capacity: 5,400 m^3/d, elevation: 137 m, raw water temperature: 2.0 - 28.4 °C, avg. 12.5 °C and raw water turbidity: 0 - 14.4 degree, avg. 1.6 degree). Both study sites are located in the mountains in Japan. Their water sources are surface water from rivers.

Measurement of the Protistan Speed

We collected a surface sand sample using a simple tool (Photo 10.1). A great variety of protists inhabit in the sand layer of SSF. We measured the movement speed of protists. Protists did not distribute uniformly in the sample. The movement behavior was not constant. It was very difficult to evaluate the average speed of movement. Measurement of movement speed was carried out using a digital microscope as follows;

1. Video recording using the digital microscope (Celestron 44341).
2. Record of a protist movement trace by photo.
3. Measurement of the movement time.
4. Calculation of the movement distance for a second.

Situation of sampling | Samples | Video recording by Celestron 44341

Photo 10.1 Situation of sampling and samples at Umagajo WTP.

RESULTS AND DISCUSSION

Sampling Conditions

Water temperature in Someya WTP was 20.8 °C. The sample was collected from the filter surface of 25 days filter run. Water temperature in Umagajo WTP decreased from 16.2 °C to 4.0 °C from October through December.

Microscope photographs of filter sand and the protists collected in Umagajo WTP are shown in Photo 10.2. The main free swimming protists were ciliates. Appearance of ciliates were not homogenous among each sand sample. A lot of ciliates were observed in the sample of 10 days filter run, then they decreased afterwards.

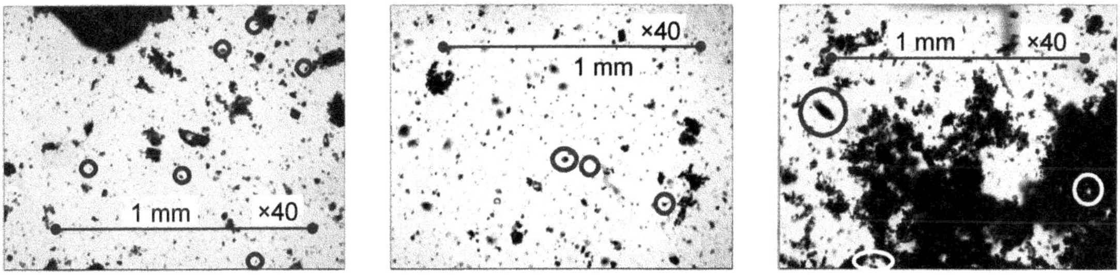

Filter run 10 days on 08, Nov. Filter run 17 days on 15, Nov. Filter run 59 days on 27, Dec.

Photo 10.2 Appearance of free swimming protists in Pond 3 of Umagajo WTP in 2013

Movement Speed of Protists

The various protists existed from a fast type to a slow type. There were different size (length) of ciliates. The protist movement was not constant. They moved quickly, slowly and stopped. We could not expect their behavior. Microscope photographs of filtration sand and the protists that collected are shown in Photo 10.3.

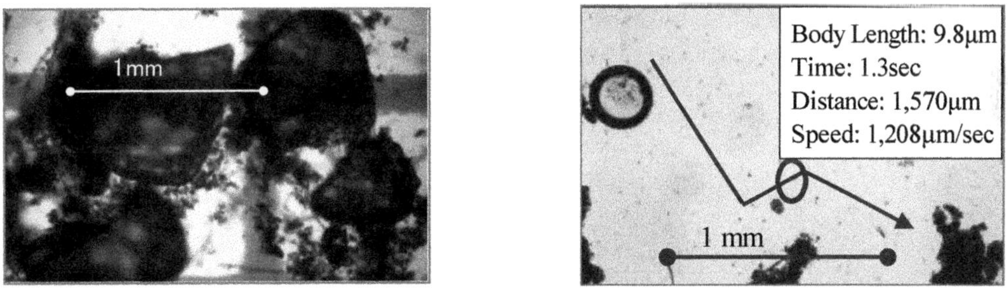

Filter sand Movement of protist

Photo 10.3 Protists among sand media and filter sand

There was a relationship between body length of protist and the movement speed (Figure 10.1). The fastest movement speed was 1,208 μm/sec. Relationship between body length of protist and movement speed / body length had negative correlation (Figure 10.2). This result showed that biological activity was higher in the small protists than large one.

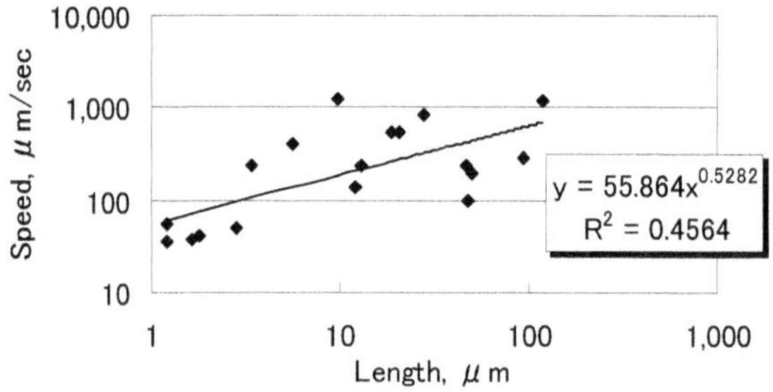

Figure 10.1 Relationship between body length of protist and movement speed

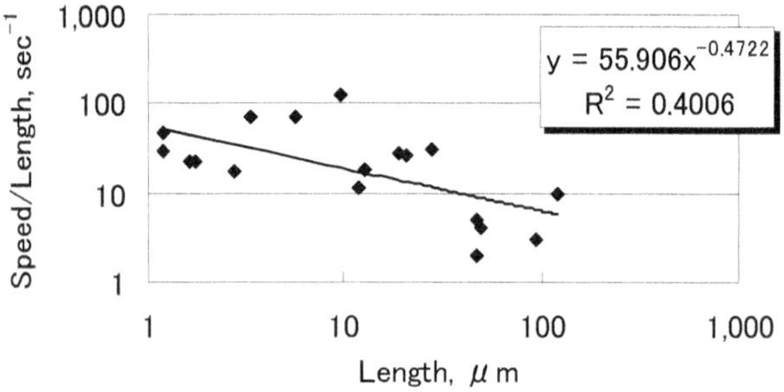

Figure 10.2 Relationship between body length of protist and movement speed / body length

Filtration rate for the Protists

Ogishima 1974 reported that *Paramecium* sp. could move in 1,200 μm/sec at 20 °C. We observed that small protists (about 1/12 length of *Paramecium* sp.) could move quickly in 1,208 μm/sec. The standard flow rate is 4.8 m/d. This rate is 56 μm/sec. If the porosity of sand layer is 40 %, the real flow rate becomes 139 μm/sec as an average among the sand grains. Protists could move 8.7 times faster speed than the standard filtration rate among the sand grains. Protists do not always swim in the center that flow is faster. They swim mainly on the surface of sand grain and behind the sand grain where the flow is slow or almost none. Our observation shows protists can move even in faster flow rate than the standard flow rate.

Lloyd (1996) showed that bacteria removal by protists decreased under faster filtration rate by a micro-chamber experiment. However, we expect that bacteria removal by protists occurs in the thick sand layer and it will be compensated by the enough thickness of the sand layer. Based on the movement speed of protists, the higher filtration rate of slow sand filter can be adopted than current standard filtration rate.

REFERENCES

Duncan, A. (1988) The Ecology of Slow Sand Filters, Slow Sand Filtration. Recent developments in water treatment technology, pp 163-180.

Nakamoto, N. (2005) *Oishii Mizu no Tsukurikata*, ISBN4-8067-1315-5

Unger, M. and Collins, M.R. (2006) Assessing the Role of the Schmutzdecke in Pathogen Removal in Riverbank and Slow Sand Filtration., Recent Progress in Slow sand and Alternative Biofiltration Processes, Proceeding of 4th Slow Sand and Alternative Biological Filtration Conference, ISBN9781843391203, pp 21-29.

Ogishima, M., Nishimura, K., Tsuchiya, Y. and Kawakubo, T. (1974) *Genseidoubutsu no Undou no Sokudobunpu*, The Physical Society of Japan, Proceeding of the spring conference, Vol.34.

Lloyd, B.J. (1996) The Significance of Protozoal Predation and Adsorption for the Removal of Bacteria by Slow Sand Filtration, Advances in Slow Sand and Alternative Biological Filtration, Proceeding of 3rd Slow Sand and Alternative Biological Filtration Conference, ISBN 0-471-96740-8, pp 129-137.

Chapter 11

Analysis of bacterial community structures in slow sand filtration bed on different region and soil depth by PCR-DGGE method

Kohei SEIKE, Masaaki EGUCH**, Kaoru UENO***

**Department of Environmental Biology, Chubu University, Japan*

***Graduate school of Bioscience and Biotechnology, Chubu University, Japan*

ABSTRACT

Slow sand filtration system is good for saving energy and reagent and supply delicious water, however managers of the systems must pay close attention to the condition of the filter bed and have original management technique based on their experience to keep stable water supply, because condition of microorganisms is changeable depend on weather and technical and historical conditions. In this study we analyzed bacterial community in some sand beds and compared different region and soil depth to find useful biological index for simple system control. We found site M (with high turbidity problem) had homogeneous community in all layers and many bands in deeper layers, but the others (without problem) had opposite results. The results were probably related to make sufficient bio film on sand beds, and the effective bacterium for make functional bio film was selected in the range of DNA marker No.2-7.

Keywords Slow sand filtration; slow sand filtration bed; Bacterial community

INTRODUCTION

In slow sand filtration method, raw water was purified by biodegradation on their soil surface biofilm and filtration effect (Huisman and Wood, 1974). The system needs little consumption of the reagent and electric, therefore, it is now being recognized again as a sustainable purification processing system in the world. However, managers of the systems must pay close attention to the condition of the filter bed and have original management technique based on their experience to keep stable water supply, because condition of microorganisms is changeable depend on weather and technical and historical conditions. Nakamura *et al.* (2010) studied on bacterial community of sands in several slow sand filter ponds, and indicated bacterial community was different from filtration plants. The mechanism and effective bacterial group to make sufficient bio filter had never been studied. In this study we analyzed bacterial community structures in some sand beds and compared different region and soil depth to find useful biological index of simple system control.

METHODOLOGY

Sampling sites and soil samples: We selected two regions and three water supply facilities using slaw sand filter system. Site M and I were in Shimane Prefecture, Site U was in Aichi Prefecture. The samples were collected each 1cm from top to 5cm of the sand bed at the day of sand scraping maintenance on November in 2012 after drainage just before the scraping. In site U, we collected 0-2mm as a top soil ("Top" in Figures) and the below to 2cm were collected and named "0-1cm". Three of replications (A, B, C) were collected at intervals of 10-12m from the center of the bed. Infiltrated water is used as raw water on site M and I, and site U intakes mountain stream water and water of a dam. Water supply were started from 2011 (site M), 1980 (site I), 1933 (site U). Turbidity of fileted water intended to be unstable in Site M, and other sites always keep stable low turbidity.

DNA extraction and PCR amplification: DNA was extracted from the soil samples using the Extra Soil DNA Kit Plus ver.2 (NIPPON STEEL & SUMIKIN Eco-Tech), by the bead-beating method. PCR amplification was carried out using the primer set F984GC and R1378, which targeted V6-8 variable region of the 16SrRNA gene of eubacteria (national institute for agro-environmental science, 2010). The PCR program had an initial denaturation at 95°C for 10min, followed by 34 cycles of denaturation at 95°C for 15s, annealing at 65°C for 30s and extension at 72°C for 30s, and a final extension at 15°C for 5min. The PCR products were observed on agarose gel (0.8%) with 6×GR Red Loading Buffer (BIO CRAFT) under UV light.

Denaturing gradient gel electrophoresis (DGGE): DGGE was performed with a Decode Universal Mutation Detection System (Bio-Rad) and the denaturant gradient of the gel was ranged from 50 to 70%. Electrophoresis was run for 18h at 58°Cat 50V.

RESULTS AND DISCUSSION

DGGE band patterns were shown in Fig. 11.1. Band patterns were homogeneous between every layers and replications in site M (Fig. 11.1a). In site I, these were homogeneous between replications, and these were heterogeneous between the sample of 0-1cm depth and the other deeper lyres (Fig. 11.1b). In this site many bans of 0-1cm samples existed in the range of DNA marker No.3-7, and in the other depth many bands were added in the range of the marker No.7-10 (Fig. 11.1b). In site U, bans were lightly marked and not clear in compare with the other sits. The band patterns of replications were also homogeneous. Many bands distributed in every layers and wide range of the marker, and the bands of top and 0-1cm depth intended to be dark and concentrated from upper marker No.1 to No.4 in compare with the other deeper layers in site U (Fig. 11.1c).

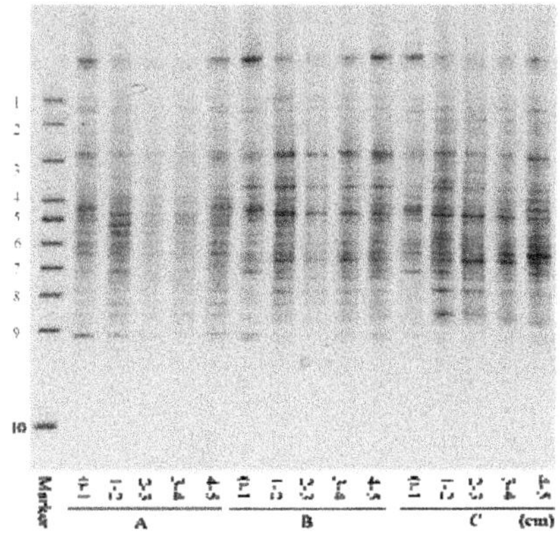

(a) Band patterns on site M.

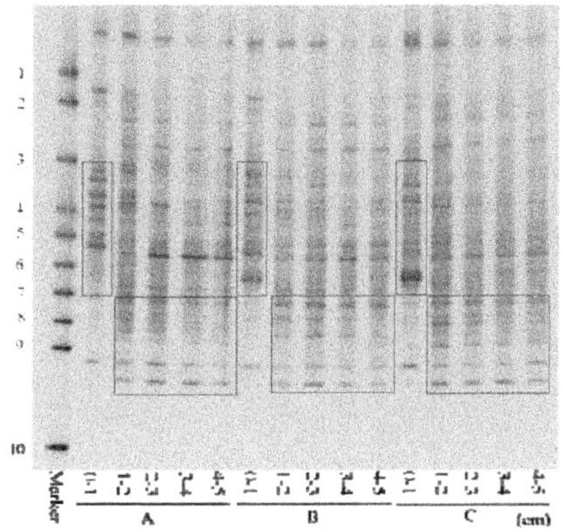

(b) Band patterns on site I.

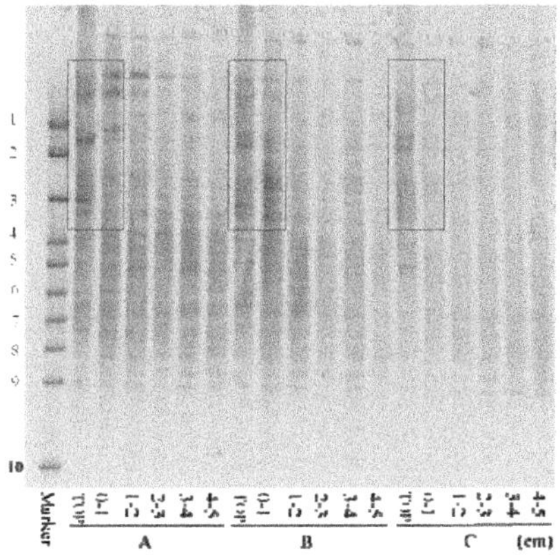

(c) Band patterns on site U.

Figure 11.1 DGGE band patterns of eubacteria of every sand depth on site M (a), I (b) and U (c). A, B and C mean replications.

For clear understandings of common and characteristic bands of the sites and depth, we analyzed the surface (top and 0-1cm depth) and 2-3cm soils of band patterns (Fig. 11.2). Band patterns were different every regions on each depth. On surface soils, many dominant bands were in the range of marker No.2-8 in site M (Fig. 11.2a; outlined-a) and were in marker No.2-7 in site I (Fig. 11.2b; outlined-b), and were in marker No.1-9 in site U (Fig. 11.2a; outlined -c). Common dominant of surface soils were found as band ①-③, ⑤, ⑥ with all sites. The band ④ and ⑦ were found in site I and U. On the other hand, in the soils of 2-3cm depth, many bands were found in the range of marker No.1-9 in site M (Fig. 11.2b; outlined-d), and found in the range of No.3-8 in site I and U (Fig 11.2.b; outlined-e). Common dominant bands were not found clearly in 2-3cm depth. On 2-3cm layers, many dominant bands were marked in site M in compare with the other sits (Fig. 11.2b).

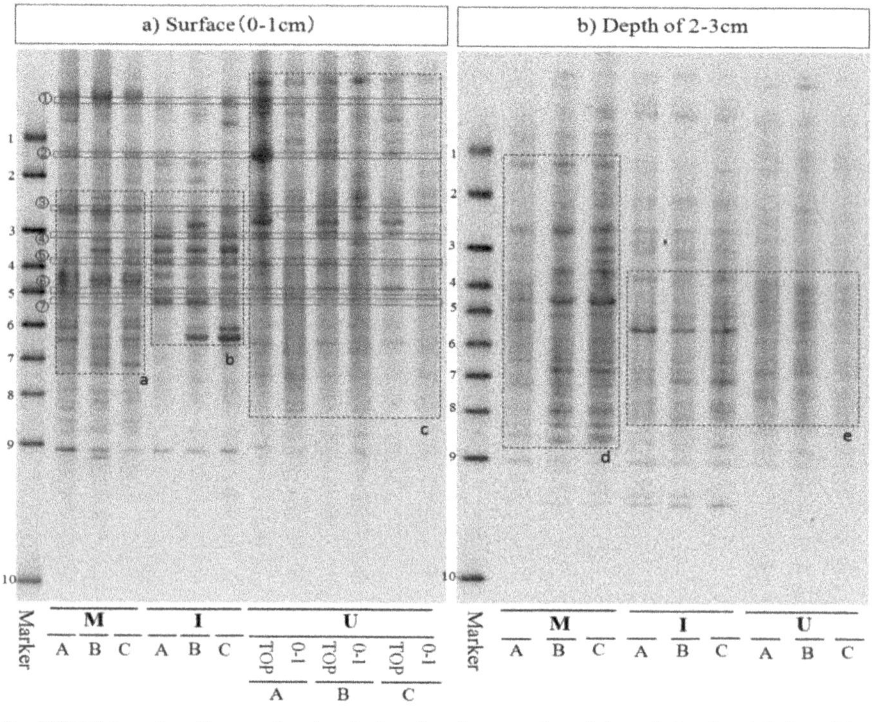

Figure 11.2 DGGE band patterns of eubacteria of soil on surface(a) and depth of 2-3cm(b) on site M, I and U. A, B and C mean replications. Surface soil of site M and I were on the depth of 0-1cm, and that of site U were collected from 0-2mm ("TOP") and directly under 2mm depth to 1cm ("0-1cm"). Outlined by dotted line (a)-(e) were the area of dominant marked in each site. Band ①-③, ⑤, ⑥were common and dominant band in all sites, ④ and ⑦were that of in site I and U.

From these results, we found the clear differences between site M and the others. Site M had homogeneous community in all layers and many bands in deeper layers, but the other sits had opposite results. Site I and U had different community between surface and the other deeper soils comparatively (Fig. 11.2), and the community of deeper layers had fewer number of bans than their surface (Fig. 11.1, 11.2). These results were probably indicated that bio films of site M was not sufficient in compare with the other sites. The film is made mainly by aerobic bacterium on surface of sand bed (Huisman and Wood, 1974). If the film was not sufficient, minute particle would leak out to the deeper layers and finally to outlet. In this case, a gap of oxygen condition would be weak and bacterial community would be similar in some depth of sand bed. Site M had turbidity problem, and the causing substances was estimated *Microcystis* sp. *Microcystis aerigompsa* dominantly occurred in the lake waters with high temperature, low DIN, DIP, and DIN: DIP ratio and high TN: DIN and TP: DIP ratios (Yoshida, 1997). Water condition of the site seemed to be in this condition frequently and also have insufficient bio film, so *Microcystis* sp. often escaped from water passing through the sand filter. Finally, the effective bacterium seemed to be in the range of marker No.2-7 (Fig. 11.2a). The range included common and dominant band with all sites or site I and site U. Particularly band ④ and ⑦ were not found in site M, so these bands had a possibility to be an effective bacterium for make sufficient film.

CONCLUSIONS

The obvious differences between site M and the others were found. Site I and U had difference community between surface and the other deeper soils, and the community of deeper layers had fewer numbers of bans than their surface. Site M had homogeneous community in all layers and large numbers of bands in deeper layers. These results in site M seemed to be an evidence of insufficient bio film on sand bed, and it seemed to become the reason of high turbidity problem. The effective bacterium for make functional bio film on sand beds were selected in the range of DNA marker No.2-7. We should identify the species and bacterial function to make sufficient bio film and avoid high turbidity in slow sand filtration system. In this study, sample soils had taken just before the scratch sands, we also should check the bacterial community in water and take soil samples before draining.

ACKNOWLEDGEMENTS

This study was supported in part by water supply facilities for sampling and Institute of Water and Environment Business for study Grant. Dr. Jun MURASE, Dr. Susumu ASAKAWA, Dr. Takeshi WATANABE and Dr. Motoyasu MINAMI gave technical instructions and advices on this study. The authors wish to acknowledge to all of them.

REFERENCES

L. HUISMAN and W.E. WOOD (1974), Slow Sand Filtration, 27-46, World Health Organization.

National institute for agro-environmental science (2010), methods for soil bacterium and fungus by PCR-DGGE ver.3.3 (in Japanese).

Taketoshi NAKAMURA, Naoko TAKEI, Motonari KAIMURA, Kenichi ABE, Takashi YAMAGUCHI, Nobuo ARAKI, Analysis of Microbial Community in Slow Sand Filters by PCR-DGGE, Kankyo to sokutei gijutsu (37), 13-17, Japan Environmental Measurement and Chemical Analysis Association, 2010 (in Japanese).

Yoichi YOSHIDA (1997), Relationship between the Dominant Occurrence of Mirocystis aeruginosa and Water Quality of Meteorological factors in South Basin of Lake Biwa, nippon suisan gakkaishi, 63(4), 531-536 (in Japanese).

Chapter 12

Gut bacterial floras of aquatic invertebrates inhabiting slow sand filter beds

J. Takeuchi, Y. Zhao**, R. Tanaka*,****

**Department of Civil & Environmental Engineering, Kure National College of Technology, Hiroshima 737-8506 Japan*

***Department of Chemistry and Chemical Engineering, Dalian University, China*

****Department of Biology, Faculty of Science, Okayama University, Japan*

ABSTRACT

Aerobic heterotrophic bacteria were isolated from the gut of aquatic invertebrates inhabiting slow sand filter beds as well as the overlying water column. Each individual was sorted with soft tweezers by hand, and only "external colonizers" on the surface of the body was partially sterelised under UV irradiation. After dispersal treatment of the animal body in an Eppendorf tube with the tube mixer, the bacteria in the suspension of a gut origin were inoculated on an agar plate and isolated as "internal colonizers". The gut floras of freshly-collected animals tended to show very uniform composition, almost a single colony type of bacteria, but depending on each host animal. This may suggest that the aquatic invertebrates can freely select their favorite food under natural conditions. Besides, most of the isolates were classified as the fermentative groups of bacteria having strong hydrolytic activities against organic substrates. Based on the current results, the gut bacterial flora and the enzymic faecal pellets themselves could contribute to the degradation of excess organics as well as possible removal of health-related microbes.

Keywords *Aeolosoma*; *Tubifex*; *Chironomus*; *Aeromonas*; internal colonizers; 16S rDNA sequencing; normal flora

INTRODUCTION

Slow sand filters (SSFs) are one of the manmade water purification systems, but they can be recognised as a part of natural ecosystems, resembling a natural pond with a stream in terms of their components and functions. Filamentous algae such as *Melosira varians* and *Spirogyra* spp. are commonly found in the SSFs as the primary produces, forming the underwater vegetation (Figure 12.1) whilst tube worms such as larva of *Chironomus* and *Tubifex* (Figure 12.2) and/or freshwater snails are representative consumers that graze the surface green mat covering the uppermost sand layer of the SSFs. Various microorganisms colonized on the "schmutzdecke" layer, eventually depending on the primary producers and consumers.

In the present report, we have focused on the internal bacterial colonizers within some tube worms' guts as well as those of planktonic crustacean, *Daphnia magna* collected from the water of the SSFs. To discuss the food-gut microbe relationships in general, the common springtails living in the terrestrial soil were also collected and compared with those aquatic lives.

Figure 12.1 Underwater algal vegetation. (Nishino Works, Mihara, Hiroshima)

Figure 12.2 Sludgeworms in the sand layer. (Noda Works, Mihara, Hiroshima)

METHODOLOGY

Aquatic invertebrates, such as *Aerosoma*, *Tubifex* and larvae of *Chironomus*, are collected with sand by a handy sieve from the uppermost layer of the filter sand beds from water works located in Kure City and Mihara City, Hiroshima Prefecture, the western part of Japan. The algal mat-sand mixture samples were transported to a laboratory to be sorted with a specially designed tweezers (Rigosha, Tokyo) by hand.

A few to several individuals of each animal were repeatedly washed by tap water, and were partially sterilized for 5 min, being stirred in a quartz glass flask placed under the UV irradiation (Figure 12.3). This treatment enabled to eliminate the bodies' surface-colonized and the surrounding water-suspended bacteria but not the inner colonizers within their gut tubes. After the partial sterilization, the whole worm's bodies were thoroughly dispersed in an Eppendorf tube by a tube mixer (Figure 12.4).

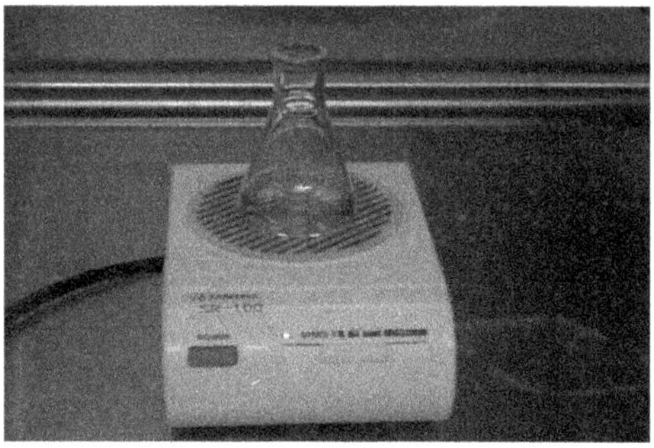

Figure 12.3 Surface sterilisation by the UV.

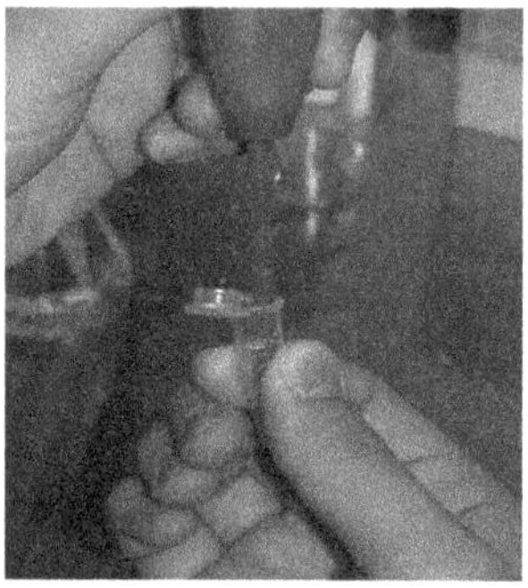

Figure 12.4 Dispersal of gut-embedded bacteria.

The resultant suspension of bacteria of a gut origin was inoculated onto a SPC (Standard Plate Count) agar (Standard Method Agar "Nissui", Tokyo) plate, and the plates were incubated in a dark place to avoid photo-reactivation by visible light for a week at ambient temperatures. The colonies were purified by streaking, and stored on the slant agar at dark/cool place. The heterotrophic bacteria of environmental origins, i.e., the sand beds and the overlying water were also isolated by an agar plate of R2A "Daigo"(Wako, Osaka), optimized for naturally-occurring bacteria. After hey were once purified, all the isolates were maintained on the same slant agar.

Differentiation of fermentative isolates from non-fermentative ones was made by the use of Hugh-Leifson medium (OF basal medium "Eiken", Tokyo) with liquid paraffin for sealing. The fermentative isolates were grouped by applying a commercial rapid identification kit, ID test EB-20 "Nissui", Tokyo; based on the test results of 20 kinds of biochemical colour changes to refer to the supplier's database.

Production of hydrolytic enzymes was also confirmed by a double layered agar plate, consisting of the "SPC" agar as the upper layer whilst the substrate-containing agar as the lower layer. 1% casein (Skim milk, Difco), 0.2% soluble starch (Wako, Osaka) and 0.4% tributyrin (Wako, Osaka) were used as each model substrate for proteins, carbohydrates, and lipids, respectively. Only the hydrolytic activity of starch was visualised by the starch-iodine reaction; those of casein and tributyrin were simply judged by presence of the enzymic diffusion as a transparent zone around a paper disc with the colony (Figure 12.5).

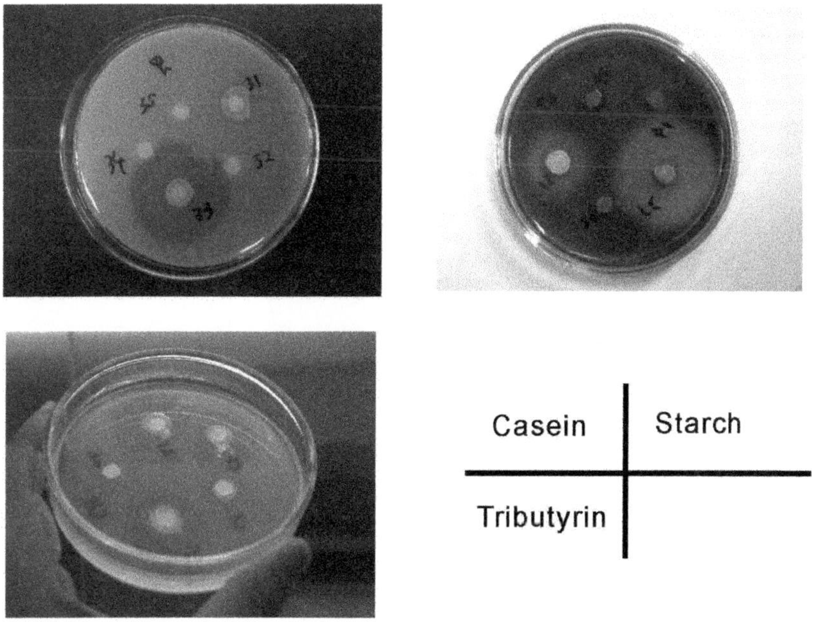

Figure 12.5 Visualisation of hydrolytic enzyme activities.
Transparent zones indicate positive reactions of each hydrolytic activity.

Each colony was suspended with a 100µL of normal saline in a 1.5mL microtube. A 20µL of the cell suspension was absorbed onto 1mm^2 of each chip of the FTA card (Whatman, UK) in a 2mL tube. The chip was dried in a desiccator containing silica gel granules, and washed with the commercially available FTA Purification Reagent (Whatman, UK) three times for 5 min. Finally, the reagent was replaced with TE buffer by a combination of vortex mixing and spinning (Eppendorf, Germany) three times for 5 min. The treated chip was then dried up, forwarding to the next PCR amplification step.

The PCR amplification from the dried FTA chips was performed in total volume of 50µL master mix containing 5µL of 10x PCR buffer, 200µL each deoxyribonucleotide triphosphate (Finnzymes, Finland) 20pM of forward (27F and reverse (1492R) primers (Invitrogen, Japan), 5 units of taq polymerase (Qiagen, Germany), 35.5µL of RO (reverse osmotic) water and the applied FTA card as a DNA template.

After a denaturation step of 5 min at 95°C, the PCR amplification was performed (Thermogen Quick Bath QB-0225AC, Nagano). This amplification process consisted of a denaturation step of 60s at 94°C, a primer- annealing step of 60s at 55°C, and an extension step of 2 min at 72°C.After 30 cycles, the final 10 min incubation at 72°C was performed. The PCR products were examined by electrophoresis on 1% agarose S (Nippon Gene, Japan) gels for 30 min at 100V. Then, the gels were stained with ethidium bromide (Nakarai, Japan) for about 10 min and soaked in tap water for 10 min. The final gel images were observed on a transilluminator UV-312-20 equipped with a Bio-Pyramid hood, MBP-01 (MeCan, Saitama), and recorded by a digital camera PowerShot G11 (Canon, Tokyo).

The PCR products were purified through a spin column direct to apply the cycle sequencing using a Genetic Analyzer ABI 3100 (Applied Biosystems, USA), which was outsourcing to MacroGen Japan at Tokyo University of Agriculture, Setagaya, Tokyo. The datesets of sequences were analysed using the Blast search system on the website of the DNA Data Base of Japan (DDBJ), Mishima, Shizuoka.

RESULTS AND DISCUSSION

Gut bacterial floras of freshly-collected aquatic invertebrates inhabiting the SSFs were characterised on the cultured plates by quite uniform composition in terms of "colony type" such as their colours and sizes, including texture description; those differ depending on their host animals as shown in Figure 12.6.

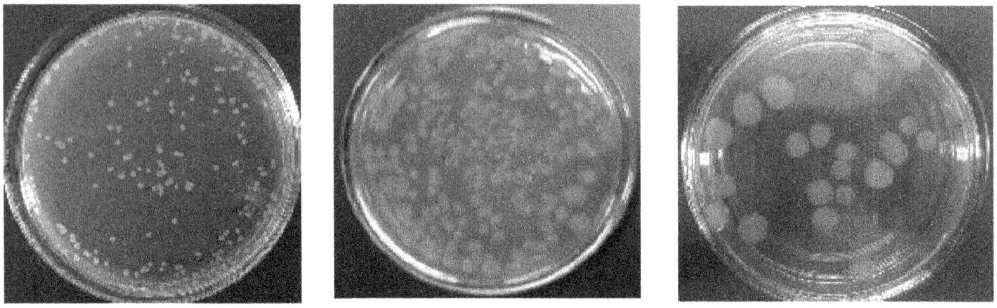

Figure 12.6 Uniform colony types appeared on agar plates.
Left: gut flora of *Daphnia magna*, Centre: that of *Aelosoma* sp., Right: *Tubifex* sp.
(Hirabara Works, Kure) (Nishino Works, Mihara) (Noda Works, Mihara)

This means that the bacteria in the gut of the invertebrates may be predominated by almost single one species or the other species close to each other within the same host animals. This strongly suggests that these wild creatures in nature can choose their favorite food suffciently available in their "open" habitats. In contrast, the gut floras may change into a bit complex composition (two or three colony types; See Figure 12.7) when the same host animals were kept in captivity in a closed space such as bottle or container for a week.

The animals were likely to eat their favorite food up, then they had to take anything else, e.g., even their dead bodies, exoskeletons or feces. In the case of springtails in the soil habitats to compare, they seemed to use limited food, possibly fallen leaves on the ground and the fungi (mushrooms) associated with the degraded litter or decayed twigs and barks. The gut flora of the springtales, as being true saprotrophic terrestrial organisms, may not be so uniform (See Figure 12.8) as compared with the aquatic counterparts in the ecosystem.

According to the OF (oxidative/fermentative) selection, almost all the gut floras were fermentative isolates, being highly adapted to their continuous culture-based conditions in the SSFs, thereby provided with exotic food sources derived from a water course or a reservor as well as self-sufficient food sources grown within the pond of the SSFs.

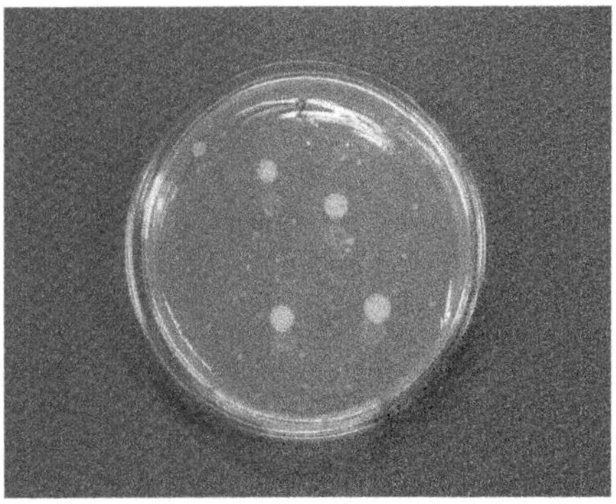

Figure 12.7 Colonies of midge larvae in captivity. (Mixed with yellow and white ones)

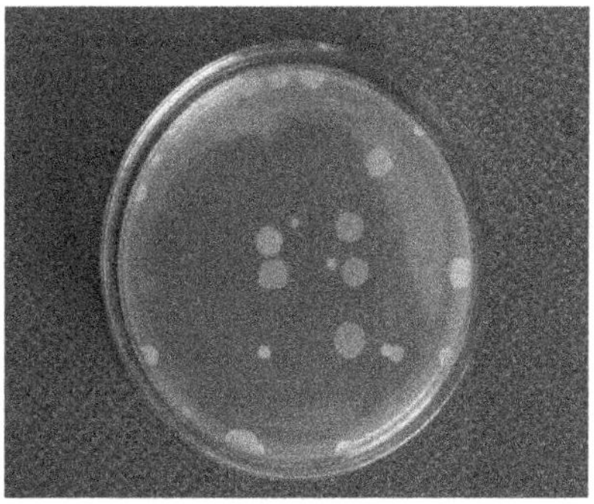

Figure 12.8 Colonies of springtails in soil. (Mixed with large and small colonies)

Fermented soybeans (Natto) and milk (yogurt), for example, are known as traditional healthy food to maintain our digestive tube and keep off harmful microbes, those accidentally leading to possible gastrointestinal infection. The episodic "Hamburg Mystery" (Nakamoto, 2005) was once reported and suggested that the SSFs can prevent the incidence of cholera from infection, possibly by removing even such a serious causative microbe in any way. The present results may suggest that the uniform floras of the native aquatic creatures found in the sand beds of the SSF purification system as a bioreactor, thereby conditioning not only qualities of tastes but biosafety as drinking water.

In addition, both floras of the gut and their relevant faecal pellets origins seemed to be almost the same in terms of hydrolytic activities (Takeuchi and Suzutoh, 2011). This may be reasonable, considering their short length of gut tubes (Figure 12.9) and the resultant quick turnovers. The present results would encourage us to make a wise decision to choose the SSF-based purification systems when employing tiny creatures as the site-specific working engineers (Wotton and Hirabayashi, 1999) without any conscious, being automatically adjustable according as the condition changes and even their "enzymic pellets" indicating strong hydrolytic activities on the agar plate (See Figure 12.5) have turned out unsanitary, regardless of our human's biased imagination derived from the warm-blooded vertebrates' defecation in general. Figure 12.9 shows the internal contents in the gut tube of a midge larva.

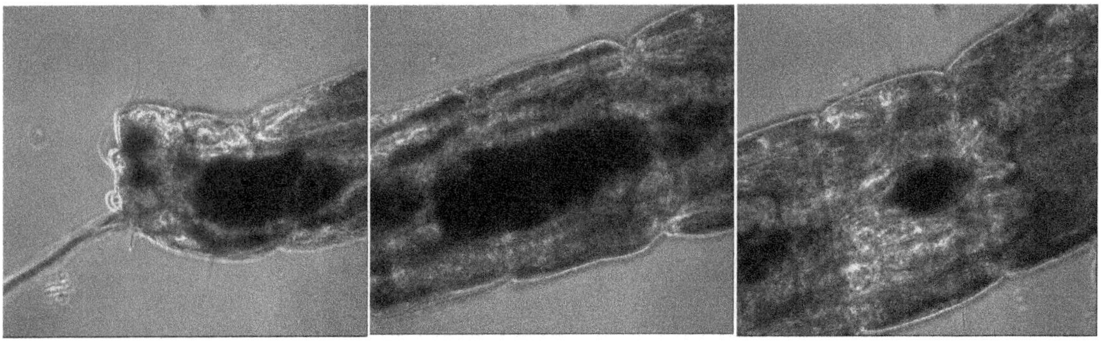

Figure 12.9 A gut tube of a midge larva and the internal contents (Phase contrast microscopy).

Left: lower part of the gut tube, Centre: middle part of the tube, Right: upper part of the tube

Table 12.1 shows the results of 16S rDNA-based molecular identification of the isolates from the animals, alongside the most probably suggested species name with some additional information based on the accumulated nucleotide sequence databases (EMBL/GenBank/DDBJ). All the isolates tested here accounted for more than 98%

similarity to the published isolates, suggesting that they would be fermentative groups of bacteria, such as the *Aeromonas* and the related groups commonly distributed in aquatic habitats. Besides being "normal flora" in the environment, a couple of isolates suggest that they could degradate natural and artificial toxic substances such as Microcystin and dioxin/PCB, respectively. Interestingly, the permanent aquiatic inhabitants, such as *Aeolosoma* and *Tubifex* spp, could possibly decompose these toxins, clearly showing the resultant contrast with the midge larva cases. As a result, gut floras of aquatic invertebrates would play an important role not only when purifying water but degrading toxic substances unfavourble for making safe and tasty water with less costly.

Table 12.1 Results of molecular identification and suggested species of the bacterial isolates.

Strain	Origin of the animals	Similarity (%)	Identification	Remarks*
#1	*Aeolosoma* sp.	99%	*A. hydrophila*	
#2	*Aeolosoma* sp.	99%	*Paucibacter* sp.	Microcystin degraders
#3	*Aeolosoma* sp.	99%	*Aeromonas* sp.	
#4	*Tubifex* sp.	98%	*Paucibacter* sp.	Microcystin degraders
#5	*Tubifex* sp.	99%	*Acidovorax* sp.	Dioxin/PCB degraders
#7	*Chironomus larva*	100%	*A. media*	
#8	*Chironomus larva*	100%	*A. veronii*	
#9	*Chironomus larva*	100%	*A. veronii*	
#10	*Chironomus larva*	100%	*A. veronii*	
#11	*Chironomus larva*	99%	*Aeromonas* sp.	
#12	*Chironomus larva*	100%	*A. veronii*	

* Information on the above remarks was the possible characteristics indirectly described in the databases.

CONCLUSIONS

The gut tube of aquatic invertebrates in the SSF beds could act as a fermentation reactor, forming a uniform bacterial flora inside and keeping other bacteria off, presumably as a result of choices of their own favorite food in nature. Such the stable "normal flora" may contribute to hygienic status of the SSF process to produce safety drinking water.

ACKNOWLEDGEMENTS

The authors wish to express special thanks to Kure Municipal Water Works and Hiroshima Prefectural Water Works for kindly providing us with fresh samples and the related information. YZ was funded by the Government of Japan and Kure-Higashi Rotary Club during her study with JT and RT in Japan in 2012.

REFERENCES

Nakamoto N. (2005) Removal of pathogens by slow sand filtration, In *How to Make Delicious Water-Rivival of Safe and Tasty Tap Water and the Supportive Technology*, 39-40, Tsukiji Shokan, Tokyo (in Japanese).

Takeuchi J. and Suzutoh S. (2011) Possible roles of midge larvae and the associated bacteria inhabiting slow sand filter beds, *Research Report of Kure National College of Technology*, 73: 19-24 (in Japanese).

Wotton R.S. and Hirabayashi K. (1999) Midge larvae (Diptera: Chironomidae) as engineers in slow sand filter beds, *Water Res.*, 33, 1509-1515.

Chapter 13

The functioning of biological slow sand filtration in relation to the presence and the role of Annelids in the schmutzdecke

Gerhard H. Wubbels, Jantinus H. Bruins*, Mieneke Bosman* and Dirk v.d. Woerdt***
**WLN (Waterlaboratorium Noord), The Netherlands*
***Waterbedrijf Groningen, The Netherlands*

ABSTRACT
At Waterbedrijf Groningen, The Netherlands for the purification of surface water to drinking water slow sand filtration is used. There are five slow sand filters, and slow sand filter three (SSF3) has different behavior in head loss (clogging). SSF3 had run for more than 25 years without renewing the schmutzdecke due to clogging. Experiments showed that certain type of annelids give the schmutzdecke a different structure than the schmutzdecke found in the other filters. The presence of the annelids had no effect on the log removal of pathogenic microorganism of the filter. Identification of the worm is doubtful. Comparison of the annelid found in the water of SSF3 and annelids found in the surrounding ground of the filters showed that the origin of the "water"-annelid lies somewhere different. The annelid is not found in the other SSF and therefore slow sand filter four is inoculated with approximately a thousand annelids. In the next years yearly the structure of the schmutzdecke is examined.

Keywords Slow sand filtration; Annelids; disinfection

INTRODUCTION

At Groningen, a city in the North of The Netherlands, drinking water is produced at the pumping Station De Punt, one of the production locations of Waterbedrijf Groningen. A part of the drinking water production in this plant is derived from surface water taken from the river the Drentsche Aa. The purification steps in this surface water plant exists of Coagulation/Sedimentation, Rapid Dual media Filtration, Active Carbon Filtration, UV and Slow Sand Filtration (SSF). In the Netherlands there is no residual chlorine usage in the distribution networks. For that reason there has to be enough disinfection capacity achieved in the water treatment system, preferably a multi barrier system. For De Punt these are the Coagulation/Sedimentation, UV and the SSF. In total there are 5 SSF units at the location De Punt and they all are built in the same period about one hundred and thirty years ago. Most SSF units have the problem of clogging after a certain period of time used. The intensity of clogging depends on velocity of filtration and the quality of the influent water mainly with respect to suspended solids and turbidity. At the five SSF units, which all have the same quality influent water, there is a significant difference of clogging between the units. Four of them are building up head loss, which is a measure for clogging, within approximately 2-5 years. After that period the "Schmutzdecke", the upper 1 to 2 cm of the filter, has to be removed or has te be raked to recondition the initial filtration capacity.

The remaining filter, SSF 3 (Figure 13.1: LZF 3), however did not clogged during the last 25 years. All this time there was no need to open the filter and to remove or rake the "Schmutzdecke". This means a significant saving in costs of maintenance. In 2007 a research was started to find the reason(s) for this unusual and different behavior of SSF3

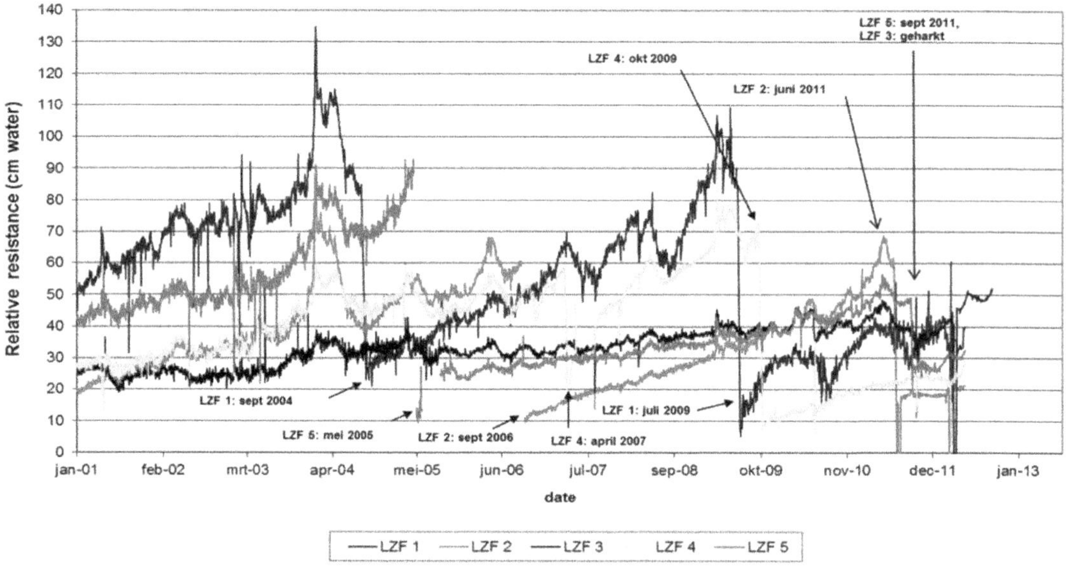

Figure 13.1 Development of (normalized) head loss of the SSF units during a period of 12 years.

METHODOLOGY

The main target for doing research at the abnormal behavior of slow sand filter three (SSF3), was the efficiency in maintenance and costs if the same behavior could be established in all five slow sand filters at the production plant. But before changing all filters and rely on the positive functioning of SSF3, there should be more knowledge about all aspects in concern.

Earlier observations during practical work on the slow sand filters (SSF) the last 15 years learned us that there was a visual difference between the filters concerning the structure of the schmutzdecke. The schmutzdecke from SSF3 was less sticky and more fluffy than the schmutzdecke from the other four filters. Secondly, there was an observation of difference in, with the naked eye visible higher organisms in the filters, looking like normal worms (Annelids). These Annelids were frequently seen in high numbers in SS3 and not in the other ones.

There were two important questions to be answered before acting on changing the other SSF's.

- Is SS3 still functioning efficiently in removing pathogenic microorganisms or is the filter more like a sieve due to the presence of Annelids.
- Are there more differences between the SSF which play a part in the clogging of the slow sand filters.
- Is it possible to change the behavior off the other filters, and if so how can we achieve in that.

For all five separate filters the logarithm removal of different indicator organisms were determined. Therefore analyses of water samples had to be done in volumes ranging from 100 ml till 1000 liter. Volume of the samples were dependent on the type of microorganism to detect.

In Table 13.1 the different microorganisms and the corresponding volumes which had to be analyzed are presented.

Table 13.1 Microbiological parameters analyzed and used methods

Microbiological parameter	Used analytical method	Volume influent (liter)	Volume effluent (liter)
Heterotrophic plate count 22°C	NEN-EN-ISO 6222	1	0.1
Escherichia coli	NEN-EN-ISO 9308-1	5	100
Spores sulphite reducing clostridia	NEN-EN-ISO 6461-2	3	100
Somatic Bacteriophages	NEN-EN-ISO 10705-2	10	1000

From all filters also the chemical composition of the effluent were analyzed. This to see if there are differenced in the polishing processes in the filters.

The concentration of the large volume samples for E.coli and SSRC were done with the MF-sampler method (Veenendaal 1996). The concentration of the large volume samples for bacteriophages were done with the Haemaflow Technique developed in the European Techneau program (Veenendaal 2007).

The next step in this project was to identify the Annelid species from SSF3 and with the knowledge of the species to explain and understand the role of this Annelid in SSF3.

The last part of this project is to seed these annelids in another filter so see if they survive and spread through the filter and alter the schmutzdecke and eventually lower the head loss.

RESULTS AND DISCUSSION

Determination of the pathogenic log-removal capacity of all the five filters was first done for the bacteriological parameters E.coli and Spores of Sulfite-Reducing Clostridia (SSRC) and a total count by Heterotrophic Plate Count (HPC). All concentrations of the micro-organisms are expressed per liter sample-volume. In Table 13.2 these results are presented. Besides the information about the bacteria also the age of the schmutzdecke and the filtration and volume characteristics are presented.

As can see in Table 13.2 in the bacterial load for all filters were very low. This is due to the former part of the purification. In the sample of the effluent ranging from approximately 40 till 130 liter no E.coli and SSRC were detected with the exception of SSF5 where a low concentration of SSRC was found. The volume of the samples was restricted to 100 liter due to clogging of the membrane filter from the MF-sampler. In SSF5 also the HPC was very high, even higher than the influent of SSF5. An explanation for this result was not found. But this was for the understanding of functioning of SSF3 also not that important. Although for E.coli and SSRC no conclusion could be drawn, the results together with the information of the HPC, which were almost equal in all filters, suggest there is no indication that SSF3 is a leaking sieve or is different from the other four filers. The second set of experiments were done for Viruses. As an indicator for enteroviruses we used somatic bacteriophages. Because of the low concentrations of bacteria it was decided to rise the volumes up to about 1000 liter.

Table 13.2 Large volume measurement (MF-sampler) E.coli, SSRC and HPC for SSF1 –SSF5.

	unit	SSF1	SSF2	SSF3	SSF4	SSF5
Age Schmutzdecke	days	550	3600	8200	3300	330
Flow filtration	m^3/h	78	98	104	99	104
Speed filtration	cm/h	8,0	7,7	8,2	7,7	8,2
E. Coli (infl)	cfu/l	< 0,2	< 0,2	< 0,2	< 0,2	< 0,2
E. Coli (effl)	cfu/l	< 0,013	< 0,013	< 0,008	< 0,013	< 0,023
Volume effluent (E.coli)	l	78	78	126	77	44
HPC22 °C (infl)	cfu/l	5200	7900	4800	5200	4800
HPC 22°C (efflu)	cfu/l	150	640	120	113	6100
Volume effluent (HPC22)	l	0,1	0,1	0,1	0,1	0,1
SSRC (infl)	cfu/l	0,6	< 0,3	<0.5	0,6	<0,5
SSRC (efflu)	cfu/l	< 0,015	< 0,013	< 0,010	< 0,016	0,017
Volume effluent (SSRC)	l	68	77	101	62	59

This could be achieved by using a new developed technique for larger volumes, the Haemaflow filtration. In Table 13.3 these results are presented. The volumes which could be analyzed in this case were not restricted to the concentration of particles but just on filtration time. 1000 liters takes about 24 hours sampling time.

Table 13.3 Concentrations of somatic coli phages in SSF1 - SSF5 sampled by Haemaflow.

	Unit	SSF1	SSF2	SSF3	SSF4	SSF5
Coliphages (infl)	pfu/l	1,2	5,7	5,7	1,2	1,2
Coliphages(effl)	pfu/l	< 0,018	< 0,034	< 0,052	< 0,08	< 0,023
Volume concentrate	ml	200	375	550	900	375
Sample Volume (effl)	l	1127	1116	1053	1112	1048

As expected from the experiments with the bacteria, the removal of bacteriophages also did not showed difference between SSF3 and the other filters. Although the concentrations in all effluent were below detection limit, also here the results suggest that SSF3 is not a sieve.

Hijnen et al (2004) showed logarithm removal of bacteriophages by slow sand filtration ranging from 1.7 to 2.2. In these experiments we show a removal in the range of more than 1.9 and 2.6.

So this is an indication that all slow sand filters are removing pathogenic viruses efficiently.

The chemical composition of all slow sand filters was almost equal (Table 13.4).

Table 13.4 Physical chemical analyses of the effluent of the slow sand filters

	Unit	SSF1	SSF 2	SSF 3	SSF 4	SSF 5
Temperatuur	°C	8,4	7,7	8,0	8,3	8,7
pH	pH	7,30	7,34	7,40	7,31	7,33
conductivity	mS/m	33,4	33,3	33,3	33,5	33,5
HCO3	mg/l	96	95	96	95	96
DOC	mg/l	5,0	5,1	5,1	4,9	5,0
UV-absorption 254 nm	M^{-1}	12,0	11,0	12,5	11,0	11,5
turbidity	FTU	< 0,1	< 0,1	< 0,1	< 0,1	< 0,1
Al	µg/l	11	11	10	11	10
Ca	mg/l	43,3	43,4	42,9	43,1	43,8
Mg	mg/l	5,55	5,55	5,45	5,50	5,55
Fe	mg/l	0,03	< 0,02	< 0,02	<0,02	< 0,02

There is no difference in the chemical composition of the effluent from the slow sand filters.

Combining all biological and chemical analyses there is no differences between the slow sand filters besides the presence of annelids in slow sand filter three.

Determination of the annelids.

About three hundred annelids were harvested from SSF3 for determination. The worms were send to four different institutes. In Table 13.4 the results of these identifications are presented. It is surprising to see that there were differences in the results and that also shows how difficult this type of identification is.

Table 13.5 Identification results Annelids SSF3

Institute	identification
Alterra Wageningen	*Eisenia fetida*
Grontmij/Aquasence	*Dendrodrilus rubidus rubrucundus*
University Lancaster	*Eisenia andrei*
National Museum History London	*Dendrodrilus rubidus*

Eisenia and Dendrodrillus are earth worms which in theory cannot survive in water. The annelid in SSF3 have already lives probably 25 years in water.

One of the possibilities of how the annelids could enter the filter could be from the outside when the filter was opened in the past. For that reason the worms found in the ground in the surrounding were determined. And five species were found (*Lumbricus terresis, Lumbricus Rubellus, Lumbrucus species, Aporrectodae caliginosa and Aporrectodae species*), but they all were different from the SSF3-Annalid.

The SSF3 annelid is also identified based on DNA-fingerprinting. Only five annelids could be used for this analyses. DNA sequencing was done from the 18S rRNA genome and a part of the 28S rRNA genome (603 bp). The 18S sequencing gave to less information for to do the determination. The 28S sequencing gave as a result the genus *Eisenia*.

Cultivating and seeding the Annelid

With all the former information it is tried to cultivate the annelid with the purpose to seed one other SSF with it and see how this filter is functioning in the time.

The cultivating did not succeeded. If the worm is outside the water column of the filter and is fed with nutrients it dies. There is a possibility that the worm is adapted to very low concentrations of nutrients in the water. As a last attempt in 2011 10 square meters of slow sand filter four was seeded with 1000 worms out of SSF3. Yearly slow sand filter four is inspected on a changed structure of the schmutzdecke.

CONCLUSIONS

The main conclusions of this work are:

(1) The presence of many annelids in SSF3 has no negative effect on the functioning of SSF3 in terms of removing of pathogenic microorganism.

(2) The annelids are probably the only reason for deflected behavior of SSF3.

(3) The annelids are responsible for a schmutzdecke which is loose and spongy so that clogging is prevented.

REFERENCES

Blackmore Robert J.(2003), Japanese earthworms (Annelida: Oligochaeta): A review and checklist of species, Org. Divers. Evol. 3, Electr. Suppl. 11: 1 - 43

Misirlioglu M.,(2007), The earthworm fauna of the Kocaeli (‹zmit) City Center (Oligochaeta, Lumbricidae), *Turk J. Zool.* **31**, 353-356

Dizer H., Grutzmacher G., Bartel H., Wiese H.B., Szewzyk, R. and Lopez-Pila, J.M.,(2004) Contribution of the colmation layer to the elimination of coliphages by slow sand filtration, *Wat. Scien. Techn* **50,** No2 211-214

Anon. (2001). Method 1602: Male-specific (F+) and Somatic Coliphage in water by Single Agar Layer (SAL) Procedure. U.S. Environmental Protection Agency, Office of Water, Washington D.C., EPA 821-R01-029.

Eighmy, T.T., Spanos, S.K., Royce, J., Collins, M.R., and Malley Jr, J.P. (1994) Microbial activity in slow sand filters. In *Slow Sand Filtration*, Collins, M.R. and N.D.J. Graham, (Eds.), AWWA, 218-240.

Haarhoff, J. and Cleasby, J.L. (1991) Biological and physical mechanisms in slow sand filtration. In *Slow Sand Filtration*, Logsdon, G.S.(Ed.), Task Committee on Slow Sand Filtration, ASCE, 19-68.

Yordanov, R.V., Lamb, A.J., Melvin, M.A.L. (1996) Biomass characteristics of slow sand filters receiving ozonated water. *In Advances in Slow Sand and Alternative Biological Filtration.* N.J.D. Graham and R. Collins (Eds.), John Wiley & Sons, UK, 107-118.

Hijnen W, Dullemont Y, Schijven J, Hanzens-Brouwer A, Rosielle M, and MedemaG. (2007) Removal and fate of cryptosporidium parvum, clostridium perfringens and small-sized centric diatoms (stephanodiscus hantzschii) in slow sand filters. *Water Research*, **41**(10), 2151–2162

Hijnen,W., Schijven J., Bonne P., Visser A. and Medema G.,(2004). Elimination of viruses, bacteria and protozoan oocysts by slow sand filtration. *Wat. Sci. Technol.* **50**, 147-154

Lloyd B.J. (1996) The significance of protozoan predation and adsorption for the removal of bacteria by slow sand filtration. *In Advances in Slow Sand and Alternative Biological Filtration*, ed. by Graham NJD and Collins MR, JohnWiley & Sons, England, 129-138.

Dullemont Y.(2008), Ingestion of Cryptosporidium oocysts and vector behaviour by rotifers in slow sand filtration, Internal Report Waternet

Wubbels,G.H.(2007), Analyse microbiologische veiligheid drinkwater productielocatie De Punt (oppervlaktewater) van waterbedrijf Groningen, Internal report WLN

Schijven J. et al, (2008), Effect filtratiesnelheid, temperatuuren korrelgrootte op de verwijdering van micro-organismen door langzame zandfiltratie., Report 330204002/2008 RIVM.

Veenendaal, H.R., en A.J. Brouwer-Hanzens (2007). A method for the concentration of microbes in large volumes of water. *Techneau.*

Veenendaal D., Hijnen W.A.M. and Veenedaal H.R., (1996), Development of a concentration method for indicator bacteria from large volumes of water, BTO 95.200

Chapter 14

The inhibitory effect of Tubificid on head loss in slow sand filtration

H. Sugino, S. Nakamasu**, T. Komatsu**, S. Takano****
**Water Purification management Department, Mizu Mirai Hiroshima Corporation, Hiroshima 739-0627, Japan*
***Water Purification Division, Hiroshima Prefectural Nutagawa Water Service Office, Hiroshima 729-0412, Japan*
****Water Works Division, Public Enterprise Bureau, Hiroshima Prefectural Government, Hiroshima 730-8511, Japan*

ABSTRACT

The degree of the rise of head loss was slower in the No.2 slow sand filter at the Noda water purification plant of Hiroshima prefecture, Japan, where tubificid had lived continuously. In the other slow sand filters, tubificid had been observed sporadically since April 2011. When tubificid was observed, the degree of the rise of head loss became relatively slower. To investigate the relation between tubificid and head loss, we put a clump of tubificid on a small slow sand filtration device. As a result, head loss became very low, after that, the degree of the rise of head loss was reduced. Thus, we confirmed the inhibitory effect of tubificid on head loss in slow sand filters. However, only in the No.2 slow sand filter, tubificid was found in the equipment downstream from the slow sand filter. Also, in a flowing continuously small slow sand filtration device, tubificid was found in the equipment downstream. So, it is clear that the amount of tubificid has to be managed as much as possible by looking in the corners of the slow sand filter, by scraping etc.

Keywords Slow sand filtration; Head Loss; Biological Filtration Treatment Techniques; Tubificid; Physidae

INTRODUCTION

In order to maintain stable water supply, management of head loss is required, because the rise of head loss in slow sand filters affects the water flowing. In past, because of the rise of head loss we experienced remarkably loss of the amount of filtered water and maintaining the water supply level was difficult.

On the other hand, when the same raw water is treated with different slow sand filters, there is a difference in the degree of the rise of head loss. On the No.2 slow sand filter at the Noda water purification plant of Hiroshima prefecture, Japan, the degree of the rise of head loss has been reduced continuously since April 2011 as tubificid has lived there. Thus, we investigated the correlation between tubificid habitats and the rise of head loss.

METHODOLOGY

Slow sand filtration at the Noda water purification plant

This site of the study was slow sand filters at the Noda water purification plant in Hiroshima Prefectural Numatagawa Water Supply Business. The plant's water resource is Numatagawa river (area:540.0km^2, the length of river:47.8km). The raw water isn't usually processed in the Hongo water intake plant and the Hongo water purification plant for industrial use which are both located upstream of the Noda water purification plant. At the Noda water purification plant, the raw water is treated by slow sand filtration and chlorine, and the treated water is supplied to Mihara and Higashihiroshima city. If the turbidity of the raw water becomes 18 degrees or more, the raw water is pretreated by poly aluminum chloride at the Hongo water purification plant. There are four slow sand filters in the Noda water purification plant. The area of one filter are about 500m^2(20m × 25m). A map of the plant was shown in Figure 14.1(a).

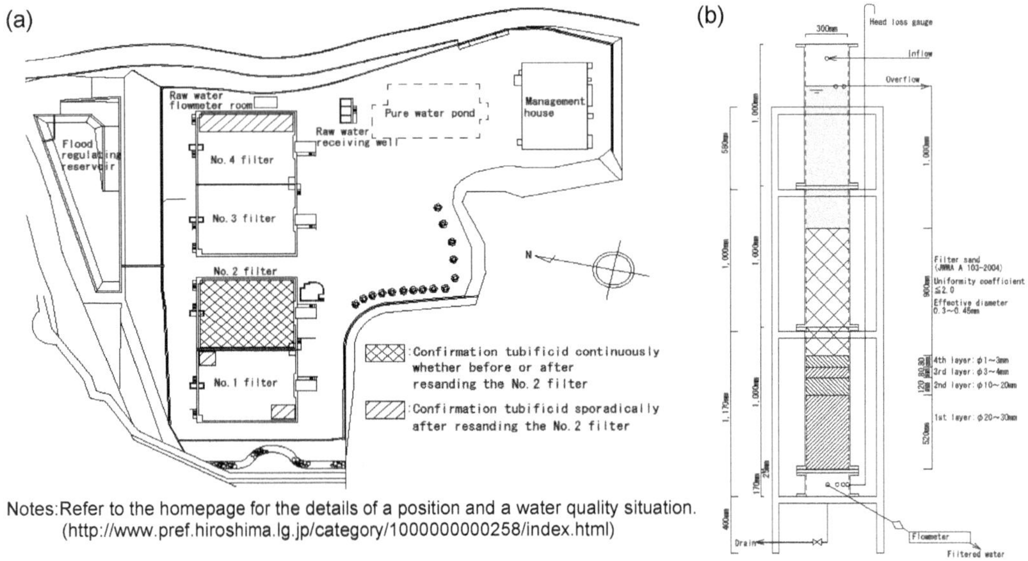

Notes:Refer to the homepage for the details of a position and a water quality situation. (http://www.pref.hiroshima.lg.jp/category/1000000000258/index.html)

Figure 14.1 Overview of subject to investigation: (a) Noda water purification plant; (b) the small slow sand filtration device.

The slow sand filters all operated the same until the No.2 slow sand filter was resanded. After then, normal operation of the two ponds differed due to the optimization of the facility scale (Table 14.1). Therefore, in this study, we analyzed their ponds separately before and after the resanding of the No.2 filter.

Table 14.1 Normal operation status of the slow sand filter at the Noda water purification plant.

The summary of normal operation	Condition of normal operation		
	Before resanding	After resanding	
		No.1 and No.2 filter	No.3 and No.4 filter
Scrap the sand	about 1cm thick	about 1.5cm thick	about 1cm thick
Start water flowing with filtration rate	1.0~1.5m/day	1.5m/day	2.0m/day
Increase filteration rate	by 10%/day	by 10%/day	by 10%/day
Keep max filteration rate	3.5m/day	2.5m/day	2.5m/day
Decrease filtration rate to lower head loss	of 30cm	of 10~20cm	of 10~20cm
Drain water slowly	for 12-18 hour	for 12-18 hour	for 12-18 hour

Small slow sand filtration device

To examine the inhibitory effect on head loss, we used a small slow sand filtration device located in the Noda water purification plant. It was an acrylic water tank. Its height is 3m and the inner diameter of about 30cm. Its sand thickness was the same as the slow sand filter of the Noda water purification plant. The filtration rate was adjusted with a needle valve attached to the area flowmeter. The overview picture of the small slow sand filtration device was shown in Figure 14.1(b).

The raw water of the small slow sand filtration device is the same as the Noda water purification plant. The filtration rate is about 4m/day (200mL/min in the area flowmeter). The adjust frequency was more than once a day. In addition, algae on the top wall exposed to sunlight was removed about once a week.

Method habitat investigation of tubificid

Tubificid habitat was observed as seen in Figure 14.2 on each scraping. Where there was a habitat, its area was approximated with the rectangle by visual observation. The rectangle area was estimated by the ratio of the area in the filter (In this report, this process is abbreviated as "visual investigation").

In addition, we investigated the filter surface using a gardening trowel (gauntlet section length: about 12cm, width: 8cm generally). The number of tubificid was confirmed by counting each cross section that was made by digging the sand (In this report, this process is abbreviated as "surface investigation"). Furthermore, the presence and the gathering status on the corners of the filter was confirmed by the same method.

While flowing, the number of tubificid was confirmed by analyzing the sand picked from the filter. A rod of 2m length with a ladle mesh of about 8.5cm diameter was used to pick the sand (In this report, this process is abbreviated as "flowing investigation").

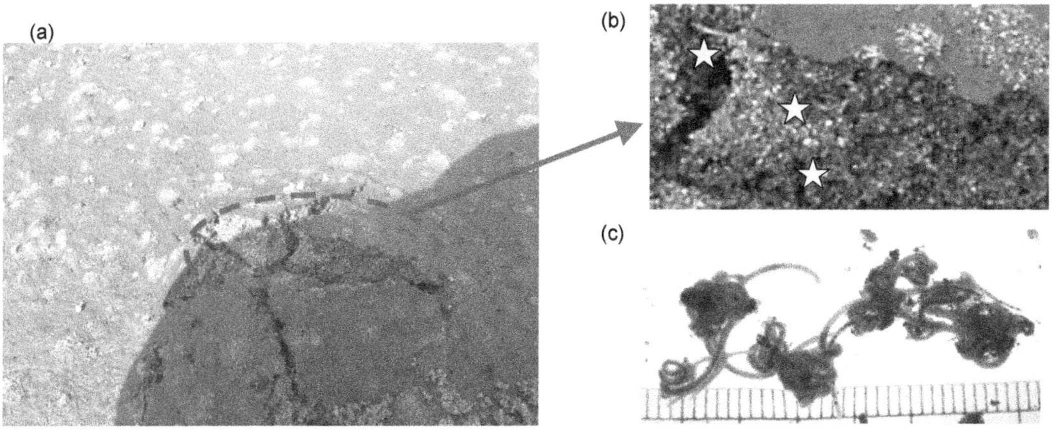

Figure 14.2 Image of habitat and tubificid: (a) 2011.4.27 habitat of tubificid on the No.2 filter while scraping. (There ware many traces of extrusive sand seen in a spotted shape.); (b) The enlarged image of habitat .(There is tubificid near the star marks); (c) Tubificid(spaces between scale marks:1mm).

Turbidity

Turbidity of the filtered water of the Noda water purification plant was measured continuously by a high sensitivity turbidity AN455 type machine (manufactured by Hitachi, Ltd.). Turbidity of the day was measured using the average value for each day. Turbidity of the filtered water on the small slow sand filtration device was measured once a day using a WA6000 type machine (utilizing an integrating-sphere electrophotometry method. Nippon Denshoku Kogyo Co.Ltd.).

Head loss

The head loss for the No.1~3 filter of the day was confirmed by measuring the difference of water level between the filter basin and the filtered water flow control well around 11 a.m. A measuring tape was used then.

As for the No.4 filter, using an immersion pressure water level gauge a ELR200-20 (0-4)-L6-AV1 (manufactured by Hitachi, Ltd.), the difference in the water level between the filter basin and the filtered water flow control well was continuously measured. And, head loss of the day was considered as the average difference for each day.

As for the small slow sand filtration device, using a head loss gauge (manometer) installed on the filtered tank, the difference in water level between the device and head loss gauge was measured using a measuring tape once a day. Head loss of the day was regarded as the difference.

In this report, to eliminate the influence of the water flow, we standardized head loss by the reference method. Specifically, using standard filtration rate (4.8m/day) of the slow sand filter, the Standard head loss was estimated by calculating the following formula.

$$c = b \times \frac{4.8}{a}$$

a : Filtration rate(m/day)
b : Head loss(cm)
c : Standard head loss(cm)

Filtered flowing

On the No.1~3 filters, while discarding water for preparing operation, filtered flow of the day was regarded as the value measured by a diving-type electromagnetic flowmeter Magne W3000FLEX (manufactured by Yamatake Corporation , Ltd. Non-Consolidated) around 11 a.m. While using the filters for water purification, the filtered flow (m^3/day) of the day was regarded as the average of the day measured continuously by electric-powered telescope (manufactured by Suidokiko Co., Ltd.).

On the No.4 filter, the filtered flow of the day was regarded as the average of the day measured continuously by diving-type electromagnetic flowmeter Magne W3000FLEX(Yamatake Corporation Ltd.).

On the small slow sand filtration device, we regarded the filtered flow (m^3/day) of the day as the value (mL/min) measured by a flowmeter RK1710-H2O-500ML/MIN (manufactured by Kofurokku Co., Ltd.).

Filtration rate

The filtration rate (m/day) was calculated using flow of filtration per day and the area of filtration. In the calculation, the filter area of the Noda purification plant was $500m^2$ and the filter area of the small slow sand filter was $0.07m^2$.

RESULTS AND DISCUSSION

Habitat status of tubificid on the No.2 filter

First, we investigated the habitant status of tubificid by surface investigation.

On the No.2 filter before resanding work, tubificid had been observed from surface to the 5cm layer in all points of filter surface since April 27th 2011 on each scraping. There were 2~3 tubificid in one digging (Figure 14.3).

On the No.2 filter during resanding work, the condition remained the same. Tubificid was not observed under the 5cm layer.

On the No.2 filter after resanding work, tubificid was observed from surface to the 2cm layer initially. However, tubificid had been observed from surface to the 5cm layer since August 29th 2012. There was one tubificid found in one digging (Figure 14.4).

By the way, when we checked the inside of the filtered flowing control well which is downstream from the filter basin, tubificid was observed only downstream of the No.2 filter in which there were a lot of tubificid. It indicates that it is important for keeping the equipment clear to manage quantity of tubificid.

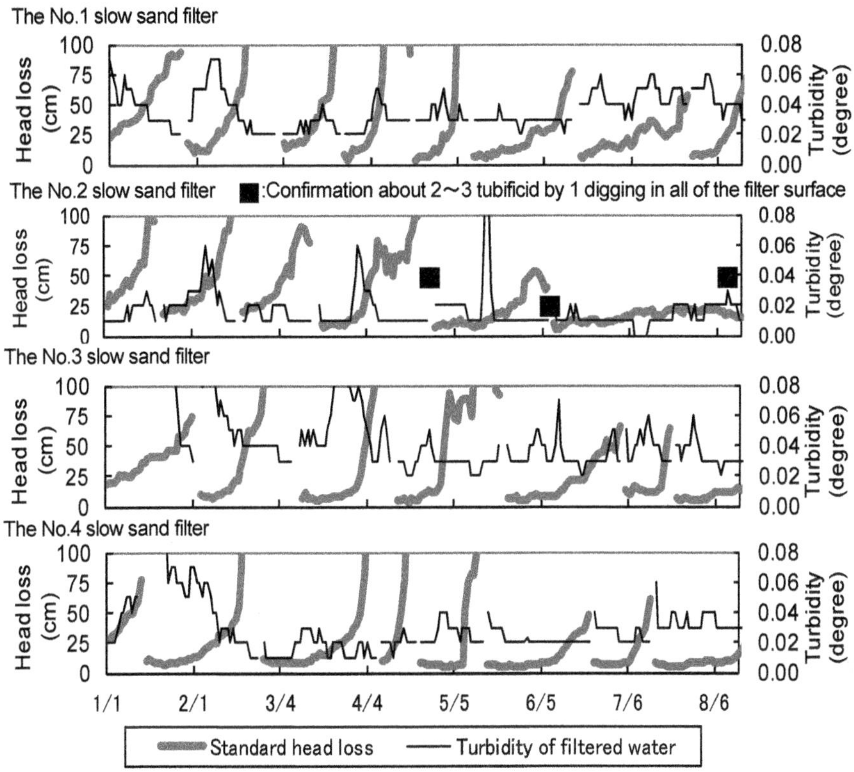

Figure 14.3 2011.1.1~8.16 Trend of standard head loss and filtered water turbidity on the No.1~No.4 slow sand filters.

Habitat status of tubificid on the other filters

Before resanding the No.2 filter, no tubificid was observed in the other filters (Figure 14.3). After resanding the No.2 filter, tubificid was sporadically observed on the No.1 and No.4 filters. However, no tubificid was observed on the No.3 filter (Figure 14.4, 14.5).

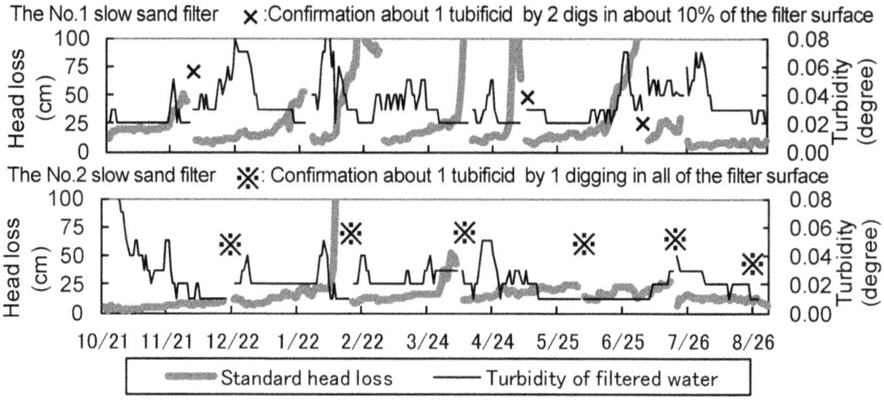

Figure 14.4 2011.10.21~2012.9.2 Trend of the standard head loss and filtered water turbidity on the No.1 and No.2 slow sand filters.

On the No.1 filter, tubificid was observed three times at the same point and was observed in about 10% of the filter surface. There was one tubificid found in two digs. That was remarkably low amount compared to the No.2 filter.

On the No.4 filter, tubificid was also observed three times at same point and was observed in 20~30% of the filter surface. In this filter, there was also one tubificid found in two digs.

Investigation of filtration status before resanding work

We investigated the influence of the tubificid on standard head loss and the turbidity of filtered water.

The trends of standard head loss and filtered water turbidity on the No.1~4 filters before resanding the No.2 filter are shown in the Figure 14.3. The term is from January 1st to August 23rd 2011.

In the No.2 filter, which had a lot of tubificid, the rise of head loss had been slower since April 2011. The filter can be operated continuously for 69 days. The sand in which tubificid was observed was relatively black. It indicates the function of the filtration become better due to the presence of tubificid. The turbidity was good compared to the others and before.

Investigation of filtration status after resanding work

In the term from October 21st 2011 to September 2nd 2012, the trend of standard head loss and filtered water turbidity on the No.1 and 2 filters, which operated the same is shown in the Figure 14.4. On the No.2 filter, tubificid had been observed continuously in all points and the head loss rise was slow. On the No.1 filter, the head loss rise was also slow when tubificid was observed in the No.1 filter. However, the rise head loss of the No.2 filter was slow compared to the No.1 filter which had less tubificid. It indicates that the difference is caused by the amount of tubificid.

In the term from October 21st 2011 to September 2nd 2012, the trend of standard head loss and filtered water turbidity on the No.3 and 4 filters were the same as shown in the Figure 14.5. The rise head loss of the No.4 filter was slow compared to the No.3 filter when tubificid was observed in the No.4 filter. The difference was caused by the presence of the tubificid because no tubificid was in the No.3 filter. However, this trend didn't differ from the No.2 filter which had tubificid continuously. It indicates that the difference in the rise is also influenced by the amount of tubificid.

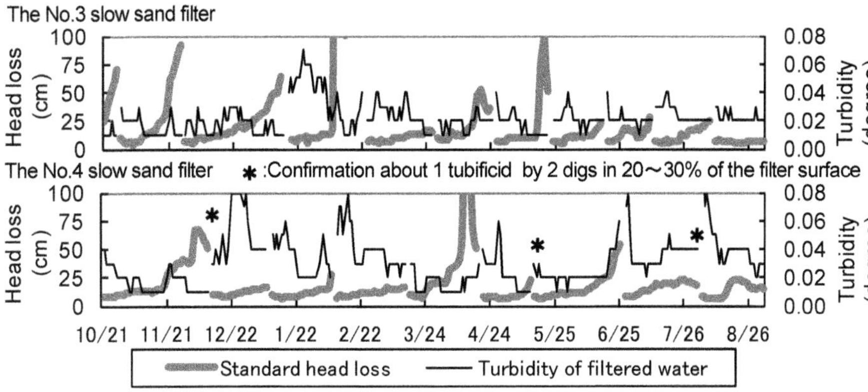

Figure 14.5 2011.10.21~2012.9.2 Trend of the standard head loss and filtered water turbidity on the No.3 and No.4 slow sand filters.

Filtered sand

It became clear using a flowing investigation that the sand in which tubificid was observed is more black than the sand in which no tubificid was.(Figure 14.6(a))

On each scraping, the sand test was carried out at surface, 0-1cm, 1-5cm, 5-10cm layer. The value of SS in the No.2 filter is remarkably higher than those of the other filters except winter. Therefore, It indicates that the habitat of tubificid effectively trapped dirt in water and this effectiveness was kept. The comparison of SS at 1~5cm layer in each filter is shown in Figure 14.6(b).

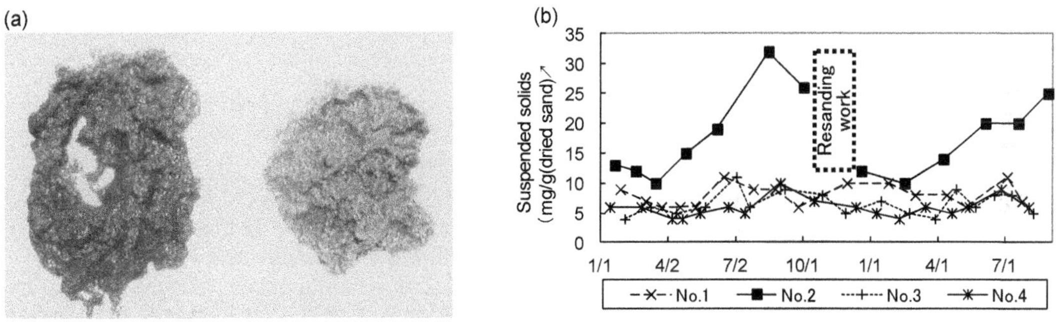

Figure 14.6 Status of filter sand : (a) 2012.7.30 filter sand picked up by flowing investigation (left : the No.4 filter which had tubificid right : the No.3 filter which had no tubicid); (b) Trend of 2011.1.1~2012.9.2 suspended solids amount at 1~5cm layer in the No.1~No.4 slow sand filters while scraping.

Inhibitory effect experiment of the head loss rise

We considered method of puting tubificid in the sand to test the inhibitory effect on head loss rise.

Specifically, we put a clump of tubificid on the sand in a bucket filled with water (Figure 14.7(a)). Though the clump became gradually smaller, the change stopped after 1 hour (Figure 14.7(b)). This condition was un changed for three days. On the other hand, some tubificid was observed on the surrounding sand.

The presence of tubificid in the sand confirmed that tubificid sink in the sand and move to the surrounding sand. It indicates that tubificid can move and live in the sand when the clump is put on the sand. So, we decided to place a clump of tubificid on the sand in our inhibitory effect experiment.

Considering this, we conducted an inhibitory effect experiment. Four clumps of tubificid were put on the surface of the filter in the small slow sand filtration device. The standard head loss and turbidity of filtered water before and after putting the tubificid is shown in Figure 14.7(c). Afterwards, the standard head loss was remarkably down and the degree of the increase became slower. There were 2~4 tubificid in one picking of flowing investigation. By scraping after the experiment, a lot of tubificid were also observed in the sand.

However, just after placing, turbidity of filtered water rose and kept a higher values. This was presumably because it was easy for water to flow in the sand after tubificid's moving. Afterwards, turbidity of filtered water was suddenly down from December 10th. On the other hand, turbidity of filtered water on the No.2 filter, which tubificid was gradually increasing in, kept low value. So, the intervention of other organisms is essential to turbidity treatment. It indicates that maturation of biota is required.

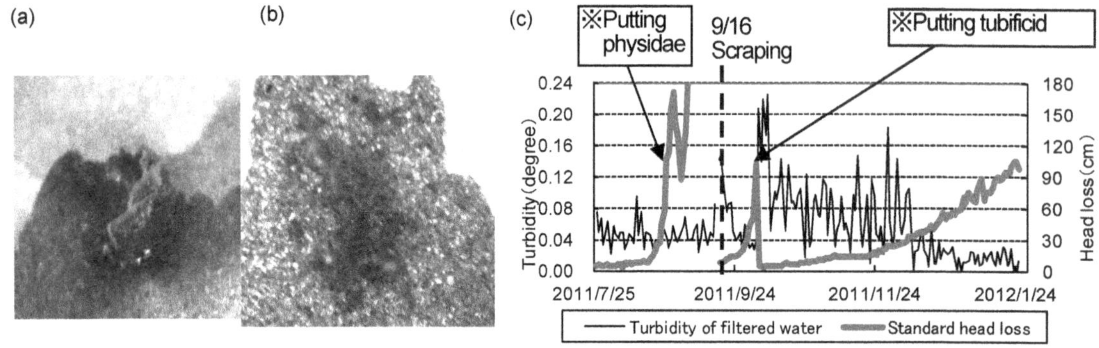

Figure 14.7 Status of Inhibitory effect experiment : (a) A clump of tubificid put on the sand in a bucket filled with water; (b) The clump of tubificid when 1 hour had passed since putting; (c) Trend of standard head loss and turbidity of filtered water on the small slow sand filtration device.

Flowing contiuously, tubificid were found around the flowmeter that was downstream from the filter basin. That indicates that tubificid pass through the filter because there were a lot more tubificid than in the slow sand filter at the Noda water purification plant. Therefore, some management is required.

Further, due for comparison, physidae which there were a lot on scraping of the No.2 filter at August 17th 2011 were also investigated. After getting physidae from the wall of the filter basin at Noda water purification plant, they were put on the sand of the small slow sand filtration device to the some extent. However, an inhibitory effect of the rise of head loss was not observed.

CONCLUSIONS

The main conclusions of this study are:

(1) It was newly found that the formation of tubificid ecosystem has an inhibitory effect on the degree of the rise of head loss;

(2) Generation and intervention of other organisms is essential to turbidity treatment and maturation of biota is required in the presence of tubificid in the sand;

(3) However, there is a risk that tubificid will be found downstream from the filter basin when too many tubificid are;

(4) Thus, amount of tubificid has to be managed as much as possible by checking in the corners of the slow sand filter, by scraping, etc.

ACKNOWLEDGEMENTS

The authors wish to acknowledge the important contribution of the Swing Corporation (business person in charge:Katsutoshi Mizuno) for providing extensive plant data, and the assistance of Limited Company Sunada Construction (site agent:Yasunobu Sunada) for scraping slow sand filters in Noda water purification plant, and the assistance of Hiroshima Environment and Health Association (identification requests contact:Shinji Handa) for the identification of organisms.

REFERENCES

Sugino, H., Hirayama, S., Nakamasu, S. (2012) The suppressive effect by tubificids on head loss increases in the slow sand filter beds. *Journal of Water and Waste*, **54**(11), 860-869. (in Japanese).

Hiroshima Prefecture (2002) Second Grade River Numatagawa River System Improvement Basic Policy (in Japanese)

Iwase, N., Kawata, K. (2001) Microbial action and the continued time of filtration on Slow Sand Filters which was used for Advanced Wastewater Treatment of Secondary Treated Water of Sewage. *Proceedings of The 38th Japan Sewege Works Association's Annual Technical Conference of Sewage Works*, 16-18. (in Japanese).

Woodword, C. A., Ta, C. T. (1988) Slow Sand Filtration Recent Developments in Water Treatment Tecnology. 349-366.

Nakamoto, N., Sakai, M (1992) Standard Head Loss and Alga as a Barometer of Filter Clogging in the Slow Sand Filtration. *Japanese journal of Water Treatment Biology,* **28**(1) 7-16. (in Japanese).

Hiroshima Prefectural Water Quality Management Center (2011) Water Quality Annual Report (2010). 5-2 - 5-9. (http://www.pref.hiroshima.lg.jp/uploaded/life/135869_207205_misc.pdf (in Japanese)).

Hiroshima Prefectural Water Quality Management Center (2012) Water Quality Annual Report (2011). 5-2 - 5-9. (http://www.pref.hiroshima.lg.jp/uploaded/life/135869_207204_misc.pdf (in Japanese)).

Chapter 15

Schmutzdecke development and treated water quality

T. Sugimoto
Nagoya City Waterworks & Sewerage Bureau , Sannomaru 3-chome, Naka-ku, Nagoya, Aichi, Japan

ABSTRACT
Seasonal schmutzdecke development is observed during filter operation using new device to collect schmutzdecke in two field scale slow sand filters while most field scale investigations observe schmutzdecke only at the end of filter runs because of difficulties to obtain schmutzdecke with supernatant on it. Detailed analysis of schmutzdecke during filter operation can improve operational management through better understanding of slow sand filter. The device featured in this paper is a portable sand bed on a mesh container (30cm x 20cm) which is set on the filter bed after cleaning between filter runs. The sampler enables collecting active schmutzdecke without disturbing it because it makes unnecessary to cut out operating schmutzdecke. This paper observes the schmutzdecke collected from eight devices set on the filter bed for each filter operation with the interval of approximately two weeks during a filter run and discusses the algae species, biomass and filter performance.

Keywords Schmutzdecke sampler; algae species; biomass; filter performance

INTRODUCTION

Slow sand filtration is a simple and low cost method for producing good quality drinking water. It effectively removes Giardia cysts and Cryptosporidium oosysts, which are highly resistant to chlorine disinfection. In slow sand filters the water purification is achieved by a combination of physico-chemical and biological processes. The investigators have indicated that the efficiency of the slow sand filtration process depends on the schmutzdecke, which is algae mat including many kinds of microorganism such as bacteria, protozoa and invertebrates and their extracellular products (Huisman and wood, 1974, Campon et al., 2002, Wakelin, 2011). Nakamoto et al. (1996) suggested that development of algal mat of filamentous diatoms in schmutzdecke has a beneficial effect on the prevention of filter clogging and the promotion of heterotrophic activity by the oxygen of photosynthesis. But biological mechanisms of water purification in slow sand filter are not fully understood (Joubert and Pillay, 2008). Several microbiologically-mediated purification mechanisms (e.g. predation, scavenging, adsorption and bio-oxidation) have been hypothesized or assumed to occur in the biofilm that forms in the filter but these have not yet been comprehensively verified (Haig et al., 2011). One of the reasons is difficulties to check the condition of the schmutzdecke during the filter operation. The new device described here allows to collect schmutzdecke as it is under water during filter operation to investigate the influence of algal schmutzdecke performance of full-scale slow sand filter.

STUDY AREA

The study was carried out at the slow sand filters of Nabeya-Ueno Purification Plant, Nagoya City Waterworks & Sewerage Bureau, in Nagoya City, located the center of Japan (Table 15.1). Thanks to good quality of the resource, the surface water of the Kiso River, its slow sand filtration is a single stage treatment without pretreatments except a plain sedimentation basin.

Table 15.1 Nabeya-Ueno Purification Plant

Rapid sand filtration	
capacity	150,000 m3/d
Slow sand filtration	
capacity	140,000 m3/d
pretreatment	Toriimatsu Plain Sedimentation Basin
Slow sand filter	14 unuts (1914-2010) 12 units (2014-)
Filter area	2,879 m2/unit
Filter rate	4 m/d
Sand depth	60-90 cm
Water depth	100 cm

NEW DEVICE TO COLLECT SCHUMITZDECKE

A new device to collect schmutzdecke was designed. It is a portable sand bed (sand depth 10cm) on a mesh container (30cm x 20cm) which is set on a filter bed after cleaning between filter runs (Figure 15.1). The device enable collecting active schmutzdecke during the filter operation without disturbing it because it is unnecessary to cut out the schmutzdecke.

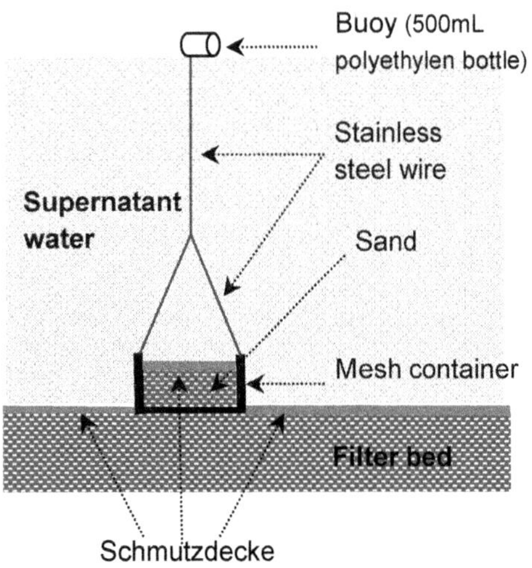

Figure 15.1 Appearance of the new device

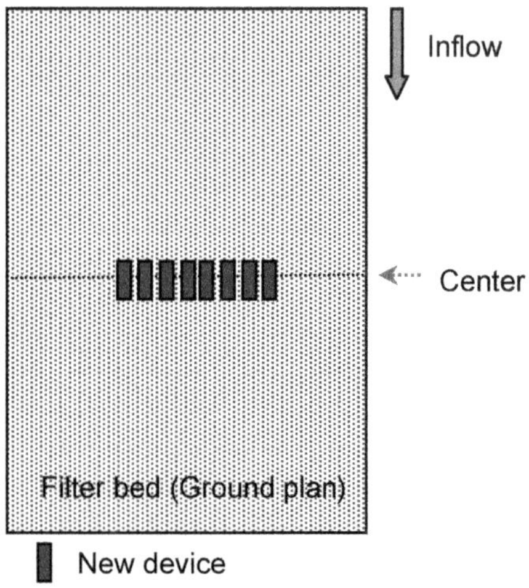

Figure 15.2 New device location

Eight devices were located on each slow sand filter bed (No.13 and 14) at each filter operation just before the water was inlet on the filter bed (Figure 15.2).

The advantages of the device are as follows:

- The schmutzdecke and 10cm surface sand of the slow sand filter can be collected at any maturation level during filter operation as it is in the water.
- The influence for slow sand filtration is small.
- According to the purpose of research, only the schmutzdecke, only the surface sand or both of them can be analyzed due to undisturbed sample condition.
- The schmutzdecke area obtained by this device is relatively wide (30cm x 20cm) and changeable.
- Aquatic insect such as chironomidae can be observed.

The algae in schmutzdecke sample obtained were identified and measured chlorophyll-a. Chlorophyll-a were measured after 30 minutes use of ultrasonics with 90% acetone and four hours extraction.

REPRESENTATIVENESS VERIFICATION

The area of the sampler is much small than that of the whole filter bed (approximatelly two 100,000th). And the schmutzdecke development is not spacially uniform on a filter bed, occasionally makes patchy distribution. Accordingly, a t-test was conducted for the verification on the representativeness of the schmutzdecke sample obtained by the new device using chlorophyll-a values.

The filter bed surfaces of Bed 13 and 14 were divided into 20 blocks, and 20cm x 20cm schmutzdecke sampling were collected from the center of each block at the end of the filter run. These 20 samples and 3 samples obtained by the new device at the end of the filter run measured chlorophyll-a. The sampling was conducted at December 9, 2003 for Bed 13, December 15, 2003 for Bed 14. For all the filter operation except the initial period, the flow rate of Bed 13 was $200m^3/H$ and Bed 14 was 400 m^3/H. Bed 14 was operated at higher filtration rate than Bed 13, and allowed much nutrient inlet to accumulated much algal biomass. In each filter bed, algal biomass of the schmutzdecke increased along the horizontal flow in the filter pond, which might have been caused by the decreasing of horizontal flow rate. Figure 15.3 and 15.4 shows the chlorophyll-a values obtained.

To know which t-test to perform, F-test was conducted for the 20 filter bed data and the three new device data. F=1.6, F-Critical=19.4(Bed 13) and F=8.3, F-Critical=19.4 (Bed 14) were obtained. For both, F<F-Critical and the difference in variance was statistically significant. According to these results, a t-test for two samples assuming unequal variances was conducted. T-statistical=-0.6, t-Critical (two-tail)=3.2 (Bed 13) and t-statistical=1.8, t-Critical (two-tail)=2.3 (Bed 14) were obtained. For both, T-statistical<t-Critical F<F-Critical and the means were not statistically significant. According to these results, the schmutzdecke obtained by the new devices can be treated as the representative samples of the filter bed.

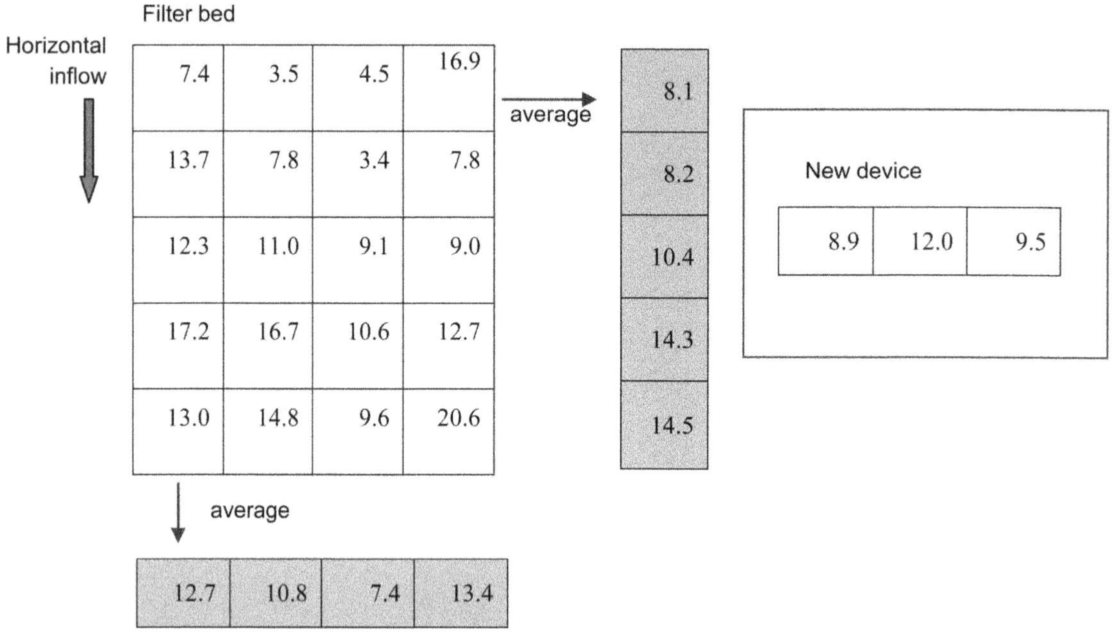

Figure 15.3 Biomass (chlorophyll-a mg/m^2) at Bed 13 and obtained by the new device

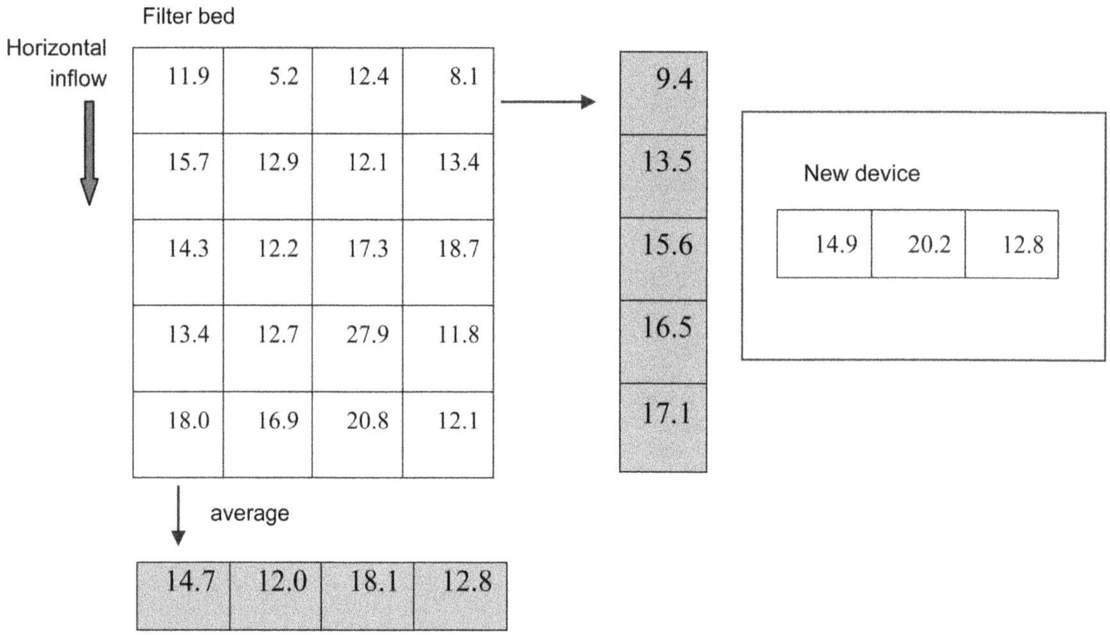

Figure 15.4 Biomass (chlorophyll-a mg/m^2) at Bed 14 and obtained by the new device

ALGAL DEVELOPMENT OF SCHMUTZDECKE AND HEADLOSS

Table 15.2 shows the change of dominant algae of the schmutzdecke collected by the new device, and Figure 15.8 shows algal biomass of it. Nakamoto et al. (1996) suggested that development of algal mats of filamentous diatom in slow sand filters has a beneficial effect on the prevention of filter clogging and the promotion of heterotrophic activity by the oxygen produced by photosynthesis. Melosira valians was dominant from middle January, but low water temperature prevented its development.

Table 15.2 Dominant algae

Date	Bed 13	Bed 14
July 22	Cymbella minuta (diatom)	
July 28	Cymbella minuta (diatom)	Cymbella minuta (diatom)
August 4		Cymbella minuta (diatom)
August 11	Cymbella minuta (diatom)	
August 18		Cymbella turgidula (diatom)
August 25	Cymbella turgidula (diatom)	
September 1		Ulothrix (green algae)
September 8	Oedogonium (green algae)	
September 16		Hydrodictyon (green algae)
September 22	Oedogonium (green algae)	
September 30		Hydrodictyon (green algae)
October 2	Oedogonium (green algae)	
October 7		Hydrodictyon (green algae)
Cleaning		
October27	Ceratoneis arcus (diatom)	Hydrodictyon (green algae)
November 12	Navicula radiosa (diatom)	Fragilaria capitellata (diatom)
Nobember28	Navicula cryptocephala (diatom)	Cymbella minuta (diatom)
December9	Nitzschia dissipata (diatom)	
December15		Navicula radiosa (diatom)
Cleaning		
January 8	**maintenance**	Achnanthes minutissima (diatom)
January22		Melosira valians (diatom)
February13		Melosira valians (diatom)
March 1		Melosira valians (diatom)
March12		Fragilaria capitellata (diatom)
March24		Melosira valians (diatom)

Filamentous algae such as Melosira valians (diatom) make sterically structured schmutzdecke which prevents from clogging and has effective ability of turbidity removal.

Figure shows the change of head loss from April, 2003 to March, 2004. Head loss (normalization) defined as equation (1) and suspended matter load as equation (2) (Nakamoto, 1992).

[Head loss (normalization) (m)] = [Head loss (m)] / [Flow rate (m3/H)] x [200 (m3/H)] (1)

[Suspended matter load (degree·H)] = Σ[Turbidity (degree)] ·[Flow rate (m3/H)]·24(H)/1000 (2)

In summer 2003, the head loss of Bed 13 and 14 rapidly increased at first according to the schmutzdecke development, then decreased due to the development of filamentous algae, Oedogonium (green algae) and Hydrodictyon (green algae). These filamentous green algae make sterically structured schmutzdecke to prevent clogging and schmutzdecke turnover to remove clogged one same as Melosira valians (diatom). Figure 15.5 shows seasonal relationship between chlorophyll-a and head loss. In winter, relatively little quantity of suspended matter load caused filter clogging. The schmutzdecke without enough developed filamentous algae did not have ability to decrease the increased head loss.

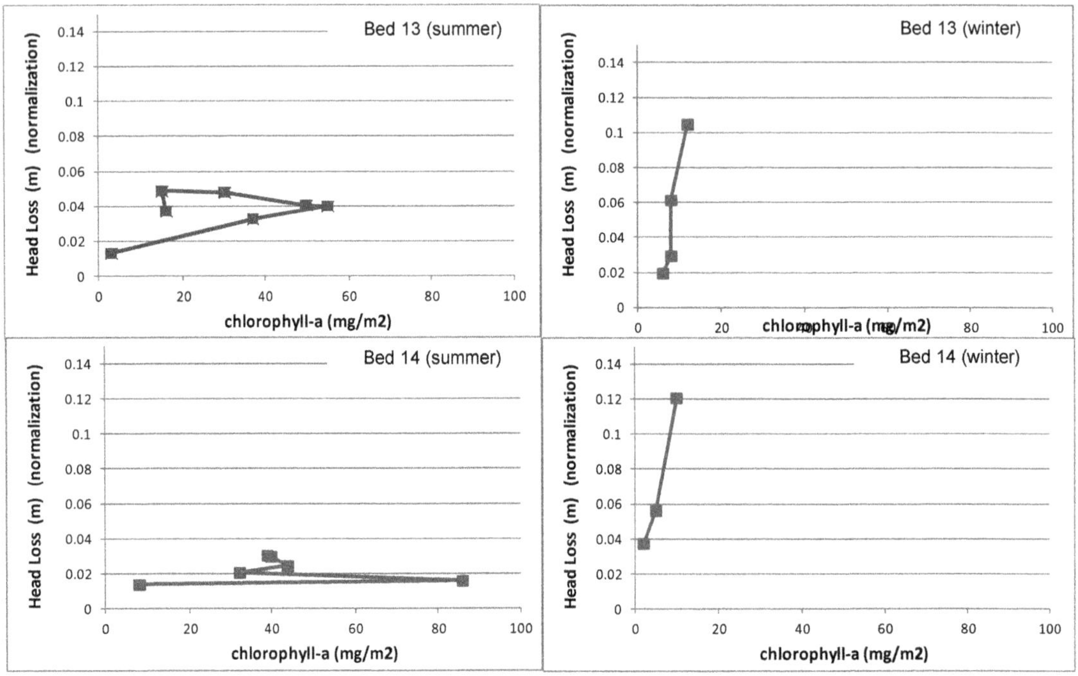

Figure 15.5 chlorophyll-a and head loss

TURBIDITY OF TREATED WATER

Continuous turbidimeter to monitor treated water of each filter was installed from 2005-2007(Bed 1,2,3,8,9 and 10). In Japan turbidity of filtration water should be 0.1 degree or less as the countermeasure for Cryptosporidium. Although turbidity kept below 0.1 degree, for several days after filtration start, turbidity was occasionally high. It is caused by immature schmutzdecke that cannot capture many particles. How long the relatively high turbidity lasts mainly depends on low water temperature.

Oosyst of Cryptosporidium is 3-5 μm. The number of this range of particle was measured in low water temperature season (January 22, March 8, march 12,2004). Figure 15.7 shows the relationship among particle removal, operation time (day) and flow rate. In low water temperature, strict turbidity management is required during first 400 hours of filtration until proper formation of schmutzdecke.

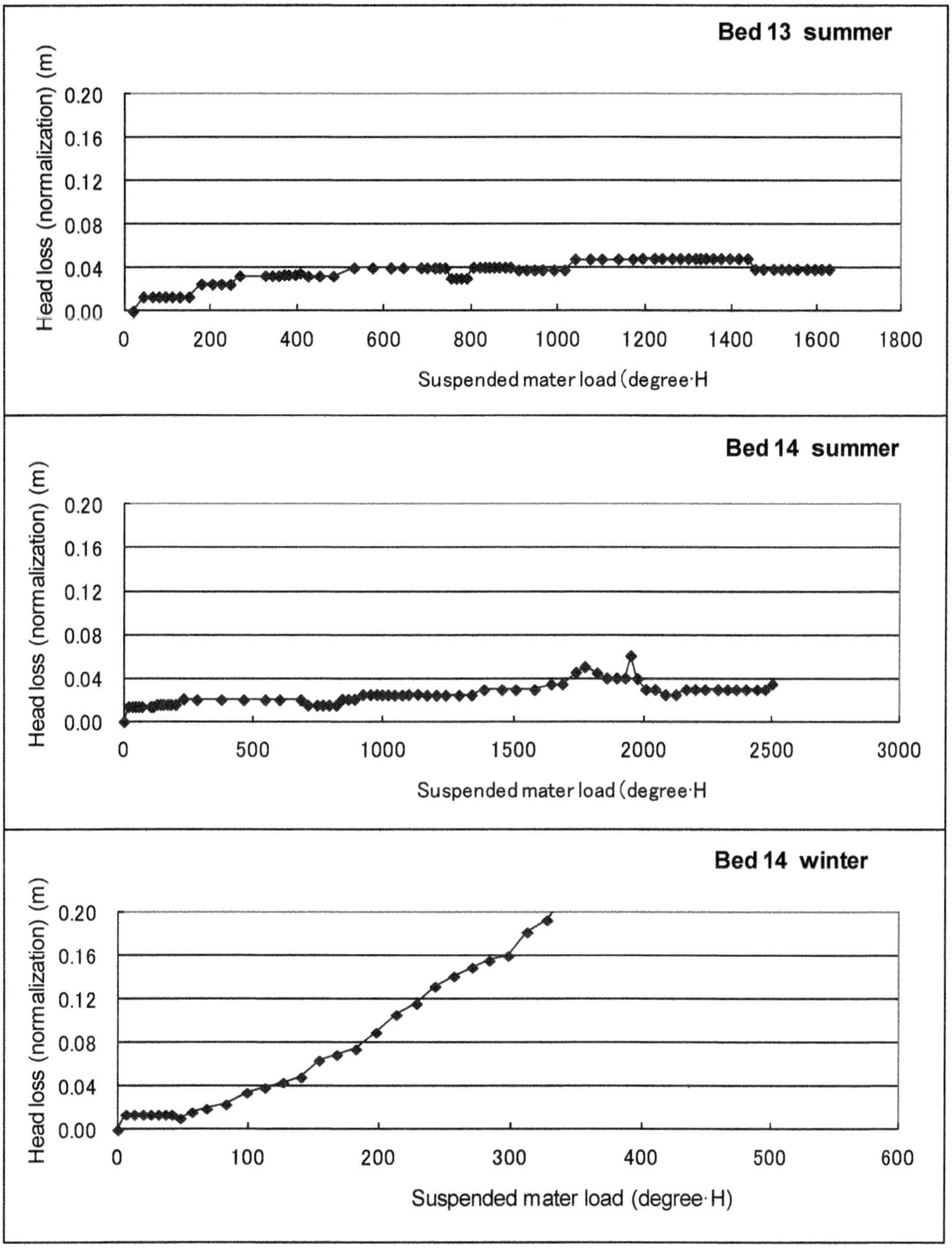

Figure 15.6 Suspended matter load and head loss

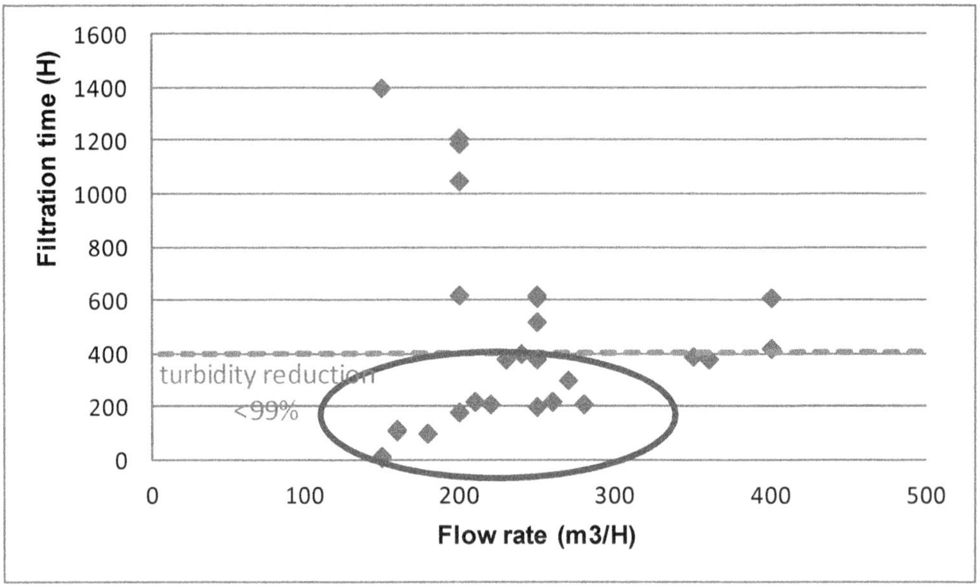

Figure 15.7 Turbidity reduction in winter

CONCLUSIONS

The main conclusions of this work are:

(1) A new to collect device schmutzdecke was designed.

(2) The advantages of the device are as follows:

- The schmutzdecke and 10cm surface sand of the slow sand filter can be collected at any maturation level during filter operation as it is in the water.
- The influence for slow sand filtration is small.
- According to the purpose of research, only the schmutzdecke, only the surface sand or both of them can be analyzed due to undisturbed sample condition.
- The schmutzdecke area obtained by this device is relatively wide (30cm x 20cm) and changeable.
- Aquatic insect such as chironomidae can be observed.

(3) The representativeness of the schmutzdecke obtained by the new devices was verified by t-test.

(4) The schmutzdecke with enough developed filamentous algae in summer had ability to decrease the increased head loss. In winter, filamentous algae did not develop enough and the schmutzdecke did not have this ability.

(5) In low water temperature, strict turbidity management is required during first 400 hours of filtration until proper formation of schmutzdecke.

REFERENCES

Campos, L.C., Su, M.F.J., Graham, N.J.D. and Smith, S.R. (2002) Biomass development in slow sand filters. *Water Research*, **36**, 4543-4551

Haig, S.N., Collins, G., Davies, R.L., Dorea, C.C. and Quince, C. (2011) Biological Aspects of Slow Sand Filtration: Past, Present and Future. *Water Science & Technology*, **11**(4), 468–472

Huisman, L. and Wood, W. E. (1974) Slow snd filtration. Genova: World Health Organization, 20-46

Joubert, E. D. and Pillay, B. (2008) Visualisation of the microbial colonisation of a slow sand filter using an Environmental Scanning Electron Microscope. *Electronic Journal of Biotechnology*, **11**(2), 0717-3458

Timms, S., Slade, J.S., Fricker, C.R. (1995) Removal of Cryptosporidium by slow sand filtration. *Water Science and Technology*, **31(5)**, 81-84Issues 5–6, 1995, 81–84.

Nakamoto, N. and Sakai, M. (1992) Normalized headloss as a clogging indicator in slow sand filter and its flora., *Japanese journal of water treatment biology*, **28**(1), 7-16

Nakamoto, N., Yamamoto, M., Nazaki, K.,Iwase, N., Yasuda, M., Kitada,T. and Hrabayashi, K. (1996) Role of filamentous diatom as an automatic purifier in a slow sand filter., advances *In slow sand and alternative biological filtration*, Graham, N. and Collins, R. (ed) John Wiley & Sons Ltd., UK, 138-149

Wakelin, S., Dillon, P., Abell, G. C. J., Gregg, A. L., Brodie, E., DeSantis, T. Z., Goldfarb, K. C. and Anderson, G. (2011) Microbial community structure of a slow sand filter schmutzdecke: a phylogenetic snapshot based on rRNA sequence analysis. *Water Science & Technology: Water Supply*, 11 (4),426-436

Chapter 16

The silicon circulation hypothesis based on slow sand filtration/ecological purification system

S. Inoue[a,b,c,d], *Y. Mochizuki*[a]
[a]*Earth-techno CO.LTD,*[b]*Meiho-techno CO.LTD,*[c]*S'tem CO. LTD,*
[d]*Hiyoshi CO.LTD,Japan*

ABSTRACT

The quantity of silicate elution increases by the function of biofilm of slow sand filtration. I suppose this phenomenon to be "microbe weathering". The silicic acid which is not included in rainwater is recognized in the mountain stream. This is conformal when I think that it is the result of the function of microbe rising in the forest soil.
It is announcement, I argue about three following items and to assume it a problem proposal.
(1) About Si circulation hypothesis caused by the microbe weathering to read from past knowledge.
(2) About the Si concentration change in slow sand filtration facilities.
(3) About significance to increase Si supplies.

Keywords Silicon; Circulation; hypothesis

INTRODUCTION

I have been involved in activities related to reproduction of natural blessings through the linkage of the forest, villages and the ocean based on technology. Since 1991, I have been paying attention to the importance of Silicon (Si). According to the analysis of spring water in the natural environment, it is considered that "Si problem" would not occur if most of the rain water be supplied downstream as spring water filtered by the soil. However as the results of urbanization, devastation of planted forests, increase of fallow fields and dam constructions, the absolute amount of Si supplied downstream is decreasing, while Nitrogen is increasing.

Decrease of Si is unfavorable for the most important diatom in microalgae which sustains fishery resourses. It is generally considered that the outbreak of blue-green algae in fresh water, and red tide or shellfish poison generations in closed bays are caused by declining diatom. On the other hand, non-diatom does not need Si. I am sure that the Si circulation will be actively discussed in the near future. But Si supply function by slow sand filtration has not attracted enough attention. I would like to refer to the important role of ecological purification system in the Si circulation, and contribute to further discussion.

The Si circulation hypothesis by microbe weathering in the past findings

A classic article, "An organism which eats stone" is included in "Silicon in a living organism". This article wrote that the *Bacillus mucilaginosus* which commonly exists in the soil, secretes mucus and dissolves silicate, the main component of sand, in order to take necessary nutrient salts into it.

Carbonic acid in the rain water dissolves silicic acid in the rock into water as ion, then the rock is disintegrated. This process is called weathering. AS the solubility of silicic acid is low, if the solubility limit drops by decreasing water temperature or evaporation, surplus of the solution is deposited. This process of recrystallization of silicic acid deposit creates the clay.

When we consider that *B.mucilaginosus* functions similarly in the soil the carbonic ion in rock weathering, a concept of microbe weathering can be conceivable, in parallel with physical or chemical weathering.

There are findings in the forestry research that Si which does not exist in rain water emerges in spring water and mountain streams in the forest (Tab. 16.1).

Table 16.1 The characteristic of nutrient salt of the water via the forest soil. (Y.Fukami (2009))

Item	DIN	DIP	Si
Rain water(μM)	8.23	0.07	0.54
Anti-rainfall magnification	4.5	8.9	350
Spring water(μM)	37	0.62	189
N:P:Si	60	1	303

Thereby,when rain water passes through the forest soil, elution of silicic acid occurs on the level of 18mg/L(by Kyoto univ.at Shiga pref.). This result supports the microbe weathering hypothesis (Fig. 16.1).

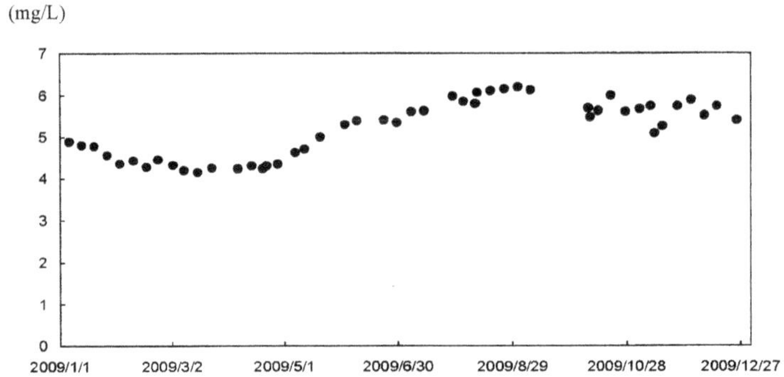

Figure 16.1 The season change of Si concentration in sampling soil water (The Si concentration rises during a period considered that the work of microbe is active) .

The changes of Si concentration in the ecological purification system

"The changes of water quality by algae biofilm in the filtration pond" by N.Nakamoto and H.Nakata in "Japanese Society of Limnology, Koshinetu branch Newsletter,1988" provided the time cource data of the receiving well, inflow water, the second pond (three days after scraping), the third pond (eleven days after scraping) and outflow of the slow sand filtration facilities (Fig. 16.2).

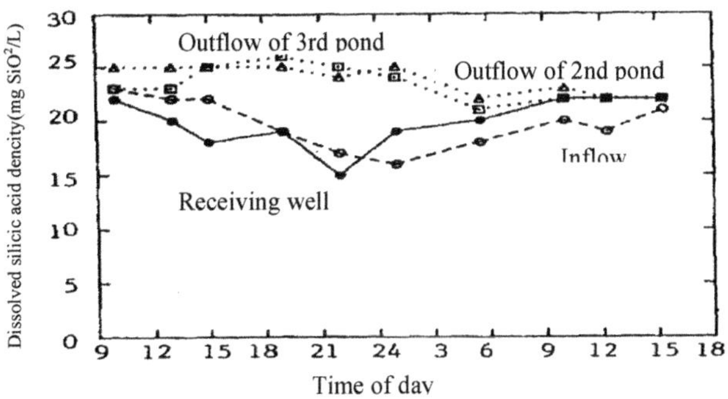

Figure 16.2 The changes of dissolved silicic acid concentration in the second pond (three days after scraping) and the third pond (eleven days after scraping).

The density of dissolved silicate in outflow water from the filrtation pond is higher than in the inflow water. This article suggests that considerable silicate elution from sand occurred in the filitration pond. It is possible to think that microbes that function similarly to *B.mucilaginosus* exist in biofilm layer of the ecological purification system. I suggest the possibility that contribution of the elution to soil ecosystem becomes the model of the silicic acid elution in the slow sand filtration (Fig. 16.3).

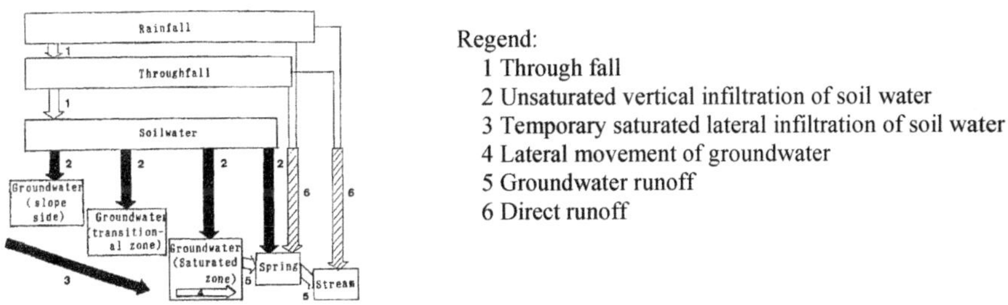

Figure 16.3 Conceptional figure to explain where the concentration of SiO^2 is formed.

Significance of increasing amount of Si supply

In the field of fisheries, it became obvious that bivalve resources decreased. This results in the decrease of water purification function by bivalves as their ecological service. This is also regarded as a social loss.Bivalves which show filter feeding habit take diatoms which have silicic acid shells. Redfield ratio of marine diatom is Fe:P:Si:N:C:O = 0.001:1:15:16:106:212. In other data, Si indicates from 15 to 50. Diatom live in different environment from sea water to fresh water and husk of fresh water diatom is thicker than sea water diatom. This reflects variety of habitat of diatom. The changes of composition of microalgae caused by comparative shortage of Si against nitrogen and phosphorus often called the silica deficiency hypothesis in the fisheries.

CONCLUSIONS

I had an opportunity to visit Lake Datonghu, Human province, China in August 2013. I observed the bivalve habitation and microbe distribution of the lake. I heard an oral tradition that the dominance of "*Bacillariophyceae* and *Dinophyceae*" meant affluence, while the dominance of "*Chlorophyceae* and *Cyanophyceae*" meant poverty. I believe that the discussion of Si circulation will soon begin also in the Asia monsoon area.

ACKNOWLEDGEMENTS

I worked on the environmental remediation of the hydrosphere as a technical expert. By the removal of nitrogen of the enrichment causative agent of the hydrosphere in particular, I had the results using a function of the activated sludge in the field of sanitary engneering.

However, I arrived at a viewpoint that the slenderest and beautiful environmental remediation is application of the normal " Food chain". That keyword was silicon with the feeling behind well-known nitrogen and phosphorus. The grobal ecosystem blessed with silicon is supported by a creature needing silicon. Above all, I knew that the significance of existence of the diatom in the hydrosphere is big, but as for the recent tendency, there is predominance of the non-diatom. It causes in the decrease of the useful marine resources. Predominance of the diatom needs supply of the silicon. Supply of silicon is necessary for becoming it predominance of the diatom. A concept called the "supply" in contadiction to the past reduction is necessary.

Merosira of the diatom is known as a main microbe of the slow sand filtration. A phenomenon that diatom grows it and the dissolved silocon in running water increase was caught by Nakamoto and otheres. When the ecosystem of the biofilm of the slow sand filtration thinks it to let you elute silicon from the sand layer, explanation does not have contradiction.

The writer who learned forestly knew that silocon supply is included in a function to have of the forest soil, but there are not many the forest people concerned noticing usefulness. However, I think that it is fact, and this function is accepted to a citizen as multiple functions of the forest in the near future.

It regards the engineer as a mission using the result of the resercher to give a result. I want to thank for all results of reserch that I reminded commonality of a biofilm and the silicon supply in the soil. In addition, I want to thank the person concerned who admitted anouncement with this title. Article presentation was anxious about, but the writer who was not proficient in English was able to complete it by the support of many acquaintances. I want to express the will of deep thanks here.

REFERENCES

N.Nakata, H.Nakamoto(1988) The changes of water quality by algae biofilm in the filtration pond. Japanese Society of Limnology Koshinetu branch newsletter.

M.Ueshima, K.Tazaki Biological weathering and clayey biomineralization. -A role of microbes during the formation of layered silicate minerals-

K.Fukami, H.Tamaki, G.Nigi(2007) Supply of inorganic nutrients to river water from forest soil and its effects on microrage in Niyodo river, Kouchi Prefecture. Kuroshio Sceience, 1, 96-104.

Y.Simada, N.Ohte, N.Tokuchi, M.Suzuki(1992) SiO2 concentration of groundwater and streamwater in a small forested watershed.J.Japan Soc. Hydol & Water Resour, Vol 5-2, 3-11.

Chapter 17

Effect of sudden changes of slow sand filtration rate on number of bacteria and particles in effluent

Willem A. Oorthuizen, Aleksandra Magic-Knezev**, Fei Cui****
**Dune water company Dunea, Zoetermeer, The Netherlands*
***Het Waterlaboratorium, Haarlem, The Netherlands*
****Delft University of Technology, Delft, The Netherlands*

ABSTRACT
The objective of this study is to measure the effects of different acceleration rates of SSF filtration velocity on total cell concentrations (TCC), adenosine triphosphate (ATP) Heterotrophic plate counts (HPC) and physical quality (turbidity and particle count) of SSF effluent.. The acceleration rate of 0.12 m/h^2 had no influence on turbidity, ATP concentration, plate count or particle count while the acceleration rate of 0.20 m/h^2 caused a slight increase of the TCC. The acceleration rate of 0.70 m/h^2 had an obvious negative influence on TCC and particle count, but not on ATP or heterotrophic plate count.

Keywords slow sand filtration; velocity, acceleration, particle counts, ATP, total cell counts

INTRODUCTION

Slow sand filtration (SSF) is one of the earliest water treatment processes used to produce microbiologically safe drinking water throughout the world. Due to the different water demand during daytime and night, many drinking water treatment plants have adjusted their water supply amount within the day. According to the forecasting of drinking water demand, water production increases in the morning and decreases at night. The increase of water flow rate during increased production in the morning generates the acceleration of SSF filtration velocity. Higher SSF filtration velocity may affect SSF effluent quality that includes particles and microbiological quality, especially if any breakthrough happens. Although the effect of different SSF filtration velocities on SSF effluent quality has already been studied (Huisman et al., 1974), there has not been any study about the effect of different acceleration rates of SSF filtration velocity on SSF effluent quality, especially particles and microbiological quality. The objective of this study is to measure the effects of different acceleration rates of SSF filtration velocity on total cell concentrations (TCC), adenosine triphosphate (ATP), heterotrophic plate counts (HPC) and physical quality (turbidity and particle count) of SSF effluent.

METHODOLOGY

Acceleration experiments were carried out at Scheveningen, the full scale drinking water treatment plant of Dunea. The SSF 7A with a surface area of 1302 m2 was used to measure the effect of acceleration events and SSF 7B, with a surface area of 1305m2, was used as the reference filter. Both SSF 7A and 7B are filled with 1 m of dune sand (d10-d90=0.3-1.2mm) on the top and 0.15m grind (d10-d90=3.4-8.0mm) on the bottom, have the same filter run time of 6 years (after previous Smutz decke removal) and operate in downstream mode with 1m water column on the top of the sand at filtration velocity of 0.2-0.3 m/h

Four acceleration events were carried out at acceleration rates of: 0.12, 0.20 (twice) and 0.70 m/h^2. For each acceleration event the TCC, ATP, HPC, turbidity and particle count were measured in effluents of SSF 7A and SSF 7B at regular time intervals. TCC was measured as described by Vital et al. 2012. TCC-intact represent fraction of cells with intact cell membrane. ATP was measured as described by van der Wielen and van der Kooij D. 2010. HPC and turbidity were measured according to syllabus of 'Het Waterlaboratorium'. Particle counts were measured online with 'MetOne PCX' Particle Counters. The smallest particles counted were down to 2 μm and the largest 15 μm. At Scheveningen, rapid sand filters (RSFs) effluent directly flows into SSFs. The influent quality is shown in Table 17.1.

Table 17.1 Influent quality into SSF 7A and SSF 7B

Parameters	Unit	Values	Parameters	Unit	Values	Parameters	Unit	Values
Turbidity [1]	FTE	0.03	Plate count (22°C) [2]	kvd/ml	3	AOC (total) [1]	μg/L	11
SS [1]	mg/L	0	ATP [3]	ng/L	4.32	Fe [1]	μg/L	8
Total hardness [1]	mmol/L	1.6	TCC-living [3]	cel/ml	184088	Mn [1]	μg/L	0.1
pH [1]		8.67	TCC-total [3]	cel/ml	209293	Aeromonas spp [2]	kvd/100ml	41

1: Average values in 2010-2011;

2: 50 percentile value of total series of number in 2010;

3: Average values of 3 experiments.

Four acceleration events were carried out according to Table 17.2.

Table 17.2 Measurement plan of four acceleration events

Acceleration event	Accelerated filter	Acceleration rate [cm/h^2]	Flow rate pattern	Duration of flow rate change [h]	Increase in flow rate [m^3/h]	Duration of acceleration
Event 1	SSF A	12	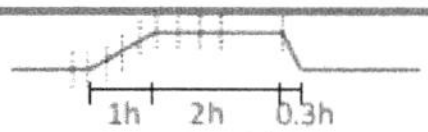	3.3	260-416	1h
Event 2	SSF A	20	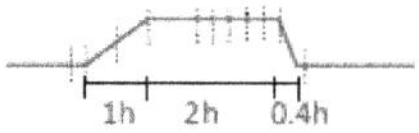	3.4	250-510	1h
Event 3	SSF A	70	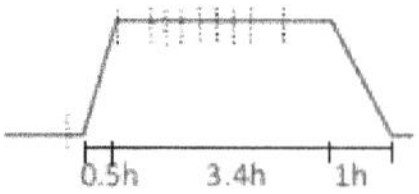	4.9	326-781	0.5h
Event 4	SSF B	20	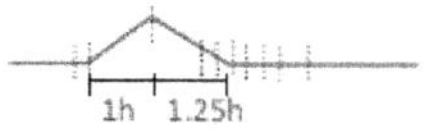	2.3	250-510	1h

Sampling was started before, during and after each acceleration event. The vertical dotted lines in the flow rate patterns in table 17.2, indicates when the samples were taken. The time between the samples varied between 15 minutes and 1 hour. Samples for bacteriological parameters were taken in sterile 250 bottles and kept on ice before analysis, and was carried out within 24 hours from the sampling.

RESULTS AND DISCUSSION

Turbidity:
For acceleration rates of 12cm/h^2 and 20cm/h^2, online and offline turbidity showed no changes. But for acceleration rate of 70cm/h^2, there was a short increase of online and offline turbidity. It is clear to see as acceleration rate increased, turbidity and TCC increased.

Plate count, ATP and TCC:
HPC and ATP stayed very stable within the range of 0-10 CFU/ml and 0-2 ng/L, respectively, for all acceleration rates tested. However, total number of cells (TTC) increased significantly after the acceleration rate of 70cm/h^2 what indicates that HPC and ATP are not sensitive enough to detect slight quality changes in the effluent when compared with TCC analysis. Besides, ATP/TCC-intact was in the range of 0.05-2.09 × 10^{-16}g/cell with all the values below 10^{-16}g/cell, indicating rather low metabolic activities (Magic-Knezev and van der Kooij, 2004).

TCC-intact:
TCC-intact accounted for 84% of TCC-total on average varying from 79% to 93% in different events. For acceleration rates of 12cm/h^2, TCC-intact stayed stable. For acceleration rates of 20cm/h^2 (Event 2), the range of TCC-intact values was within 22% and there was a slight increase during the period of stable high filtration velocity. For acceleration rates of 70cm/h^2, the range of TCC-intact values is within 64%. So there was an obvious negative influence on TCC-intact due to an acceleration rate of 70cm/h^2.

TCC:
For acceleration rates of 12cm/h^2, TCC even decreased as time passed by. For acceleration rates of 20cm/h^2 (Event 2), the range of TCC values was within 32% and there was a slight increase. For acceleration rate of 70cm/h^2, the range of TC values was within 72% and TCC had the same trend as TCC-intact. So there was a clear influence of acceleration rate of 70cm/h^2 on TCC values.

Particle count:
After comparing of all the particle count results, only the smallest particles in the range of 2-3 μm showed obvious variation. The results with acceleration rates of 12- and 20cm/h^2 made clear that the particle counts in the range 2-3 μm of SSF 7A and SSF 7B were almost the same. For acceleration rates of 70cm/h^2 there was a big increase of particle counts in effluent of SSF 7A (Figure 17.1). Even until the next day the particle counts were still higher than in the reference filter, which proved that particles are released for a long time after the acceleration rate of 70cm/h^2.

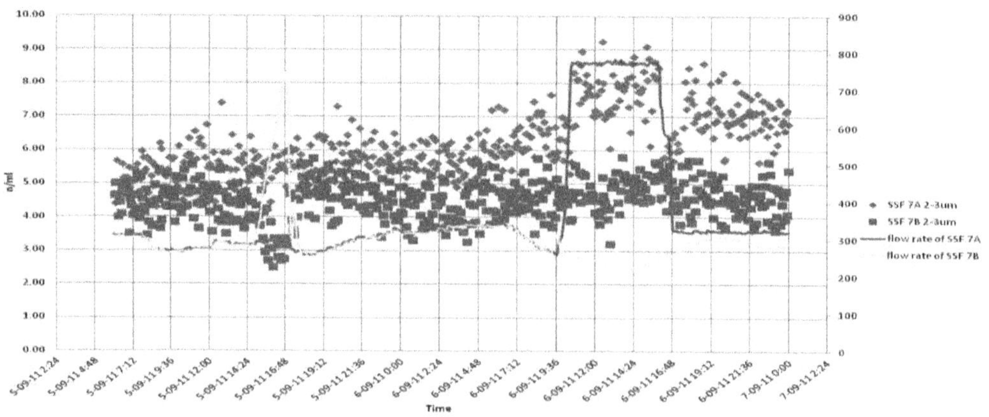

Figure 17.1 Flow rate of SSF 7A (red line) and SSF 7B (yellow line) and particle counts (2-3 m) in effluent SSF 7A (blue diamonds) and 7B (red squares) at acceleration rate of 0.70m/h

CONCLUSIONS

1. The acceleration rate of 0.12 m/h^2 had no influence on turbidity, ATP concentration, plate count or particle count. TCC decreased slightly after acceleration, probably because the top layer (Smutz decke) of the SSF became more compacted due to the acceleration of filtration velocity.
2. The acceleration rate of 0.20 m/h^2 had no influence on turbidity, ATP, HPC or particle count, but a slight increase of the TCC was observed.
3. The acceleration rate of 0.70 m/h^2 had an obvious negative influence on TCC and particle count, but not on ATP or HPC.

The results of this research indicate that acceleration rate in a SSF to a maximum of 0.12 m/h^2 does not affect the quality of effluent. In accordance, Dunea set the operating limit for acceleration rate in a SSF to a maximum of 0.12 m/h^2 evenly distributed over one hour.

REFERENCES

Magic-Knezev A., van der Kooij D. (2004) Optimization and significance of ATP analysis for measuring active biomass in granular activated carbon filters used in water treatment. *Wat. Res.*, **38**(18), 3971-9.

Hammes F., M. Berney, Y. Wang, M. Vital,O. Koster, T. Egli. (2008). Flow-cytometric total bacterial cell counts as a descriptive microbiological parameter for drinking water treatment processes *Wat. Res.*, **42**, 269 – 277

L.Huisman. W.E. WOOD, F.I.C.E. 1974. Slow sand filtration. WHO, Geneva. http://whqlibdoc.who.int/publications/9241540370.pdf

Van der Wielen P.W., Van der Kooij D. (2010a). Effect of water composition, distance and season on the adenosine triphosphate concentration in unchlorinated drinking water in the Netherlands. *Wat. Res.* **44**(17), 4860-4867.

Part III

Slow Sand Filtration – Treatment Performance

Chapter 18

Slow sand filtration process model for removal of microorganisms

Jack F. Schijven[a,b], Harold J.J. van den Berg[a], Michel Colin[c], Yolanda Dullemont[c], Wim A.M. Hijnen[d], Aleksandra Magic-Knezev[e], Wim Oorthuizen[f], Gerhard Wubbels[g]

[a]National Institute of Public Health and the Environment, Bilthoven, The Netherlands
[b]Utrecht University, The Netherlands
[c]Waternet, Amsterdam, The Netherlands
[d]KWR Watercycle Research Institute, Nieuwegein, The Netherlands
[e]Het Waterlaboratorium, Haarlem, The Netherlands
[f]DUNEA, The Hague, The Netherlands
[g]Waterbedrijf Groningen, The Netherlands

ABSTRACT

A new Slow Sand Filtration Process Model for removal of microorganisms has been developed into an interactive computational tool. Operational conditions, such as filter bed depth, filtration rate, water temperature, grain size of the sand and age of the Schmutzdecke, can be set to predict removal of bacteriophage MS2 and *E. coli* as process indicators for removal of enterovirus and *Campylobacter*. These predictions can be conducted as a function of time, accounting for time-variable operational conditions. A Beta distribution is fitted to the fractions of microorganisms that pass slow sand filtration. It is demonstrated how the Beta distribution parameters can be used as integral part of Quantitative Microbial Risk Assessment for the production of potable water from surface water with the computational tool QMRAspot.

Keywords Slow sand filtration; microorganism removal; model; risk assessment

INTRODUCTION

Following the preventive management framework for safe drinking water of The World Health Organization (WHO) Guidelines for Drinking Water (WHO 2011), the Dutch Drinking Water Act (Anonymous, 2001) requires that the drinking water companies demonstrate compliance with a maximum risk in infection of one per ten thousand persons per year from drinking water consumption. It is required to demonstrate this for four index pathogens: Enterovirus, *Campylobacter*, *Cryptosporidium* and *Giardia.*

Compliance with the risk level is done by means of a Quantitative Microbial Risk Assessment (QMRA). QMRA entails determining pathogen concentrations in the source water and determining efficiency of drinking water treatment to remove pathogens from the water on the basis of indicator organisms. By default, bacteriophages and *E. coli* are used as the process indicators for removal of enterovirus and *Campylobacter*, respectively.

At six drinking water production locations in the Netherlands, slow sand filtration is applied indoors as the final treatment step, so receiving almost finished drinking water. Because of that, concentrations of indicator organisms in the influent and effluent of the slow sand filters are usually below detection limit, hindering proper assessment of the efficiency of the slow sand filters. Therefore, refuge was taken to using pilot plant slow sand filters that were run in parallel with the full-scale filters under exactly the same conditions. The pilot filters could be challenged with sufficiently high concentrations of indicator organisms to determine removal efficiency (Dullemont et al., 2006).

Based on the results of thirteen such pilot plant experiments that were conducted under various operational conditions, Schijven et al. (2013) developed a Slow Sand Filtration Process Model that can be used to predict removal of bacteriophage MS2 and of *E. coli* by slow sand filtration as a function of temperature, grain size, filtration rate and the age of the Schmutzdecke. The latter is a biologically active slime layer that gradually develops on top of a sand filter. The Schmutzdecke is known to improve microorganism removal, but every once and a while it needs to be scraped of when the filter bed resistance becomes too high (*e.g.* Amy et al., 2006).

Having developed a Slow Sand Filtration Process Model, the next step to take was applying it as part of the legislatively required QMRA. For conducting QMRA, the computational tool QMRAspot has been developed (Schijven et al., 2011).

The objective of the current study was to develop the Slow Sand Filtration Process Model into an interactive computational tool that can be used by operators to set and manage the proper operational conditions as well to provide values that can be used in QMRAspot, thus integrating the Slow Sand Filtration Process Model into QMRA, accounting for all variability in removal efficiency over a period of several years.

METHODOLOGY

In brief, the Slow Sand Filtration Process Model by Schijven et al. (2013) is cited here:

$$\ln\frac{C}{C_0} = -\frac{3}{2}\frac{1-n}{d_c}\left(\alpha\eta_{0,d_p,d_c,u,T,n}z + f_0T\left(1 - e^{-\alpha f_1 a}\right)\right) \tag{1}$$

Attachment was described by colloid filtration theory (Tufenkji and Elimelech,2004), where α is the sticking efficiency, η_0 is the single collector efficiency, d_p is the size of the microorganism, d_c is the grain size, u is the flow rate, T is the temperature, n is the porosity and z is the filter bed depth. The interaction with the Schmutzdecke increases with a, the age of the Schmutzdecke, the sticking efficiency and temperature and decreased with grain size. The Schmutzdecke develops at a specific rate and reaches a maximum, corresponding to logistic growth, therefore scale factor f_0 and shape factor f_1 were introduced.

Schijven et al. (2013) used likelihood ratio tests for combining parameter values for sticking efficiency α, f_0, and f_1 between experiments. They found that the sticking efficiency α for MS2 and *E. coli* WR1 was filter specific, but f_0, and f_1 were only microorganism specific. The correlation coefficient between predicted and observed removal for both MS2 and *E. coli* WR1 was 98%. Cross-validation showed that the model was able to predict log removal of MS2 and *E. coli* WR1 within ±0.6 $\log_{10}$.

An interactive computational tool of the Slow Sand Filtration Process Model (equation 1) was coded in Mathematica language (version 9.0.1, Wolfram Inc., Champaign, Illinois) and saved in Computable Document Format. It can be run using Mathematica or Mathematica's Player Pro with full functionality. It can also be run with Mathematica's free CDF Player, however, in that case it cannot read time series data from a spreadsheet.

Mathematica provides a collection of commands that allows fast development of user-friendly, interactive, dynamic tools.

The tool contains an interactive part for setting single values for operational conditions, and by reading time series data from a spreadsheet file, removal of bacteriophage MS2 and *E. coli* WR1 can be predicted as a function of time, accounting for the time variability of water temperature, age of the Schmutzdecke, and filtration rate. The current version of the tool applies to a single slow sand filter bed.

Variability of the fraction of microorganisms that passes treatment can be described by a Beta distribution (Teunis et al., 2009). Therefore, the predicted log removal values as function of time were transformed into fractions to which a Beta distribution was fitted using maximum likelihood estimation.

As output, the tool plots the time dependent values of filtration rate, age of the Schmutzdecke, water temperature and log removal of both indicator organisms. In addition, the parameters of the Beta distributed fractions of indicator organisms that pass treatment are given. These parameter values can be used interactively in QMRAspot (Schijven et al., 2011), thus coupling the outcomes of the Slow Filtration Process Model to QMRA.

RESULTS AND DISCUSSION

Figure 18.1 shows the screen print of the tool after reading data from the spreadsheet. The left part provides input fields for single values of the operational conditions that are part of model (see equation 1 in the introduction). Under "Production location", a number of production locations can be selected for which sets of operational conditions are incorporated in the code of the tool. These values can also be set by loading data from the spreadsheet or by the user's input.

Obviously, operational conditions are location specific. The values of the sticking efficiencies of the process indicators are location specific as well and need to be determined for each production location.

Under "Removal $-\log_{10}C/C_0$", the single log removal value is given for the single operational conditions that were set.

Under "Times series data", an Excel spreadsheet can be selected and, subsequently, its data can be loaded. At fixed cells in the spreadsheet the same single values for the operational conditions are given, and below those a list of dates with filtration rates and water temperatures for any time period. Also a list of dates within this time period when the Schmutzdecke was scraped off is read. In the tool these dates are converted into a list of day numbers. Note that the single value for the age of the Schmutzdecke is used as the age of the Schmutzdecke at the start of the period of concern.

On the bottom left the $\log_{10}$ of the mean C/C_0 (fraction) is given as well as the 95-percentile. And below those, the parameter values of the Beta distribution that describe removal variability.

On the right side, graphical output is presented. The top plot represents the filtration rate. This is an important operational parameter. Note that the model is not developed to predict microorganism removal under intermittent operation, *i.e.* including periods with zero flow rate. The second and third plot represent the age of the Schmutzdecke and the measured water temperature, respectively.

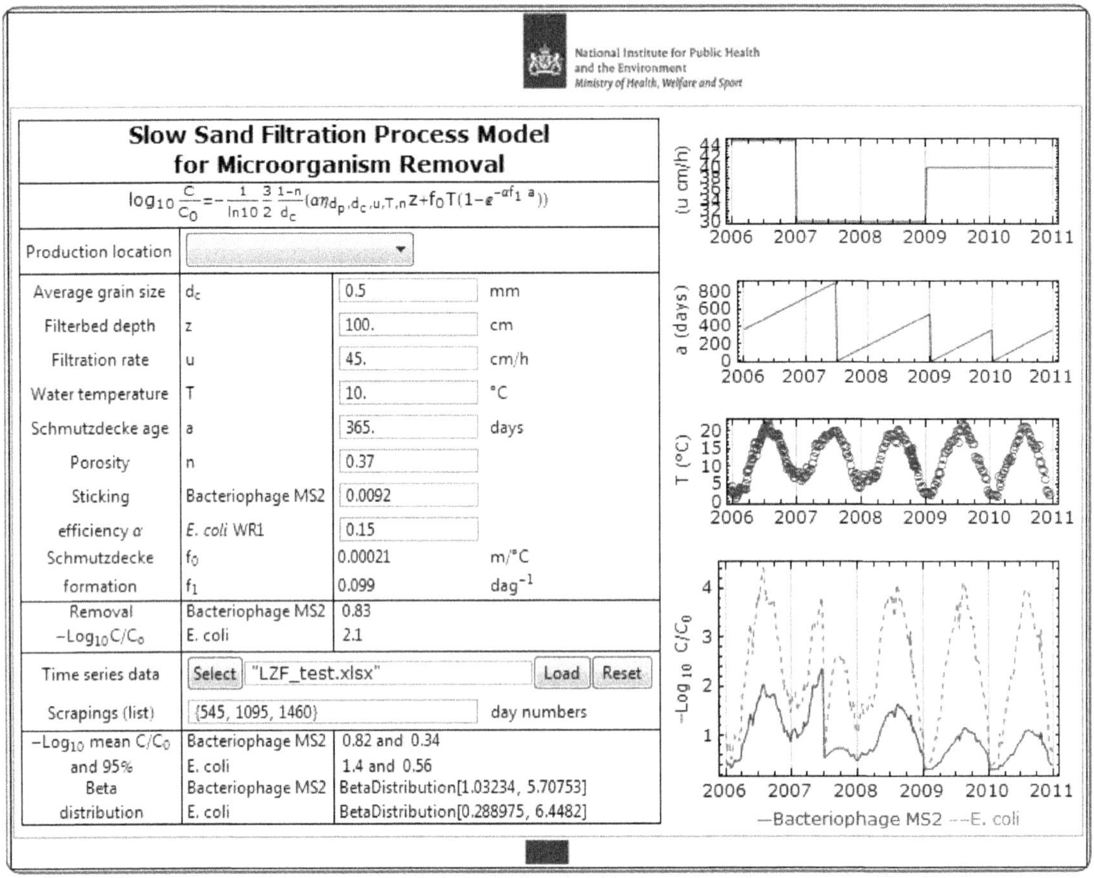

Figure 18.1 Screen print of the Slow Sand Filtration Process Model Tool with input and output.

Finally, the plot at the bottom right shows the predicted log removal of bacteriophage MS2 and *E. coli* as a function of time. Because of its interactive and dynamic properties the tool can easily be used to explore the effects of the operational conditions on the removal of microorganisms. It will be clear that finer sand will significantly increase removal. Obviously, a higher age of the Schmutzdecke is beneficial as well up to a certain maximum, mainly determined by f_0 and the water temperature (equation 1). This maximum may not be reached if filter bed resistance becomes too high and necessitates scraping the Schmutzdecke off from the sand filter.

Commonly, several filter beds are operated in parallel. It is straightforward to extend the tool for the removal by each of the filter beds and for the combination of the filter beds. It is common practice that the filter beds will not be scraped all at the same time. Here, the tool can be used to determine when a filter bed that has been scraped can be taken into operation again.

Figure 18.2 shows the screen "Option-Treatment-Enterovirus" of QMRAspot, where the parameter distribution values can be given. Figure 18.3 shows for bacteriophage MS2 the corresponding histogram of fractions from Monte Carlo simulation as part of the QMRA. It should be noted that variability of treatment in time is randomized in the current version of QMRAspot. Often, pathogen concentrations may be higher in winter and lower in summer, whereas in winter slow sand filtration performs less well than in summer. The randomization may therefore lead to some overdispersion of risk estimates.

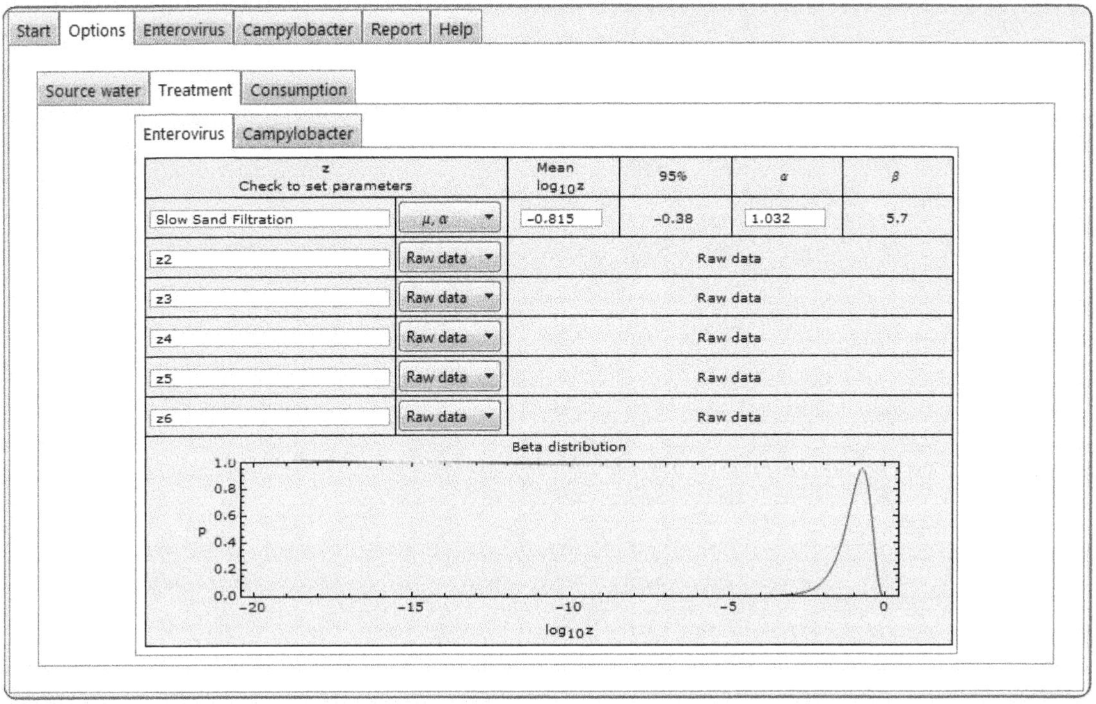

Figure 18.2 "Options-Treatment-Enterovirus" screen in QMRAspot (Schijven et al., 2011), where the Beta distribution parameter values for Slow Sand Filtration are set.

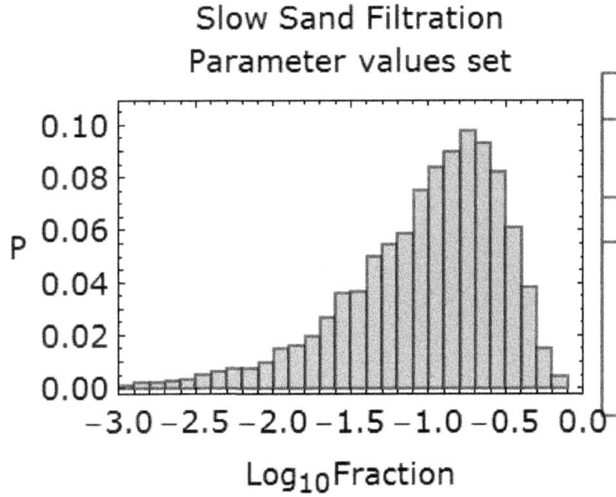

Unpaired beta model		
α	1.	
β	5.7	
	x	logx
Mean	0.15	−0.82
5%	0.0098	−2.
Median	0.12	−0.92
95%	0.41	−0.39

Figure 18.3 Histogram of fractions (Monte Carlo simulation data) of the bacteriophages MS2 that pass slow sand filtration on the basis of the predicted Beta distribution parameters from the Slow Sand Filtration Process Model. The histogram is generated by QMRAspot (Schijven et al., 2011) as part of QMRA.

CONCLUSIONS

The main conclusions of this work are:

(1) A simple to use interactive computational tool has been developed to predict removal of bacteriophage MS2 and *E. coli* as a function of variable process conditions.

(2) The Beta distribution parameter values generated by the Slow Sand Filtration Process Model describe variability in removal of microorganisms by slow sand filtration and are direct input in QMRAspot for risk assessment;

(3) The Slow Sand Filtration Process Model Tool can be used to set optimal operational conditions.

REFERENCES

Amy, G., Carlson, K., Collins, M.R., Drewes, J., Gruenheid, S., Jekel, M. (2006) Integrated comparison of biofiltration in engineered versus natural systems. *In recent progress in slow sand and alternative biofiltration processes*, Gimbel, R., Graham, J.D., Collins, M.R., Eds., IWA, Mülheim, 1, pp 3-11.

Anonymous. (2001) Besluit van 9 januari 2001 tot wijziging van het waterleidingbesluit in verband met de richtlijn betreffende de kwaliteit van voor menselijke consumptie bestemd water. *Staatsblad van het Koninkrijk der Nederlanden*, **31**:1-53.

Dullemont, Y., Schijven, J. Hijnen, W., Colin M., Magic-Knezev A. and Oorthuizen W. (2006) Removal of microorganisms by slow sand filtration. *In 'Recent Progress in Slow Sand and Alternative Biofiltration Processes*, ed. R. Gimbel, N.J.D. Graham and M.R. Collins, IWA publishing, Cornwall, UK.

Schijven, J. F., van den Berg, H. H., Colin, M., Dullemont, Y., Hijnen, W. A., Magic-Knezev, A. and Wubbels, G. (2013). A mathematical model for removal of human pathogenic viruses and bacteria by slow sand filtration under variable operational conditions. *Water Res*. **47**(7), 2592–2602

Schijven, J.F., Teunis, P., Rutjes, S., de Roda Husman, A.M. (2011) QMRAspot: A computational user-friendly interactive tool for quantitative microbial risk assessment from surface water to drinking water. *Water Res,* **45**, 5564-5576.

Teunis, P.F.M, Rutjes, S.A., Westrell, T. and de Roda Husman, A.M. (2009) Characterization of drinking water treatment for virus risk assessment. *Water Res*, **43**(2):395-404.

Tufenkji, N., Elimelech, M. (2004) Deviation from the classical colloid filtration theory in the presence of repulsive DLVO Interactions. *Langmuir* **20**, 10818-10828.

WHO (2011). Guidelines for Drinking-water Quality. Fourth edition. *World Health Organisation*, Geneve, Zwitserland. http://whqlibdoc.who.int/publications/2011/9789241548151_eng.pdf.

Chapter 19

Removal of anti-inflammatory compounds by ecological filtration

C. M. Erba, E. P. Tangerino**, W. D. Isique**, L.C. Campos****
**Universidade de São Paulo, EESC, Brazil*
***Universidade Estadual Paulista "Julio de Mesquita Filho" FEIS, Brazil*
****University College London, UK.*

ABSTRACT
The use of the slow sand filtration, better known as *ecological filter* was tested to removal of anti-inflammatory compounds from water, (i.e. Diclofenac, Naproxen and Ibuprofen). The treatment system consisted of pre-filtration followed by ecological sand filter and the mixture of the three different compounds was added the afluent of the ecological filter. The filtration system operated during dry season (June-November 2012) in Brazil. The pharmaceutical compounds were determined by SPE (solid phase extraction) followed by HPLC (high performance liquid chromatography). In addition, turbidity, apparent color, true color, total and thermotolerant and coliforms were measured to further evaluate the water quality. The ecological sand filters showed to be an efficient process to remove Diclofenac, Naproxen and Ibuprofen in this study.

Keywords Slow sand filtration; anti-inflammatory compounds; ecological filtration

INTRODUCTION

The increasing contamination of water bodies by pharmaceutical compounds and the frequent occurrence in drinking water sources have raised concerns about the impact on public health and fauna and flora (Tundisi, 2005).

These compounds are highly persistent environmental micropollutants and difficult to be removed by conventional water treatment processes (Melo et al., 2009). They also have high potential for bioaccumulation and low biodegradability, which exacerbates the adverse effects in the longer term (Kummerer, 2004).

The use of slow sand filtration, better known as *ecological filter* (Nakamoto, 2011), seems a cost-effective treatment technology because it does not require application of chemicals and it can efficiently remove pathogenic microorganisms such as viruses, bacteria, cryptosporidium, algae, as well as dissolved organic matter after pre-oxidation (Jenkins et al., 2011; Graham, 1999).

METHODOLOGY

The aim of this work was to evaluate the removal of three anti-inflammatory compounds from water by ecological filtration. The raw water used for the testing was taken from Ipê Lake – Bairro do Ipê (Ilha Solteira, Sao Paulo state, Brazil).

The water treatment system developed (Figure 19.1) for this research consisted of the a prefilter (PFVA) followed by Ecological Filter (FEco). The PFVA was built in steel with cylindrical shape, diameter of 0.80 m, area of 0.50 m^2 and height of 2.95 m. The filter media was composed by four sub-layers of gravel (25.4mm to 19mm; 15.4mm to 9.6mm; 6.4mm to 3.2mm; 3.2mm to 1.4mm) with decreasing grain size in the direction of flow.

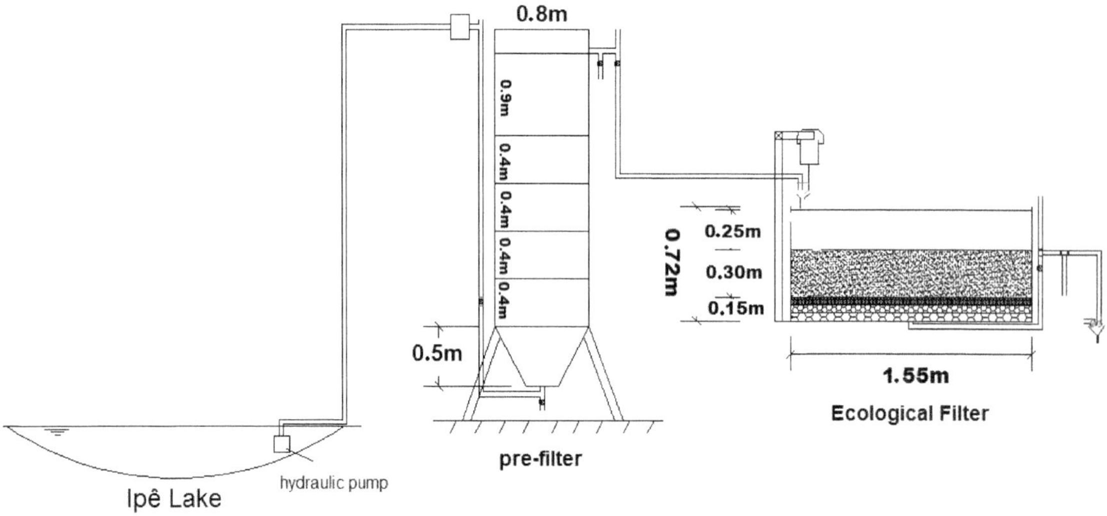

Figure 19.1 The ecological filtration system.

The FEco consisted of a tank with 1.00 m wide, 1.55 m long and 0.72 m depth. The support layer of the filter media was formed by three layers of gravel with 15 cm of total height and particle size of 12.5 mm to 1.41 mm . The sand bed had 30 cm, and a particle size between 1.0 and 0.08 mm, coefficient of uniformity between 2 and 3 and the effective diameter of 0.25 mm. The water depth above the sand surface was equal to 25 cm.

The filtration rate of the FEco was abour 3m^{-3}m^{-2}d^{-1}. The affluent water feed was made by a PVC pipe with diameter of 20 mm, and mounted in a cross shape, with holes spaced to evenly distribute the water over the filter.

A solution containing a mixture of three different anti-inflammatory compounds (i.e. Diclofenac, Naproxen and Ibuprofen) was added weekly to the ecological filter influent (FEco1) during June and November 2012 (6 months). This contaminant solution was prepared with concentration of 2µg.l^{1} by diluting the anti-inflammatory compounds in 50 ml of 100% methanol and adding them to 2 liters of deionized water for homogenization. The solution was then placed in a sealed bucket to prevent degradation by sunlight. This bucket had a dripping system linked to the water filter inlet and the solution dosage was set at 10 ml.min^{-1}.

In the FEco, samples of 300 ml were collected daily from the supernatant water (i.e. 25cm above the sand surface) and from the effluent (FEco 2) being the first sample was taken 24 hours after the first dosing of the contaminants. The anti-inflammatory compound concentrations were determined through Solid Phase Extraction (SPE) on a C18 cartridge and High Performance Liquid Chromatography (HPLC-DAD) using a modified method developed by Nebot et al., (2007).

For fecal and total coliforms, Petrifilm/3M plates were used followed by Greenhouse culture/Fanem/A LT-502, while the true color was determined by membrane filtration (0.45 mM) followed by spectrophotometry - HACH DR5000. The bacteriological tests and the true color were measured weekly.

The determination of apparent color and turbidity were performed three times a week. The Nephelometric method (i.e. 2100A-HACH) was used for turbidity measurement, while apparent color was measured spectrophotometrically using a HACH DR5000. The monitored points were the raw water Ipe Lake water, i.e. PFVA affluent (AB), the ecological filter affluent (FEco 1) and the effluent of ecological filter (FEco 2).

RESULTS AND DISCUSSION

Results showed that Diclofenac, Naproxen and Ibuprofen compounds were efficiently removed by ecological filtration with average removal of 94.9%, 87.3% and 76.4%, respectively (Figure 19.2). These results are in agreement with those found in sand filters by Matamoros et al. (2007) to removal some PPCPS (Pharmaceutical and Personal Care Products). Table 19.1 shows the average removal of each compound during each month (4 contaminations per month). Ibuprofen was the anti-inflammatory compound that presented the least removal (76.4%), probably because it has more complex and isomeric molecules. The average temperature in the period of this study was 21.5 °C.

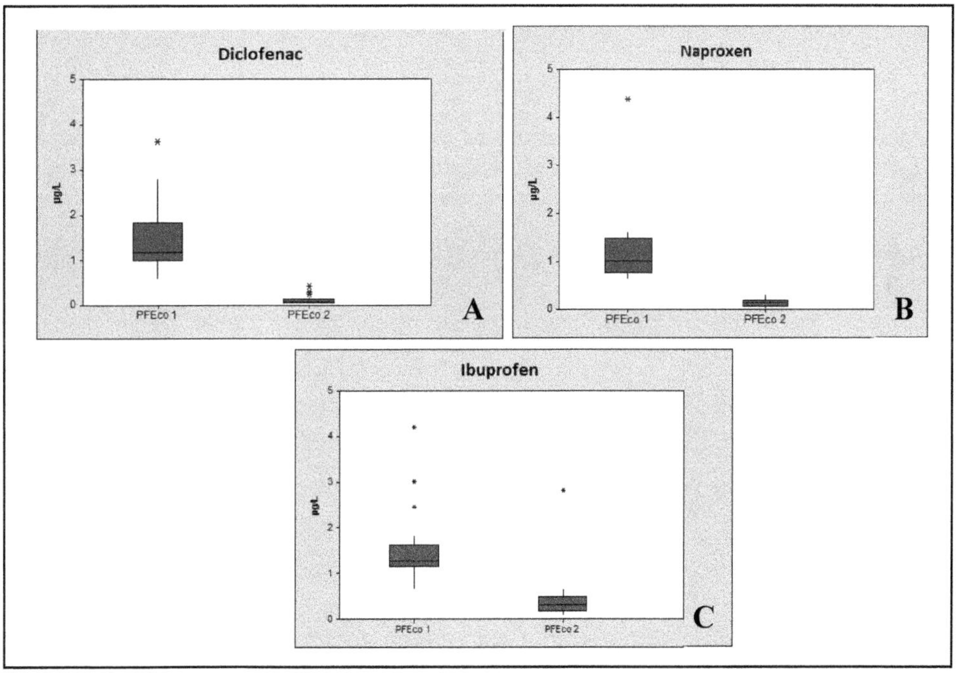

Figure 19.2 Boxplots of average removal of anti-inflamatory drugs by ecological filter during a 5 month filtration run. A: Diclofenac; B: Naproxen; C: Ibuprofen. (PFEco1 = filter influent; PFEco2 = filter effluent).

Table 19.1 Average values per month for each anti-inflamatory compound applied.

Month	Compound	FEco1 (μg mL^{-1})	FEco2 (μg mL^{-1})
June	Diclofenac	1.26	0.02
	Naproxen	1.04	0.04
	Ibuprofen	1.19	0.31
July	Diclofenac	1.87	0.1
	Naproxen	1.26	0.19
	Ibuprofen	2.11	0.19
September	Diclofenac	1.14	0.07
	Naproxen	1.22	0.14
	Ibuprofen	1.23	0.47
October	Diclofenac	0.92	0.13
	Naproxen	1.27	0.09
	Ibuprofen	1.04	0.38
November	Diclofenac	1.02	0.14
	Naproxen	0.73	0.10
	Ibuprofen	1.30	0.19

It is difficult to state if the final concentrations of the pharmaceutical compounds achieved by the ecological filter in this study were at safe values because of the lack of information about the potential human health risks from exposure to trace concentrations of these compounds in drinking water.

Turbidity had removal of 68.0% (Figure 19.3) and effluent turbidity remained below 1 NTU and apparent color had removal of 61.6 % (Figure 19.4a) and true color removal was 47.97 % (Figure 19.4b).

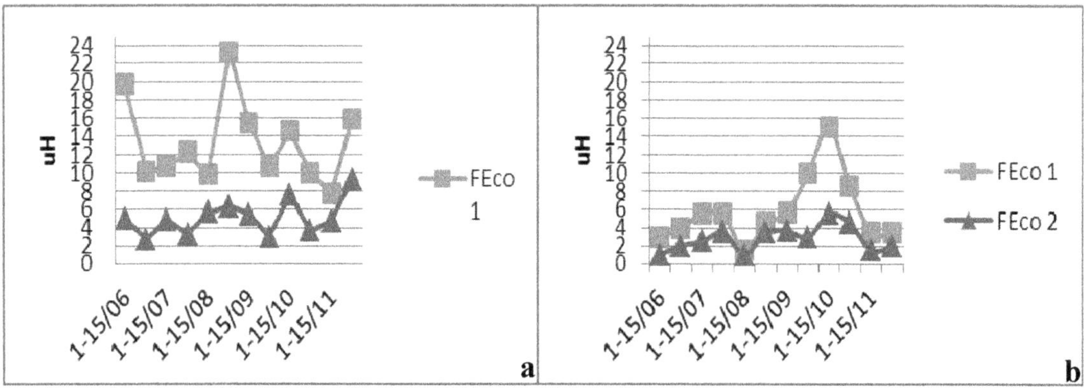

Figure 19.4 Removal of (a) apparent color and (b) true color by the ecological filter.

Total coliforms removal, as expected, was above 99% (Figure 19.5) during the whole period even during the occurrence of highest coliforms peak in the second half of August (dry season). Fecal coliforms were not detected in all samples during the whole study period.

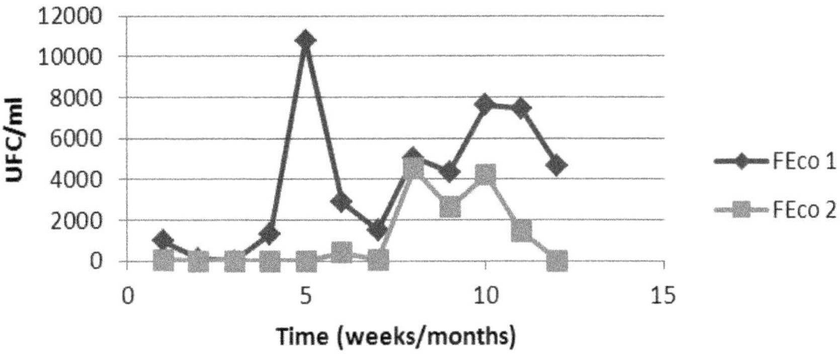

Figure 19.5 Total coliforms variation in the ecological filter.

A wide variety of algae and cyanobacteria were also identified in the ecological filters but these are not discussed in this paper.

CONCLUSIONS

The main conclusions of this work are:

(1) The ecological filter has apparently removed efficiently the three anti-inflamatory compounds studied (i.e. Diclofenac, Naproxen and Ibuprofen);
(2) Investigation on potential effect of trace concentrations of pharmaceutical compounds on human health is necessary;
(3) Turbidity, colour and total coliforms concentration in the filter effluent were found to be withn the Brazilian standards of potability (Brasil, 2004).

REFERENCES

Brasil. 2004. Ministério da Saúde. Portaria nº 518, 25 de março de 2004. Estabelece os procedimentos e responsabilidades relativos ao controle e vigilância da qualidade da água para consumo humano e seu padrão de potabilidade, e dá outras providências. Brasília/DF.

Graham, N.J.D. (1999). Removal of humic substances by oxidation/biofiltration processes – a review. *Wat.Sci.Tech.*, **40**:9:141-148

Jenkins, M.W., Tiwari, S.K., and Darby, J. (2011). Bacterial, viral and turbidity removal by intermittent slow sand filtration for household use in developing countries: Experimental investigation and modeling. *Water Research*, **45**:6227-6339.

Kummerer, K. (2004).. Resistance in the environment. Journal of Antimicrobial Chemotherapy, Oxford, **54**:2:311–320.

Matamoros, V., Arias, C.; Brix, H. and Bayona, J.M., (2007) Removal of Pharmaceuticals and Personal Care Products (PPCPs) from Urban Wastewater in a Pilot Vertical Flow Constructed Wetland and a Sand Filter. *Environ. Sci. Technol.*, **41**:23:8171–8177.

Mello, S.A.S.; Trovó, A.G.; Bautitz, I.R. and Nogueira, R.F.P. (2009). Degradação de fármacos residuais por processos oxidativos avançados. Quimica Nova, São Carlos, v. 32, n. 1, p. 188-197.

Nakamoto, N. (2009). Produza você mesmo uma água saborosa – sistema de purificação ecológica - revendo a tecnologia de produção de água potável. São Paulo: Ferrari. 210p.

Nebot, C; Gibb, S.W., and Boyd, K.G. (2007). Quantification of human pharmaceuticals in water samples by high performance liquid chromatography– tandem mass spectrometry. *Analytica Chimica Acta,* 598, 87–94.

Tundisi, J. G. (2005). Água no século XXI: enfrentando a escassez. 2. ed. São Paulo: Rima, 248 p.

Chapter 20

Removal of bromophenols by slow sand filtration

Keiko Katayama-Hirayama, Hirofumi Suzuki,* Naoki Toda,* Akihiko Tauchi,** Atsushi Fujioka,** Tetsuya Akitsu,* Hidehiro Kaneko,* and Kimiaki Hirayama**

**Interdisciplinary Graduate School of Medicine and Engineering, University of Yamanashi, 4-3-11 Takeda, Kofu, 400-8511, Japan*

***Technology Development Division, Toshiba Lighting and Technology Corporation, Asahimachi 5-2-1, Imabari, Ehime, 794-0042, Japan*

ABSTRACT

We tried to remove bromophenols with a slow sand filter and compared the degradability and the structure. The phenols examined are phenol, catechol, and mono-bromophenols (MBPs), namely 2-bromophemol (2-BP), 3-bromophenol (3-BP) and 4-bromophemol (4-BP); and di-bromophenols (DBPs), namely 2,4-dibromophenol (2,4-DBP), 2,6-dibromophenol (2,6-DBP), and 3,5-dibromophenol (3,5-DBP). Bromophenols (BPs) with low removal rates were also examined for removal under ultraviolet lamp (UV) irradiation and the relationship between degradability and molecular orbital properties of the phenols was estimated. The order of removability with a slow sand filter of examined phenols was catechol > phenol ≥ 4-BP > 3-BP > 2-BP > 2,4-DBP, 2,6-DBP > 3,5-DBP. After 5 min of UV irradiation, the initial concentrations of MBPs and DBPs of 20 mg per liter were almost removed. The decrease in the dissolved organic carbon concentration suggests mineralization of MBPs and DBPs. Bromide ions were released along with the decrease of BPs examined. Molecular orbital calculation suggests that the electrophilic frontier density and HOMO energy may be related to the degradability of BPs under UV irradiation.

Keywords Bromophenols; slow sand filtration; UV

INTRODUCTION

A slow sand filter is the easiest water purification system to build and operate to obtain tap water. This method is recommended for developing countries. Contaminants are biologically removed at the upper layer of the filter. Also adsorption and physical filtration are effective. The International Reference Centre for Community Water Supply and Sanitation published a brochure (WHO 1987), in which slow sand filtration was reported as a suitable technique for water supplies in developing countries. Industrial development in newly developed countries is expanding rapidly and the risk to tap water is a concern, including the risk of the leakage of industrial products in the disposal process.

In the previous study, we examined removal of tetrabromobisphenol A (TBBPA) by a slow sand filter, and 2,6-dibromophenol was identified as a decomposition product of TBBPA by filtration (Katayama-Hirayama et al. 2010). TBBPA is one of the flame retardants widely used in industrial products such as plastic polymers, electronic circuits and fabrics, and it has the potential to enter the water environment after the disposal of these products. The World Health Organization reported that irritation, abnormal swimming behavior, and reduced survival and growth have been observed in fish exposed to TBBPA (WHO 1995). Many researchers have examined the fate of TBBPA in the environment (Ronen and Abeliovich 2000; Oeberg et al. 2002; Erikson et al. 2004; Greeke et al. 2006; Arbeli et al. 2006) and in animals (Meerts et al. 2000; Fujitani et al. 2006; Zalko et al. 2006; Antignac et al. 2009) because of the evidence of toxicity (WHO 1995; Hakk et al. 2003; Ogunbayoet al. 2008; Lilienthal et al. 2008).

2,6-Dibromophenol is also used as a flame retardant after polymerization and has the potential to enter the water environment through the disposal of industrial products. Several marine benthos such as Enteropneusta have been reported to produce mono- and di-bromophenols, sometimes causing off-flavor in fish as well as accumulation in coastal waters (CICAD 2005; Iida et al. 1997).

We tried to remove bromophenols (BPs) with a slow sand filter and compared the degradability and the structure. BPs with low removal rates were also examined for removal under ultraviolet lamp (UV) irradiation and the relationship between degradability and molecular orbital properties of the phenols was estimated.

MATERIALS AND METHODS

The phenols examined are shown in Figure 20.1: phenol, catechol, the mono-bromophenols (MBPs), 2-bromophemol (2-BP), 3-bromophenol (3-BP) and 4-bromophemol (4-BP); and the di-bromophenols (DBPs), 2,4-dibromophenol (2,4-DBP), 2,6-dibromophenol (2,6-DBP), and 3,5-dibromophenol (3,5-DBP).

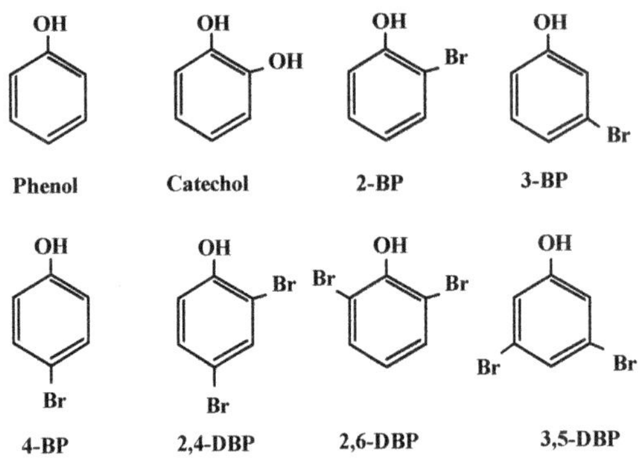

Figure 20.1 Phenols examined.

A slow sand filter was prepared using a 10 cm × 10 cm × 30 cm plastic bottle, completely submerged and operating in down-flow mode (Figure 20.2). The filter column was filled with filter sand of an effective diameter of 0.43 mm and uniformity coefficient of 2.00. The filter was operated at filtration rates of 3 to 4 m per day. Prior to the experiment to remove of phenols', local river water was applied to the sand filter to develop biofilms. After biofilm development, we filtered each of the phenols added to river water at the concentration of 20μg per liter. The concentration of BOD in the river water was below 2 mg per liter.

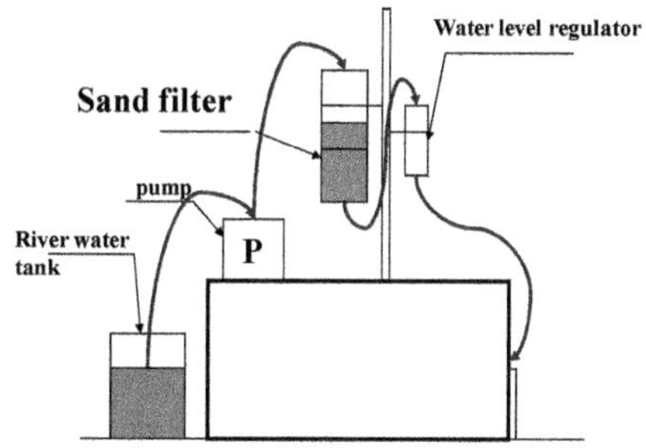

Figure 20.2 Experimental set-up of a slow sand filter.

The concentration of DBPs was determined by high-performance liquid chromatography (HPLC). The conditions for HPLC were as follows: column, XBridge C18 5 mm, 4.6 mm i.d. × 250 mm (Waters, Mass., USA); eluent, methanol/water = 70/30, pH 4.0, adjusted with phosphoric acid; flow rate, 1 ml min^{-1}; detector, Hitachi L-2455 Photo Diode Array Detector; temperature, 40°C. The concentration of bromide ion was determined by ion chromatography (IC). The conditions for IC were as follows: column, Hitachi #2710-SA-IC (Hitachi Co. Tokyo) 4 mm i.d. × 150 mm; eluent, 0.75 mM potassium biphthalate; flow rate, 1 mL min^{-1}; detector HITACHI L-3700 Conductivity Detector; temperature, 40°C. Sodium bromide was used as the standard sample. Dissolved organic carbon (DOC) was analyzed by a TOC analyzer (Shimadzu TOC Analyzer 5000). The semi-empirical molecular orbital calculation using a PM5 Hamiltonian was performed in MOPAC Version 6 with a CAChe package (Fujitsu Co., Tokyo).

Figure 20.3 shows the experimental set-up, and Figure 20.4 shows the spectrum of the UV lamp. The lamp emitted UV light of wavelengths of 200–400nm. The properties of the UV lamp are as follows: Input voltage 120±10 V, electric energy 1500W, Irradiance at 280mm under the lamp UV-25 (240–280nm), UV-31 (280–360nm) and UV-35 (300–390nm), 41.8, 37.5 and 46.2 mW cm^{-2} , respectively. The reaction mixture was 300 mL of initial BP concentration of 20 mg L^{-1} in a glass beaker, and it was continuously mixed with a magnetic stirrer. The initial pH of reaction mixtures was adjusted to 7. Sample water was taken periodically. The distance between the water surface and the UV lamp was kept at 280mm by changing the height of the magnetic stirrer.

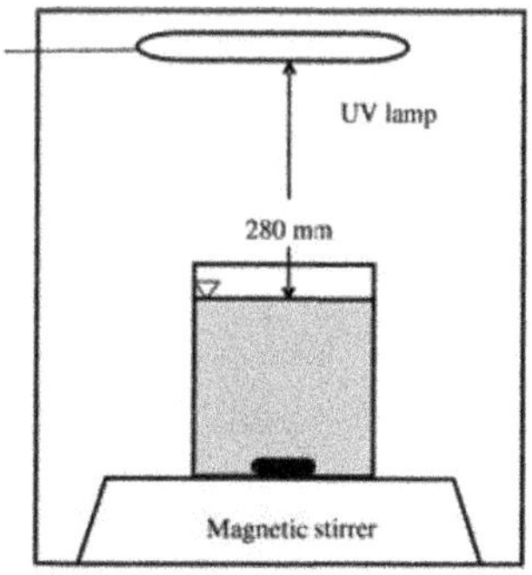

Figure 20.3 Experimental set-up of UV irradiation.

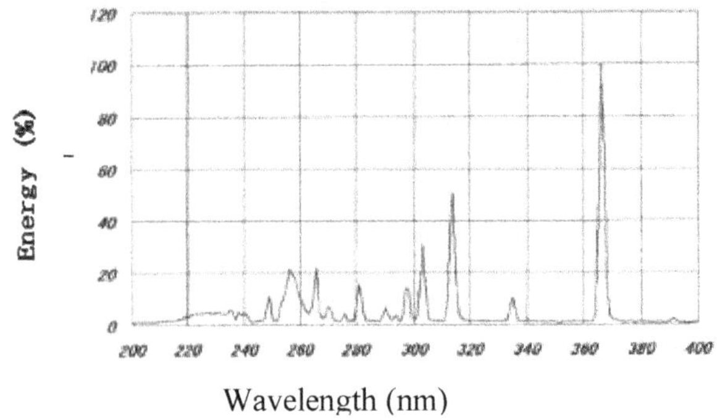

Figure 20.4 The spectrum of the UV lamp.

RESULTS AND DISCUSSION

Figures 20.5 and 20.6 show the removal rates profiles of phenol and catechol, and the removal rates profile of BPs by a slow sand filter during experimental periods, respectively. Table 20.1 summarizes the removal rates of phenol, catechol and BPs during experimental periods.

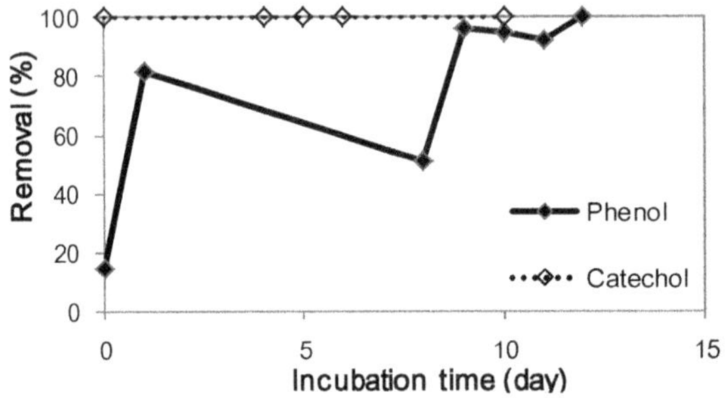

Figure 20.5 Removal rates profile of phenol, catechol by a slow sand filter.

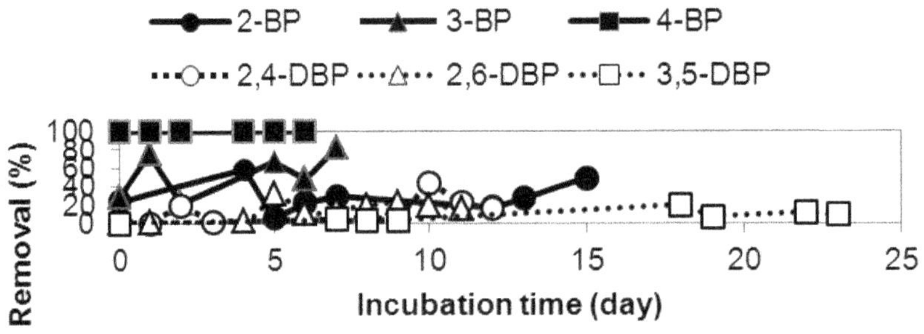

Figure 20.6 Removal rates profile of BPs by a slow sand filter.

Table 20.1 Removal rates of phenol, catechol and bromophenols by a slow sand filter.

Phenols	Removal rate (%)
Phenol	100
Catechol	100
2-bromophenol	<50
3-bromophenol	<80
4-bromophenol	100
2,4-dibromophenol	<30
2,4-dibromophenol	<40
3,5-dibromophenol	<20

Catechol shows full removal during filtration periods. Phenol and 4-BP were removed well also. Removal rates of MBPs were higher than DBPs, suggesting a relationship between the removal rates and the attached numbers of bromine atoms. The removal rate of 2-BP was lower than 3-BP and 4-BP. An attached group next to a hydroxyl group may inhibit phenol hydroxylation because hydroxylation of a phenol ring to make catechol is an early step in microbial aromatic degradation (Cain et al. 1968; Cook & Cain 1974; Katayama-Hirayama et al. 1992; Katayama-Hirayama et al. 1994: Katayama-Hirayama et al. 2009). Catechol showed the highest removability while catechol is the intermediate of phenol degradation (Katayama-Hirayama et al. 1992).

The removal rate of 3,5-DBP was improved to 40% when the initial concentration was lowered to 10μg L^{-1} (graph is not shown). The order of removability of examined phenols was catechol > phenol ≥ 4-BP > 3-BP > 2- BP > 2,4-DBP, 2,6-DBP > 3,5-DBP.

BPs of low removal rates were also examined for removal under ultraviolet lamp (UV) irradiation. The removal of BPs under a UV lamp system was successfully performed in an aqueous solution of BPs.

Table 20.2 summarizes the removal rates (%) of BPs, removal rates (%) of DOC and the released % of bromide ion after 5 and 60 min of UV irradiation.

Decrease of BPs was immediate. After 5 min of UV irradiation, the initial concentrations of 20 mg L^{-1} 2-BP, 3-BP, 4-BP, 2,4-DBP, and 2,6-DBP were 100% removed. Removal of 3,5-DBP of 20 mg L^{-1} was 98.2%. The decrease of DOC in Table 20.2 suggested mineralization of BPs. Bromide ions were released along with the decrease of BPs. The mineralization may occur after release of bromide ions because the decrease of DOC was slower than the release of bromide ions (figure is not shown).

Table 20.2 Removal rates of BPs, removal rates of DOC and released % of bromide ion after 5min and 60 min of UV irradiation.

Bromophenol	Removal % BP		Removal % of DOC		Released % of bromide ion	
	5min	60min	5min	60min	5min	60min
2-BP	100	100	22.2	94.8	52.5	60.0
3-BP	100	100	16.2	97.0	65.0	67.5
4-BP	100	100	9.6	93.0	63.8	70.3
2,4-DBP	100	100	10.5	99.7	42.0	64.8
2,6-DBP	100	100	15.5	96.9	42.9	65.0
3,5-DBP	98.2	99.3	10.5	86.5	41.3	64.6

After 60 min of UV irradiation, removal (%) of DOC was almost 100%. BPs were successfully degraded and mineralized by UV irradiation. Bromide ions released (%) were around 65%. Other bromine (ca. 35%) might exist as molecular bromine and/or hypobromous acid as reported in the literatures (Encyclopedia of Chemistry 1960; von Gunten and Oliveras 1997; von Gunten 2003).

Electrophilic frontier density is a measure of the susceptibility of the substrate to attack by an electrophile such as hydroxyl radical (JSSP, 1983). The UV irradiation process produces ozone from O_2 in air and water. Hydroxyl radical is generated from degradation of ozone by UV as follows (JOA 2009):

$$O_3 + H_2O + h\nu\ (\lambda = 254\ \text{nm}) \rightarrow H_2O_2 + O_2$$
$$H_2O_2 + h\nu\ (\lambda = 254\ \text{nm}) \rightleftarrows 2HO\cdot$$

Electrophilic frontier density reveals reactive sites based on the electron distribution of active orbitals near the HOMO (highest occupied molecular orbital). In nucleophile such like phenols, larger absolute value of HOMO energy indicates lower reactivity (JSSP, 1983).

Table 20.3 summarizes the electrophilic frontier density of the Br atm and HOMO energy. The magnitude of decomposition of BPs by UV may be related to the electrophilic frontier density of Br atom and HOMO energy. The Electrophilic frontier density of Br atom in 2-BP, 3-BP and 4-BP was 0.318, 0.164 and 0.297, respectively. The absolute value of HOMO energy of 2-BP, 3-BP and 4-BP was 9.15, 9.20 and 9.05, respectively. The Electrophilic frontier density of Br atoms in 2,4-DBP, 2,6-DBP, and 3,5-DBP was 0.175 and 0.236, 0.293 and 0.118, and 0.130 and 0.123, respectively. The absolute value of HOMO energy of 2,4-DBP, 2,6-DBP, and 3,5-DBP was 9.15, 9.24 and 9.41, respectively.

Table 20.3 The Electrophilic frontier density of Br and C atoms of DBPs.

Bromophenol	Atom No. of Br	Electrophilic frontier density on Br	HOMO energy (eV)
2-BP	Br7	0.318	-9.15
3-BP	Br7	0.164	-9.20
4-BP	Br7	0.297	-9.05
2,4-DBP	Br7	0.175	-9.15
	Br8	0.236	
2,6-DBP	Br7	0.293	-9.24
	Br8	0.118	
3,5-DBP	Br7	0.130	-9.41
	Br8	0.123	

The degradability of 3,5-DBP and release of bromide ions from 3,5-DBP were a little lower than 2,6-DBP and 2,4-DBP. Molecular orbital calculation suggests that the electrophilic frontier density might be related to the degradability because the density of Br atoms in 3,5-DBP was lower than the other two DBPs. The absolute value of HOMO energy of 3,5-DBP was the largest in three DBPs. Removal of BPs by UV irradiation is partly explained on the basis of electrophilic frontier density of Br atoms of BPs and HOMO energy.

CONCLUSIONS

The main conclusions of this work are:

(1) The order of removability of examined phenols of 20μg L^{-1} was catechol > phenol ≥ 4-BP > 3-BP > 2- BP > 2,4-DBP, 2,6-DBP > 3,5-DBP with a slow sand filter.

(2) An attached group next to a hydroxyl group may inhibit phenol hydroxylation in sand filtration.

(3) Removal rates of MBPs were higher than DBPs, suggesting a relationship between the removal rates and the attached numbers of bromine atoms.

(4) The removal rate of 3,5-DBP by a slow sand filter was improved when the initial concentration was lowered to 10μg L-1.

(5) The removal of BPs was successfully performed in an aqueous solution under a UV lamp.

(6) Removal of BPs by UV irradiation is partly explained on the basis of electrophilic frontier density of Br atoms of BPs and HOMO energy.

REFERENCES

Antignac J-P., Cariou R., Zalko D., Berrebi A., Cravedi J-P., Maume D., Marchand P., Monteau F., Riu A., Andre F. and Le Bizec B. (2009) Exposure assessment of French women and their newborn to brominated flame retardants: Determination of tri- to deca- polybromodiphenylethers (PBDE) in maternal adipose tissue, serum, breast milk and cord serum. *Environmental Pollution*, **157**, 164-173.

Arbeli Z., Ronen Z. and Diaz-Baez M. C. (2006) Reductive dehalogenation of tetrabromobisphenol-A by sediment from a contaminated ephemeral streamed and an enrichment culture. *Chemosphere*, **64**, 1472-1478.

Cain, R. B., Bilton, R. F. & Darrah, J. H. (1968) The metabolism of aromatic acids by micro-organisms. *Biochemical Journal*, **108**, 797–828.

Cook, K. A & Cain, R. B. (1974) Regulation of aromatic metabolism in the fungi: metabolic control of 3-oxoadipate pathway in the yeast *Rhodotorula musilaginosa*. *Journal of General Microbiology*, **85**, 37–50.

CICAD (Concise International Chemical Assessment Document). (2005) 66 2,4,6-TRIBROMOPHENOL AND OTHER SIMPLEBROMINATED PHENOLS, World Health Organization, Geneva.

Encyclopedia of Chemistry. (1960) Editing Committee for Encyclopedia of Chemistry, Kyoritsu Shuppan, Tokyo, 4, 652-654.

Erikson J., Rahm S., Green N., Bergman A. and Jakobson E. (2004) Photochemical transformations of tetrabromobisphenol A and related phenols in water. *Chemosphere*, **54**, 117-226.

Greeke A. C., Giger W., Hartmann P. Heeb N. V. Kohler H.-P. E., Schmid P. Zennegg M. and Kohler M. (2006) Anaerobic degradation of brominated flame retardants in sewage sludge. *Chemosphere*, **64**, 311-317.

Fujitani T., Toda Y., Takahashi H., Yano N., Ando H., Kubo Y., Yuzawa K., Nagasawa A. Ogata A. and Kamimura H. (2006) Placental transfer of tetrabromobisphenol A and excretion to milk in mice. *Annual Report of Tokyo Metropolitan Institute of Public Health*, **57**, 361-365.

Hakk H. and Letche, R. J., Metabolism in the toxicokinetics and fate of brominated flame retardants—a review. (2003) *Environment International*, **29**(6), 801-828.

Iida, H. Yamashita, Y. and Okada, M. (1997) Detection of some bromophenols in the tissues of Mexican Croker, Umbrina coroides and U. rncador, in the gulf of California. *Bulletin of the National Reserch Institute of Fisheries Science,* No.9, 1-10.

JSSP (Japan Scientific Society Press) (1983) Frontier Orbital Theory and Kenichi Fukui, Chemical Review No.38, Japan Chemical Society, Tokyo.

JOA (Japan Ozone Association) (2009) Advanced Oxidation Processes, Ozone Handbook, Tokyo, 96-98. Katayama-Hirayama, K., Tobita, S. & Hirayama, K. (1992) Aromatic degradation in yeast *Rhodotorula rubra. Water Science and Technology,* **26**(3–4), 773–781.

Katayama-Hirayama, K., Tobita, S. & Hirayama, K. (1994) Biodegradation of phenol and monochlorophenols by yeast *Rhodotorula glutinis*. *Water Science and Technology*, **30**(9), 59–66.

Katayama-Hirayama K., S. Arai, T. Kobayashi, H. Matsuda, Z. Luo, M. Tachibana, H. Kaneko, T. Akitsu and K. Hirayama (2009) Removal of bisphenols by slow sand filtration, *Water Science & Technology: Water Supply,* **9**(3),263-268.

Katayama-Hirayama, K., Tobita, S. & Hirayama, K. (1992) Aromatic degradation in yeast *Rhodotorula rubra. Water Science & Technology*. **26**(3–4), 773–781.

Katayama-Hirayama, K., Suzuki, A. Mukaiyama, S. Kaneko, H., Hirayama,K. and Akitsu, T. (2010) Removal of tetrabromobisphenol A by slow sand filtration and a high-performance UV lamp system. *Sustainable Environment Research*, **20**(4), 221-225.

Lilienthal H., Verwer C. M., Van der Ven L. T. M., Piersma A. H., Vos J. G. (2008) Exposure to tetrabromobisphenol A (TBBPA) in Wistar rats: Neurobehavioral effects in offspring from a one-generation reproduction study. *Toxicology*, **246**(1), 45-54.

Meerts I. A. T. M., van Zenden J. J., Luijks E. A. C., van Leeuwen-Bol I. Marsh G. Jakobson E., Bergman A. and Brouwer A. (2000) Potent competitive interactions of some brominated flame retardants and related compounds with human transthyration in vitro. *Toxicological Science*, **56**, 95-104.

Oeberg K., Warman K. and Oeberg T., (2002) Distribution and levels of brominated flame retardants in sewage sludge. *Chemosphere*, **48**, 805-809.

Ogunbayo O. A., Lai P. F., Connolly T. J., Michelangeli F. (2008) Tetrabromobisphenol A (TBBPA), induces cell death in TM4 Sertoli cells by modulating Ca2+ transport proteins and causing disregulation of Ca2+ homeostasis. Toxicology in Vitro, 22(4), 943-952.

Ronen Z. and Abeliovich A., (2000) Anaerobic-Aerobic Process for Microbial Degradation of Tetrabromobisphenol A. Applied and Environmental Microbiology, 66(6), 2372-2377.

von Gunten Urs and Oliveras, Y. (1997). Kinetics of the reaction between hydrogenperoxyde and hypobromous acid: Implication of water treatment and natural systems, Water Research, 31(4), 900-906.

von Gunten Urs (2003). Ozonation of drinking water: Part II. Disinfection and by-product formation in presence of bromide, iodide or chlorine, *Water Research* **37** (7), 1469–1487.

WHO (World Health Organization) Collaborating Centre (1987). Slow Sand Filtration for Community Water Supply, The Hague, The Netherlands.

WHO (World Health Organization) (1995) In: Van Esch GJ, ed., Tetrabromobisphenol A and Derivatives. Geneva: WHO International Programme on Chemical Safety, Environmental Health Criteria, 172, 25–47.

Zalko D., Prouillac C., Riu A., Perde E., Dolo L., Jouanin I., Canlet C., Debrauwer L. and Cravedi J.-P. (2006) Biotransformation of the flame retardant terrabromo-bisphenol A by human and rat sub-cellular liver fractions. *Chemosphere*, **64**, 318-327.

Chapter 21

The iron and manganese removal process: Its microbial habitats and functions

*H. Kaya**,**, J. Takeuchi*,****
**NPO member,Community Centre for Water Supply Support, Tokyo, Japan*
***Daiichi Consultants and Technology Co. Ltd., Hiroshima 731-1142 Japan*
****Department of Civil and Environmental Engineering, Kure National College of Technology, Hiroshima 737-8506 Japan*

ABSTRACT

The biological iron/manganese removal process is a variant of the slow sand filtration (SSF) process; both mimic the natural ecosystems, thereby employing the naturally-occurring microorganisms without any chemicals and all the more friendly to the environment, eventually leading to the sustainable lifestyle. We have focused on possible microbial seeds and the resulting colonization, comprehensively to compare environmental conditions between natural and artificial habitats, including the other relevant implications. The seeds of Fe/Mn oxidizers could be originated from the groundwater, whilst other co-existing organisms may be accidentally transported from different places, most likely via an airborne route.

Keywords Iron/manganese (Fe/Mn) removal; consortium; water purification facilities; diapauses; natural ecosystems

INTRODUCTION

The slow sand filtration (SSF) process is one of the applied techniques to produce drinking water, alongside the activated sludge process to treat wastewater, both using a part of natural ecosystems' components in the human-built facilities, thereby a wide variety of naturally-occurring organisms are involved in. However, the iron and manganese (Fe/Mn) removal process basically employs the similar process but uses limited, specialised consortia of microorganisms capable of metabolizing inorganic substances of a groundwater origin. We would therefore like to discuss the Fe/Mn removal process and the related issues in natural habitats, bridging the mainstream and the surrounding ecological processes including a possible source of microorganisms and the relevant interference and/or association with other organisms sharing the same ecological niche.

METHODOLOGY

Field surveys were carried out mainly in Hiroshima Prefecture, ranging from on-site underground water purification facilities. The operational parameters of the facilities were monitored and the water samples were collected by ourselves and officially analysed by a company using the approved methods.

Apart from the indoor facilities, water samples were collected from a natural lake named "Budo Ike" located on campus of Hiroshima University (Higashi-Hiroshima, Japan). Rain water dropped from the canopy was collected using a trap set above the floor of a coastal forest in Kure, Hiroshima. A special specimen of the Fe-containing deposits was also collected from a rain drain pipe at Kure Kosen (College).

The collected samples were observed using a phase contrast microscope BX-2700TPH, and the images were captured with a WrayCam G130 (Wraymer, Osaka). Underwater photos were taken using an Olympus's waterproof digital camera "μTough" to describe the intact of biofilms formed on the sand beds.

RESULTS AND DISCUSSION

Water purification equipment. Table 21.1 shows typical cases of different conditions in terms of the sizes, and Fe/Mn contents, and purposes for use of shallow wells as raw water. The site F was once abandoned, and finally replaced by the public water supply. The rest two sites have still been under operation.

Table 21.1 Specifications and the removal rate of each equipment.

Study site	Purposes	Fil. areas	Flow rate *2	Mn, initial	Fe, initial	Removal *3
Site F	Food indust.	$3m^2$	24 m/day	0.94 mg/L	0.18 mg/L	>99.9%
Site N	Care facility	$4m^2$	22 m/day	1.30 mg/L	0.0 mg/L	>99.9%
Site K	Public hall	$0.5m^2$ *1	29 m/day	0.29 mg/L	4.5 mg/L	>99.9%

*1, two step -wise reactors ; *2. calculated figures, as each site runs intermittently per a day;
*3, the rate was calculated by both Fe and Mn removal.

The raw water samples contain various iron (0-4.5 mg/L) and manganese (0.29-1.3 mg/L) concentrations, but the final removal rates are all more than 99.9%; thus satisfactory levels.

The two step-wise reactors (site K) indicated the first tank acted for iron (Fe) removal, whereas the second one acted for polishing up as well as manganese (Mn) removal, showing dark granules distributed on the sand surface as evidence of manganese deposits (Figure 21.1). The equipment was designed to remove the specific solutes of high Fe/Mn content derived from groundwater. Thus, the whole equipment is to be shaded or partly darkened; this is useful for the microbial communities to reduce their biological productivity and diversity, leading to predominance of the enriched Fe/Mn-removing bacterial groups.

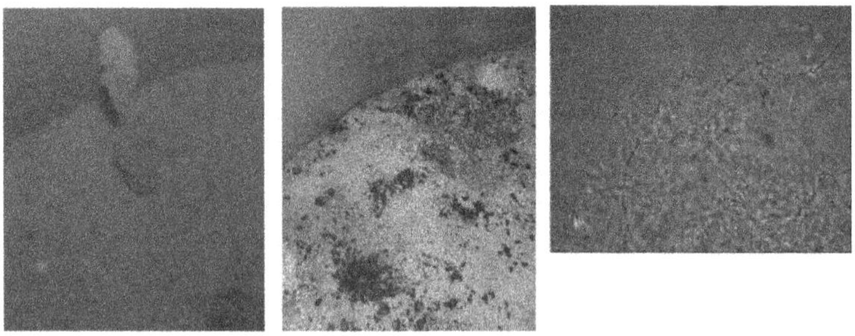

Figure 21.1 Submerged sand beds, showing the 1st reactor (left) and the 2nd reactor (centre) and the *Gallionella* type of iron oxidising bacteria (right) found in the 1st reactor.

Hiroshima University "Lake Budo" is located in the centre of the University's campus. Two types of iron bacterial colonies were observed (Figure 21.2), whether the iron-rust deposits on the lake bottom of the running water or oil-like membrane flouting on the surface of the stagnant water. These are typical natural habitat status of the *Leptothrix* group of iron oxidising bacteria (Figure 21.3) commonly found in such a natural habitat, implying a possibility of inflowing ground water of higher iron contents.

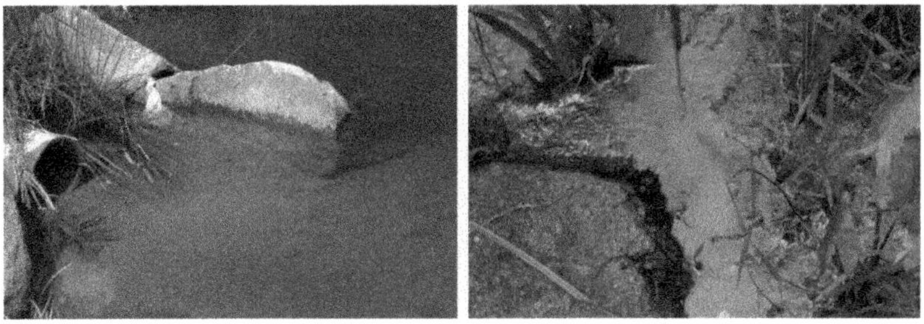

Figure 21.2 Two growth modes of iron bacteria deposited (left) or forming an oily membrane (right).

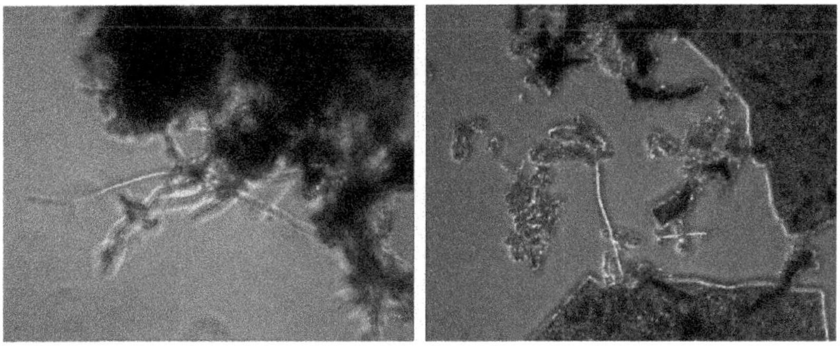

Figure 21.3 The *Leptothrix* type of iron bacteria in a lake at Hiroshima Univ. with the rusty deposits.

Kure College An ecosystem was found on campus; the lower end of a rain drain pipe (See Figure 21.4).

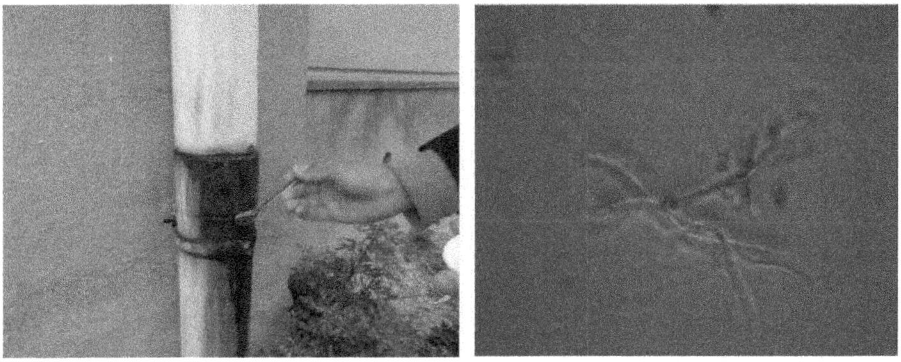

Figure 21.4 A rusty rain drain pipe (left) and the associated iron-oxidizing bacteria (right).

Based on microscopic examination, we found iron bacteria and the other microbes such as protozoa and rotifers. In addition, we were surprised at knowing that there exists a large *Hypsibius* (water bears) population, forming an iron-depending ecosystem, being equivalent to the sulphur-depending ecosystems often reported from hot springs and the associated black smoky chimney pipes on the deep seafloors.

Figure 21.5 shows a possible mechanism of establishing the iron-depending ecosystem; this would strongly suggest that microbial seeds could be widely distributed everywhere in a dormant state and possibly colonize onto any suitable place. Water bears and other related protists are known for their strong resistance against the fatal UV irradiation and drying as well; these factors enable them to survive longer during travelling in the air, often being "diapauses" status.

Daphnia, nematode, rotifers were such representative invertebrates as being revived from the rain water harvested under the canopy of a forest. These results suggest that they can transported from a place to another in the air, and they could accidentally come to a water purification facility, always searching for a niche. Thus, we should be ware of accidental contamination into the Fe-Mn removal reactors, where autotrophic iron bacteria act as a substitute for the primary producers, usually photosynthetic algae, to provide the other heterotrophic consumers with organic matter.

Most of microorganisms responsible for SSFs, including the Fe/Mn removal, are also cosmopolitan members widely distributed on the earth, thus those are available from the air without any cost, being environmentally sustainable.

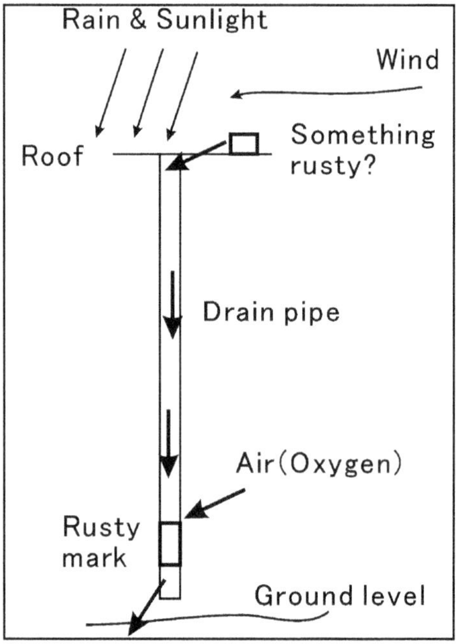

Figure 21.5 The iron-depending ecosystem, showing a possible scenario at a Kure case.

CONCLUSIONS

Biological iron and manganese removal is a variant of slow sand filters, employing more specific microbes of the environmental origin. However, some of the aquatic invertebrates may invade the water purification system, and they could occupy such a special niche, via the air as a result of their unexpected happening due to their diapauses.

ACKNOWLEDGEMENTS

The author (JT) wishes to express his thanks to Dr. Norio Shimizu, Hiroshima University Museum, to guide the nature reserve area along the walking trail when collecting the iron bacteria-related samples. Mr. Yuki Ito, a student at Kure Kosen, informed JT his findings on the sink of iron-oxidizing bacteria.

REFERENCE

Alekseev, V.R., De Stasio, B., Gilbert, J.J. (2010) *Diapause in Aquatic Invertebrates: Theory and Human Use*, Springer, The Netherlands, 257pp.

Chapter 22

Biological iron removal from community water source - Corriverton, Republic of Guyana

H. Nakazono, P. Thapa*, A. Takechi*, S. Barran***

**Technical Group, TEC International Co., Ltd., Japan, 7-1 Kasumigaseki 3-Chome Chiyoda-ku Tokyo Japan*

***Guyana Water Incorporated, Guyana*

ABSTRACT

After completion of a project with grant aid from JICA in 2010 quality of supplied water has significantly improved in Corriverton, Republic of Guyana. Prior to this project, Corriverton used to be supplied raw water directly from deep wells without any disinfection. The raw water had iron concentrations exceeding the WHO standard and faecal coliforms were often detected in it. To address these problems a slow sand filter was constructed by the project which provided treatment for both excess iron and biological contamination in raw water. Average iron removal rate in the newly-constructed WTP is 88%, and the average iron concentration of treated water is 0.23mg/L, compared with 1.87mg/L of raw water during February to July 2013. Frequency of sand scraping is every 2 to 3 weeks, and the filters resume operation after a few hours of scraping due to fast breeding of iron bacteria. In this project, the slow sand filter has been successfully used for removing iron with iron bacteria in addition to its conventional purification capacity for other physical and biological contaminants. Due to easy operation and maintenance of this filter, mainly consisting of sand scraping, which is carried out adequately,, the customers actually felt the benefit of improvement in water quality.

Keywords biological iron removal; slow sand filter; ground water

BACKGROUND

The Co-operative Republic of Guyana (hereinafter referred to as Guyana) is situated in the northeast part of the South American Continent facing the Atlantic Ocean with population of 795 thousands (Bureau of Statistics Guyana, Population and Housing Census 2012) and area of 215 thousands km2. In Guyana, as most of population is concentrated along the coastal line, 90% of water sources rely on groundwater from rich aquifers of the coastal plain while only 10% rely on the surface water.

Development of the water supply system in Guyana is rather easy by digging wells in populated areas and laying down the distribution pipes along the trunk roads within a few kilo meter width. Water supply coverage has already reached to 83% in national average and nearly 90% (as of 2006) in the coastal areas. However, it has problems in quality of the supplied water, such as the risk of the bacterial contamination and unacceptability due to unpleasant taste, odor and coloring by high concentration of iron.

The Government of Guyana adopted in 2001 the National Development Strategy 2001 – 2010 and Poverty Reduction Strategy Paper (PRSP) as a national development policy. The median term plan of PRSP included the promotion of water treatment in the coastal areas, of which Corriverton is one of the prioritized areas. (IMF, Guyana Poverty Reduction Strategy Paper, 2001)

Under the circumstances, the project was launched to develop water supply facility, including water treatment plants (WTPs), in order to improve water quality and supply situation in Corriverton.

This paper introduces a case of slow sand filter WTP designed to remove iron biologically, and discusses the treatment performance as well as operation and maintenance (O&M).

TARGET AREA

The population of Corriverton was 30,942 people in 2002 (Bureau of Statistics Guyana, Population and Housing Census 2002). The water supply service in Corriverton is operated by Guyana Water Incorporated (GWI), which was established as a consolidated water supply service company in Guyana, and the water supply coverage has almost reached 100% in Corriverton.

Prior to construction of WTPs in 2010 with grant aid support from the Government of Japan, Corriverton used to be supplied raw water directly from 8 deep wells without any disinfection. The characteristics of water quality were as follows.

i) As it was confirmed that iron concentrations from 8 existing wells in targeted area exceeded the WHO standard (less than 0.3mg/L) and the faecal coliform were often detected from 7 wells, WTP(s) for the iron removal and disinfection are required.
ii) To consider the installation of slow sand filtration with iron bacteria as an iron removal method, a simple test to check the existence of iron bacteria in raw water was conducted, and their existence was confirmed. Detail of this test is shown in next section.

In addition, there are 2 existing WTPs of GWI operation in New Amsterdam and Rose Hall around Corrivertion. The present situations of the facilities and O&M are shown as follows.

i) The existing 2 WTPs employ rapid sand filtration with pre-treatment for pH control. In addition, as an iron removal method, New Amsterdam WTP employs air oxidation by cascade method, and Rose Hall WTP employs air oxidation by compressor.
ii) There are not any pH meters, no jar testers, no coagulant and no pH control chemicals in existing 2 WTPs.

IRON BACTERIA

(1) Simple Test for Iron Bacteria

The test was carried out by setting up a simplified slow sand filter. After setting up, raw well water was introduced to filter continuously, and the surface conditions of the sand media was observed visually. After 3, 4 days, the sand media turned to yellow or brown. After 1 week, sand surface was scraped as a sample and brought to Japan for the microscopic observation.

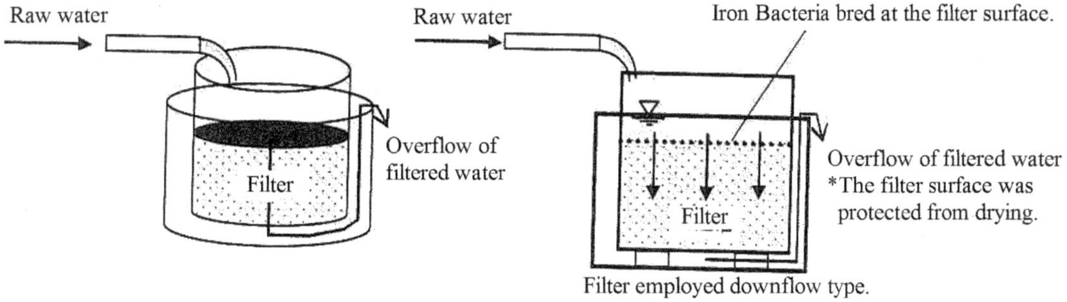

Figure 22.1 Simplified Slow Sand Filter for Iron Bacteria Test

As the result of the microscopic observation, the following 3 kinds of iron bacteria were identified.
Siderocapsa: dominant species
cf. Leptothrix sp.:
cf. GAllionella ferruginea:

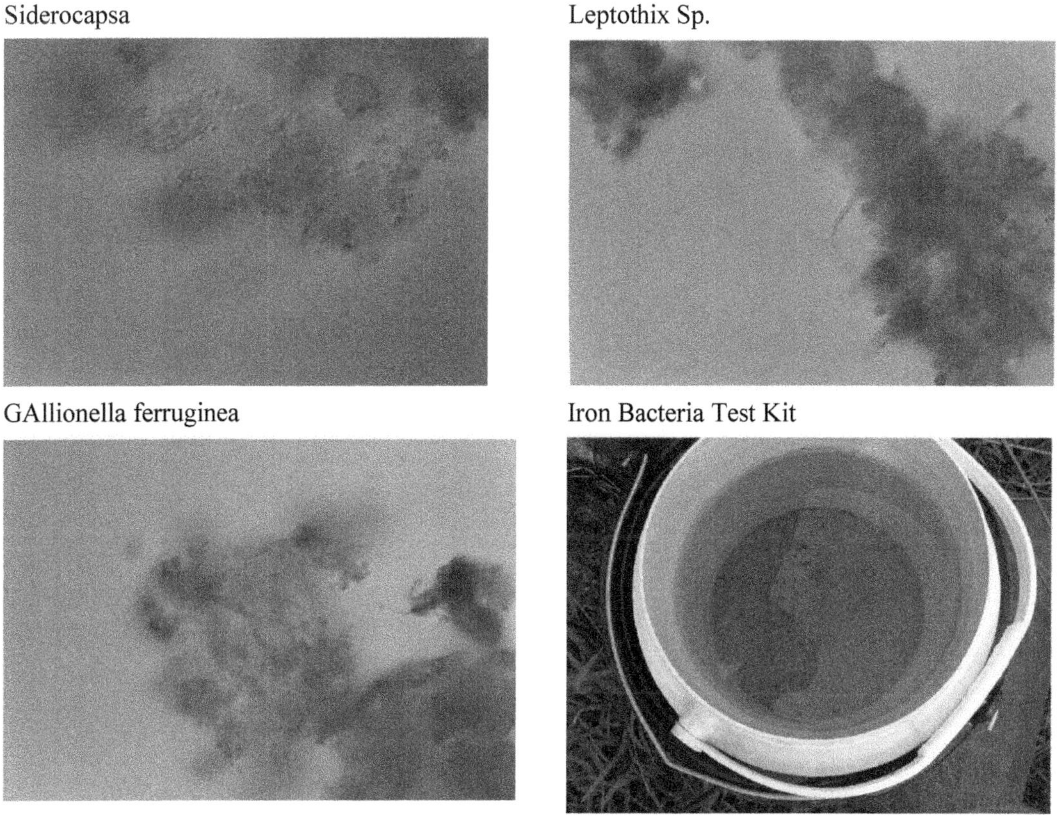

Figure 22.2 Identified iron bacteria, together with iron bacteria test kit.

(2) Variety of Iron bacteria

It is described that iron bacteria reacts with ferrous iron ($Fe2+$) through an oxidation process to form ferric hydroxide ($Fe(OH)3$) and has functions to store and precipitate. According to standard method for the examination of water, iron bacteria are classified as follows in terms of morphometric characteristic (JWWA, 2011).

i) filamentous form
* Toxothrix
* Leptothrix
* Crenothrix
* Clonothrix
* Gallionella

ii) flocculated form
* Siderocapsa
* Siderococcus
* Siderocystis

iii) special form
* Planktomyces

As reported by Dr. Kojima, Siderocapsa is minim coccoid, which is just 0.2 to 1.0 μ diameter. The smaller bacteria are, the larger surface area is. Therefore, Siderocapsa is more efficient in iron removal (Dr. Kojima, 1972). The dominant species in Corriverton is the genus Siderocapsa, that is why, it is good condition to produce the slow sand filtration with iron bacteria in Corriverton.

DESIGN CONDITION

(1) Slow sand filtration with iron bacteria

In consideration of present O&M in New Amsterdam and Rose Hall WTPs and the results of simple test for the presence of iron bacteria, the slow sand filtration with iron bacteria was adopted for new treatment method.

As GWI had no experience to operate and maintain slow sand filtration basin, technical transfer was required for plant operation including iron bacteria control.

(2) Facility system

In consideration of well performances, site of WTP (dimension, access, ground condition, land acquisition, environmental impact) and future service population, the facility system was selected as follows.

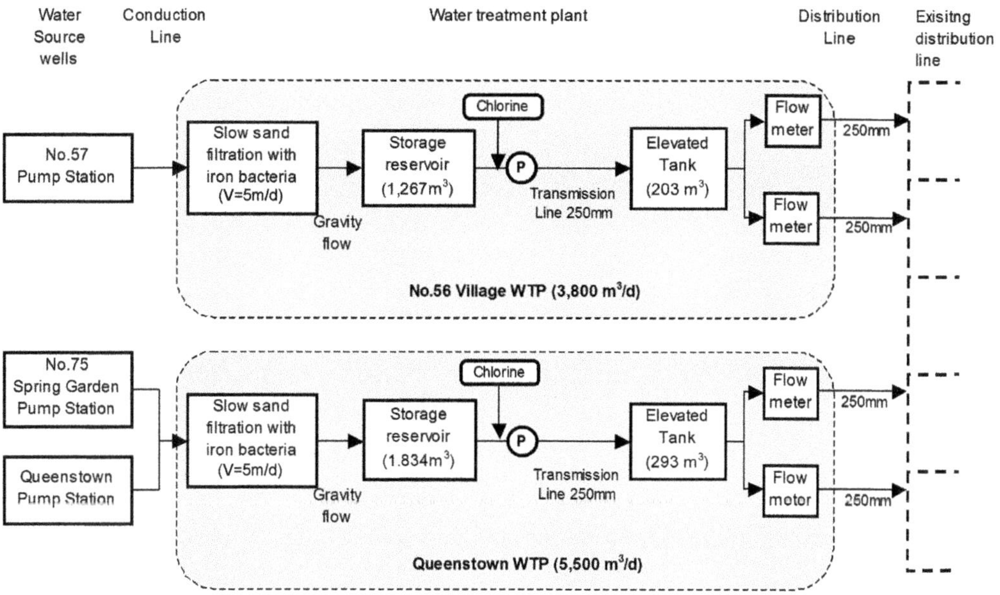

Figure 22.3 Facility System of the Project

RESULTS

No.56 Village WTP and Queenstown WTP completed to be constructed on March 2008 and October 2010. The results of present O&M in No.56 Village WTP are described here based on recent data in 2013.

(1) Water Quality

As the result, iron removal rate in No.57 Village WTP is 88%, and the average iron concentration of treated water is 0.23mg/L, compared with 1.87mg/L of raw water during February to July 2013. In addition, technical transfer of O&M for slow sand filtration with iron bacteria was intensively carried out through seminar, manual and on-the-job training in the project. Therefore, the filtration basins have been operating well.

Table 22.1 Water Quality Data in No.56 Village WTP

Date	pH		Turbidity (NTU)		Iron (mg/L)	
	Raw water	Treated water	Raw water	Treated water	Raw water	Treated water
WHO drinking water guidelines	-	-	-	5.00	-	0.30
26/Feb./2013	5.98	ND	10.9	ND	1.61	ND
19/Mar./2013	5.76	ND	14.0	ND	2.02	ND
7/May/2013	6.01	ND	9.2	ND	1.98	ND
4/Jun./2013	5.83	6.61	20.3	2.90	ND	0.39
18/Jun./2013	5.77	6.61	14.7	0.60	ND	0.13
3/Jul./2013	5.64	6.34	12.6	1.30	ND	0.18
17/Jul./2013	5.77	6.22	20.1	0.90	ND	ND
Average	5.82	6.45	14.5	1.43	1.87	0.23

(2) Sand Scraping

Frequency of sand scraping was carried out every 2 to 4 weeks during January to August 2013, as shown in Table 22.2. As reported by Dr. Kojima, iron bacteria layer should be scraped out every few days to few weeks, because the thickness of the layer reached 15cm on the sand surface by fast breeding of iron bacteria and water was not fully filtered due to increasing in filtration resistance (Dr. Kojima, 1972).

And, the filters resume the operation after a few hours of scraping due to fast breeding of iron bacteria in Corriverton. As reported by Dr. Kojima, the growth rate of iron bacteria is very quick under favorable conditions. In case of slow sand filtration basin of paper-manufacturing company in Takamatsu city, after scraping out the sand surface, iron bacteria bred in just few hours, as if red-brown cotton was spread on the sand (Dr. Kojima, 1972).

Table 22.2 Frequency of Sand Scraping in No.56 Village WTP (January to August 2013)

Filter No.			Jan.				Feb.				Mar.					Apr.				May				Jun.					Jul.					Aug.	
1				x			x			x			x			x			x				x					x			x				
2	x				x				x			x			x			x			x			x					x					x	
3		x						x			x		x				x			x					x		x								x

*Check-mark is placed in the week of sand scraping.

TOWARD SUSTAINABLE OPERATION

As shown in Figure 22.4, water pressure was poor, that is 5 m or less, in 73.2% of service blocks, before project implementation. But, after the project, it was achieved to supply treated water to all customers by adequate water pressure; more than10 m.

In addition, the leakage reduction project, funded by DFID, was assisted in this project to provide 8,400 water-meters for GWI. GWI installed the water-meters and established the structure of leakage monitoring and repair, in parallel with applying to metered tariff system. Finally, GWI achieved the reduction in leakage volume, and GWI has been creating a virtuous cycle for sustainability of water supply operation.

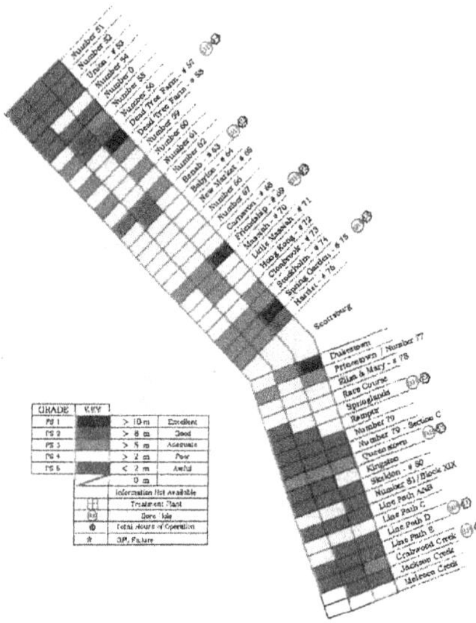

Figure 22.4 Water pressure distribution in Corriverton before the project

CONCLUSION

In this project, slow sand filter with iron bacteria was developed, and iron could be removed. And, GWI continues to carry out adequate operation and maintenance including top sand scraping of slow sand filtration basin as required. Consequently, the customers actually felt the benefit of improvement inwater quality.

REFERENCES

Bureau of Statistics Guyana, The Co-operative Republic of Guyana, Population and Housing Census 2012

Bureau of Statistics Guyana, The Co-operative Republic of Guyana, Population and Housing Census 2002

IMF, Guyana Poverty Reduction Strategy Paper, 2001

Dr. Kojima, Iron Removal by Iron Bacteria, Journal of Water and Waste, Vol.3 No.10, pp45-52, 1961

Dr. Kojima, Iron and Manganese Removal Treatment by Bacteria, Journal of Water and Waste, Vol.14 No.6, pp59-65, 1972

JWWA, Standard Method for the Examination of Water, 2011

Chapter 23

Case studies of slow sand filtration/ecological purification system in small scale water supply

K. Sanda, M. Seno***

**Sankyo Consultant Corporation, 4-122, Hanajiri-midorimachi, Kita-ku, Okayama 701-0131, Japan*

***Community Water Supply Support Center of Japan, 655-96, Hirano, Kita-ku, Okayama 701-0151, Japan*

ABSTRACT

We developed five small water treatment plants in rural area, Okayama, Japan. Capacity of water demand in each plant ranged 10 to 20 m^3/day. Upper limit of a burden per family was 1,200 thousand yen. Filtration rate of upflow roughing filters was set to 12 to 20 m/day. Filtration rate of slow sand filter was set to 4 to 5 m/day. All facilities achieved the quality of drinking water criteria of Japan.

Keywords slow sand filter, pre-treatment, roughing filter, small scale water

INTRODUCTION

There are many small scale water supply systems in Japan. The system is only piping system which raw water is derived from a surface water of a river or ground water. Non treated water is directly supplied or the water is only disinfected by chlorine in such system. These systems have been maintained by the rural people. When heavy rain attacks in the catchment area, the supply water from a river becomes easily turbid water. Then people cannot use this water. This is a common serious problem in rural area where the people does not receive any authorized water service in Japan.

We developed five small water treatment plants in such rural area. We made a plan to construct slow sand filter plant in cooperation with local residents and local governments. We contributed fully as a consultant from the start-up of the projects; conception, planning, design, construction, construction management, and consensus among local governments and residents.

Our experiences are useful information for planners who intend to develop a new small scale water system in a rural area.

OUTLINE OF THE PROJECTS

The outline of the projects is summarized in Table 23.1. Capacity of water demand in each plant ranged 10 to 20 m^3/day. Upper limit of a burden per family was 1,200 thousand yen. Okutsugawa plant was a remarkable small burden per family than other plants. This plant was able to use the existing facilities. Fundamental items of the plants are shown in Table 23.2.

Table 23.1 Outline of the Projects

	Unit	Miyama	Okutsu-gawa	Sagari kaya	Kamiyokono kami-No.1	Kamiyokono kami-No.2	Subtotal
Population	Person	73	90	41	57	56	317
Household Number	Family	17	45	12	15	15	104
Supply Capacity	m^3/day	20.0	20.0	10.3	14.3	14.0	78.6
Total Project Cost	10^3 yen	50,890	25,904	47,805	41,504	62,188	228,291
Subsidy	10^3 yen	30,459	14,813	33,405	23,504	44,188	146,369
Local Burden	10^3 yen	20,400	11,091	14,400	18,000	18,000	81,891
TPC per Family	10^3 yen / family	2,994	576	3,984	2,767	4,146	2,195
Burden per Family	10^3 yen / family	1,200	246	1,200	1,200	1,200	787

Table 23.2 Fundamental Items of the Treatment Facilities

Water Source	Surface water of a high land river
Expected Water Quality	Acceptable level of the criteria of water supply act in Japan
Treatment Flow	Intake → Sedimentation → 3 Steps of upflow roughing filter → Slow sand filter → Distribution tank

SPECIFICATIONS OF FACILITIES

Specifications of each facility are shown in Table 23.3. The design of RF referred to Wegelin (1996). The design of other facilities was based on Water Facilities Design Guideline of Japan Water Works Association. Filtration rate of upflow roughing filters were set to 12 to 20 m / day. Filtration rate of slow sand filters were set to 4.0 to 5.0 m / day. For an example, general layout and cross-section of Sagarikaya are shown in Figure 23.1.

Table 23.3 Specifications of Facilities

	unit	Miyama	Okutsugawa	Sagarikaya	Kamiyokonokami No.1	Kamiyokonokami No.2
Sedimentation Pond Retention time	hour	24	8	3	6	6
Roughing Filter rate	m / day	14.4	19.2	12.0	12.0	12.0
Slow Sand Filter rate	m / day	4.0	4.0	5.0	4.0	4.0

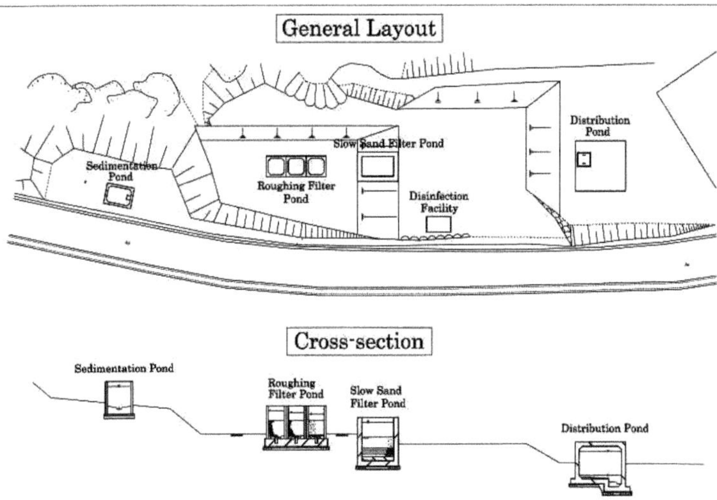

Figure 23.1 General Layout and Cross-Section of Sagarikaya (for an example)

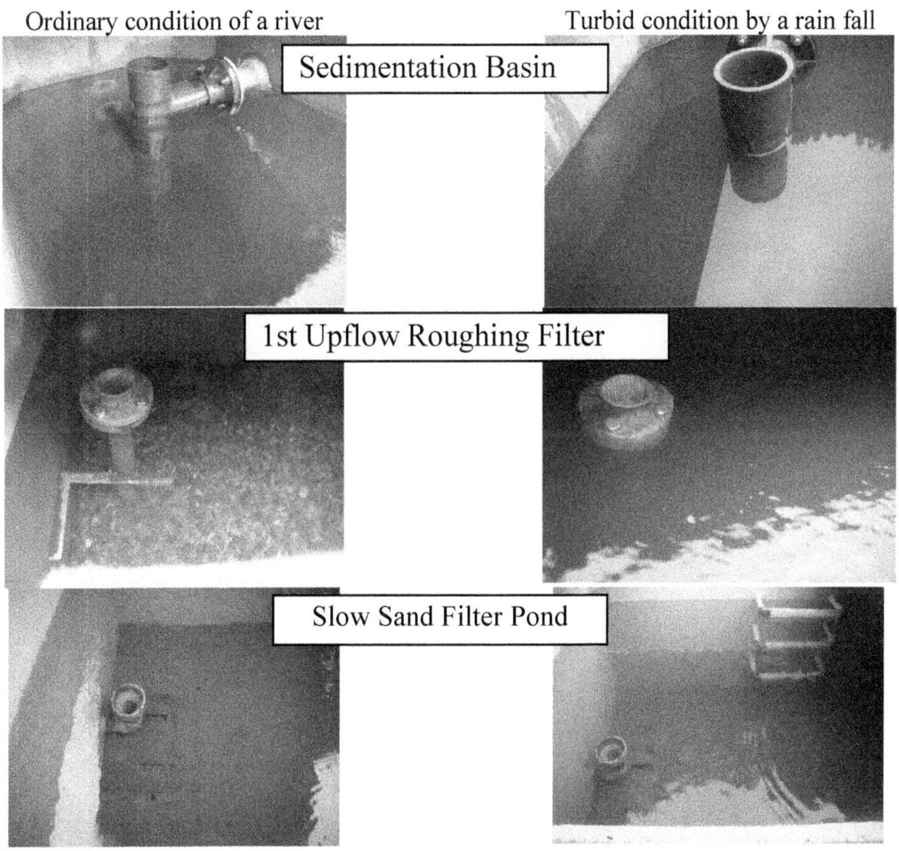

WATER QUALITY

Main items of water quality of five facilities are shown in Table 23.4.

Table 23.4 Main Items of Water Quality in Five Facilities

		General bacteria	E. Coli	TOC	Color	Turbidity
Unit		n / ml	-	mg/L	degree	degree
Drinking Water Quality Standards in Japan		<=100	Negative	<=3mg/L	<=5	<=2 degree
Miyama	Inlet of Sedimentation Tank	21	Positive	0.8	3.3	0.7
From Sept. 2010.	Outlet of SSF	0	Negative	0.7	< 1	0.1
Okutsugawa	Inlet of Sedimentation Tank	110	Positive	1.4	9.3	4.2
From March 2012.	Outlet of SSF	0	Negative	0.8	1.3	0.1
Sagarikaya	Inlet of Sedimentation Tank	530	Positive	0.5	3.3	2.1
From May 2012.	Outlet of SSF	0	Negative	1.1	1	0.2
Kamiyokonokami No.1	Inlet of Sedimentation Tank	12	Positive	1.1	2.8	1.8
From Nov 2013.	Outlet of SSF	12	Negative	0.8	0.7	0.2
Kamiyokonokami No.2	Inlet of Sedimentation Tank	12	Positive	1.1	2.8	1.8
From Nov 2013.	Outlet of SSF	2	Negative	1.2	0.6	0.2

CONCLUSIONS

The main conclusions of this work are:

(1) Upper limit of a burden per family was 1,200 thousand yen, and the local government bore the excess money over the burden.

(2) The newly built WTP was composed with a sedimentation pond, 3 steps of upflow roughing filter and a slow sand filter (ecological purification system).

(3) Filtration rate of URF was from 12 to 20 m/day, and the rate of SSF (EPS) was from 4 to 5 m/day.

(4) The quality of all of 3 operating facilities meets the criteria of Japanese drinking water.

(5) Upflow roughing filters can be removed high turbidity enough for the slow sand filters.

(6) Slow sand filter had not clogged and it runs without any scraping the filter sand over the years after completion the plant.

REFERENCES

Wegelin, M. (1996) Surface Water Treatment by Roughing Filters., A Design, Contracture and Operation Manual, p IX-10, ISBN: 3-908001-67-6

Japan Water Works Association, Water Facilities Design Guideline

Chapter 24
Water supply of a small community with a small slow sand filtration/ecological purification system unit

F. Kazaki
NPO Oita Water and Life Association Japan

ABSTRACT

Many people live to where there is no public water services.
There are many small villages can not be used to clean water.
Profitability is not taken population is small, the management of water services is impossible.
It is the same even in the privatization even in public.
So they have to operate the water on their own.
We focused on Slow sand filtration/ecological purification.
And, so that anyone can build a water treatment plant easily, we have devised a system unit that is made of FRP.
Then, in a region that contains iron and manganese in the water, we are doing support residents to clean the water.

Keywords Slow sand filtration/ecological purification; iron; manganese; system unit

INTRODUCTION

In order to contribute to ensuring the "clean water" in a small community located in mountainous areas where a public water supply is not in place yet, NPO "Oita Water and Life Association" have proposed a small-scale water supply suited to the local circumstances, We will help with the maintenance and management.

This paper will introduce the case of small-scale water supply in District K of Bungotakada, Oita Prefecture.

DISTRICT K OVERVIEW

District K is a small village. There are 14 households in the village, most of the residents are elderly people. The village is located near the center of Kunisaki Peninsula.

In the vicinity of this village, there is a volcano. This region has volcanic topography and much of the ground is covered with rock. So in this region, water resources, such as shallow groundwater and surface water is poor. On the other hand, the deep groundwater is relatively plentiful, but it contains a lot of iron and manganese. Water (deep well) co-wells that have been used in the area also contained a lot of iron and manganese. It was murky, black brown water. It was an inconvenient situation for residents, not only for drinking water, but for bathroom and laundry too.

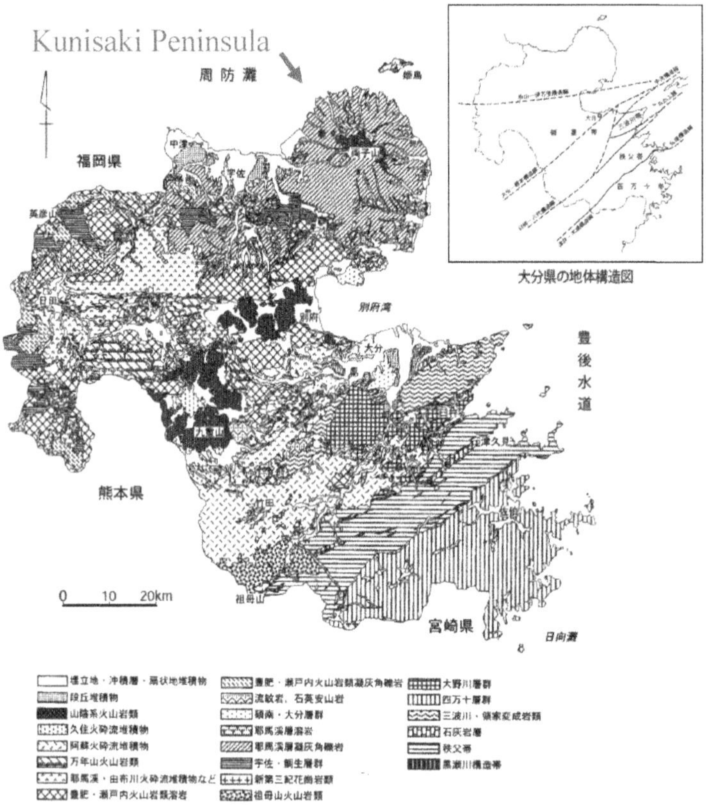

Figure 24.1 The volcanicity topography of Oita

REGIONAL CHARACTERISTICS, AND NEEDS OF DISTRICT RESIDENTS

- Source of income for most of the elderly people living in the region is the pension. Therefore, There is no burden of facility development. (U.S. $ 1,000 per household is the limit)
- Maintenance tasks, such as "hard work", "dealing with chemicals", "complex tasks", is difficult for aging region.
- On the other hand, older people are relaxing at home, enjoying a life of retirement. So, if maintenance work is simple and light work, it is possible for them to do it, even if there is a high frequency of maintenance
- Water quality of water sources currently in use are good, except for manganese and iron.

PROPOSED MEASURES FOR IMPROVEMENT, VALIDATION AND EXPERIMENT

In this district, in order to ensure clean water, we have proposed adding a slow sand filtration on the current facility. The benefits are as follows.

- It is simple; it is a water purification reliable method
- It requires frequent maintenance (scraping of sand due to filtration blockage), but it is light work, and suited to the elderly.
- Does not use chemicals to clean the water.
- Construction costs and maintenance costs are low.

In order to realize this proposal, cost reduction, and easy maintenance, we experimented with prototype units container made of FRP. (L: W: H = 1.0 × 0.5 × 1.0 (m))

The results; the ability to remove iron and manganese was good. Maintenance of the prototype testing was also high.

Therefore, it was decided to develop a real machine based on this.

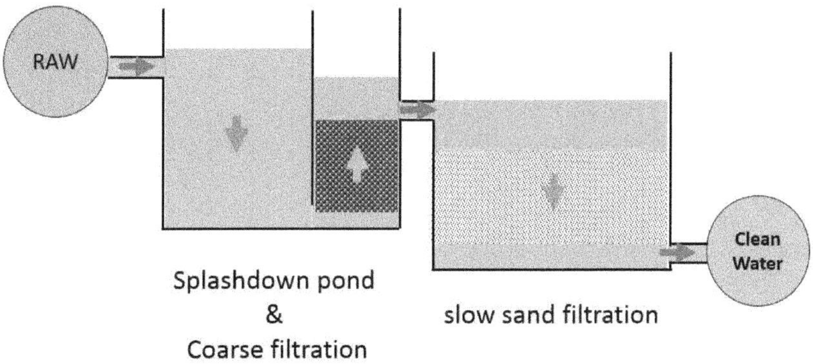

Figure 24.2 Experimental device

Table 24.1 The ability to remove iron and manganese

	Raw	Clean	unit	standard
general bacteria	0	16	CFU/ml	100≧
E. coli bacteria	N.D.	N.D.		N.D.
Zinc	0.1	0.05>	mg/l	1.0≧
Iron	1.5	0.03>	mg/l	0.3≧
Manganese	0.24	0.005>	mg/l	0.05≧
PH	7.2	7.6		5. 8−8. 6
Colour	14	0.5>		5≧
Turbidity	4.6	0.1>		2≧

FULL-SCALE INTRODUCTION

Through the course of the appeal, it was possible to obtain the trust of local government and local residents.

Then, it was possible to expand the facility below.

- Slow sand filtration with 4 units of using the prototype ($8m^3$/day water treatment capacity)
- Reservoirs made of SUS ($4m^3$)
- Distribution pump

Figure 24.3 (left): Splashdown pond, coarse filtration, Slow sand filtration×4. (right): Reservoirs made of SUS ($4m^3$), distribution pump, deep well

RESULTS

The facility began operation in April 2011.

Residents are willing to say "clean water is obtained from the tap" he said.

Now also a good operation that meets the criteria of water quality is maintained in the hands of the residents themselves as follows.

- The day-to-day management of filtration flow rate
- Coarse filtration pond cleaning every 2 weeks
- Removal of blockage of filtration pond every 2 months (Scraping of sand)

In our NPO,

- Status monitoring by one visit per month
- Support of water quality testing
- Research for more convenient maintenance and improvement of facilities, In addition to continuing maintenance support of small scale water ways, we are acquiring "know how" regarding small scale water facility operations.

Chapter 25

Refocus on slow sand filter in Kakuda city and Marumori town after Great East Japan Earthquake

Masahide NANBU and Kazuhiro YAMAKI***

**Water Works Office of Kakuda City, Miyagi Prefecture, Japan*

***Marumori Town office, Miyagi Prefecture, Japan*

ABSTRACT

Small water supply system had a plan to close in rural area in north east of Japan. After the great east Japan earthquake, small water supply system of slow sand filter plant was refocused as more reliable system in rural area. Slow sand system is reconfirmed that is simple and easy maintenance.

Keywords Slow sand filtration; refocus; earthquake; small system

INTRODUCTION

Kakuda city is a small city with the population of 31,500 and 147.6 km^2 in Japan. It is located in southern Miyagi Prefecture, and is surrounded by the Abukuma mountain region in the circumference. The underground water in this area has a taste of iron. The concentration often exceeds 0.3 mg/l of the Japanese drinkable standard. This water is not good for drinking water. Oda water purification plant (Oda WPP) was built in 1934. The surface water of a river has been treated by slow sand filtration. The treatment capacity was 2,880 m^3/day using 6 filter ponds. And the present water service coverage is 95.9% of residents in the city.

Safe and stable water supply system had been thought in Japan after 1960s. A large water supply system by rapid sand filtration system has been rercommended in the most of Japan. New plant of Edano water purification plant (Edano WPP) was built in 1985 in this city. New plant is rapid sand filtration system which capacity is 1,970 m^3/day. Miyagi prefecture government built a huge water purification plant by rapid sand filter system in 1990. The treatment capacity is 279,000 m^3/day to cover a large area in the prefecture. Kakuda city also had to use this water. As the result, present contribution of water supply is covered only 6% in the city by Oda WPP of slow sand filter. The Edano WPP in the city covers 14% by rapid sand filtration and the remaining 80% of water demand is supplied by the Sen-nan and-Sen-en bulk water supply scheme of the Miyagi Prefecture Water Enterprise Bureau.

Marumori town is a neighbor town of Kakuda city with a population of 15,800, which is also surrounded by mountains and houses are broadly scattered in the town. The water service coverage rate still remains 68.9% by Ishiba water purification plant (Ishiba WPP). This plant has two systems of slow sand filter and rapid sand filter.

In March 2011, the great east earthquake was attacked in this area. The large water supply system of rapid sand filter system received a big damage (mainly distribution system) and the water supply stopped for long period. However, Oda WPP of slow sand filter had no damage. This plant used natural gravidity system. And a small rapid sand filter system of Edano WPP received small damage. This plant has an automatic power supply system with an enough capacity in case of stopping the public power supply.

After the great earthquake, Kakuda city and Marumori town were refocused the slow sand filter system as more safe and reliable water supply system.

SLOW SAND FILTER OF ODA WPP IN KAKUDA CITY

The original Oda WPP had 3 slow sand filters in 1934. Area of each slow sand filter basin is 108 m^2. The source water is derived from Oda River using natural gravidity. This plant was not necessarily any electric power supply. The distribution tank (988 m^3) was built in 1934. Natural gravidity was used in whole system. This system is really energy save system. After 1945, the chlorination began to filtered water in every water supply system in Japan. Present plant has 6 filter basins (Figure 25.1 and 25.2). The filtration capacity is 2,880 m^3/d for 6 filters.

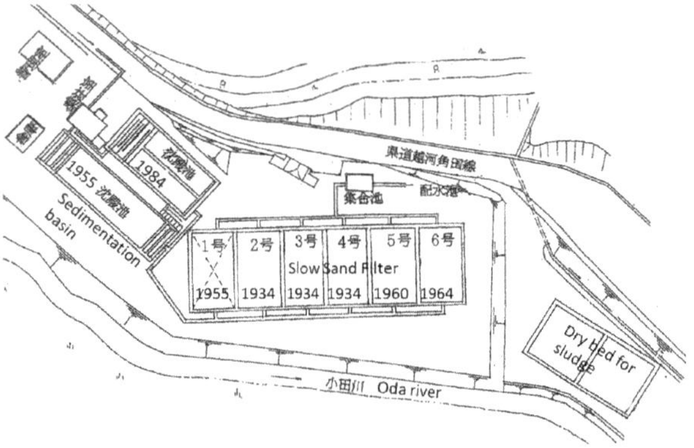

Figure 25.1 Outline of Oda WPP

Figure 25.2 Slow sand filter of Oda WPP

The design flow rate was 4.4 to 5.3 m/d. This filter rate is the standard filter rate of the English filter. At the present time, one filter basin is not used due to the leakage of water. Moreover, the sedimentation basin (182 m^2) was installed in 1955. Another sedimentation and precipitation basin (153 m^2) was installed with deterioration of quality of raw water since 1984.

DAMAGE OF EARTHQUAKE IN KAKUDA CITY

The great east Japan earthquake was attacked in this area on 11th of March, 2011. The large water supply system of rapid sand filter system received a big damage (mainly distribution system) which usually covers 80% of the water demand in Kakuda city. And the water supply was cut for long period. However, Oda WPP of slow sand filter had no damage. This plant used natural gravidity system. And a small rapid sand filter system of Edano WPP received also small damage. This plant has an enough capacity of an automatic power supply system in case of stopping the public power supply. The residents in Kakuda city could get enough amount of safe water after the earthquake. These two plants covered temporarily the supply water by the Miyagi Prefecture Water Enterprise Bureau.

SLOW SAND FITER IN MARUMORI TOWN

Slow sand filter of Ishiba WPP was built in Marumori town in 1970 (Figure 25.3). There are 4 filter beds. An area of a filter bed is 109.2 m^2. The capacity of the plant is 1,200 m^3/d. Rapid sand filter system was built in this WPP in 1990. The capacity of the plant is 4,440 m^3/d. The design plan was a combined use of slow sand filtration and rapid filtration. However, the rapid sand filter system was mainly used and the slow sand filter system became a compensational use after 1990. Treated water by slow sand filter became 255 m^3/d which was about 21% of the capacity of slow sand filter. At that time the amount of treated water by rapid sand filter was 2,898 m^3/d which was 65.3% of the capacity. The contribution of the rapid sand filter was 92% of the total demand. The slow sand filter system was renovated in 2000 and the treated water by slow sand filter system became 800 m^3/d. And the treated water by rapid sand filter system became 2,353 m^3/d. The contribution of the rapid sand filter reduced to 75% of the total demand, and the contribution of slow sand filter increased to 25% of the total demand.

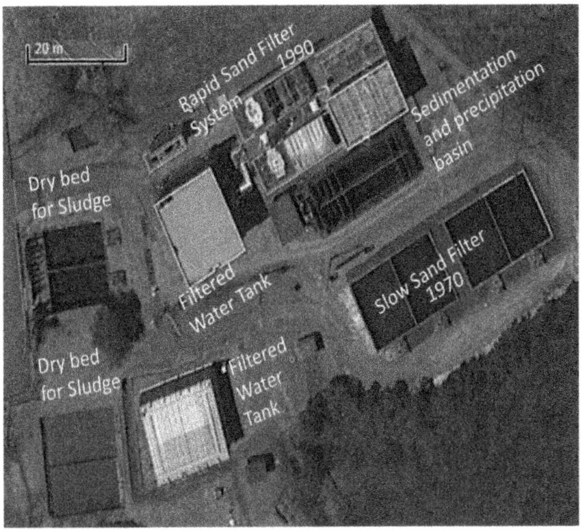

Figure 25.3 Oda WPD in Marumori Town

Since both purified waters of slow sand filtration and rapid filtration were blended in the pipe to a distributing reservoir, the extended project (rapid filter machinery etc.) was undertaken in 1992 in order to an increase of the amount of purified water by rapid filtration. There was a tendency that most clean water was made from slow sand filtration. We understood that major contribution of rapid sand filter in this plant was not wise selection. The rapid filtration system of the town has an ability to provide the whole quantity of water demand. Morumori town office got a useful information by Asahi newspaper on 26 April, 2008 that Misato town in Miyagi prefecture built a new slow sand filter plant to make more safe and delicious water by natural purification system. Marumori town office reconfirmed the water quality of slow sand filtration is valued, and present this system should be used effectively also in this town. With these circumstances, the town started improvement work. Repair project of the both systems of Ishiba WPP and new blend system to the pipe had started. Although the whole use of the slow sand filtrated water of 1,200 m^3/d was not completed in 2013, water use of slow sand filter was increased from 255 to 800 m^3/d by the repair. Since slow sand filter provides water of stable water quality, water quality does not deteriorate rapidly even when rapid filtrated water has abnormalities. Cascade mode system to chlorine addition was used in this rapid sand filter plant. A problem of the pre-chlorination pouring system into raw water was found in this plant. Cascade mode system must be changed in near future. In addition, the damage of the slow sand filter plant did not arise in the great earthquake in March 2011.

REFOCUS ON SLOW SAND FILTER

After the great earthquake, Kakuda city and Marumori town were refocused the slow sand filter system as more safe and reliable water supply system. Since the Oda WPP had passed 80 years after the construction, abolition of the Oda WPP was considered before the great earthquake due to its remarkable superannuation. But the plan was changed after the great earthquake. Our offices were decided to be preserved the slow sand filter system and a small water supply system as a precious self-water source within the region.

Since the Oda WPP arranges the facilities along a physiographic vertical interval, it is an environment-friendly system without using additional power. Kakuda city is due to carry out operation management of Oda WPP from now on, finding out new additional values as a place of the education of a historical structure and water service institution.

Kakuda city and Marumori town, Miyagi prefecture, will continue to use slow sand filtration in the future, taking advantage of a good experience of operation and management for a long time.

Chapter 26

Small slow sand filter system for rural area in Bolivia

Toshiki HORIE, Yoshinori FUKUSHIMA**, Grover Calichoc*** and Chika TAKAHATAKE*****

**Japan Overseas Cooperation Volunteer (JOCV) (Currently, Japan Techno. Co. Ltd,) Flos Bldg.19-7-2, Hatagaya, Shibuya-ku ,Tokyo, Japan*

***JICA expert (Currently, SUIGEN S.R.L, Bolivia) Miguel Rodriguez,No.39 Santa Cruz de la Sierra,Bolivia*

****Santa Cruz Prefecture PROASU-JICA,Bolivia,(Currently,SUIGEN S.R.L,Bolivia) Miguel Rodriguez, No.39 Santa Cruz de la Sierra Bolivia*

*****Assistant representative of Japan International Cooperation Agency (JICA) Bolivia Office (Currently, JICA headquarter), 5-25 Niban-cho,Chiyoda-ku, Tokyo Japan*

ABSTRACT

Slow sand filtration is sustainable technology for drinking water treatment in small communities. The purpose of this study is to prove feasibility of adequate technology for drinking water supply system in rural area in developing country. We introduced our pilot project of small slow sand filter system in Bolivian rural area. The results of water analysis and our observations about sustainability of slow sand filter are described.

Keywords small slow sand filter system; rural water supply; adequate technology

INTRODUCTION

JICA has conducted master plan study, grant aid and technical cooperation project for rural water supply in Bolivia since 1994. Japan donated two sets of machineries for well drilling to two prefectures including Prefecture of Santa Cruz in 1998. In this period, prefectures were responsible for groundwater development and the municipalities were for construction of water supply system by Bolivian law. However, around 2004 the following problems were observed; there were communities with wells drilled, but without water distribution system, and in case of the communities with water distribution system, some communities could not maintain the system very well. These problems were caused by insufficient and inadequate coordination among prefectures, municipalities, and communities. And also there was insufficient capacity of municipalities and communities to construct and maintain the water distribution system. In this circumstance, an approach for improvement of maintenance capacity of communities called Productive Community Development was proposed and was legislated by Ministry of Water Resources. In this approach, the responsibility of prefectures was extended to, not only groundwater development as water source, but also technical assistance to municipalities for construction of water distribution systems and assistance to communities for its maintenance.

At that period, prefecture had two major problems: lack of technicians capable to design the water distribution system, and of knowledge on appropriate technologies for water distribution system. Then, Santa Cruz prefecture requested JICA technical assistance by dispatch of JOCV on design water distribution system and promotion of appropriate technologies for Santa Cruz prefecture. The first author worked at Santa Cruz prefecture from 2007 to 2009.

Santa Cruz prefecture implemented the pilot project of small slow sand filter system as one of possible appropriate technologies. This small slow sand filter system was composed by only Bolivian material and it is not necessary to use coagulation and chlorine to maintain the slow sand filter system.

TARGETED AREA

Target area was San Antonio de Lomerío community, Ñuflo de Chavez municipality, in Santa Cruz prefecture. It was located north east 416 km from Santa Cruz de la Sierra city which is center of Santa Cruz prefecture. It took approximately ten hours from Santa Cruz de la Sierra by car. The location is indicated in Figure 26.1. The population of the community was about six thousands and most of the inhabitants make their living in agriculture such as cultivation of cassava, corn, banana or rice. Electricity supply for the community was by a generator, and only for two hours from 7 pm to 9 pm.

There were several hand pumps utilizing underground water through wells. Since some well water contained iron, the water had yellow-colored and with metallic taste. Therefore, the inhabitants didn't want to use such kind of water.

Figure 26.1 Targeted area in Bolivia

MATERIAL AND METHODS

We constructed small sand filter system (Figure 26.2, Photograph 26.1). It was composed of raw water tank (500 L), three gravel filter tanks (200 L), one sand filter tank (200 L) and treated water tank (2,000 L). Each tank was made of polyethylene. Tank sizes of gravel filter and sand filter, grain sizes and filter depth are shown in Table 26.1. Three gravel filters were up-flow. And there was drain valve in lower part of each tank. It was easy to wash the gravel filter. And sand filter was down-flow. A natural gravity flow was used for the whole system. The diameter of slow sand filter tank was 600 mm and line velocity of sand filter was 5 m/day. Therefore, quantity of water was 1,400 L/day. The water demand per person was 20 L/day. This system could provide water for about 70 persons per day. The water depth on the sand surface of sand filter was only 50 mm. Biological community became active due in the shallow depth. And a continuous flow system was important to keep active aerobic biological community.

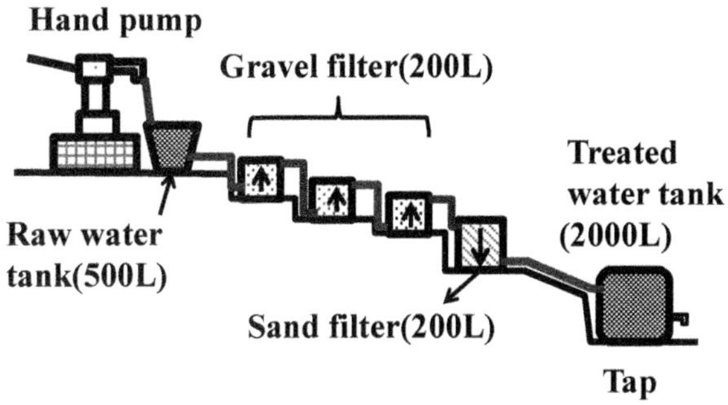

Figure 26.2 Small slow sand filter system

Photograph 26.1 Small slow sand filter system

Table 26.1 Tank size, grained filter, depth

	Tank size	Grain diameter	Depth
Gravel filter	L2000m×W500×H200mm	20mm-50mm	120mm
Sand filter	Φ600m×H1000mm	0.5mm-1.0mm	600mm

RESULTS AND DISCUSSION

After seven days of operation, color of treated water was clearer than raw water. The comparison of water color is shown in Photograph 26.2. The sample of after gravel filter became nearly transparent. And also metallic taste was not detected.

Photograph 26.2 The comparision of water color

Table 26.2 showed the result of water analysis. Iron concentration decreased from 5.64 mg/L to 0.13 mg/L after gravel filters. Its removal rate was 96.8%. And manganese concentration decreased from 0.3 mg/L to 0.1 mg/L after sand filter. This data showed that iron and manganese can be treated effectively. Especially, iron concentration became under Bolivian standard level after gravel filters.

Red sediment was observed on the surface of the gravel filters. The authors consider that red sediment was iron bacteria. We also observed the biofilm on the surface of sand filter. Algal growth was observed in the sand filter. Since small bubbles were observed on the algal mat, it was considered algal photosynthesis was significantly active. When algal mat floated to the water surface in the sand filter, we removed by hand.

It was necessary to maintain every day to check the condition of the sand filter. We considered that it was important to keep line velocity to activate biological activity. Therefore, quantity of water should be checked every day. And also back washing of gravel filter and scratch sand layer should be implemented as necessary.

Table 26.2 The result of water analysis

Item	Unit	Raw water	After gravel filter	Treated Water	Bolivian standard
PH		6.18	7.12	7.59	6.5-8.5
Color	TCU	76	16	14	Under 15
Turbidity	NTU	25	6	4	Under 5
TH	mg -$CaCo_3$/L	73	119	116	500
Fe	mg/L	5.64	0.18	0.03	0.3
Mn	mg/L	0.3	0.2	0.1	0.1
Cu	mg/L	0.11	0.01	0.01	1
Cr	mg/L	0	0	0	0.05
Cl	mg/L	28	23	22	250
S	mg/L	5	2	2	400
NO_3	mg/L	6	5.6	5.4	45
NO_2	mg/L	0.011	0.044	0.025	0.1
NH_4-N	mg/L	0.13	0.08	0.07	Under 0.1
S	mg/L	0.009	0.007	0.005	Under 1
EC	µS/cm	416	495	496	1000
E-coli	CFU	0	0	0	0

CONCLUSIONS

The main conclusions of this work are:

(1) The water quality was improved by this system under Bolivian standard. And this system can be made by local material. It is not necessary to use coagulation. Therefore it is adequate water treatment system in rural area of developing country.

(2) It is important for this system to maintain well. Therefore, the maintenance training should be conducted.

ACKNOWLEDGEMENT

The authors appreciate the San Antonio de Lomerío community for their contribution in the construction of the small sand filter system.

REFERENCES

Huisman, L. and Wood, W.E. (1974) Slow sand filtration, World Health Organization, Geneva

Nakamoto, N. (2002) Tasty raw tap water, Tsukijisyokan, pp.30-49, Japan

Nakamoto, N. (2005) How to make tasty water, Tsukijisyokan, pp.10-52, Japan

Visscher, J.T., Paramasivam, R., Raman, A., and Heijnen, H.A. (1987) Slow sand filtration for community water supply, IRC Technical Paper Series No.24. IRC, The Netherlands

Part IV

Slow Sand Filtration – Process Developments

Chapter 27

Switching on pesticide degraders in biological filters used in drinking water production

*C. A. Rolph, B. Jefferson, R. Villa**

Cranfield University, Bedfordshire, MK43 0AL, UK

ABSTRACT

This study investigated the use of biological processes for the removal of the pesticide metaldehyde from drinking water. Analysis of data from a full scale site, alongside lab scale batch and column experiments identified some of the conditions required to obtain biological removal of metaldehyde. Metaldehyde removal appeared to be most effective in slow sand filters following a period of exposure to high concentrations of the pesticide (above 0.2 $\mu g\ L^{-1}$). This was demonstrated in laboratory batch experiments where metaldehyde removal was accelerated following a period of dosing at higher concentrations than expected levels (0.5 $\mu g\ L^{-1}$). Similarly, in column experiments significant metaldehyde removal was achieved only when the bioreactors were dosed with high metaldehyde levels, after which removal could be sustained for several days.

Keywords Slow sand filtration, pesticides, metaldehyde

INTRODUCTION

Pesticides are a threat to drinking water quality in many countries. In the EU levels of pesticides are strictly regulated and must not exceed 0.1 $\mu g\ L^{-1}$ in treated drinking water (*Council Directive 98/83/EC*). Conventional chemical and physical treatments are not always effective at removing pesticides to these levels (Hall et al., 2011). In particular, in the UK, the pesticide metaldehyde has been regularly found in treated drinking water at elevated levels and was responsible for 232 out of 253 pesticide failures in 2012 (DWI, 2013). Metaldehyde is a molluscicide used to control populations of slugs and snails. It is a polar compound with a low log K_{ow} value (0.12) and a low molecular weight (176.21 g mol^{-1}) and is therefore difficult to remove by adsorption processes.

Biological processes such as slow sand filtration proved to be effective for the removal of a range of compounds including organic matter, bacterial particles and trace organics including recalcitrant compounds. Zearley and Summers (2012) investigated biological technologies for the removal of a range of micropollutants including pesticides and pharmaceuticals and showed that removals were highly variable. However, under the right conditions biodegradation is possible for any compound (Montagnolli et al., 2009; Peterson, 2008).

To our knowledge there is no published work on metaldehyde degradation in water but degradation of metaldehyde in soil and aquatic sediments indicates it is amenable to biodegradation (Coloso et al., 1998; Bieri, 2003). In addition, data from a full-scale water treatment works in the UK has revealed good removal of metaldehyde in slow sand filters, further indicating its biodegradability. This work investigated the possibility of capturing and enhancing the mechanisms of this treatment to provide the benefits of slow sand filtration at a faster rate. This paper reviews the current ability of a full scale biological treatment works to remove metaldehyde and examines some of the conditions necessary to achieve metaldehyde removal at lab scale.

METHODOLOGY

Materials and reagents

Sand with an active biofilm was collected from an operational water treatment works in the UK. The main treatment stages are GAC roughing filters followed by slow sand filtration. Biologically active sand was collected by scraping the surface of a sand filter with a seeding tool. A portion of the sand was inactivated by heating at 105°C overnight. Clean sand was also collected after it had been through the onsite mechanical cleaning system. Raw water was obtained from the outlet of the roughing filters which feed directly on to the slow sand filters.

10 mg of metaldehyde (Fisher scientific, UK) was dissolved in 1 litre of ultrapure water for spiking. HPLC grade methanol and dichloromethane (Rathburn chemicals, UK) were used for the extraction of metaldehyde from water. Metaldehyde d_{16} (Qmx laboratories, UK) was dissolved in methanol (50 mg L^{-1}) for use as an internal standard.

Batch experiments

Batch experiments were undertaken using different media and spiked raw water at concentration of 10 µg L^{-1} and 1 µg L^{-1}. Four types of sand were used for batch experiments; *Active:* sand with an active biofilm which had not been exposed to metaldehyde concentrations above 0.17 µg L^{-1} for more than 6 months. *Acclimated:* sand with an active biofilm exposed to metaldehyde concentrations of 0.5 µg L^{-1} for 10 days in a downflow column in the laboratory. *Inactive:* sand with an inactive biofilm. *Clean*: clean sand with no biofilm. 100 g of each media were added to conical flasks with 300 mL of spiked raw water. The samples were shaken at 150 RPM for 72 hours and samples were collected regularly for immediate analysis.

Column experiments

Two lab scale columns were set up using a 50 cm glass column with an I.D. of 2.5 cm. The columns were filled with a 5 cm gravel support and 30 cm of sand with an active biofilm. The columns were operated in upflow mode with most of the water recycled to obtain extended contact times. A further pump fed raw water from the inlet into the recycle container and from the recycle to the outlet where samples were collected for analysis (Figure 27.1).

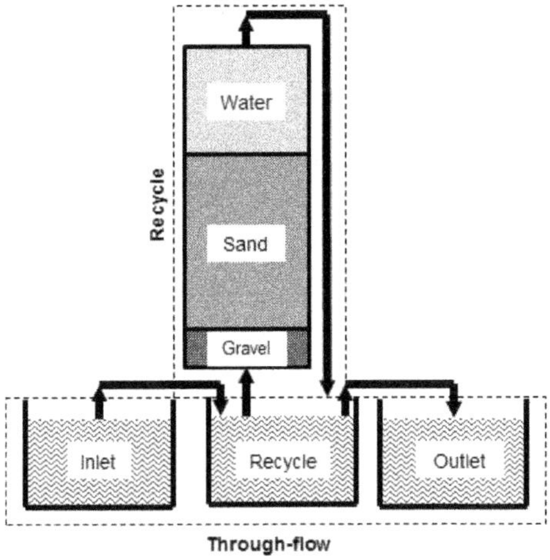

Figure 27.1 Lab scale column set-up

Columns were operated in upflow mode to maximize surface area and therefore contact with the biofilm. They were operated for 1 month with raw water (average metaldehyde concentration 0.04 μg L-1) prior to the commencement of the experiment. Following this, the columns were fed with raw water spiked with different concentrations of metaldehyde. The pump speed was adjusted to achieve different contact times and the system was allowed to stabilize for 24 hours following a change in contact time.

Analytical methods

The determination of metaldehyde in aqueous samples was performed using solid phase extraction (SPE) followed by analysis with gas chromatography-mass spectrometry (GC-MS). SPE cartridges (styrene-divinylbenzene cartridges, Strata, 200 mg/ 3mL, Phenomenex, UK) were conditioned using 10 mL of methanol followed by 2 mL of ultrapure water. Between 10 and 150 mL of sample, spiked with metaldehyde d_{16} as an internal standard, was passed through the cartridge. The cartridge was rinsed with 2 mL of ultrapure water and left to dry under vacuum for 45 minutes. The samples were then eluted using 2 mL of dichloromethane before being evaporated down to 0.5 mL. Quantification was achieved using an Agilent 6890N gas chromatograph coupled with an Agilent 5673 mass selective detector with a limit of detection of 0.05 μg L^{-1} (GC-MS + SPE).

RESULTS AND DISCUSSION

Metaldehyde removal by full scale biological filters

At full scale, negligible removal of metaldehyde was observed through primary GAC roughing filters. The hydrophilic, low molecular weight structure of metaldehyde makes it difficult to remove through adsorption processes. Furthermore, the short retention time in the roughing filter is unlikely to be sufficient for biodegradation. Good removal of metaldehyde (up to 98%) can be achieved through biological slow sand filtration (Figure 27.2a). 85% of samples taken from this site had compliant metaldehyde levels whilst only 50% of samples were compliant at a site using a O_3-GAC pesticide treatment despite the same average influent concentration of 0.2 μg L^{-1}.

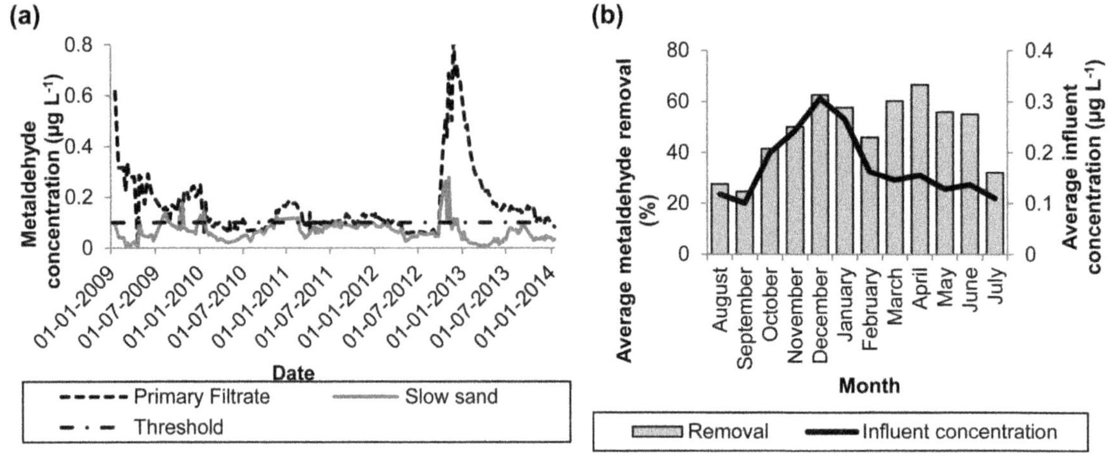

Figure 27.2 **(a)** Metaldehyde removal by slow sand filters at an operational water treatment works (2009 – 2014) **(b)** Average monthly metaldehyde influent and removal 2008 - 2013

Influent concentrations of metaldehyde peak between October and January when they average above 0.2 µg L^{-1}, following the main metaldehyde spreading period by farmers between August and December. Despite a fall in influent concentrations between February and June, high metaldehyde removal is still achieved (>45%). Removal efficiency begins to drop between July and September (25 – 32% average removal) despite influent concentrations remaining relatively stable from May to September (Figure 27.2b). It could be that exposure to these high concentrations for several months results in an acclimated microbial population that continues to successfully degrade metaldehyde. Alternatively, reduced performance in the degradation of herbicides during the summer months was observed by Woudneh et al. (1996) who speculated this could be due to a reduction in oxygen availability or an increase in nutrients which are preferentially degraded.

Metaldehyde removal by slow sand filtration appears to be particularly effective when metaldehyde levels are high such as at the end of 2012 when 0.6 µg L^{-1} of metaldehyde was regularly removed. There is significant correlation ($P < 0.05$) between concentration of the primary filtrate and metaldehyde removal (Figure 27.3). When influent concentrations are above 0.2 µg L^{-1} average removal is always greater than 40% and the average removal is 82%.

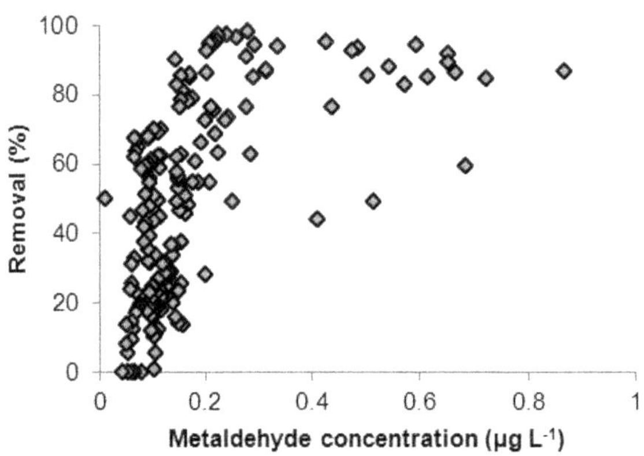

Figure 27.3 Relationship between metaldehyde concentration and removal in a slow sand filter

Metaldehyde removal in batch experiments

Metaldehyde removal was observed in batch experiments at different concentrations to establish the effect this had on removal (Figure 27.4). Compliant water was achieved after 48 hours in the 1 μg L^{-1} experiment. Increasing the concentration to 10 μg L^{-1} increased the time to compliant water by 24 hours to 72 hours. No significant removal of metaldehyde was observed using inactive or clean sand indicating a biological removal mechanism. Following this experiment the active sand was shaken vigorously with clean water for several hours and no desorption was observed.

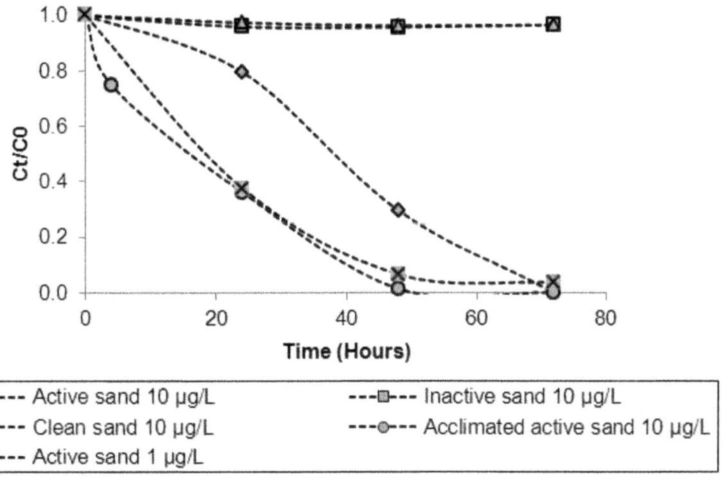

Figure 27.4 Metaldehyde removal in batch experiments

The acclimated active sand achieved drinking water compliance in 48 hours, significantly faster than non-acclimated sand. The results suggest that exposure to higher levels of metaldehyde for a short period of time was sufficient for an acclimated community to develop and produce more efficient degradation.

Metaldehyde removal in lab scale columns

Lab scale columns were established to investigate necessary conditions for metaldehyde removal. Initial trials at high concentrations (50 μg L^{-1}) using the column in recycle mode showed that metaldehyde could be effectively removed (compliant drinking water levels) within 4 days (Figure 27.5a). A through-flow column was then established to identify whether metaldehyde could be removed with an extended contact time (16 hours) at lower concentrations. Influent concentration below 0.5 μg L^{-1} resulted in metaldehyde removal greater than 50% however compliant water was not achieved (Figure 27.5b).

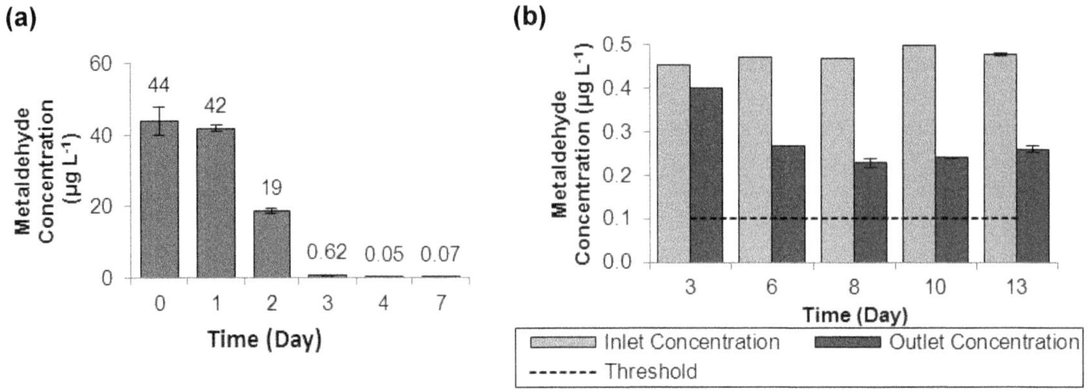

Figure 27.5 **(a)** Metaldehyde removal in an upflow sand column operated in recycle mode **(b)** Metaldehyde removal in an upflow sand column operated in recycle mode

Adjusting the contact time by increasing the flow rate had only a small effect when inlet concentrations were high (10 μg L^{-1}). Reducing contact times from 16 hours to 3.2 hours had a negligible effect on metaldehyde removal showing the importance of optimising filter efficiency (Table 27.1). Further increases lead to a drop in performance. More work will be required to identify the optimal contact time for the removal of metaldehyde removal at different concentrations.

Table 27.1 Column operating parameters

Experiment	Recycle flow rate (mL/min)	Through-flow flow rate (mL/min)	Time in contact with the sand (Hrs)	Average Removal (%)
1	156	0.2	16	70
2	156	0.8	3.5	67
3	156	1.5	1.7	45

Additional experiments were undertaken looking at the impact of the addition of a carbon source, to increase overall biomass, or phosphorous, identified as a limiting factor in the biodegradation of organic substances (Nishijima et al., 1997). In both cases there was no significant improvement in performance following these additions. However, by spiking the influent with a high level of metaldehyde (50 μg L^{-1}) for one week and then returning to an effluent with a concentration of 0.5 μg L^{-1} compliant water was achieved and high levels of removal were sustained for more than 20 days.

CONCLUSIONS

Biologically active filters are suitable for the removal of metaldehyde and can be highly effective at achieving water that is compliant with drinking water regulations under the correct conditions. A high concentration of metaldehyde leads to the switching on of pesticide degraders and therefore leads to an efficient degradation which was observed at both lab and full-scale.

ACKNOWLEDGEMENTS

This work was undertaken as part of the STREAM IDC and was supported by the EPSRC, Anglian Water, Yorkshire Water and the WRc.

REFERENCES

Bieri, M. (2003). The environmental profile of metaldehyde. In: World Agriculture, British Crop Protection Council, Symposium Proceedings 80. pp. 255–260.

Coloso, R., Borlongan, I.G. and Blum, R.A. (1998). Use of metaldehyde as a molluscicide in semi-commercial and commercial milkfish ponds. Crop Protection. 17 (8). p.pp. 669–674.

DWI (2013). Drinking water 2012. London; DWI.

Hall, T., Holden, B. and Haley, J. (2011). Treatment for metaldehyde and other problem pesticides, 4th Developments in Water Treatment and Supply conference. Cheltenham, 7th-8th June 2011.

Montagnolli, R.N., Lopes, P.R.M. and Bidoia, E.D. (2009). Applied models to biodegradation kinetics of lubricant and vegetable oils in wastewater. International Biodeterioration and Biodegradation. 63 (3). p.pp. 297–305.

Nishijima, W., Shoto, E. and Okada, M. (1997). Improvement of biodegradation of organic substance by addition of phosphorus in biological activated carbon. Water Science and Technology. 36 (12). p.pp. 251–257.

Peterson, H. (2008). The Power of Biology in Drinking Water Treatment. Canadian Water Treatment. (February). p.pp. 18–20.

The Council of the European Union (1998). Council Directive 98/83/EC of 3 November 1998 on the quality of water intended for human consumption. p.p. L 330/32 – 330/52.

Woudneh, W.B., Lloyd, B.J. and Stevenson, D. (1996). Removal of Herbicides by Biological Filters. In: N. Graham and R. Collins (eds.). Advances in slow sand and alternative biological filtration. Chichester: Wiley, pp. 211–221.

Zearley, T.L. and Summers, R.S. (2012). Removal of Trace Organic Micropollutants by Drinking Water Biological Filters. Environmental Science and Technology. 46. p.pp. 9412–9419.

Chapter 28

Bioaugmentation reduces negative effect of estrogens on coliform removal in slow sand filters

S. J. Haig, C. Quince*, G. Collins**, R. L. Davies*** and C. Gauchotte-Lindsay**

**School of Engineering, University of Glasgow, UK*
***School of Natural Sciences, National University of Ireland, Galway, Ireland*
****Institute of Infection, Immunity and Inflammation, University of Glasgow, UK*
(e-mail: s.haig.1@research.gla.ac.uk)

ABSTRACT

Both the effect of natural estrogens (estone, estradiol and estriol), and the ability to improve estrogen removal using bioaugmentation with three estrogen-metabolizing bacteria, was examined in slow sand filters (SSFs). Concentrations of the natural estrogens determined by gas chromatography coupled with mass spectrometry (GC-MS) revealed that augmented filters removed significantly more estone and estradiol than non-augmented filters. Additionally, a positive correlation was found between coliform retention and estrogen concentration in non-augmented filters. This was explained by the toxic inhibition of protozoa, suggesting the functional implications (impaired coliform removal) that high estrogen concentrations might have in SSFs. Consequently, we suggest that high estrogen concentrations could impact significantly on water quality production, and in particular on pathogen removal. Monitoring and adapting the microbial community in SSFs will become an important element of water production and quality assurance in the future.

Keywords: Slow sand filter; estrogen; coliform removal; protozoa; water quality; bioaugmentation

INTRODUCTION

Slow sand filtration is one of the oldest and most effective means of treating drinking water, storm water and horticultural irrigation water by controlling microbiological contamination. Despite their effectiveness in producing high-quality water, slow sand filters (SSFs) are still operated as "black boxes", where purification is attributed to naturally-occurring biochemical processes (e.g., predation and bio-oxidation). However, these processes have never been comprehensively verified (Haig *et al.*, 2011).

Thus, the true breadth and extent of contaminant deactivation, transformation or removal capabilities of SSFs are either unknown or poorly understood.

Previously, we compared laboratory-scale filters to full-scale SSFs using phyla-specific qPCR primers, and both 454 and Illumina sequencing (Haig *et al.,* 2014b). This revealed that the microbial communities underpinning SSFs are extremely complex, and phylogenetically and metabolically diverse. Additionally, we have found through metagenomic analysis that there are several key phages and protozoan genera, which are important for coliform removal (Haig *et al.,* 2013). However, from a functional perspective, determining whether removal of "emerging" contaminants (e.g., estrogen) can be achieved and/or enhanced through bioaugmentation of SSF microbial communities is important. This is especially important as the list of contaminants being added to water quality guidelines expands. Particularly, there is increasing concern over estrogenic effects on human health and wildlife, and over the rising levels of natural estrogens in waterways; therefore, their impact on existing treatment systems and removal mechanisms should be investigated (Yu *et al.,* 2013).

Our aim was to understand the functional effect of estrogen exposure on SSFs, and to determine whether bioaugmentation can improve estrogen removal and overall filter performance.

METHODOLOGY

Eight experimental SSFs (four augmented with three enriched estrogen (estone, estradiol and estriol)-metabolizing bacteria (Table 28.1) and four non-augmented filters) were used to treat river water over 10 weeks at a flow rate of 0.15 $m^3 m^{-2} h^{-1}$. Bioaugmentation was done at the beginning of the experiment, following the method of Kim *et al.* (2004), using exponentially grown bacteria at a density of 10^8 cfu/ml. Each laboratory-scale SSF comprised a PVC tube (2.5m x 54mm) with eight sampling ports (11mm diameter) down the length (0,5,10,15,20,30,45,70cm) of the sand bed (sterilized from industrial SSF), which were 1m high with 6.5cm of under-drainage (Haig *et al.* 2014b). Influent and effluent from each filter was taken weekly to analyze water quality indicators, including estone, estradiol and estriol, using GC-MS. GC-MS analysis of natural estrogens was performed following the procedure outlined by Quintana *et al.* (2004). Briefly, one litter of effluent was concentrated using 60mg Oasis HLB solid phase extraction cartridges, (Waters, UK). Resulting analytes were derivitized by *N*-methyl-*N*-(trimethylsilyl)trifluoroacetamide (MSTFA) and analyzed using GC (Agilent 7890A) system equipped with a HP-5MS capillary column coupled to a single quadrapole (Agilents 3157 5975c Inert XL EI/CI MSD). Compounds were separated using the following oven program: 1min at 50◦C, first ramp at 20◦C/min to 220◦C (held for 27min), second ramp at 20◦C/min to 250◦C (held for 20min). The GC-MS interface temperature was set to 250◦C. Quantification of each of the estrogens was achieved by using single ion wavelengths (estone = 342 m/z, estradiol = 416 m/z and estriol = 504 m/z). All GC-MS runs were performed in triplicate alongside 6 standards, negative controls and blanks.

Table 28.1 Overview of the isolated and enriched estrogen degrading bacterial species used for bioaugmentation

Degrading Estrogen	Closest match[a] (accession no)	Classification Based on RDP query	Percentage Similarity
Estone (E1)	*Acidovorax sp.* KKS102 (CP003872.1)	Beta-Proteobacteria	99
Estradiol (E2)	*Pseudomonas sp.* UW4 (CP003880.1)	Gamma-*Proteobacteria*	97
Estriol (E3)	*Pseudomonas fluorescenes* Pf0-1 (CP000094.2)	Gamma-*Proteobacteria*	97

Additionally, sand samples were retrieved from every sampling port to determine microbial community composition. To assess SSF effectiveness in removing high concentrations of estrogens, filters were spiked with estrone (50ng/L), estradiol (12ng/L) and estriol (39ng/L) at weeks 5 and 8. Concentrations were chosen based on averages from Chen et al. (2010) and simulate agricultural or wastewater pollution events.

RESULTS AND DISCUSSION

Results indicate that augmented filters removed significantly more estone and estradiol than non-augmented filters (Table 28.2). However, there was no significant difference in estriol removal. Estrogen concentration positively correlated with coliform retention in non-augmented filters (correlation = 0.34, p-value = 0.0379) (Fig. 28.1), which appears to be due to inhibition of coliform-grazing protozoa (Fig 28.2). Referring more closely to Figure 28.2, a reduction in the number of eukaryotes (i.e., protozoa) present during the two spiking periods of estrogen can be seen in both augmented and non-augmented filters. However, this reduction was significantly lower in augmented filters (55% and 53% respectively) compared to non-augmented filters (93 and 85% respectively). This alongside the interesting finding that coliform removal in augmented filters was less affected by estrogen concentration than non-augmented filters, implies that augmentation reduces the toxic effect of estrogen on the coliform-grazing, protozoan community (Haig *et al.* 2014a).

Table 28.2 Average percentage removal of estrogens in augmented and non-augmented filters. Significant differences tested using Wilcoxon tests, negative percentages mean an increase in concentration (i.e., no removal)

	Average percentage removal by filter type		
Chemical	**Augmented**	**Non-augmented**	**p-value**
Estone	20.1	-130	* 0.00784
Estradiol	0.2	-90	* 0.02285
Estriol	-69	-282	0.104

Positive percentages correspond to removal. Negative percentage correspond to an increase in the concentration of estrogens i.e. no removal.

This study underscores the importance of investigating the effects of emerging contaminants on the micro- and macro-community of SSFs, and further shows that bioaugmentation of SSFs is feasible and beneficial for both chemical and microbiological removal. Moreover, this is the first study, to our knowledge, to demonstrate the negative effect of estrogens on pathogen removal.

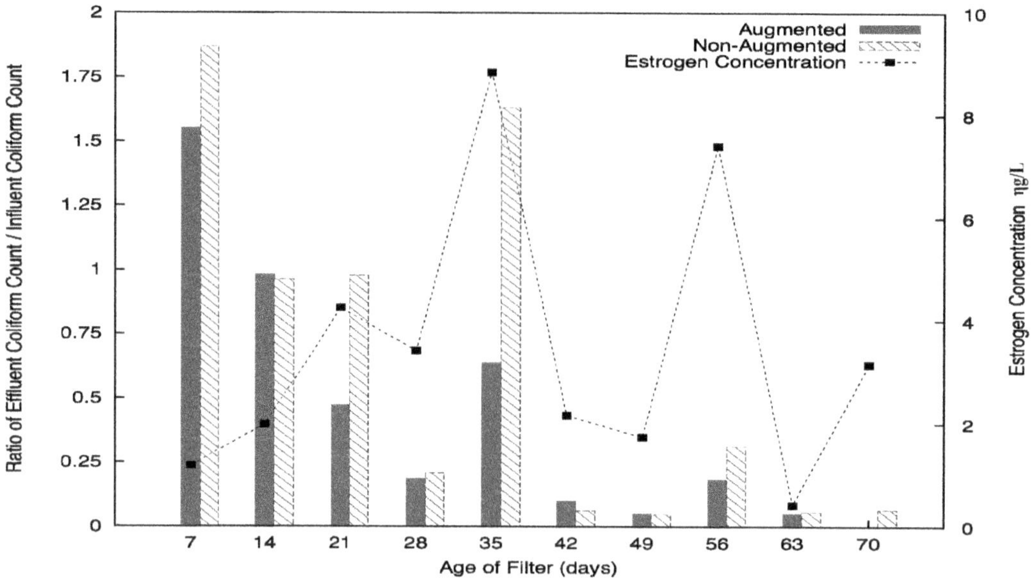

Figure 28.1 Coliform retention in augmented and non-augmented lab-scale slow sand filters, alongside the corresponding combined estrogen (estone, estradiol and estriol) concentration.

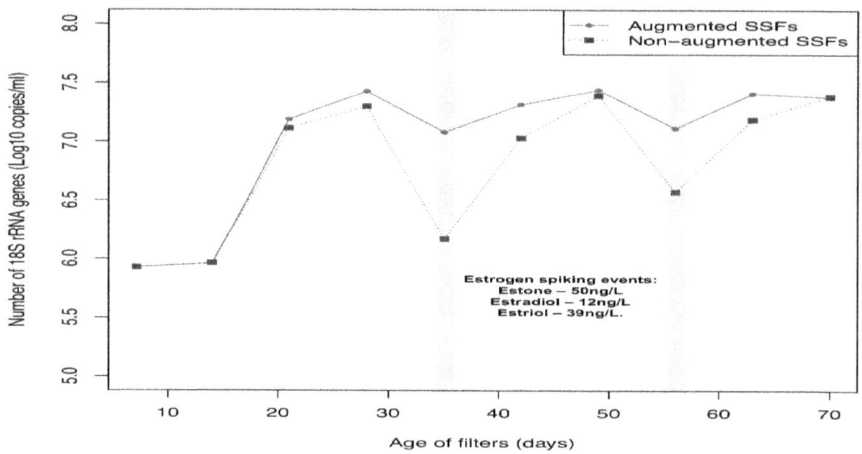

Figure 28.2 Number of copies of the 18S rRNA gene (used to detect eukaryotes) in augmented and non-augmented slow sand filters. Yellow highlighted columns correspond to estrogen spiking time points.

CONCLUSIONS

The main conclusions of this work are:

(1) Bioaugmention is possible and significantly improved the removal of estone and estradiol in experimental slow sand filters;

(2) Estrogen concentration is positively correlated with coliform retention in non-augmented filters (i.e., the higher the estrogen concentration, the lower the coliform removing capability);

(3) High estrogen concentrations cause reduced protozoan growth and death resulting in reduced coliform removal. However, these effects were significantly lower in augmented filters compared to non-augmented filters.

To summarize, bioaugmentation helped to reduce the negative effects of natural estrogens on protozoan populations. This in turn prevented subsequent protozoan death, allowing for optimal coliform removal.

In terms of current and future implications, the discovered negative effects of estrogen exposure on coliform removal in SSFs poses serious questions for the performance of SSFs and other biological water filters (both waste and drinking water). These include understanding the effects of natural estrogens on the performance and micro- and macro- community of such biological filters. This is of paramount importance especially due to the increase in intensive farming and weather phenomena, all of which result in elevated estrogen levels reaching various water sources. This study is the first to discover that estrogen exposure induces a negative effect on coliform removal in SSFs. Further, this work highlights the potential serious consequences of unmonitored estrogen exposure, not only for water quality production but for the survival and health of aquatic ecosystems in water sources such as reservoirs and rivers.

ACKNOWLEDGEMENTS

Sarah Haig is supported by a Lord Kelvin / Adam Smith Research scholarship from the University of Glasgow. A special thanks to the GU68 Engineers Trust, Ian Scouller, Tim Montgomery and Robert Boyd without whom this work would not have been possible.

REFERENCES

Chen T-S, Chen T-C, Yeh K-J C, Chao, Liaw E-T, Sieh C-Y, Chen K-C, Hsieh L-T, and Yeh Y-L. (2010) High estrogen concentrations in receiving river discharge from a concentrated livestock feedlot. *Science of the Total Environment*, **408**(16):3223–3230

Haig S.J, Collins G, Davies R.L, Dorea C, and Quince C, (2011) Biological aspects of slow sand filtration: past, present and future. *Water Science and Technology: Water Supply*, **11**(4):468–472

Haig S.J, Collins G, Davies R.L, and Quince C, (2013, August) Exploring Coliform Removal in Slow Sand Filters using DNA-SIP Coupled with Metagenomics. Presented at Microbial Ecology in Water Engineering, Ann Arbor, USA

Haig S.J, Quince, C., Collins G, Davies R.L, and Gauchotte-Lindsay C. 2014a (In submission) Differential toxicity of estrogens to protozoa species. *Ecotoxicology*

Haig S.J, Quince, Davies R.L, Dorea C.C. and Collins G. 2014b (In submission) Replicating the microbial community and water quality performance of full scale filters in laboratory-scale filters. *Water Research.*

Kim H, Jaffe P.Ř, and Young L.Y. (2004) Simulating biodegradation of toluene in sand column experiments at the macroscopic and pore-level scale for aerobic and denitrifying conditions. *Advances in water resources*, **27**(4):335–348

Quintana J. Carpinteiro J. Rodrıguez I. Lorenzo R. Carro A. and Cela R. (2004) Determination of natural and synthetic estrogens in water by gas chromatography with mass spectrometric detection. *Journal of chromatography A*, **1024**(1):177–185.

Yu C-P, Deeb RA, and Chu K-H. (2013) Microbial degradation of steroidal estrogens. *Chemosphere,* **91**:1225–1235

Chapter 29

Reduction of water hardness with a compact slow sand filtration system

R. Miyahira, K. Yajima**, J. Nakagawa**, M. Koseki**

**Tokyo Kasei University, Japan*

***Oyalox Clean Servis Co., Ltd, Japan*

ABSTRACT

Hardness has a marked effect on the taste of water and it is very important in the manufacture of food. Consequently, water treatment plants in regions with high levels of water hardness typically treat the water chemically before filtering and processing it for further use. We examined the effectiveness of slow sand-filtration for reducing water hardness. Specifically, we examined whether the method, which is safe, simple, inexpensive, and does not employ chemicals, could be used to soften very hard water and produce water that is suitable for producing food. We devised a small, pilot-scale, slow sand-filtration system consisting of three gravel filters and one sand filter connected in series. First, we circulated the water samples through the three gravel filters only, and then measured the hardness of the water by chelatometry. The results showed that water hardness was markedly reduced by filtration through the gravel filters only. The amount of reduction in water hardness was found to correspond to the temporary hardness component of the water samples. By investigating the structure of coarse filter columns in use at water purification plants, it may be possible to soften hard water and produce high-quality water suitable for both food production and drinking without the need for chemical treatment.

Keywords Slow sand filtration, biological purification, food manufacture, hardness, reduction

INTRODUCTION

Water hardness is a function of the calcium and/or magnesium concentration in the water. It is classified as being temporary when the hydrogen carbonate concentration is high, and permanent when the sulfate, nitrate or chloride concentration is high (The Pharmaceutical Society of Japan, 2010). Hard water occurs due to the presence of calcium and magnesium in the ground; as water percolates through the ground, these minerals dissolve in the water (The Pharmaceutical Society of Japan, 2010). The longer the water remains underground under such conditions, the harder it becomes. However, since the calcium and magnesium content of soil varies between regions, so too does the hardness of the water (The Pharmaceutical Society of Japan, 2010). The World Health Organization (WHO) characterizes water with a hardness of 0-60 mg/L as soft, 60-120 mg/L as moderately hard, 120-180 mg/L as hard, and >180 mg/L as very hard (World Health Organization,

2011). While most regions in Japan have soft water, some have hard water. Water hardness affects its taste and makes it difficult to boil high-protein foods (Egawa and Komaki, 2001). In addition, drinking hard water can cause diarrhea (The Pharmaceutical Society of Japan, 2010). Thus, since water hardness can affect quality of life, the maximum national water quality standard was set at <300 mg/L, and preferable in the range 10 to 100 mg/L (Ministry of Health, Labour and Welfare in Japan, 2003). Water treatment plants in areas with hard water typically treat the water chemically before filtration, but such treatment is expensive.

We therefore examined whether a small, slow sand-filtration system that is safe, simple, inexpensive, and which employed biological filtration with no chemicals, could be used to soften very hard water and produce water that was suitable for food production.

METHODOLOGY

I. FILTRATION SYSTEM

We devised a pilot-scale, slow sand-filtration system consisting of three gravel filters and one sand filter connected in series (Fig. 29.1). The first of the three gravel filters contained 1.5 L of pumice as a filter medium (particle diam.: approx. 13 mm), and the other two each contained about 1.5 L of gravel (particle diam.: approx. 5 mm). The sand filter consisted of an elongated polyethylene terephthalate (PET) bottle containing approximately 3 L of fine sand (particle diam.: 0.3–0.45 mm) as the filtration medium (Katayama-Hirayama, 2009). Groundwater was circulated through this array of filters for approximately one month to promote the development of a biofilm, which was monitored by conducting viable bacterial counts and monitoring water physicochemistry. The filtration rate set flow from 5 to 6 m/day. (Nakamoto, 2008)

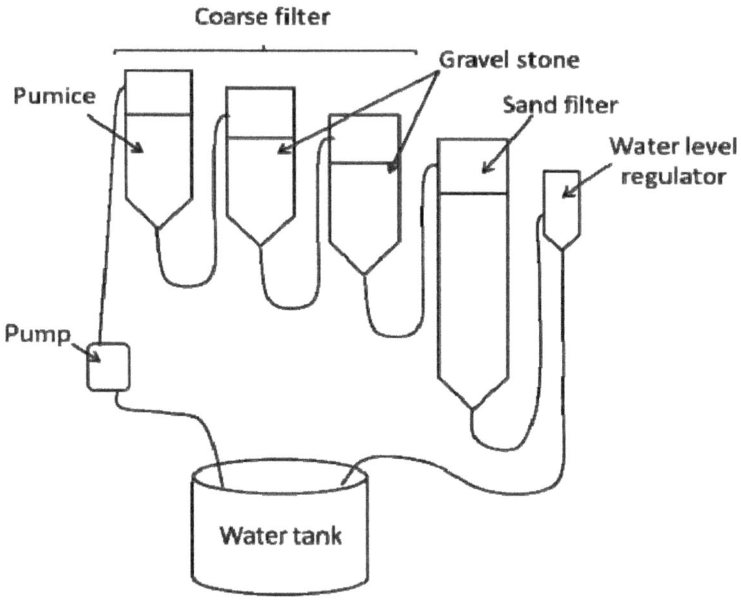

Figure 29.1 Schematic diagram of the slow sand-filtration system.

II. EXPERIMENTAL METHODS

Confirmation of biofilm development by viable bacterial counts

To confirm whether a biofilm had formed in the system, three water samples were collected and viable bacterial were counted: before the filter array, after the gravel filters, and after the entire filter array. Each of the water samples (1 ml) was mixed with sterilized phosphate buffered saline (9 ml), plated on Petrifilm™ Aerobic Count Plates (3M Co., USA) and incubated at 35°C for 2 days, after which colonies were counted.

Removal of dissolved chemical components

After confirming the existence of a biofilm, we examined the changes in the physicochemical characteristics of river water filtered using the filtration system. Water samples were collected from the Shakujii River in Itabashi Ward, Tokyo, Japan, and the following 16 chemical parameters were measured: total organic carbon (TOC), fluorine, chloride ions, nitrite ions, nitrate ions, sulfate ions, sodium ions, magnesium ions, calcium ions, magnesium hardness, calcium hardness, total hardness, potassium ions, ammonium ions, iron and manganese. TOC was analyzed using a total organic carbon analyzer (Shimadzu, TOC-V), and the other parameters were analyzed by ion chromatography (TOSHO,IC2010 and Shimadzu, LC10) using the assay conditions shown in Table 29.1.

Table 29.1 Analysis conditions for ion chromatography

	Anion analysis	Cation analysis
Model	TOSHO IC2010	Shimadzu LC10
Column	TSKgel Super Anion HS (diam.) 4.6 mm × 100 mm	Shim-pack IC-SC1 (diam.) 4.6 mm × 150 mm
Moving phase	3.8 mM/L sodium bicarbonate 3.0 mM/L sodium carbonate	7 mM methanesulfonic acid solution
Capacity	1.5 ml/min	1.0 mll/min
Column temperature	40°C	40°C
Suppressor	Ion exchange resins	———
Detector	Electrical conductivity	Electrical conductivity
Injection sample volume	30 µl	50 µl

Softening commercial mineral water

Mineral water (Evian, Groupe, Danone, France), which has a hardness of approximately 310 mg/L, was used to assess whether the slow sand-filtration system could reduce the hardness of bottled drinking water. Three filter array configurations were examined: (i) three gravel filters and one sand filter in series, (ii) three gravel filters only, and (iii) one sand filter only. Ten liters of mineral drinking water was then circulated through each filter configuration. Samples were collected before filtration, and following 1, 3, 5, 7 and 9 days of filtration.

Measurement of temporary and permanent hardness in mineral water

Four-hundred milliliters of mineral water was boiled continuously while adding decarbonized water for an hour. After boiling, 400 ml of the mineral water/decarbonized water mixture was placed into a measuring cylinder. The boiled water was then passed through filter paper (Advantec® 5c; 110 mm) and the temporary and permanent harnesses were determined by measuring the difference in hardness before and after boiling (The Pharmaceutical Society of Japan, 1995).

Softening hard water containing aqueous solutions of calcium sulfate and calcium chloride

Two aqueous solutions with a hardness of 310 mg/L were prepared using calcium sulfate and calcium chloride, both of which contribute to permanent hardness. These aqueous solutions were filtered through the gravel filter array only. Samples were taken before filtration and following 1, 3, 5, 7 and 9 days of filtration, and the water hardness was measured.

Softening very hard water for potable water supply

Groundwater with a high hardness was collected from the Sodeyama Filtration Plant in Miyakojima, Okinawa Prefecture, Japan. Hardness measurements were taken before filtration and following 1, 3, 5, 7 and 9 days of filtration.

Measurement of temporary and permanent hardness of raw groundwater at the Sodeyama Filtration Plant

Four-hundred milliliters of mineral water was kept on the boil while adding decarbonized water for an hour. The temporary and permanent hardness was measured in the same manner as for the mineral water (see ***Softening commercial mineral water***).

Hardness measurements

Water hardness of the collected samples was measured by chelatometry using a standard solution EDTA solution (0.01 M) (The Pharmaceutical Society of Japan, 2010). In addition, subsamples of this water were subjected to ion chromatography analysis (The Pharmaceutical Society of Japan, 2010).

RESULTS AND DISCUSSION

Confirmation of biofilm formation by viable bacteria count

Figure 29.2 shows the observed decrease in viable bacterial count and the increase in filtration efficiency (solid line) associated with filtration. The viable bacteria count decreased by approximately 89% due to filtration.

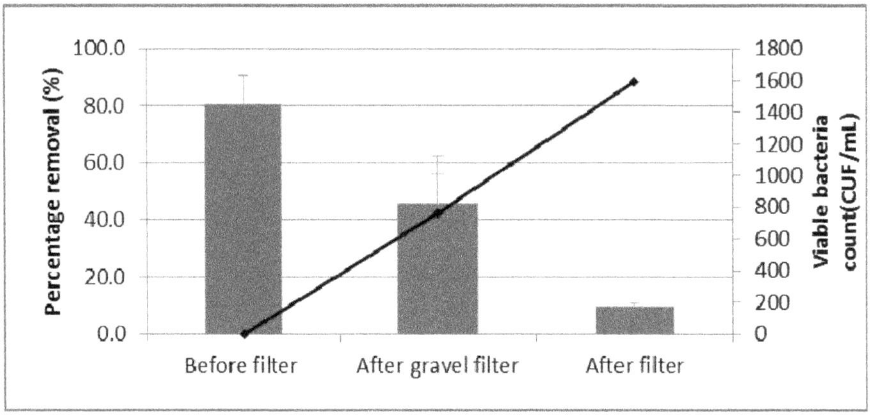

Figure 29.2 Average viable bacterial count in Shakujii River water in Itabashi Ward, Tokyo, Japan. (assays were performed in duplicate)

Removal of dissolved chemical components

Figure 29.3 shows the changes in Shakujii River water quality due to filtration. Of the chemical parameters examined, TOC, iron, and manganese levels decreased over the course of filtration. These results confirmed that a biofilm had been formed in the filter array. In addition, five chemical parameters – magnesium ions, calcium ions, magnesium hardness, calcium hardness, total hardness – also decreased, implying that the slow sand-filtration system was able to soften the water. The levels of fluorine, nitrite ions and ammonium ions were below detectable limits, implying that they were also below detectable limits in the river water sample. Ions of chloride, nitrate, sulfate, sodium and potassium did not decrease and it appears that these ions occurred at levels that were too low for the microbes in the biofilm to remove. However, the items for which a decrease was not observed meet the water-quality standards set by the Ministry of Health, Labour and Welfare, Japan – see web site (http://www.mhlw.go.jp/stf/seisakunitsuite/bunya/topics/bukyoku/kenkou/suido/kijun/kijunchi.html) - and the WHO (2011). From the above results, the presence of a biofilm was confirmed.

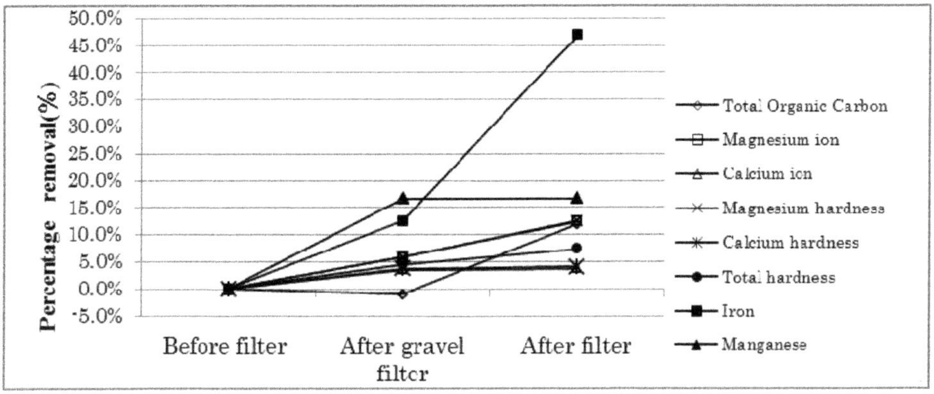

Figure 29.3 Changes in physicochemical parameters over the course of filtration (assays were performed in duplicate)

Softening of mineral drinking water

The hardness of the sample water decreased after being circulated through the complete filter array (i.e., three gravel filters and one sand filter in series). The water hardness decreased from about 310 mg/L before filtration, to about 216 mg/L on day 1, to 98 mg/L on day 9 of filtration (see Fig 29.4).

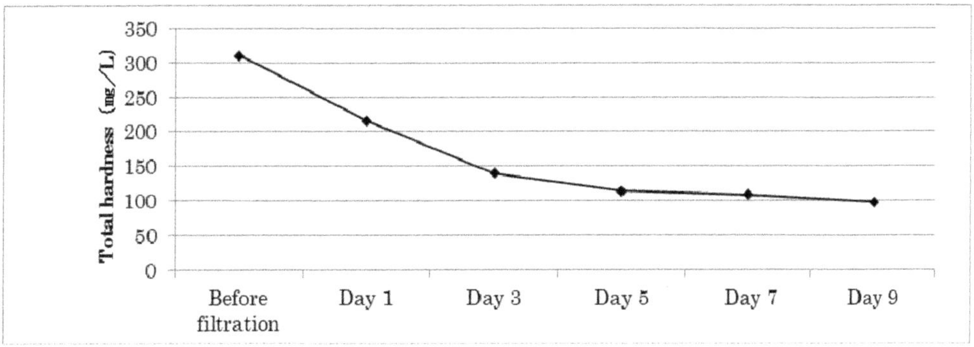

Figure 29.4 Change in commercial mineral water hardness after filtering through the complete filter array.

When the commercial mineral water sample was circulated through the three gravel filters only, the hardness decreased from about 311 mg/L before filtration, to about 178 mg/L on day 1, and 82 mg/L on day 9 (see Fig. 29.5). These results are comparable to those obtained using the complete array of three gravel filters and one sand filter.

In filtration experiments using only the sand filter, the hardness of the sample water decreased from about 315 mg/L before filtration, to about 292 mg/L on day 1, and 183 mg/L on day 9 of filtration (see Fig. 29.6). Of the three filter configurations tested, filtering the commercial mineral water sample through the sand filter only was the least effective for reducing water hardness. The effectiveness of the gravel filters is considered to be due to deposition of magnesium and calcium by the biofilm.

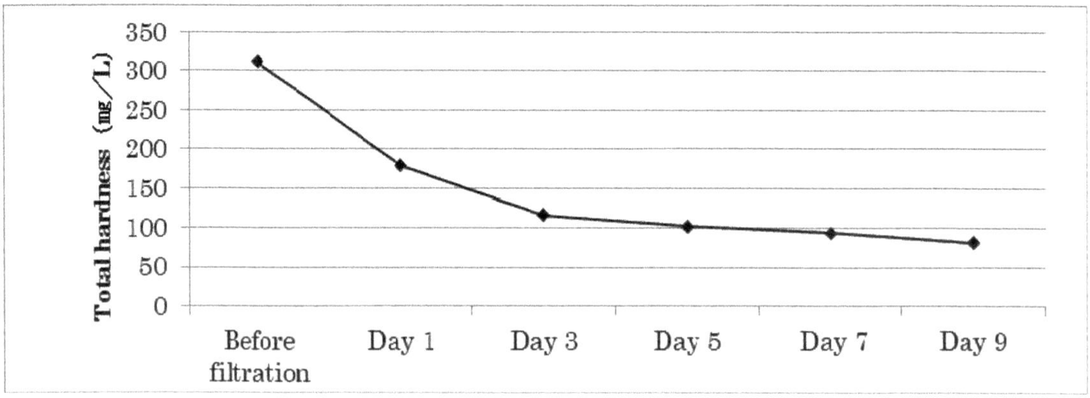

Figure 29.5 Change in commercial mineral water hardness after filtering through the three gravel filters only.

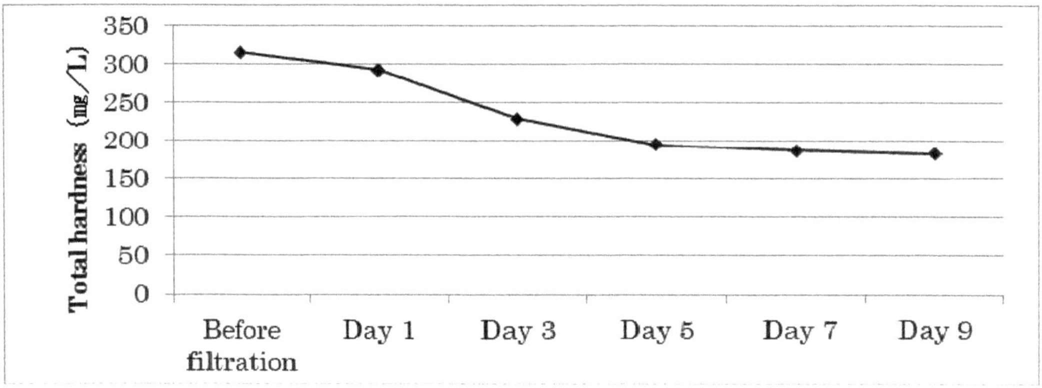

Figure 29.6 Change in commercial mineral water hardness after filtering through the sand filter only.

Temporary and permanent hardness of sample mineral water

The hardness of the unboiled mineral water was 313 mg/L, while that of the boiled mineral water was 81 mg/L. These findings imply that the permanent hardness of the mineral water sample is 81 mg/L and the temporary hardness is 232 mg/L. Thus, the temporary hardness of the mineral water decreased due to gravel filtration.

Softening aqueous solutions containing calcium sulfate and calcium chloride

The hardness of the calcium sulfate aqueous solution before and after filtration was 312 and 304 mg/L, respectively. The hardness of the calcium chloride aqueous solution before and after filtration was 308 and 294 mg/L, respectively (Fig. 29.7). The slight decrease in the hardness of both aqueous solutions after filtration indicated that gravel filtration is better suited for treating temporary hardness.

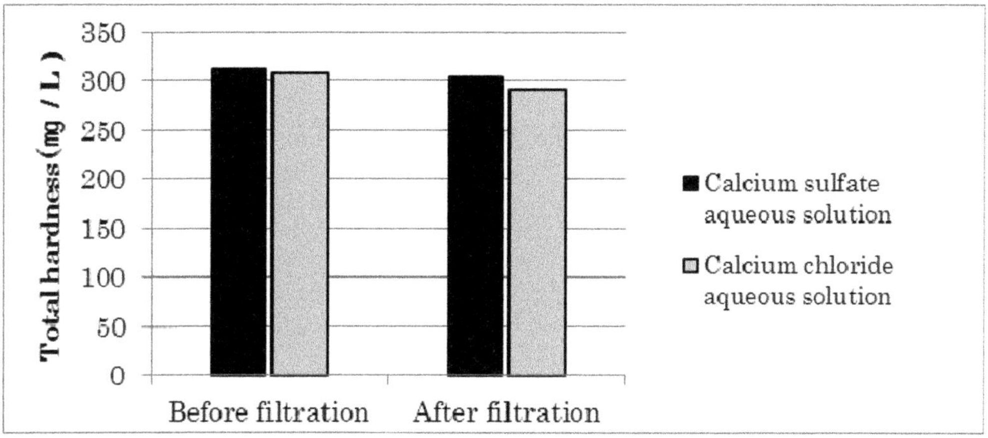

Figure 29.7 Change in permanent hardness before and after filtration through the gravel-only filter array.

Softening very hard water for potable water supply

The hardness of the raw groundwater samples decreased from about 283 mg/L before filtration, to about 215 mg/L on day 1, and 139 mg/L on day 9, respectively, of filtering through the gravel-only array (see Fig. 29.8). Thus, even after 9 days, the decrease was only moderate.

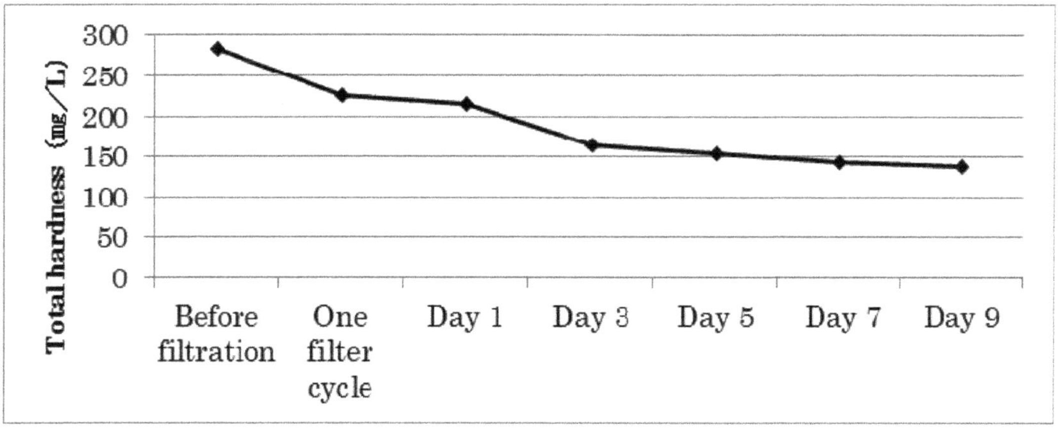

Figure 29.8 Change in hardness of very hard water used for potable water supply.

Temporary and permanent hardness of raw water from the Sodeyama Filtration Plant

The hardness of unboiled and boiled water samples from the Sodeyama Filtration Plant was 283 and 65 mg/L, respectively, which implies that the permanent hardness of the mineral water was 65 mg/L and the temporary hardness was 218 mg/L. These findings show that the gravel filter array was unable to completely soften the sample water from the Sodeyama Filtration Plant, possibly because of the loss of the green algae biofilm in the gravel filters. This is considered to be because this experiment was performed after those on the commercial mineral water, and the aqueous solutions of calcium sulfate and calcium chloride. It is thought that the algal component of the biofilm had changed from predominantly green algae to brown algae due to the low temperatures and short days in wintertime. It is considered likely that if this had not occurred, the gravel filters could completely soften the raw water from the Sodeyama Filtration Plant.

General discussion

The results of the experiments showed that the hardness of water could be reduced by filtering the water through an array of gravel filters. In addition, the slow sand-filtration system was shown to be capable of removing temporary hardness in water. However, a total of 9 days of filtration was required to soften the water, which is problematic as the algal component of the biofilter changes according to the season. To decrease the processing time required, we propose that both the water circulation rate and the size of the gravel filter should be increased, as this would increase the surface area available for algal attachment and, consequently, facilitate bioprocessing of the water. In addition, to maintain a healthy population of algae, the water and filter temperatures need to be carefully controlled. We are currently examining ways in which the filtration time can be reduced while maintaining sufficient plant growth. In addition, we will examine whether increasing the size of the gravel filter system in an actual filtration plant is practicable.

CONCLUSIONS

The conclusions of this study are as follows:

(1) In the slow sand-filtration system, the gravel filter array was more effective than the sand filter. Specifically, the gravel filter array was as effective for softening water as the gravel filter array *and* the sand filter in series. The sand filter was the least effective filter for reducing water hardness.

(2) Only the gravel filter was capable of reducing the temporary hardness component of the water. The reduction in the hardness of the mineral water (evian) was found to be equal to its temporary hardness component. Thus, the gravel filter cannot decrease the permanent hardness component of the water.

(3) The ability of the filters to reduce hardness was affected by the condition of the algae colonizing the filters. It is considered that green algae are the most effective, because when brown algae formed, the temporary hardness of raw water from the sodeyama filtration plant could not be completely eliminated.

ACKNOWLEDGEMENTS

The authors wish to acknowledge the important contribution of The Towa Foundation for Food Research, the very kind advice of Dr. N. Nakamoto and kind gift of water samples from Mr. K. Kajiwara, Mr. H. Nohara and Mr. K. Miyaguni of Miyakojima-city.

REFERENCES

The Pharmaceutical Society of Japan (2010) Methods of Analysis in Health Science 730-731.

A. Egawa and M. Komaki (2001) On Water Hardness. *Journal of Human Environmental Engineering*, 3(1), 106-109.

K. Katayama-Hirayama, S. Arai, T. Kobayashi, H. Matsuda, Z. Luo, M. Tachibana, H. Kaneko, T. Akitsu and K. Hirayama (2009) Removal of bisphenols by slow sand filtration. *Water Science & Technology: Water Supply,* **9**(3), 263-268.

The Pharmaceutical Society of Japan (1995) Standard Methods of Analysis for Hygienic Chemists – With Commentary – authorized by the Pharmaceutical Society of Japan. 970-973.

N. Nakamoto (2008) How to make delicious water, Ecological Purification System-trump technology of revival of delicious water supply. Tsukiji Shokan. Tokyo, Japan. 6-17, 43, 52

Homepage of Ministry of Health, Labour and Welfare in Japan, http://www.mhlw.go.jp/stf/seisakunitsuite/bunya/topics/bukyoku/kenkou/suido/kijun/kijunchi.html

World Health Organization (2011) *Guidelines for drinking-water quality.* Fourth edition.

Chapter 30

Installing a DIY-based home slow sand filter - Is it possible to make it without speciality?

*H. Hukuoka**, **
**Self-employment plumber, **NPO member, CWSC, Tokyo, Japan*
Kawasaki 3-19-29,Shunan-shi, Yamaguchi Prefecture 746-0001 Japan

ABSTRACT
A trial of installing small-scale water purification systems has so far been performed, since the biological purification process is easy to go without academic training in water engineering. Two cases have been introduced herein as evidence of private facilities made by the author; this suggests that practice of slow sand filters could be actualized even on a DIY private basis. A series of my practice has been carried out, taking a stance on "wait and see" the natural life. The results of bacteriological hygienic examination were remarkably satisfied in spite of simple procedures of scarce maintenance after initial installation. One of the cases was targeting mountain runoff as a raw water source was resulted in usual outputs, but the other case applied to municipal tap water was resulted in unusual ones, thereby influenced by chlorine residue in the water. The whole process was occupied by chlorine-resistant fungi having gelled polysaccharide, leading to an "extreme environment" accidentally appeared in a manmade specific niche.

Keywords Water purification system; private use; handmade facilities; chlorine-resistant fungi; extreme environment; human-induced stress; niche occupation

INTRODUCTION

The slow sand filtration (SSF) process to produce drinking water has a simple structure, but a high efficiency in terms of cost-benefit balance (Nakamoto, 2005). This encouraged me to build up such a system using DIY (Do-it-yourself)-related daily materials. These experience-based techniques could easily be applicable for less-populated remote areas and/or in emergency situations after natural disasters.

However, lack of speciality in water engineering, in other words, being free from the fixed notion, would lead me to the unknown outcome, hence resulted in the predominance of chlorine-resistant fungi. Through the present field investigation, daily monitoring by unaided eyes and outsourcing bacteriological examinations were constantly introduced, together with the NPO-based advisory members using an epi-fluorescence microscopy and/or specific staining techniques to differentiate the mixed populations.

METHODOLOGY

All the materials, such as plastic buckets, pipes and sand, were basically obtained from a home-centre nearby my house. However, second-handed sand gifted from a municipal water purification plant was mixed with natural sand collected from the bed of a water course. Two types of raw water were fed to each DIY-SSFs; one was municipal tap water and the other was mountain runoff, as shown in Figure 30.1.

Figure 30.1 Two types of DIY-slow sand filters. Left (a): tap water-fed Right (b): mountain runoff-fed

Table 30.1 shows specification details of these two SSFs, their running conditions in details included. Routine monitoring on both the SSF systems has been carried out by unaided eyes on an everyday basis. Biosafety of the filtrate as the final effluent was intermittently been monitored by counting ordinary SPC (Standard Plate Count) and coliforms, both being outsourced to a nearby public testing organisation.

Table 30.1 Specifications of two types of DIY-slow sand filters.

Case/Run time	A (May 2002-present)	B (July 2012-present)
Raw water	Tap water (clear, chlorinated)	Mountain runoff (usually clear)*[1]
Pre-treatment	A pond removing chlorine residue, Water plant community , Cascading water courses	Runoff surface water intake, A transit pond with gravels, A net/cover to trap litters
Main body	Polypropylene barrel (300mm diameter)	Polypropylene bucket (450mm diameter)
Additional tank	A storage tank for filtrate	An original stabilizer for water flow
Filter media	Second-handed sand/gravels	Natural and commercial sand mix
Flow rate	Initially, 4m/day Currently, 2-4m/day	Initially, 15m/day Meantime, 4-27m/day Currently, 13-13.5m/day

*[1] A sedimentation sink used for separating suspended mud particles from the runoff water.

Microbiological observation of the samples was made by a phase contrast microscope (Wraymer BX-2700TPH, Osaka) plus a WrayCam G130. Cyanobacterial cells were detected by an epi-fluorescence microscope (Olympus BX-60, Tokyo) with a digital camera (Canon EOS "Kiss" DigitalX, Tokyo).

RESULTS AND DISCUSSION

Bacteriological indices of the filtered water have intermittently been monitored as shown in Table 30.2.

Table 30.2 Results of bacteriological examinations.

Case A			Case B			
Date	SPC bacteria	Coliform	Date	SPC	*E. coli* counts	F. rate
11 Jun 2002	14,000*	+	10 Dec 2012	0	Not detected	15m/d
10 Sep 2002	790	Not detected	04 Mar 2013	1	Not detected	25m/d
17 Dec 2002	2	Not detected	28 May 2013	1	Not detected	25m/d
08 Mar 2005	12	Not detected	SPC stands for Standard Plate Count, namely, the fast-growing heterotrophic bacterial group.			
14 Jun 2005	13	Not detected				

*All the bacterial counts (/mL) were provided from a nearby public organisation outsourced.

Filamentous fungi and alga were observed on the sand surface of Case A and B, respectively. The latter was commonly observed in the other SSFs as well, whilst the former was an unusual outcome, induced under a stress by chlorine residue existing in the tap water, suggesting "extreme environment".

Macro-colonies of the fungi floating on the water covered almost the whole area, and the fungi tended to form mushroom-shaped colonies around the inflowing point, creeping up the side walls (Figure 30.2). The texture of the colonies was somewhat jellied, showing evidence of the translucent capsular materials stained with Indian ink, alongside the mycelia/spores stained with lactophenol cotton blue (See Figure 30.3).

Figure 30.2 Macro-colonies of chlorine-resistant fungi. Left (a): Floating on the water Right (b): Forming a mushroom-like shape

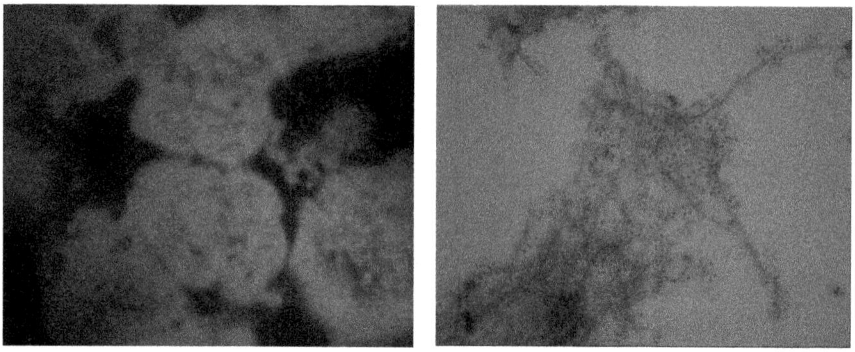

Figure 30.3 Chlorine-resistant fungi, showing a stained specimen. Left (a): Indian ink stained (phase contrast) Right (b): Lactophenol cotton blue stained

None of the filamentous diatoms were found herein this extreme habitat under artificial stress with the chlorine contamination of a tap water origin. Then, who was responsible primary production there? Figure 30.4 shows the colonization of photosynthetic prokaryotes (possibly cyanobacteria) direct onto sand granules was observed by autofluorescence under the blue excitation of an epi-fluorescence microscopy. Thus, these photosynthetic populations may play an important role in producing organic matter to support the heterotrophic growth of fungi in a tap water containing scarce organic nutrients.

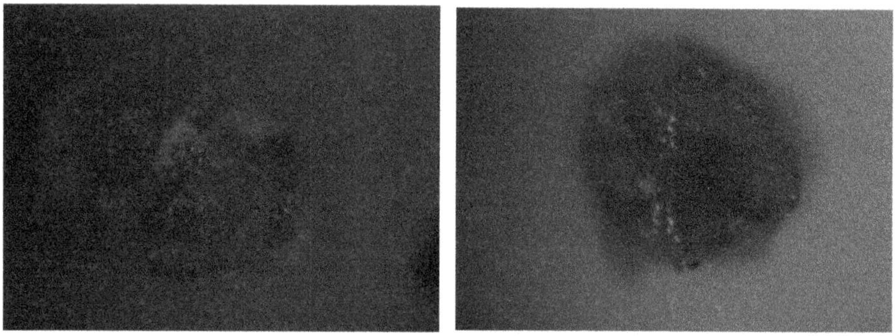

Figure 30.4 Epi-fluorescence microphotographs of photosynthetic bacteria on the sand granules, their autofluorescence signals were detected under blue excitation.

Application of biological SSFs to the chlorinated water from a tap may sound non-sense from the view of engineering. However, the author and his adviser eventually reached at a new discovery unexpected.

CONCLUSIONS

The specialist group usually tends to take their already known fixed routes, but the non-specialist group tends to choose an alternative way without notice; this may invite us to the unveiled world. Thus, diversity of community consisting of both specialist and non-specialist mix would be of necessity, hopefully helping each other to find a better solution or failure-based learning towards sustainable lifestyles in the end. We have learnt an emergence of chlorine resistant fungi in an extreme environment in quite closely-related to our surrounding living areas.

ACKNOWLEDGEMENTS

The author wishes to express his special thanks to Dr. J Takeuchi for his kind help when preparing the English manuscript and capturing microphotographs in his laboratory at Kure Kosen College, Hiroshima. Thanks are also due to Mecan Imaging Inc., Saitama, for lending JT a camera adaptor to take photos.

REFERENCES

Nakamoto, N. (2002) *Nama de Oishii Suidosui* (Natural slow sand filters), Tsukiji Shokan, Tokyo.
Nakamoto, N. (2005) *Oishii Mizu no Tsukurikata* (Biofiltration technique), Tsukiji Shokan, Tokyo.

Chapter 31

Development of an unmanned operate system for a slow sand filter/ecological purification system

T. Hirose
Takemi corporation Co., Ltd., Sakai-city, Osaka-prefecture, Japan

ABSTRACT
In order to meet various needs, the authors have developed moderate speed filtration facilities and their remote monitoring operation system that can become independent for energy demands only with electric power supplied by photovoltaic power generation. As a result of having performed an operation experiment using a pilot plant for 60 days, the system that was ecological superior in economy was demonstrated.

Keywords Water supply, biological treatment, slow sand filtration, photovoltaic power generation, remote control.

INTRODUCTION

Electric power supply system is essential for the modern water supply system. When the plant is attacked by a big earthquake or a severe disaster, the electric power supply is sometimes stopped for long period. Then the plant cannot produce the drinking water. The water supply in this area should be covered temporally from other purification plant. When the recovery of the plant from the disaster takes long time, life of the residence becomes hard. Secure of an enough supply of the safe water is essential for the residence.

Further, the number of engineer staff members engaged in water supply business tends to decrease every year. Security and cultivation of human resources for small business entities in short of staff members, of which management burden on their staff members is large in particular, are a serious problem. Therefore, the authors focused on small purification plants scattering in suburbs of various parts of Japan and have developed a system that enables slow sand filtration being unmanned, even without power supply. A pilot plant was used for the experiment. Operation adjustment by a teleoperator and scraping of filter surface were performed with electric power generated by photovoltaic power generation. As a result, it has reached the level at which the system can be operated economically in actual applications and therefore we report the achievement in this paper.

OUTLINE OF THE SYSTEM

A pilot plant was installed in the existing small scale water purification plant. raw water was taken by natural flow from the river. Internet by a phone line was used for communication of the supervisory control. Remote supervisory control was performed with a desktop PC installed in our office 21km away from the pilot plant. Illustration of the pilot plant is shown in Figure 31.1. Specifications of the pilot plant are shown in Table 31.1.

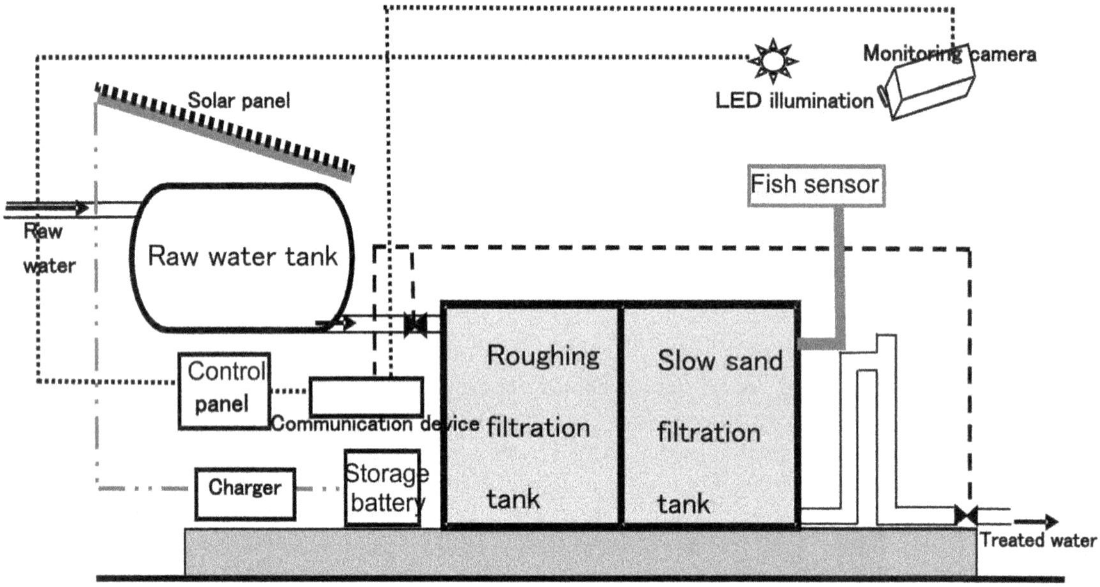

Figure 31.1 Outline of the pilot plant

Table 31.1 Specifications of the pilot plant

Raw water tank	Polyethylene tank
Polyethylene tank	Stainless steel 1,200mm in width, 800mm in length, 1,200mm in height
	Slow sand filtration x 1 tank (filter medium: gravel 12mm - 20mm, 600mm in depth)
	(filter medium: effective diameter of the slow sand filtration and 0.41mm, Uniformity coefficient 1.8, thickness 500mm, water depth 450mm)
Filtration rate	Coarse filtration: 10m/day, Moderate speed filtration: 5m/day
Surface scraping	Automatic siphon-type scraping device
Photovoltaic power generation	Maximum power 420W, electricity accumulation system
Monitoring device	Monitor CCD Camera x 2 (with illumination)
Valves	Electric valve
Control panel	Valve control equipment, Internet device, and Telecontrol device incorporated

EXPERIMENT AND RESULT

RAW WATER AND TREATED WATER

The experiment period was 60 days. Although the flow rate at an early stage of flowing was raised straight to 5m/day, it did not affect the treated water. Average of the raw water turbidity during the period was 11.3 degrees. The treated water turbidity was 1.47 degrees on Day 0 of flowing though it gradually lowered to 0.16 degrees on Day 4. The water quality management target value of Japan (0.1 degrees or less) was obtained on Day 16. Although raw water of more than 50 degrees flowed due to rise in raw water turbidity by rainfall, it did not affect the treated water. Since the loss head rose to 35cm after the rise in raw water turbidity (Day 40), scraping of the slow sand filtration outer layer was performed.

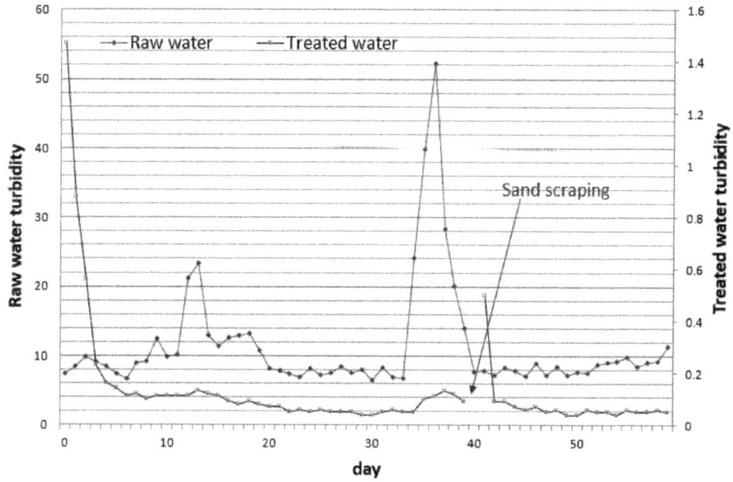

Figure 31.2 Daily change of raw water turbidity and treated water turbidity

TELECONTROL

Filtration rate adjustment was performed by telecontrol while watching images from a surveillance camera. Increase of the loss head was confirmed with a surveillance camera. Cleaning of the roughing filtration tank and scraping of the slow sand filtration outer layer were performed automatically by telecontrol. The upflow type was chosen for the roughing filtration tank, having a structure by which sedimented mud deposits in the lower part. For cleaning of the roughing filtration tank, the drain valve was opened and sedimented mud was discharged. For scraping of the outer layer, the scraping device was operated without draining the water in the slow sand filtration tank. A siphon is generated by an ejector and seaweed and sand on the surface were discharged by the stream. The scraping device moved in the tank by motor drive. A series of operation was automatically performed by sequence. Valve operation was performed by telecontrol so that treated water with high turbidity could not flow into the pure water reservoir at the time of starting and restarting filtration.

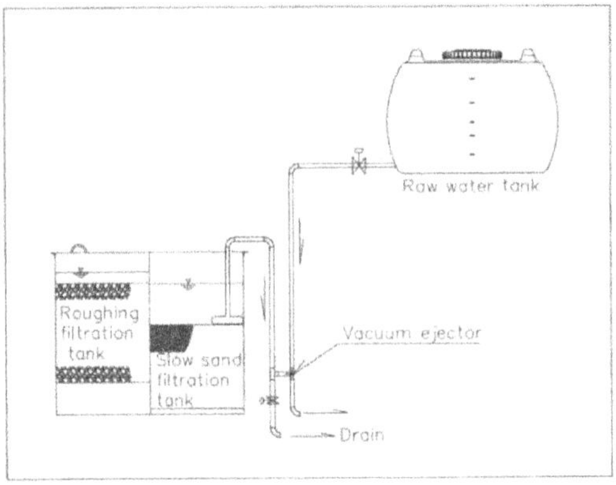

Figure 31.3 Outline of the scraping device (1)

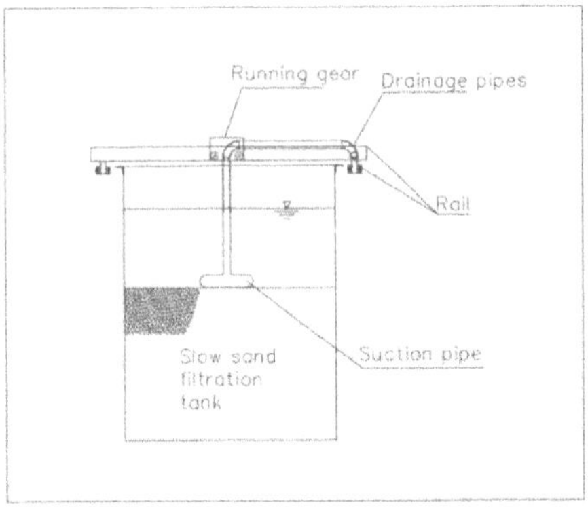

Figure 31.4 Outline of the scraping device (2)

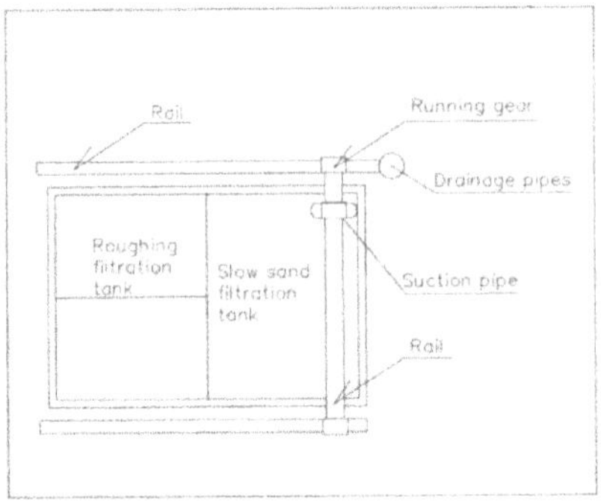

Figure 31.5 Outline of the scraping device (3)

RESULT

In a 60-day operation experiment, filtration facilities were successfully operated only with electric power generated by photovoltaic power generation. Further, frequency of the check was reduced by performing a remote control. The burden on staff in the operation management can be reduced in this way.

ECONOMICAL EFFICIENCY

In this experiment, we performed remote operation of one place of facilities. In actual operation, plural facilities are to be operated. In that case, having one engineer who operate facilities from a remote place enables to reduce plural staff members of each facility required for daily management and cost required for scraping. Further, fully independent operation can be performed in facilities by photovoltaic power generation in a water purification plant and therefore no electric cost is required.

Photo 31.1 View of the Experiment Plant

CONCLUSIONS

The main conclusions of this work are:

(1) Scraping work by telecontrol is achieved.

(2) Plural water purification plants can be managed by one engineer.

(3) Most of the labor cost for daily management and scraping cost are reduced.

(4) Since operation of facilities can be performed by photovoltaic power generation, no electric cost is required.

Chapter 32

Process Optimization of Solar Energy Based Slow Sand Filtration

Ronjon Chakrabarti, Martin Jekel*, Pankaj Kumar Roy**, Manisha Banik**, Gourab Banerjee**, Somnath Pal**, Asis Mazumdar**,*

**Department of Water Quality Control, Technical University Berlin, 10623 Berlin, Germany*

***School of Water Resource Engineering, Jadavpur University, Kolkata 700032, India*

ABSTRACT

The biological performance of the Slow Sand Filtration (SSF) process is influenced by aerobic conditions in the filter media. The dissolved oxygen levels of the raw water as well as continuous filtration rates are necessary for achieving constant aerobic conditions. Solar Energy Based Slow Sand Filtration have the disadvantage to provide their energy output for a limited daily time period which in addition changes accordingly with varying climatic conditions and yearly seasons. In order to run a SEBSSF, various process modifications and additional plant equipment need to be setup. The presented paper deals with the constructional design of the plant components and the optimization of the SEBSSF at a pilot site in the rural area of West Bengal, India

Keywords Optimization of slow sand filtration, sustainable solar water treatment

INTRODUCTION

The Solar Energy Based Slow Sand Filtration (SEBSSF) is setup in a rural community which is suffering from arsenic groundwater contamination. The approach of the SEBSSF based water treatment plant is to substitute the current drinking water source and use surface water which has very low arsenic concentrations and does not exceed the Indian Drinking Water Quality Standards. The presented solution has been developed on the basis of an intense needs assessment and participatory planning process (Chakrabarti, 2013). The pilot project is taken up and evaluated for its performance in the framework of the DST, EC funded ECO-India project (EC Grant No. 308467).

The jointly developed water supply consists of two treatment plants which are located near to the village Jyotsujan in the Murshidabad district, around 200 km north from Kolkata (Figure 32.1). One of the treatment plants of the community based water supply is sourcing its raw water from a rain-fed village pond. The basic treatment process consists of catchment area management for pollution prevention of rainwater harvesting, a silt trap, sedimentation pond, horizontal roughing filter (HRF), slow sand filter (SSF) and activated carbon filter (ACF). The basic processes are adopted from Gimbel, 2004 and Wegelin, 1996. In addition, optional treatment steps for improving the water quality are developed by European ECO-India partner like aeration equipment, activated filter media and disinfection by mixed oxidants. The SEBSSF system runs with two 1 hp pumps. The raw water pump lifts the pond water into an elevated raw water tank (RWT) from where the water is fed into the gravity based filtration system which discharges it treated water into a clear water tank (CWT). The clear water pump lifts the water from the ground water tank to an overhead tank (OHT) from where the water is distributed via a pipeline to the consumers.

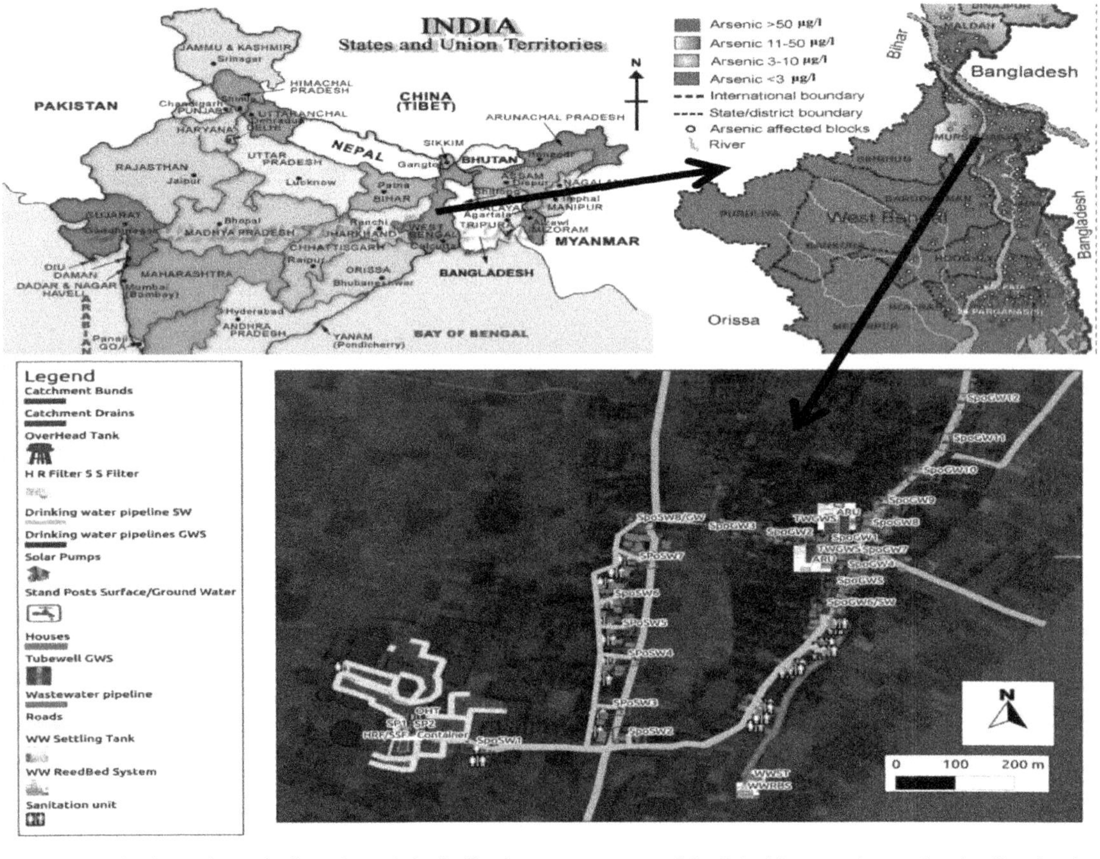

Figure 32.1 Location of pilot site: a) in India (source: maps of India); b) arsenic contamination in the area (source: SOES JU); c) water supply scheme in Jyotsujan

The biological performance of the Slow Sand Filtration (SSF) process is influenced by aerobic conditions in the filter media, the oxygen activates biodegradation in the filter bed. This can be provided by algae which can grow on uncovered filters (Noriyasu IWASE, 2006). Uncovered filters in tropical regions can though also lead to intense algae grow which can clog the filter quickly and the algae can consume oxygen in the night time. Further uncovered filters which are exposed to direct sunlight tend to heat up and increase the water temperature which is not desirable for the consumers. In covered filters not only dissolved oxygen (DO) levels of the raw water but also continuous filtration rates are necessary for achieving constant aerobic conditions (J.T. Visscher, 1987). The pilot plant in Jyotsujan is shaded by the photovoltaic cells and can be considered as covered. Even if the raw water from the pond has been measured to have a DO level between of 3.3 and 7 mg/l according to climatic season and time of the day, the challenge for a solar based system lies in the continuous filtration rate. The PV system provides energy output during a limited daily time period which in addition changes according to varying climatic conditions and yearly seasons. Thus in order to ensure aerobic conditions in the slow sand filter bed various process modifications and additional plant equipment have to be setup related to prolonging the availability of the power source as well as increasing the tank capacity for the raw water storage at an elevated location.

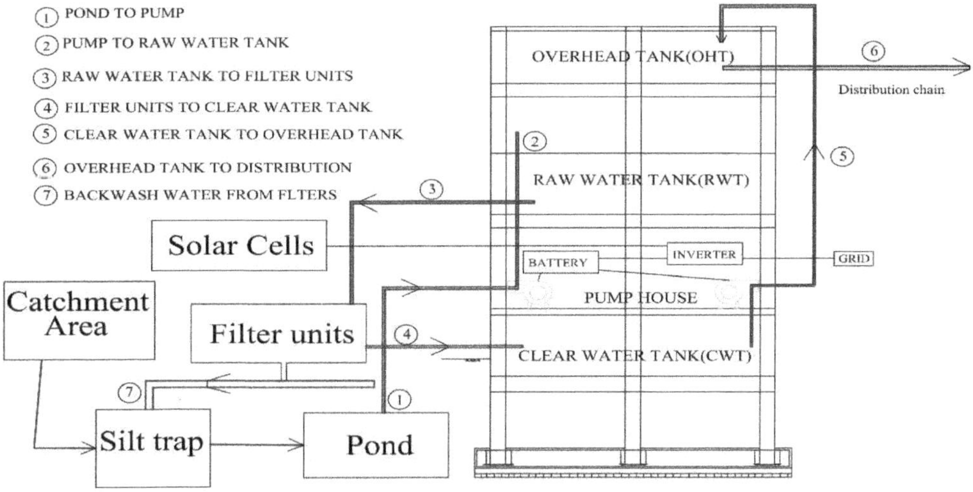

Figure 32.2 Schematic drawing of treatment components and their basic processes

METHODOLOGY

The design of the overall system for the pumping and raw water tank is primarily based on the filtration rate. The SSF process is designed for a flow rate in between 0.1 to 0.2 m/hr (Hendricks, 1991), (Okun, 1984). The water supply is designed to be a 24*7 uninterrupted supply and based on the assessed demand. The total filter area of two times 2.95 m² supplies 14 m³ per day which can be achieved at a constant flow rate of 0.1 m/hr. The higher flow rate of 0.2 m/hr is used in case one of the units has to interrupt its process due to maintenance like scraping, resanding or ripening. The setup considering uninterrupted power supply serves as a basis for comparing it with the process changes. For the solar power availability the buffer tank capacities, flow rates of the pumps, and battery backup are looked into in order to be able to provide the demand throughout the day. In addition aeration equipment, filter bed heights, grain sizes, direct exposure to sun radiation of the Schmutzdecke and further process arrangements are designed for the intermediate power availability. In the basic case with continuous available grid power the tanks are considered to be of a minimum size of 5m³ each. The pumps are considered to have a maximum flow rate of appr. 2m³/hr and can run according to demand. The day is divided into 3h periods for which tank levels and pumping flow rates are calculated.

In the case of solar power the pumps can run only during the peak sun period. For simplification full running time is considered between 8:30 and 15:30. They thus supply 1 m³ between 6-9 h and 15-18 h, and 12 m³ between 9-15 h. In order to keep a sufficient buffer for possible demand fluctuation and pumping timing modification a minimum of 1250 L is considered to have to be in the tank in any of the 3 h periods. For covering the period in which power is unavailable from the PV cells a battery bank is installed. On the other hand in order to cope with the solar dependent pump flow rate the tank size is increased. As an optimized solution the two measures are combined.

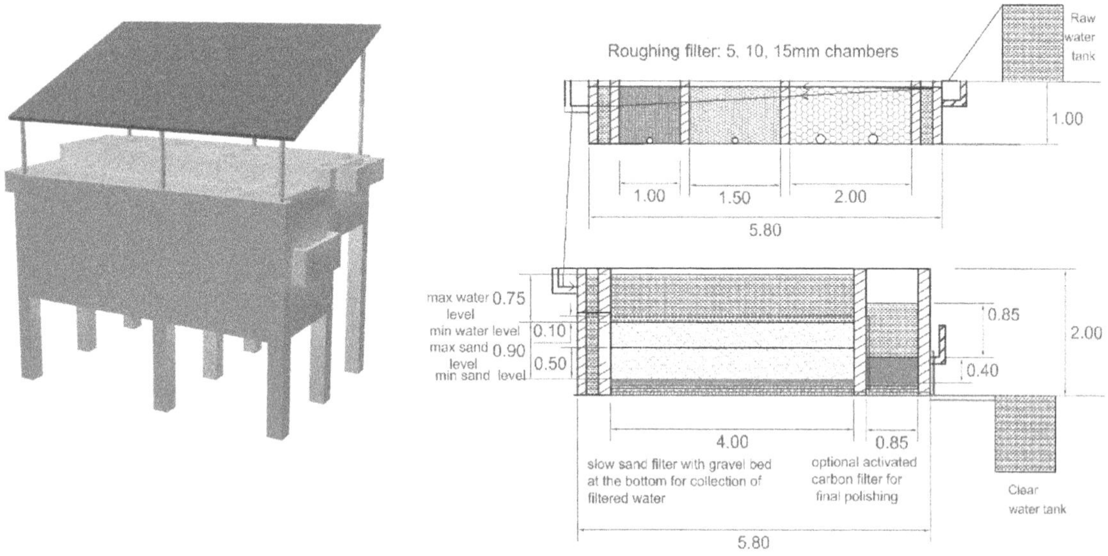

Figure 32.3 3D view of filter units and units with filter media depths

RESULTS AND DISCUSSION

The standard treatment plant design for the SSF system has to be modified when running as SEBSSF. Intermediate filtration processes can be a serious risk for ensuring the overall performance of the plant to produce drinking water quality. Intermediate processes can further lead to large amounts of waste water which has to be discarded when reinitiating the filtration process due to remobilisation of contamination in the filter media during anaerobic conditions. Power backups, like batteries, can be a solution in order to provide energy during radiation poor time periods, but it may be expensive and maintenance intensive. A grid connected SSF system can run the pumps at any time required. As shown in Figure 32.4 the pump extracting raw water from the lake can have the flow rate exactly required for the SSF process. As the pump will have a higher capacity to cater situations other than the standard flow rate the pumping process will not be continuous but intermittent using a raw water tank. Considering the display in 3 h periods the raw water tank has the same water level throughout the day and thus is theoretically unnecessary. According to the continuous filtration process the clear water tank is filled at a constant rate. As the demand varies throughout the day (Garg, 2012): The CWT is used as a buffer tank and the pump is run exactly according to the demand making the OHT theoretically unnecessary. In order to have at least 1250 L of buffer the CWT has to have a size of 5000 L. This system can provide 14 m^3 of water along the assessed demand curve throughout the day and provide a 24*7 water supply.

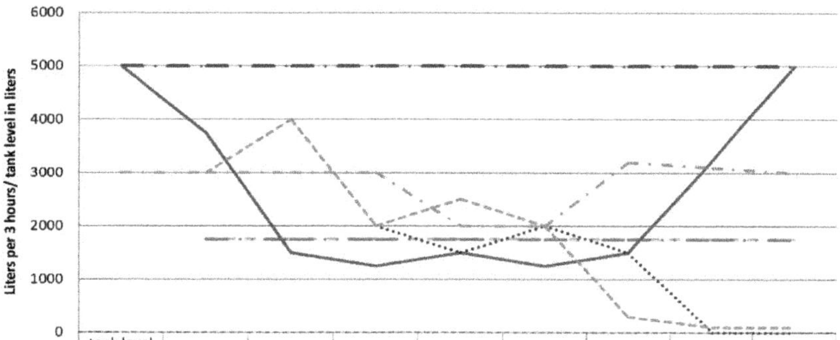

	tank level in the morning 6:00	6-9	9-12	12-15	15-18	18-21	21-24	00-03	3-6
RW pump (RWT in) [L/3h]		1750	1750	1750	1750	1750	1750	1750	1750
Raw Water Tank level [L]	5000	5000	5000	5000	5000	5000	5000	5000	5000
HRF SSF Filter (RWT out / CWT in) [L/3h]		1750	1750	1750	1750	1750	1750	1750	1750
Clear Water Tank level [L]	5000	3750	1500	1250	1500	1250	1500	3250	5000
CW pump (CWT out / OHT in) [L/3h]		3000	4000	2000	1500	2000	1500	0	0
OverHead Tank level [L]	3000	3000	3000	3000	2000	2000	3200	3100	3000
demand (OHT out) [L/3h]		3000	4000	2000	2500	2000	300	100	100

Figure 32.4 SSF system with 5000 L tanks and uninterrupted grid power

The purely solar based system with same tank sizes of 5000 L (Figure 32.5) makes the raw water tank overflow in the timespan 12-15h. The RWT also runs dry in the 0-9h. Thus the SSF will not have a continuous water supply which could influence the quality of water and can also only filter 10500 L instead of 14000 L. The next problem is that the CWT overflows in the night time 0-3 as it cannot be emptied into the OHT due to lack of power for the pumps. The demand cannot be met between at night and in the morning between 18 and 9 h. In total only 10250 L instead of 14000 L can be provided to the consumers. In contrast to the grid connected solution the full capacity of the RWT (5000 L) and a minimum of 2500 L of the OHT are used as buffer and tanks run dry thus 1250 L of reserve are not maintained.

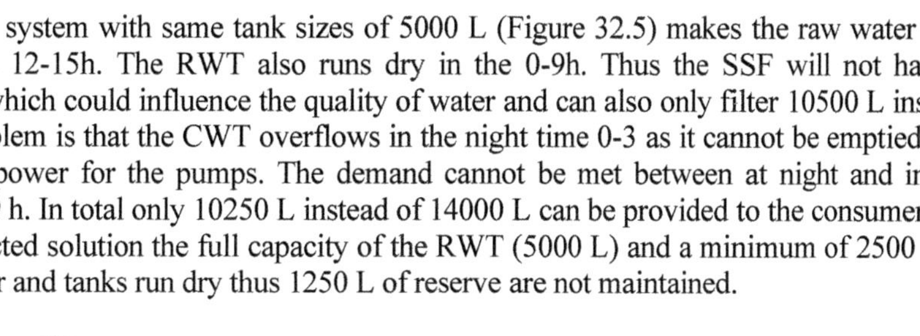

	tank level in the morning 6:00	6-9	9-12	12-15	15-18	18-21	21-24	00-03	3-6
RW pump (RWT in) [L/3h]		1000	6000	6000	1000	0	0	0	0
Raw Water Tank level [L]	0	0	4250	5000	4250	2500	750	0	0
HRF SSF Filter (RWT out / CWT in) [L/3h]		1000	1750	1750	1750	1750	1750	750	0
Clear Water Tank level [L]	5000	5000	750	0	1000	2750	4500	5000	5000
CW pump (CWT out / OHT in) [L/3h]		1000	6000	2500	750	0	0	0	0
OverHead Tank level [L]	0	0	2000	2500	750	0	0	0	0
supply (OHT out) [L/3h]	1000	4000	2000	2500	750	0	0	0	
demand [L/3h]	3000	4000	2000	2500	2000	300	100	100	

Figure 32.5 SSF system with 5000 L tanks and PV power only

The problem can be solved by increasing the buffer tank size (Figure 32.6a)). The full flow rate of the solar PV system can be used if the RWT has a size of at least 9750 L, the CWT has a size of at least 9750 L and the OHT has a size of 7250 L. These values include a buffer of 1250 L for each tank. The SSF will always receive sufficient water and the full demand can be met. As can be seen in Figure 32.5 the bottle neck of the system is the non-availability of energy in the evening and early morning hours when the demand is pretty high. Thus instead of increasing the tank size a battery pack can be installed which can provide energy to the pumps at times when no solar energy is available. The pump has to pump additional 750 L between 6-9 h. It should pump only 500 L less between 9-12 h, 2000 L instead of 6000 L between 12-15h, additional 750 L between 21-24h, additional 1500 L between 0-3h and 1750 L between 3-6 h. In total the raw water pump has to be provided with almost 2.5 h backup. The Clear Water Pump has to pump an additional 1000 L from 6-9h, 750 L less from 9-12h, 4250 L less from 12-15 h, 500 L more from 15-8 h, 750 L more from 18-21 h and 21-24 h and 1000 L from 0-3h and 3-6h each. In total the Clear Water Pump has to run an additional 2.5 hours. This solution considers a minimum buffer of 1250 l in each tank to any 3 h timespan.

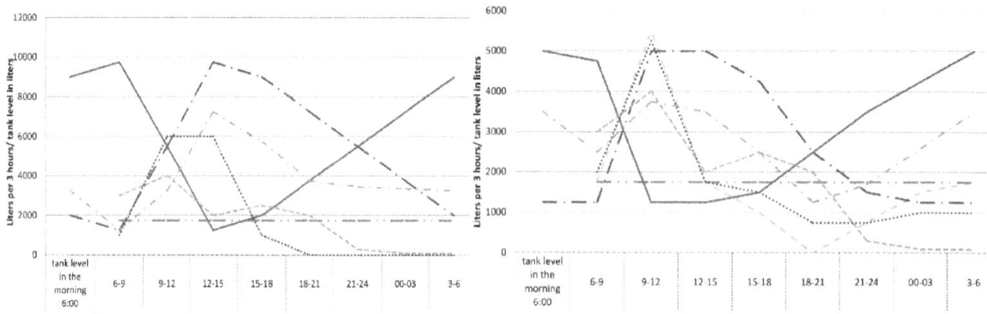

Figure 32.6 a) SSF system with increased tank size on PV power only; b) SSF system with 5000 L tanks on PV power with 2.5 hours battery backup

A combination of increasing tank size and installing a battery backup is looked into as to provide two buffer options at the same time (Figure 32.7). The raw water and clear water pump provide additional 1000 L each from 3-6 h and 6-9 h provided by a one hour backup for each. Both pumps provide 2000 L less from 12-15 h. The RWT thus needs only a size of 8000 L instead of 9750 L. Similarly the CWT has only 8000L instead of 9750 L. The maximum size of the OHT concludes to 5250 L instead of 7250 L.

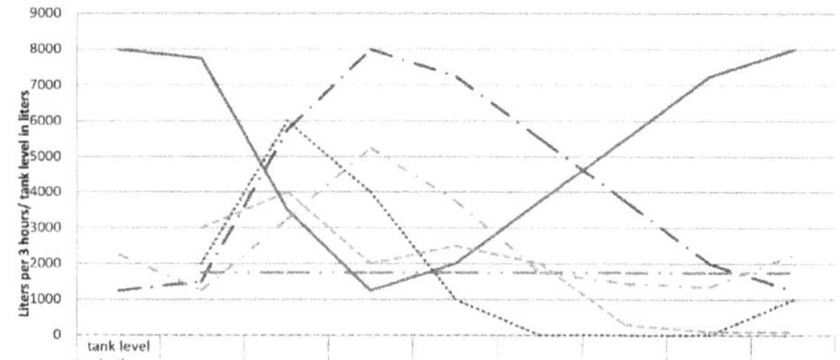

	tank level in the morning 6:00	6-9	9-12	12-15	15-18	18-21	21-24	00-03	3-6
RW pump (RWT in) [L/3h]		2000	6000	4000	1000	0	0	0	1000
Raw Water Tank level [L]	1250	1500	5750	8000	7250	5500	3750	2000	1250
HRF SSF Filter (RWT out / CWT in) [L/3h]		1750	1750	1750	1750	1750	1750	1750	1750
Clear Water Tank level [L]	8000	7750	3500	1250	2000	3750	5500	7250	8000
CW pump (CWT out / OHT in) [L/3h]		2000	6000	4000	1000	0	0	0	1000
OverHead Tank level [L]	2250	1250	3250	5250	3750	1750	1450	1350	2250
demand (OHT out) [L/3h]		3000	4000	2000	2500	2000	300	100	100

Figure 32.7 SSF system with slightly increased tank size, PV power and 1 hour battery backup

Other approaches to increase the available oxygen in the SSF system are additional aeration steps increasing the dissolved oxygen levels in the treatment process which can make up for slower filtration rates during the unavailability of solar power. In the presented structure the HRF can be constructed with a 30cm elevation enabling a spill over between the HRF and SSF which increases the DO level in the SSF feed water. This solution of course only works at times when flow rates are available and is no solution for longer filtration breaks. Shading of the Schmutzdecke among other factors (L. Campos, 2006) leads to less algae growth and longer ripening periods but can play a major impact on the oxygen consumption of the algae during the night time, enabling slower flow rates during the unavailability of power and thus further decreasing the size necessities for buffer tanks. The solar panels themselves can be used for shading the filter as shown in Figure 32.3. Although the measures taken up in the presented design aim at providing continuous constant filtration rate it has to be mentioned that current studies have also come up with suitable performance characteristics of intermittent small scale SSF operation (M. Mansoor Ahammed, 2013)

CONCLUSIONS

SEBSSF can be a viable solution for rural areas which lack from a reliable grid power connection or for environmentally friendly project approaches intending not to make use of any fossil energies. Depending on the energy economics the approach can be cost efficient. The developed pilot plant shows a very promising approach which should be further pursued. The increase of the buffer tanks has comparatively higher investment due to the battery option, but less maintenance necessity.

The optimization of the processes needs further research. The presented solution has been developed in an iterative approach. By application of minimization formulas there is further scope for optimization regarding the optimal tank size and battery size. In addition varying environmental conditions have a significant impact on the biological efficiency of the SSF and cannot be studied in a short period. Long term studies will help in identifying sufficient buffer capacities and process modifications for a reliable uninterrupted drinking water supply and minimizing the necessity of chemical post treatment. For the evaluation of the performance of the setup provisions are being made for measuring dissolved oxygen, turbidity and temperature. These parameters are measured online at the raw water tank, the supernatant of the SSF as well as from the treated water. It is expected that the retrieved water quality data under varying process condition can provide further suggestions for optimization of the treatment process.

REFERENCES

Daniel A.Okun, C. A. (1984). Surface water treatment for communities in developing countries. USA: Water and Sanitation for Health Project.

Garg, S. K. (2012). Water Supply Engineering. New Delhi: Khanna Publishers.

Hendricks, D. (1991). Manual of Design for Slow Sand Filtration. USA: AWWA.

J.T. Visccher, R. P. (1987). Slow Sand Filtration for Community Water Supply. The Hague: IRC International Reference Center for Community Water Supply and Sanitation.

L. Campos, S. S. (2006). Deterministic-Based Model of Slow Sand Filtration. I: Model Development. J. Environ. Eng., S. 132(8), 872-886.

M. Mansoor Ahammed, K. D. (2013). Influence of Operating Parameters on the Performance of Household Slow Sand Filter. Conference on Small Water & Wastewater Systems and Sludge Management. Harbin: IWA.

Noriyasu IWASE, S. K. (2006). Role of algal growth and photosynthesis in slow sand filters as an advanced wastewater treatment. Recent Progress in Slow Sand and Alternative Biofiltration Processes. IWA Publishing.

Rolf Gimbel, M. Jekel (2004). Wasseraufbereitung - Grundlagen und Verfahren. München: DVGW Deutsche Vereinigung des Gas und Wasserfaches e.V.

Ronjon Chakrabarti, M. J. (2013). Sustainable Integrated Water Supply for Rural Communities. Conference of Small Water & Waste Water & Sludge Management. Harbin: IWA.

Wegelin, M. (1996). Surface Water Treatment by Roughing Filters. Duebendorf: SANDEC.

Chapter 33

Performance of ecological purification system in Samoa

J. Tamasese*, E. Fuimaono**, S. Lautua***

*Engineer Lucky Construction, Samoa Former Urban Operation & Maintenance Engineer, Samoa Water Authority

**Urban Operations and Maintenance Water Engineer, Samoa Water Authority

***Asset Management Specialist, Samoa Water Authority

ABSTRACT

Slow Sand Filtration was introduced to Samoa in the late 1980's when the principles of Ecological Purification System (EPS) were yet to be completely understood. The physical design and operational procedures that came with the EPS Water Treatment Plants have been the primary causes of the many issues and failures on water supply services often experienced. Naturally the Slow Sand Filtration technology lost its appeal within Samoa's water supply industry. In the recent years however, EPS has again have been positively accepted owing largely to the strong EPS capacity building relationship between Samoa and Miyakojima. Inadvertently another contributing factor has been the immensely poor performance of the rapid sand systems that were introduced into Samoa from 2007 to 2010. In saying that however, the transition from the older version of slow sand filtration of to the new and improved EPS technology has been somewhat slow, as decision makers and funding restrict the necessary modifications to improve their performance. One treatment plant where progress has been made however is the Malololelei EPS WTP, where overflows were lowered to decrease the tanks water depth to around 500 mm. This paper discusses the issues, challenges and future direction for EPS in Samoa. It also looks at how the modifications to the Malololelei system are changing attitudes and what more needs to be done to bring the Samoa Water Authority systems into the 21st century.

Keywords Slow sand filtration; ecological purification system, Samoa Water Authority

INTRODUCTION

Water supply in the Samoan islands are supplied by Samoa Water Authority and 17% of the countryby village independent water schemes. The Upolu islands where majority of Samoan population inhabit, 90% of water supply is from surface water. (SWA, 2013) This paper looks into the operational performances of the Slow Sand Filter Water Treatment Plants of two (2) out of the three (3) water treatment plants that supply water to the capital city of Apia, and the peri-urban area. Alaoa and Malololelei WTP both have the same German design introduced in the 1980s, were commissioned in 1986. Both systems are located in the same river catchment of Apia. Their slow sand filters (SSF) consist of a bed of sand through which water is passed in a downward direction. The sand layer is 1m thick and is supported on a 0.9m deep gravel layer. The supporting gravel consists of three 0.3mm thick graded layers, 2 - 4mm (top), 4 - 10mm (middle) and 10 - 25mm (bottom).

In 2010, a three (3) year Technical Capacity Building relationship was formalized between the SWA and Miyakojima City Council through a Japan International Cooperation Agency (JICA) funded initiative. One key learning from this initiative was that, shallow water depths improve the SSF filtration efficiency by allowing for maximised penetration of sunlight onto the sand bed where photosynthesis is a key component for an active aerobic community. In 2011 the water depth for Malololelei was lowered to 500mm and it was observed that at a shallower water depth, shorter algae dominated with a much greater rate of algal uplift. After five months it was noted that scrapping of the sand layer had not been required since the lowering of the water levels in the SSF. From 2011to 2013 the Alaoa SSF's were overhauled although the water depth was maintained at approximately 1m. In 2013 Malololelei SSF's were also overhauled with the water level kept at 500mm. The analysis discussed in the results are taken from operational records and water quality results of 16 months, from July 2012 to November 2013, showing the differences in performance of the two water treatment plants at contrasting water level conditions.

METHODOLOGY

For the evaluation discussed in this paper, comparison of the operational maintenance of the two water treatment plants of Alaoa and Malololelei were observed and reported. SSF perfomance is evaluated by the analysis of available water quality data from June 2012 to November 2013, as well as the frequency of algal layer removal (schmutzdecke layer) following an increase in headloss and filter resisitance. Performance of the two plants is mostly impacted on by the raw water characteristics. Recently, significant changes have been observed owing largely to deterioration of water catchment areas. Turbidity is identified as one major parameter determining performance with a huge impact on the Alaoa Plant. Malololelei settling tank inflow cannot currently be adjusted due to problems with the raw water transmission line with the settling tank overloading frequently. The physico-chemical and microbiological parameters are tested weekly by SWA's Water Quality laboratory (WQ Unit). Turbidity data are obtained from these reports are analyzed for trend(s). Where values are used, a derived monthly average is used for comparable analysis over time.

Figure 33.1 Alaoa Water Treatment Plant showing high turbidity at intake (a) and settling tank (b)

Analysis of filter performace efficiency by comparing turbidity data prior and post overhauling of filters at Malololelelei treatment plant. Filters were overhauled within the month of February 2013.

Data were recorded for frequency of filter loss resuting in removal of dirty layer. Frequency are observed as due to high loading matters due to poor quality of inflow raw water (see Figure 33.1) or poor performance of plant operation.

RESULTS AND DISCUSSION

Turbidity

Comparison prior and post overhauling of Malololelei

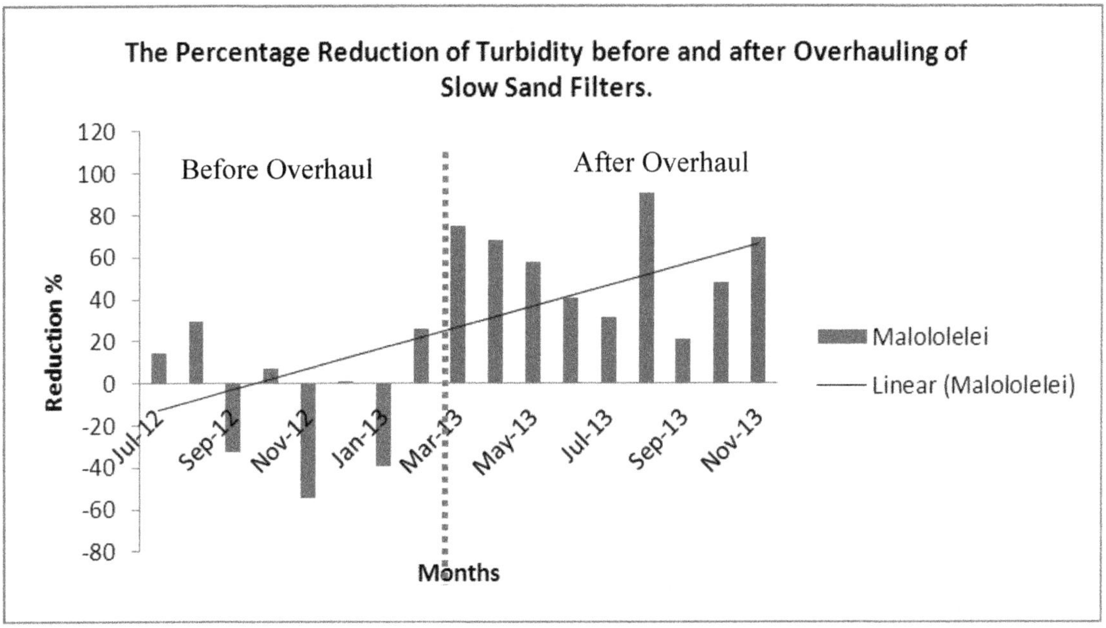

Figure 33.2 The percentage reduction of turbidity before and after overhauling of SSF. Improved performance is shown in the period following the overhaul of the Slow sand filters at Malololelei where there is a significant increase in percentage of turbidity reduction as the water passes through the sand medium to the balancing tank.

High turbidity at the Alaoa Treatment Plant

Susceptibility of Alaoa Plant to high turbidity in heavy rainfall periods has been most challenging for the Authority. Turbidity levels at intake can be as high as 655 NTU. As such, filter run is short and the high loading rates often result in more frequent scraping of top algal layer.

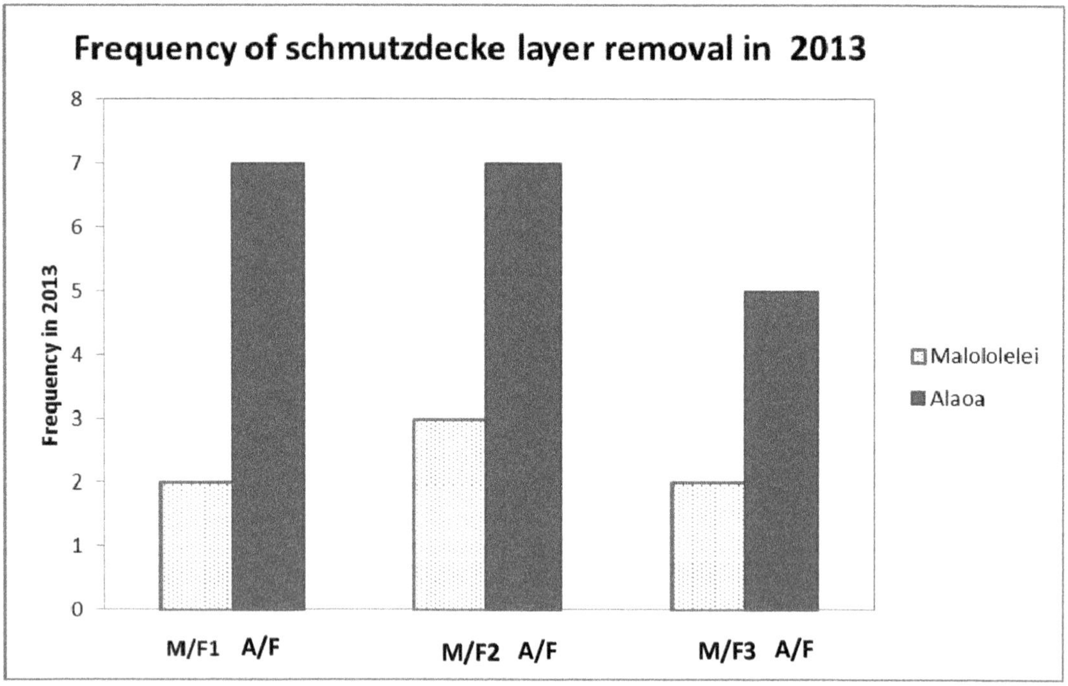

Figure 33.3 Comparison of the frequency of algal layer removal for Malololelei WTP and Alaoa WTP

Figure 33.3, shows the comparison in number of times the SSF in Malololelei and Alaoa have been scraped in the 12 month period for 2013. SSF sand depth was averaging 100mm before the overhaul. After the overhaul the sand depth was increased to 500mm allowing for greater sediment retention in the SSF's

The scraping frequency of Malololelei WTP was similar to Alaoa WTP before the reduction in SSF water level. Malololelei scraping frequency is now observed to be significantly less in comparison to Alaoa WTP scraping frequency which remains unchanged where the water depth was still retained at 1m. Historical operational data is very difficult to come by as it has only until recently become a priority to collect it.

Justification of Overhauling

Figure 33.4 Images during overhauling of the Malololelei WTP (February 2013)

Overhauling of the treatment plant SSF's was an initiative by the SWA Urban Operations Manager, Ekiumeni Fauolo in 2011 as the WTP had been in operation for over 25 years with little to no maintenance having ever been carried out. Sand layers had been scraped without replenishment and most tanks were already showing exposed areas of the underlying aggregate material.

There was also little to no information on the tank designs as most of it had been lost in a fire that ravished the main office where these documents were kept in the early 90's. During the overhaul each layer of aggregate was removed and noted for thickness and size. When the final aggregrate layer was removed perforated pipes were exposed and cleaned. These pipes were heavily clogged with thick muddy sediment to between 60% - 90% of each pipes internal diameter.

After reinstating the Alaoa tanks to their original design specifications and water depth of close to 1 meter, it was noted that flows had improved through the tanks although quantifiable measurements were not recorded. It was also observed however that the sand layer still became clogged over a similar time period to before the overhaul, also scrapping frequency still remains largely unchanged.

Figure 33.5 Image of the scraping of the schmutzdecke layer of the Alaoa WTP

At the Malololelei Treatment Plant there has been great improvement in slow sand filtration. Having an adequate sand layer (500mm) has allowed for improved filtration (in terms of quality water produced) and reduced filter loss adjusting water depths (allowing for a maximum of ~500mm. Increased uplifting of the algal layer was observed a few weeks following overhauling, several occasions have seen the immediate ripening of the new algal layer within 4 days of the last uplifting. Whereas a detailed study is required to further identify this new behavior of SSFs, it is speculated that the adjusted levels of water depths allow for favourable conditions for rapid growth of algal layer and hence better filtration.

In the shallow water it was also noted that a shorter algae would develop. Often there were contrasting colours from light brown in one tank, opaque green/blue in another and yellowish/green in the third. The shorter algae uplift was more frequent with small patches of 20mm to 40mm diameter on average rising to the surface constantly. When turbidity loadings into the SSF's was low, tank scraping was not required. Due to a problem controlling the flow through the settling tank, high turbidity loads do still enter the SSF's on a periodic basis and it has been observed that scrapping has only been required after these heavy turbidity loadings.

CONCLUSIONS

- The results show that a shallower water depth in the SSF increases the efficiency of the system.
- Lowering the water depth in the SSF causes increased algae uplift with sediment, creating a natural self cleansing process that enables the schmutzdecke layer to free itself of clogging material and thereby removing the need for scraping.
- The reduction in SSF scraping frequency has also meant a reduction in manpower resource required to scrape the tanks allowing for refocus to other areas of the operational schedule.
- The capacity building relationship between Samoa and Miyako Jima has been of great help improving conceptual approaches and hands on experience for some of SWA personnel aiding a change in more efficient operational attitudes
- Other challenges are related to the vulnerability of the WTP (aging infrastructure) , plant operating over its design capacity and various understanding of what is 'deemed' best operating practices for slow sand filtration

ACKNOWLEDGEMENTS

The authors wish to acknowledge the important contribution of the Samoa Water Authority Urban Operation and Maintenance Division for the operational data , experience and observation. The Water Quality Unit for the water testing results. Miyako Jima City Council for the lessons learnt in the capacity building resulting in the attitude change in operating the SSF in Samoa.

REFERENCES

Miyako Jima City Council, Capacity Building 2010
Samoa Water Authority, Water Quality Report, June2012, Vaitele, Samoa
Samoa Water Authority, Water Quality Report, July 2012, Vaitele, Samoa
Samoa Water Authority, Water Quality Report, August 2012, Vaitele, Samoa
Samoa Water Authority, Water Quality Report, September 2012, Vaitele, Samoa
Samoa Water Authority, Water Quality Report, October 2012, Vaitele, Samoa
Samoa Water Authority, Water Quality Report, November 2012, Vaitele, Samoa
Samoa Water Authority, Water Quality Report, January 2013, Vaitele, Samoa
Samoa Water Authority, Water Quality Report, February 2013, Vaitele, Samoa
Samoa Water Authority, Water Quality Report, March 2013, Vaitele, Samoa
Samoa Water Authority, Water Quality Report, April 2013, Vaitele, Samoa
Samoa Water Authority, Water Quality Report, May 2013, Vaitele, Samoa
Samoa Water Authority, Water Quality Report, June2013 Vaitele, Samoa
Samoa Water Authority, Water Quality Report, July2013, Vaitele, Samoa
Samoa Water Authority, Water Quality Report, August 2013, Vaitele, Samoa
Samoa Water Authority, Water Quality Report, September 2013, Vaitele, Samoa
Samoa Water Authority, Water Quality Report, October 2013, Vaitele, Samoa
Samoa Water Authority, Water Quality Report, November 2013, Vaitele, Samoa
Samoa Water Authority, Urban Operation and Management Division Maintenance Records, Vaitele, Samoa
Samoa Water Authority, Annual Report 2012-2013, Apia, Samoa

Chapter 34

Household bio-purification device

Shengzhe Jin

Jiangsu Comfity Power Technology Co.,Ltd, CHINA

ABSTRACT

Since 2008, household bio-purification devices using concrete pipes have been installed in China. The design, construction and maintenance details of these household devices are presented in the paper.

Keywords slow sand, algae, photosynthesis

INTRODUCTION

The design was done under the guidance of Professor Nakamoto of Japan Shinshu University, based on his theory (algae, photosynthesis, small animals, micro-organisms, ecosystem structure). It is very convenient to use concrete pipes as it allows for simple and economical construction of the device, which can be installed in a short period of time. It is also easy to obtain concrete pipes in China.

Concrete pipes used are mostly 1m in diameter (there was one case where the diameter was 0.6m). Amount of water purified is roughly 4 tons per day (the flow rate is 5m per day).

DESIGN

Figure 34.1 shows the layout schematic of the household bio-purification device. Valve 1 is for cleaning. When pouring in water, valve 3 closes when water reaches the sand surface. Valves 4, 5 and 6 are used to regulate the amount of water being purified.

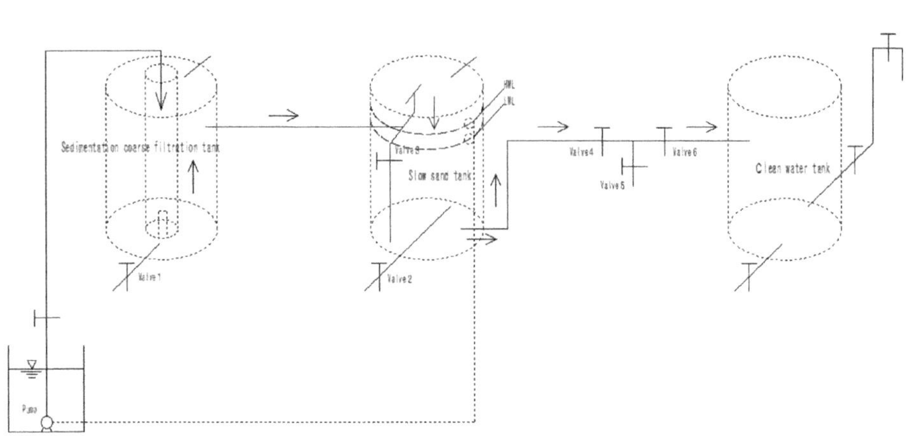

Figure 34.1 Schematic of household bio-purification device

Figures 34.2 and 34.3 show the elevation and ground plan respectively. The circulation tank in the figure is used based on site conditions, such as where raw water is drawn from a deep well or where water is in shortage. As such, it is also possible to achieve energy saving.

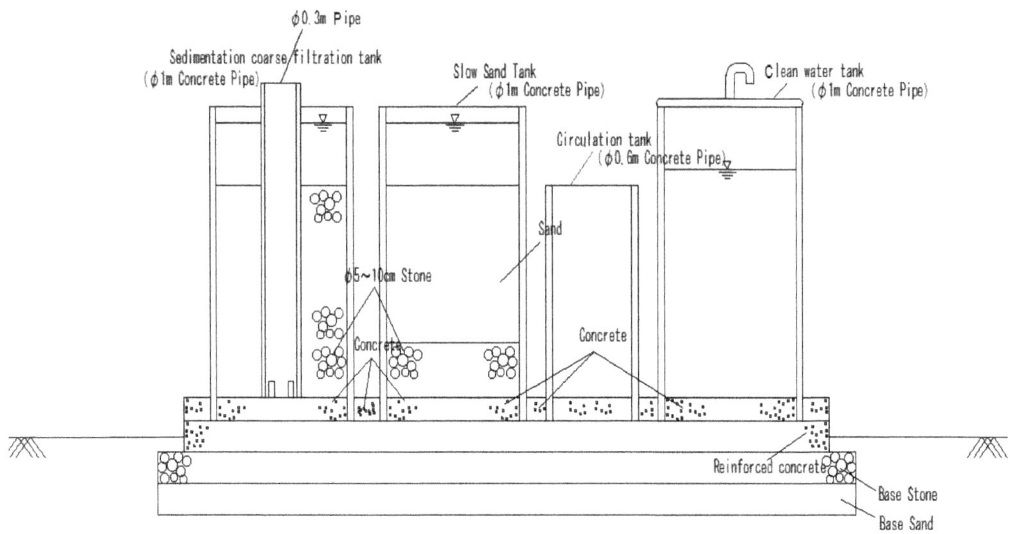

Figure 34.2 Elevation view of household bio-purification device

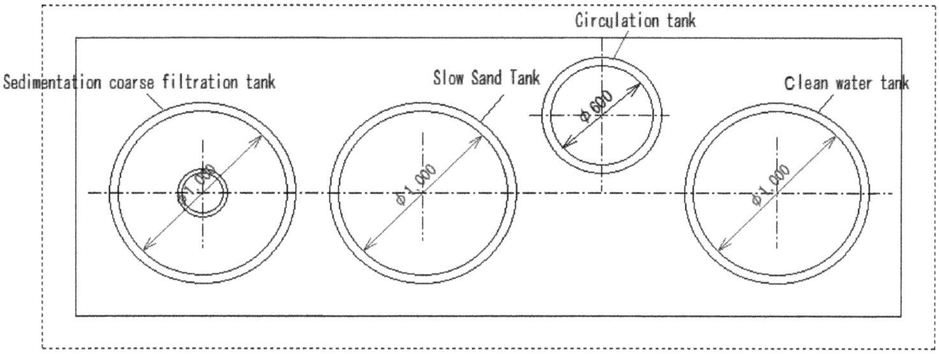

Figure 34.3 Plan view of household bio-purification device

CONSTRUCTION

Construction takes about 7 to10 days. First, create a reinforced concrete foundation, and then mount the concrete pipe after about 3 days. As the pipe weighs more than one ton, it would be convenient to have some heavy equipment such as cranes on hand. To prevent water leakage, secure the pipe to the base with about 5cm of mortar and repeat one more time when hardened. Use coarse sand with diameter of about 1mm and wash off the soil with water before loading.

CONTROLLING THE PUMP

Water abstraction is done using a pump. The pump's on/off is controlled by water level (HWL, LWL), which is detected by either electrode or a float. Water level control is performed in the slow sand tank (Figure 34.1), but can also be done in the sedimentation coarse filtration tank.

MAINTENANCE

Inspection of the electric equipment (pump) and cleaning of the Sedimentation coarse filtration tank (about 10 minutes of water drainage) should be performed periodically. During summer time when algae growth may be vigorous, manual extraction of algae is advisable to prevent it from rotting in the tanks should aging algae and waste fail to be fully drained from the slow sand tank and the sedimentation coarse filtration tank.

Photos 34.1~4 exhibit a device under construction at an orphanage in a Chinese prefecture.

Photo 34.1

Photo 34.2

Photo 34.3

Photo 34.4

In this region, temperature drops below freezing point in wintertime, so a prefabricated hut was used as counter measure.

CONCLUSIONS

As described above, the construction of Household Bio-Purification Devices using concrete pipes is easy and economical. Additionally, maintenance is also simple. Inevitably, this device causes the complete ecosystem cycle by naturally inducing the food chain. Raw water contaminated by algae, small animals, and microorganisms is purified to clean water at the end. The purified water is of course clean and the taste is round and delicious. Due to site constraints, water quality tests were performed at limited locations, where the following trends were observed.

a. Common bacteria: below 30/ml
b. E. coli: none
c. pH value: 7.5-8.5
d. Odor and taste, chromaticity, turbidity: no issue
e. Hardness: appears to be less than raw water to some extent
f. Sulfate content: appears to be less than raw water to some extent

ACKNOWLEDGEMENTS

I would like to thank the Program Committee of the 5th International Slow Sand and Alternative Biological Filtration Conference for the acceptance of this paper. I would also like to take this opportunity to express my gratitude to Professor Nakamoto of Shinshu University who has given me guidance and support for many years.

REFERENCES

Nobutada Nakamoto (2006) How to make delicious water- biological purification method.

Design Standard of Concrete Structures GB50010-2002,China

Design Standard of Basis Building Foundation GB50007-2002,China

Design Standard of Power Supply System GB50052-95,China

Chapter 35

A study to further reduce the turbidity level of slow sand filtered water in Tanigahara water purification plant, Japan

I. Tateno, M. Sekiyama*, N. Kobayashi*, S. Arii**

** Tanigahara Water Purification Plant,Kanagawa Prefectural Public Enterprises Agency, 2-6-1 Tanigahara, Midori-ku,Sagamihara 2520113, Japan*

ABSTRACT

In Tanigahara water purification plant (TWPP), the source of the water used in slow sand filtration (SSF) is from artificial lakes (Sagami Lake, Tsukui Lake) and river-bed water. Compared to river-bed water, eutrophic artificial lake water contains higher levels of algae. Some algae in water from artificial lakes sometimes results in an increase in the turbidity in the filtered water in SSF. *Nitzschia acicularis* and *Pandorina morum* were the cause of the filtered water turbidity increase during our regular water production. We therefore decided to investigate the relationship between the algae and turbidity and to work towards further reducing the turbidity. To reduce the cause of filtered water turbidity we studied SSF full scale plant while using only river-bed water. Two runs were conducted; in May and in December of 2012. In both runs, the results varied from basin to basin. The species observed during our regular water production were also observed in filtered water , while *Melosira varians* and *Fragilaria* spp. predominated in the schmutzdecke. In May (Run1), *Melosira varians* or *Cymbella* spp. predominated, while in December (Run2), *Melosira varians* initially predominated but later changed to *Scenedesmus* spp. when *Cymbella* spp. and *Scenedesmus* spp. predominated in the schmutzdecke, Under these conditions, that the filtered water turbidity could be kept at less than our taget of 0.05 degrees.

Keywords Slow sand filtration, algae, turbidity, schmutzdecke

INTRODUCTION

Presentation of Tanigahara water purification plant

Kanagawa Prefectural Public Enterprises Agency (KPPEA) supplies water to 12 cities and 6 towns in Kanagawa Prefecture as of March 2013, Japan. The population served is approximately 2,790,000. Tanigahara Water Purification Plant (TWPP) is one of the two major purification plants in KPPEA. The population served by this plant is approximately 440,000. Two water treatment systems are used in this plant: Slow sand filtration (SSF) and Rapid sand filtration (RSF) each with a production capacity of 32,800m^3/day and 210,000m^3/day, respectively. The source of the water used in SSF is from artificial lakes (Sagami Lake, Tsukui Lake) and river-bed water. The amount of river-bed water use is more than that of lake water in raw water. The treatment process of SSF includes flocculation(in case of high turbidity raw water), sedimentation (3 basins), filtration (8 basins) and final disinfection by chlorine.

The problems of SSF in TWPP

The Japanese guidelines for the Cryptosporidium treatment in waterworks stipulate that the turbidity of filtered water be kept at 0.1 degrees or less (MHLW 1996, 2007). The guidelines also order that the SSF rate is less than 5m/day. In our plant, the target is to reduce the turbidity to 0.05 degrees or less while the filtration rate is kept at about 2m/day. Due to the stricter turbidity target, some plant operation problems have arisen. For example, in order to reduce the algae, cause of turbidity in the filtered water, we used to reduce water retention period in the basin by increasing the filtration rate. However, increasing the filtration rate results in an increase in turbidity in the filtered water in some cases. Furthermore if the filtered water turbidity is higher than 0.05 degrees, we have to drain the water and reduce the filtration rate, thus reducing the volume of water produced and the treatment process is inefficient.

The algae observed in filtered water in SSF

During our regular water production, when the filtered water turbidity increased to more than 0.05 degrees, we investigated the cause. Most frequently observed algae in filtered water were that of *Nitzschia acicularis* of Bacillariophyceae family. The other species were those of flagellate species *Pandorina morum*, *Eudorina* spp., *Chlamydomonas* spp., *Carteria* spp. of Chlophyceae family and *Cryptomonas* spp. species of Cryptophyceae family and *Peridinium* spp. of Dynophyceae family. We considered these species to be the cause of high turbidity.

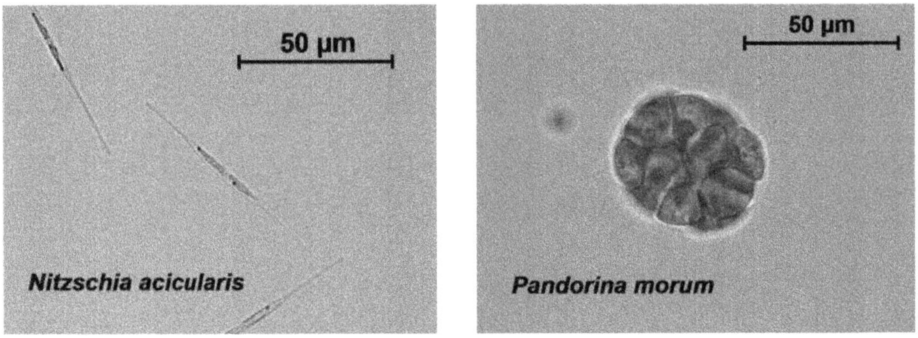

Figure 35.1 The algae causing filtered water turbidity in SSF

The purpose of this study

The purpose of this study was to reduce the filtered water turbidity to 0.05 degrees while using as high a filtration rate as possible and using only river-bed water. The mat of biologicall matter that forms on the sand surface, known as the "schmutzdecke" (Huisman, 1974). Previous work has shown that compound algae species in algal mat(schmutzdecke) on sand filters were changed by using a different source of water (Nozaki, 1993). We therefore decided to investigate between the algae in pre-filtered water and in the schmutzdecke on sand filters and in the filtered water and their relationship to the filterd water turbidity.

The water quality of two water sources in SSF

Under normal conditions river-bed water turbidity was below 0.1 degrees and rarely contained algae. However, artificial lake water contained high levels of algae. The level of algae was from 10^2 to 10^4cells/mL and the predominant species were almost planktonic diatoms, in summer Cyanobacteria also predominated. The river-bed water algae levels were very low despite water containing enough nutrient to support algae growth. Water quality is shown in Table 35.1.

Table 35.1 Average raw water quality in SSF

		River-bed water	Lake water
Temperature	℃	15.6	14.8
Turbidity	degrees	0.6	8.3
pH	-	7.4	7.8
TOC	mg/L	0.3	0.9
Phospate ion	mg/L	0.08	0.16
Ammonium nitrogen	mg/L	<0.01	0.03
Nitrate & Nitrite	mg/L	1.6	1.1

Average of 2003-2012 data

METHODOLOGY

The period of study

Run 1 was conducted for 35 days from 14 May to 17 June in 2012 and Run 2 was conducted for 34 days from 6 December in 2012 to 8 January in 2013 while using only river-bed water. During other studies outside the forementioned river-bed water and atificial lake water was used in SSF.The experiments started after scraping the schmutzdecke on sand filters. At the end of the study, we continued using the filtration basin in which the filtered water turbidity and algae cell numbers were kept low.

Sampling point and analysed water quality

Two filtration basins were investigated during each study (Run 1, Run 2). Sampling point and analysed water quality are shown in Table 35.2. Water quality,temperature, turbidity, pH and biological examination were investigated. These methods were according to the Standard methods for the examination of water (JWWA, 2011). Turbidity measurement was obtained using online laser turbidity analyser Yokogawa TB600G (Yokogawa electric Co., Japan). The turbidity unit was expressed in degrees according to the Japanese drinking water quality standard (0.1 NTU = approximately 0.07 degrees).

Table 35.2 Sampling point and analysed water quality

	Tem.	Turbidity	pH	Bio.exam.
Sampling point	℃	degrees		cells/mL
River-bed water	O	O	O	O
Sedimentation basin water	–	O	O	O
Pre-filterd water	O	O	O	O
Schmutzdecke	–	–	–	O
Filterd water	–	O	O	O

Cell numbers of schmutzdecke were not quantitatively analysed

Biological examination

The samples were fixed by 25% glutaraldehyde. The algae in these samples were observed, identified and counted by phase microscopy. The schmutzdecke on sand filters were scraped using a scoop and the predominant algae species were observed and identified using ordinary microscopy. The microscope was AixioImager 2 and the camera was AxioCam ICc 3 (Carl Zeiss Co., Germany).

The operation of SSF during the studies

The filtered water turbidity was kept at 0.05 degrees and the filtration rate was increased to as high a rate as possible while keeping it under 5m/day. The headloss was until 500mm.

RESULTS

Investigation of river-bed water and sedimentation basin water

The results are shown in Table 35.3. Algae cell numbers and turbidity increased in sedimentation basin in both Run 1 and Run 2. The algae species observed in sedimentation basin water were *Cyclotella* spp. & *Stephanodiscus* spp. of Bacillariophyceae family.

Table 35.3 Average water quality of river-bed water and sedimentation basin during the study

		Tem.	Turbidity	pH	Bio.exam.
Run	Sampling point	°C	degrees		cells/mL
Run 1	River-bed water	14.6	0.10	7.2	5
	Sedimentation basin water	-	0.43	-	850
Run 2	River-bed water	13.5	0.08	7.3	<1
	Sedimentation basin water	-	0.27	7.5	100

The result of Run1

Basin1:The experiment in this basin ran for 24 days. The average water temperature in the filtration basin was 16.9 C. Filtered water turbidity increased to 0.03 degrees after day 9 and contained algae cell numbers of 252cells/mL (Figure 35.2). Observed algae in filtered water were *Nitzschia acicularis*, *Nitzschia palea* and *Cryptomonas* spp. After day 23 filtered water turbidity increased again. The cell numbers were 272cells/mL and were predominantlly *Pandorina morum*. These species was also observed growing in the pre-filtered water. Due to high turbidity, we had to drain the filtered water (Figure 35.2:days 10-11, days 13-14, after day 24). The predominant species in schmutzdecke were *Melosira varians*, *Fragilaria* spp., *Nitzschia palea* and *Cryptomonas* spp. (Figure 35.4). Maximum filtration rate was 1.0m/day (Figure 35.2).

Basin2:The experiment in this basin ran for 65 days. During the first 5 days and after day 41 the raw water contained lake water. Except for the aforementioned days, only river-bed water was used. The average water temperature in the filtration basin was 16.4 C. Filtered water turbidity increased to 0.03 degrees after day 12 and contained algae cell numbers of 23 cells/mL after day 15 (Figure 35.3). Observed algae in filtered water were *Nitzschia acicularis* and *Cryptomonas* spp. and the filtration rate was decreased to 0.5m/day (Figure 35.3). The predominant species in schmutzdecke were *Melosira varians* and *Fragilaria* spp. until day 24, but later changed to only *Cymbella* spp.(day 30) and predominated until the last day (Figure 35.4). *Cymbella* spp. predominated in the schmutzdecke and under these conditions filtered water turbidity was kept low. After day 45, *Cryptomonas* spp. increased to 4,100cells/mL in the pre-filtered water. However this species was observed in the filtered water in very low numbers (1 cell/mL). Therefore we could increase the filtration rate to a maximum rate of 4.9m/day (Figure 35.3).

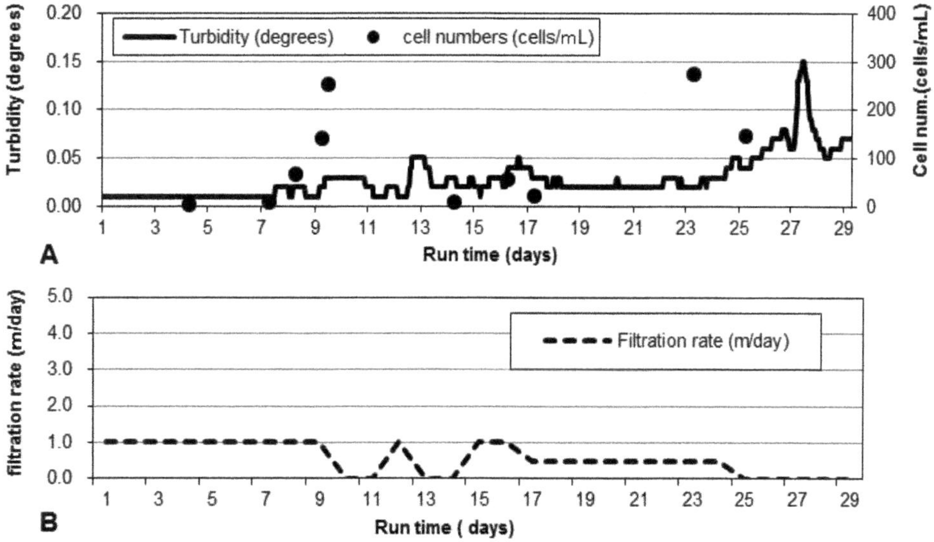

Figure 35.2 Run1 Basin1
A : The relationship between turbidity and cell number in filtered water and filtration rate.
B : Filtration rate during the study. 0m/day means the basin was drained.

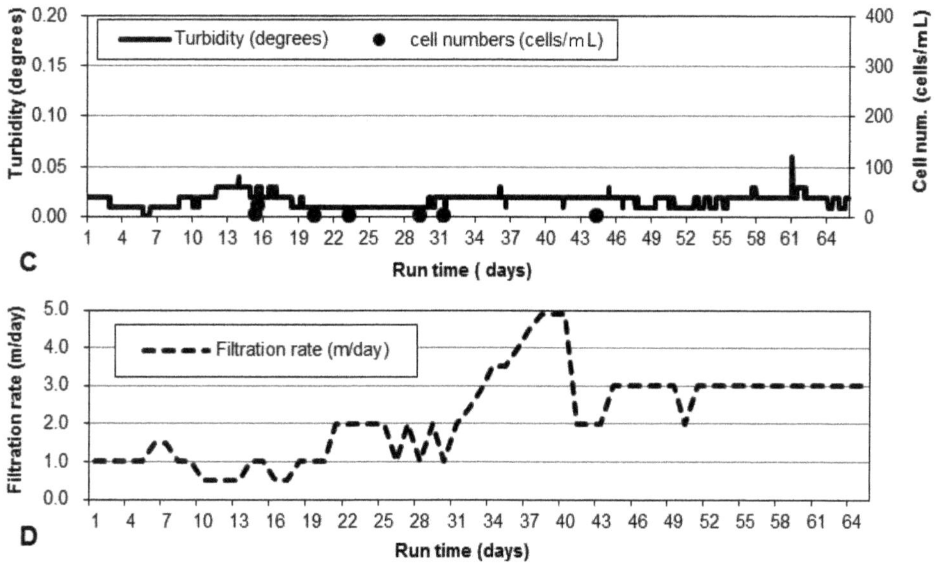

Figure 35.3 Run1 Basin2
C : The relationship between turbidity and cell number in filtered water and filtration rate.
D : Filtration rate during the study. 0m/day means the basin was drained.

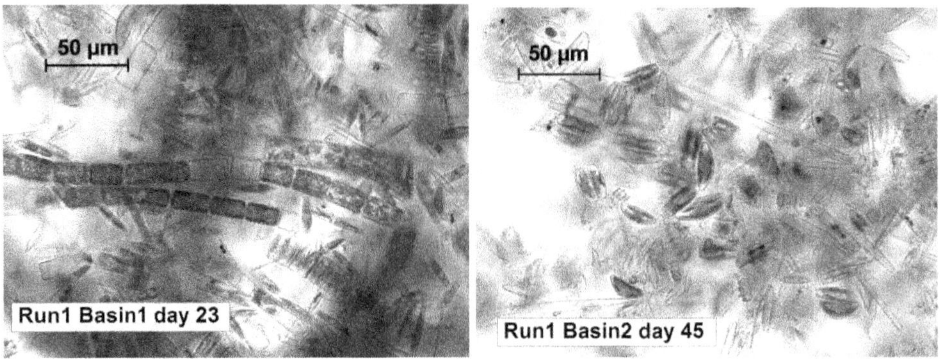

Figure 35.4 Schmutzdecke Run1 Basin1:day23 (4a, left), Run1 Basin2:day45 (4b, right)

The result of Run2

Basin1: The experiment in this basin ran for 49 days .The average water temperature in filtration basin was 11.6 C. After day 35, the raw water contained lake water. Filtered water turbidity increased to 0.02 degrees after day 22 and contained algae cell numbers of 340 cells/mL. The filtration rate was 2m/ day. Observed algae in filtered water were *Nitzschia acicularis*. Due to a high level of algae , we had to drain the filtered water and use the filtration rate of 1.0m/day after day 26. The predominant species in schmutzdecke were *Melosira varians* and *Fragilaria* spp. and auxospore formation in *Melosira varians* was observed (Figure 35.5). After day 33 almost all the diatoms cells became empty and *Scenedesmus* spp. of gleen algae predominated in the schmutzdecke (Figure 35.5). Under these conditions, the filtered water turbidity was kept low and the filtration rate was 2m/day. The average filtration rate was 1.3m/day and the maximum was 2.0m/day.

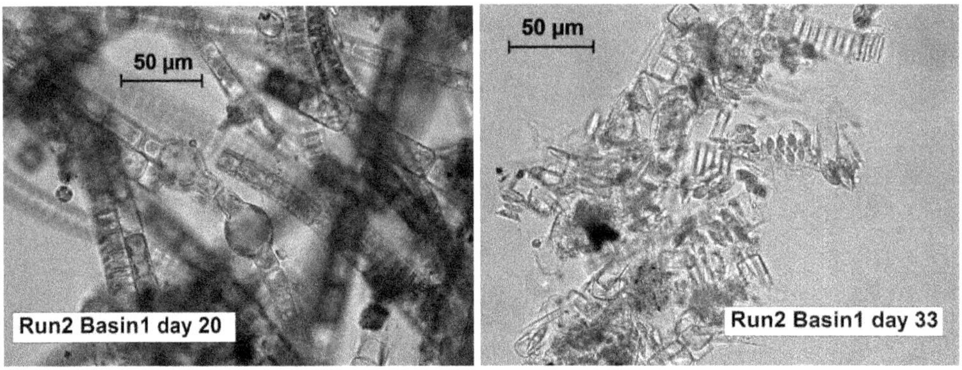

Figure 35.5 Schmutzdecke Run2 Basin1: day20 (5a, left) and day33 (5b, right)

Basin2: The experiment in this basin ran for 45 days .The average water temperature in the filtration basin was 11.3C. The raw water contained lake water after day 30. But the filtered water turbidity was kept low. *Cryptomonas* spp. was growing in pre-filtered water after day 11 (840 cells/mL) but could only be detected in small numbers (24 cells/mL)in filtered water after day 18. The filtered water turbidity was kept low and the filtration rate was 1m/day. The predominant species in schmutzdecke were *Melosira varians* , *Fragilaria* spp. and *Scenedesmus* spp. (Figure 35.6). A large number of midge larvae was growing in the basin after 25 days. After that diatoms cells became empty and *Scenedesmus* spp. predominated in the schmutzdecke (Figure 35.6). When *Scenedesmus* spp. predominated in the schmutzdecke, *Nitzschia acicularis* was observed in the filtered water (30 cells/mL) after day 25 but the turbidity was kept at 0.01degrees. We could therefore increase the filtration rate from 1.0m/day to 2.0m/day. *Chlamydomonas* spp. was observed in pre-filtered water (520 cells/mL) after day 41 but it was not observed in filtered water. The average filteration rate was 1.4m/day and the maximum was 2.0 m/day.

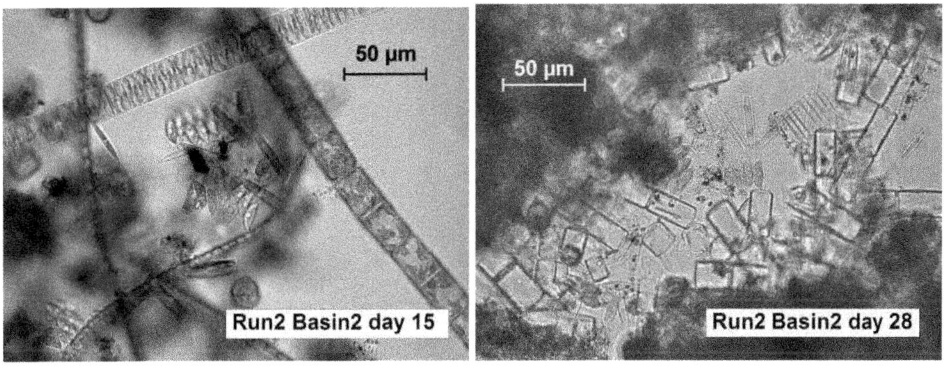

Figure 35.6 Schmutzdecke Run2 Basin2: day15 (6a, left) and day28 (6b, right)

Table 35.4 The basins condition comparison between the study and regular water production

Run	Basin	Water Tem.	Priod	Filtration rate (m / day)		Algae in filtered water (cells/mL)
		°C	day	Max.	Average	Max.
Run 1	Basin1	16.9	24	1.0	0.7	275
	Basin2	16.4	65	4.9	2.2	23
Run 2	Basin1	11.6	49	2.0	1.3	340
	Basin2	11.3	45	2.0	1.4	30
Regular water production	Best condicion basin in the year	22.6 *	71	4.0	2.0	-

* This water temperature was raw water temperature.

DISCUSSION

In sedimentation basin the algae species observed growing were almost all of the planktonic diatoms *Cyclotella* spp. & *Stephanodiscus* spp. which are unrelated to the filtered water turbidity. The species such as *Pandorina morum*, *Chlamydomonas* spp., *Carteria* spp. and *Cryptomonas* spp. that cause high turbidity, were observed in pre-filtered water and in filtered water. We confirmed these species growing in the filtration basin. During our study, *Nitzschia acicularis* was observed in the filtered water (Run1 basin1 and Run2 basin1) while the predominant species in schmutzdecke were *Melosira varians* , *Fragilaria* spp.. We inferred that the reason for the increase in *Nitzschia acicularis* in filtered water from river-bed water is due to enough nutrient and the low turbidity of river-bed water which allows a high sunlight irradiation. Therefore the sunlight effect on *Nitzschia acicularis* increased in the schmutzdecke. However *Cymbella* spp. Predominated in the schmutzdecke and under these conditions filtered water turbidity was kept low (Run1 basin2). In spite of *Cryptomonas* spp. increase to 4,100 cells/mL in the pre-filtered water, very few cell numbers were contained in filtered water. The result suggested that

Cymbella spp. has an important role in the schmutzdecke as a filter. We inferred that *Cymbella* spp., being a periphytonic diatoms species, attached onto the bed by their mucilage stalk and therefore filled the role of a filter. During Run2, the predominant species in schmutzdecke were *Melosira varians* , *Fragilaria* spp. in both basins. But after a month, diatoms cells became empty and *Scenedesmus* spp. predominated. Previous study (Nakamoto, 2006) reported that midge larvae was grazing filamentous daitoms. We also observed a large number of midge larvae in the basins, and therefore inferred that the predominant species changed in schmutzdecke by the grazing activity. Under these conditions, *Scenedesmus* spp. predominated in the schmutzdecke and the turbidity was kept low in both basins (Run2 basin1,2). The reason for these results is that empty diatoms cells, together with *Scenedesmus* spp., an immotile type of algae, settled and filled the role of a filter. The best basin condition in the year was shown during regular water production (Table 35.4) while using raw water. These results suggested that the filtration rate could be raised and the turbidity kept low even when lake water is used. Since the period of our study was limited and lake water was used during part of the study (thus influencing the results), we did not see clear advantages of using only river-bed water. In future, we would like to study a SSF system that would not just remove algae in filtered water, thus reducing turbidity, but would also be a more efficient treatment process.

CONCLUSIONS

The main conclusions of this study are:

(1) The algae observed in filtered water in SSF were *Nitzschia acicularis*, *Nitzschia palea*; exhibiting gliding movement species and *Pandorina morum*, *Eudorina* spp, *Chlamydomonas* spp., *Carteria* spp. *Cryptomonas* spp. and *Peridinium* spp.; flagellate species.

(2) When using only river-bed water in SSF, *Nitzschia acicularis* and *Pandorina morum* were observed in filtered water and identified the cause of turbidity, while the predominant species in schmutzdecke were *Melosira varians* , *Fragilaria* spp..

(3) The best basin condition during the study was when *Cymbella* spp. predominanted in the schmutzdecke. We could raise the filtration rate while keeping the filterd water turbidity low.

(4) This study clarified the relationship between the algae in the schmutzdecke and the filtered water turbidity. However, since the period of our study was limited and lake water was used during part of the study ,we did not see clear advantages of using only river-bed water in SSF.

REFERENCES

K. Nozaki,T.Yoshioka,M,Sakai,N.Nakamoto.(1993).Alge in slow sand filters connected with a lake. *Ann. Environ. Sci.*,Shinsyu University, Japan.**15** 89-94.

Ministry of Health, Labour and Welfare (1996).The tentative guidelines for cryptosporidium treatment in waterworks (Revised in 1998 and 2001), Japan.

Ministry of Health, Labour and Welfare (2007) The guidelines for cryptosporidium treatment in waterworks, Japan.

L.Huisman,W.E.Wood,F.I.C.E.(1974) *Slow Sand Filtration*,Word Health Organization,Geneva.

Japan Water Works Association (2011) Standard Mthods for the Examination of Water,Japan.

N.Nakamoto,H.Kato (2006) Development pattern of filamentous diatom and its condition rerated with midge larvae in slow sand filter. In: *Recent progress in slow sand and alternative biofiltration processes* (IWA),68-73.

Chapter 36

The countermeasure against the increase of disinfection by-products concentration by humic substances at a water purification plant using a slow sand filtration system

Yuki Kado, Kenichiro Takao, Kenji Hashiwata
Waterworks Bureau of the City of Hiroshima, 6-1-1 Ochiai Minami, Asakita Ward Hiroshima City, Japan

ABSTRACT

Dichloroacetic acid (DCA) was detected in high concentration in the tap water supplied by a water purification plant using a slow sand filtration system. DCA is a kind of disinfection by-product considering that it was derived from the humic substances due to the fallen leaves.

The correlation between some factors (DCA, TOC, chromaticity) was investigated and the appropriate operation and maintenance method at a water purification plant with a slow sand filtration system when DCA is in high concentration were established.

Keywords Slow sand filtration; disinfection by-products; dichloroacetic acid; humic substances; chromaticity

INTRODUCTION

Tsubakidani Water Purification Plant is located in a mountainous region. This water purification plant was taken over by the management of our water utility accompanying the incorporation of a small town into Hiroshima City. The outline of this water purification plant is shown as follows.

- Number of reservoirs: 3
- Supply capacity: 225 cubic meters per day
- Scale of distribution reservoir: 197 cubic meters
- Design population supplied: 570 persons
- Population supplied: 389 persons (161houses)
- Average daily supply: 96 cubic meters per day

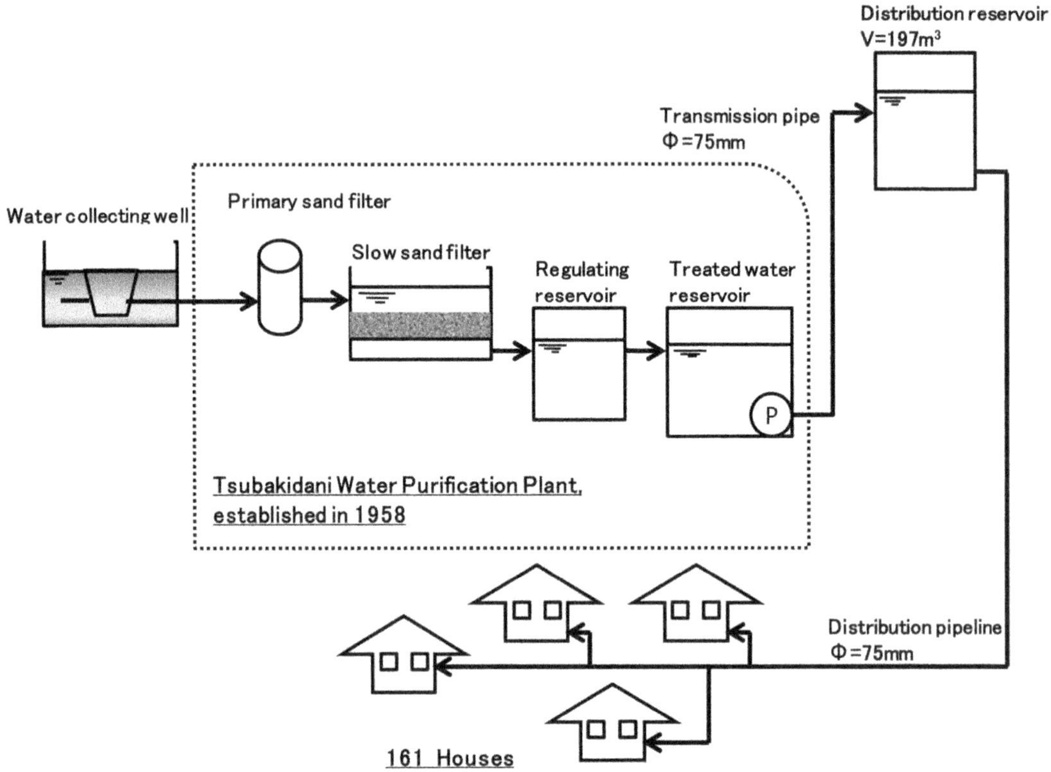

Figure 36.1 The flowchart of Tsubakidani Water Purification Plant

It is a small plant utilizing a slow sand filtration system with a water resource of river water in small quantities. The water resource hardly has any pollution by humans. Therefore its water quality is good.

However in June 2008, dichloroacetic acid (DCA) was detected with a concentration of 0.037mg/L in the tap water supplied by the Tsubakidani Water Purification Plant. DCA is a kind of disinfection by-product (DBP) and the inspection result showed that it had risen to 90 percent of DCA's maximum allowable level that was 0.04mg/L. A slow sand filtration system is a water purification method depending on capture and oxidative decomposition effects by microbes in the sand layer. Organic matter in the raw water is changed into inorganic matter by this microbial film. DBPs concentration in the purified water is generally lower than that by rapid sand filtration. Increase of DBPs concentration had not occurred at water purification plants using slow sand filtration in Hiroshima City.

This work reports the cause and the countermeasure of an increase of DCA concentration in Tsubakidani Water Purification Plant.

INVESTIGATION OF THE CAUSE

Many fallen leaves and leaf mold had piled near the plant's water source which contained organic matter called humic substances which produce DBPs by a reaction with chlorine. It was considered that humic substances flowed in the water source by rain, because high concentration of DBPs was detected after rain.

It is widely recognized that there is a high correlation between humic substances and chromaticity. The chromaticity of raw water was continuously measured to investigate the correlation with rain. Fig. 36.2 shows this result. Shortly after it rained heavily on the evening of 15th July 2008, the chromaticity rapidly increased. Then, the chromaticity gradually decreased and returned to the normal value in half a day. This investigation clarified that a large quantity of humic substances temporarily flowed in the water source due to hard rain.

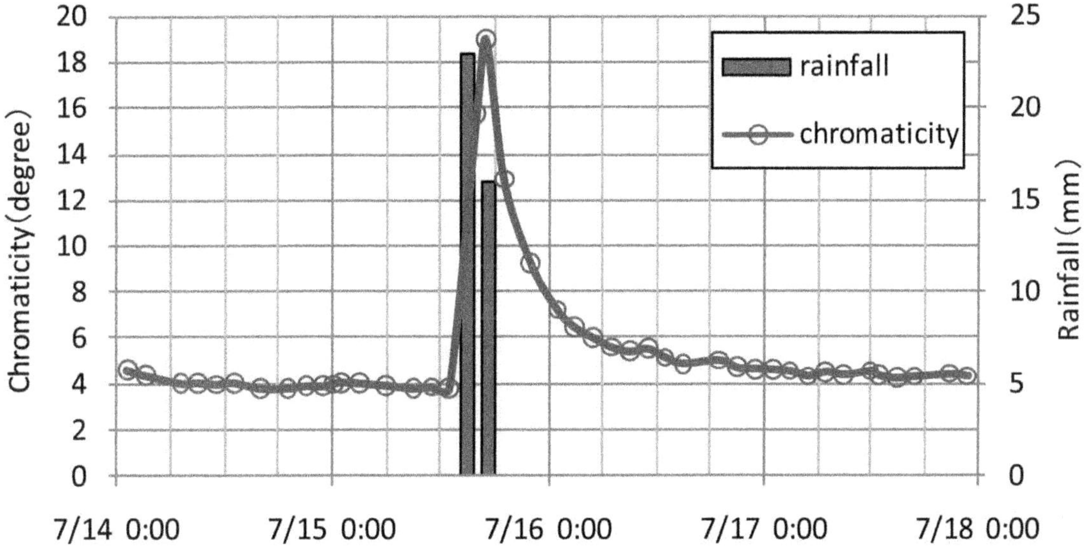

Figure 36.2 A correlation between amount of rainfall and chromaticity of the raw water in July 2008

COUNTERMEASURE AGAINST HUMIC SUBSTANCES

It is well known that humic substances cannot be removed by a slow sand filtration system because humic substances are produced by decomposition of plants by microbes. To solve this problem, active carbon or ozone oxidation treatment operations were proposed as countermeasures, but those operations were considered difficult due to the maintenance and the cost effectiveness. Therefore, it was proposed to suspend water intake because a large quantity of humic substances temporarily flowed in after hard rain.

Firstly, the timing of suspending water intake was discussed and it was determined to operate so that 75 percent of DCA's maximum allowable level (0.03mg/L) may not be exceeded. Concentration of DCA in raw water can't be monitored. So chromaticity of raw water, which has high correlation with humic substances and can be monitored, was used as an index. However, to observe the correlation between chromaticity of raw water and DCA of tap water is difficult because it is necessary to consider the time lag from raw water to tap water. Therefore the past data had been analyzed and it was found that total organic carbon (TOC) had high correlation with chromaticity of raw water and DCA of tap water. Therefore TOC concentration being equivalent to 0.03mg/L of DCA was firstly investigated and the chromaticity of raw water being equivalent to the TOC concentration was secondarily investigated and the chromaticity of raw water being equivalent to 0.03mg/L of DCA was lastly calculated. Fig. 36.3 shows the results of this investigation. It was found that TOC concentration being equivalent to 0.03mg/L of DCA was about 1.5mg/L and the chromaticity of raw water being equivalent to this was about 5 degrees.

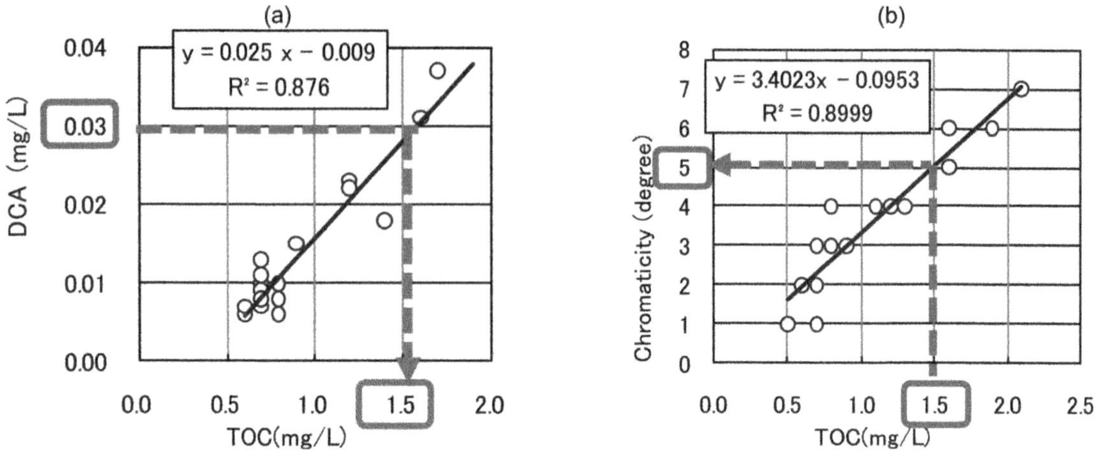

Figure 36.3 Result of investigation about correlation between : (a) DCA and TOC in tap water; (b) TOC and chromaticity in the raw water.

It is considered that the operation of the plant while suspending intake water is possible when the chromaticity of raw water increases, because focusing on the transition of the chromaticity at a rainy time in Fig. 36.2, the time while chromaticity exceeds 5 degrees is 23 hours and the possible time of suspending intake water at Tsubakidani Water Purification Plant is up to 36 hours. That is the plant's reservoir can supply water to customers for 36 hours without being replenished.

It was decided to counteract the inflow of humic substances by temporarily suspending intake as the index of chromaticity in the raw water reaches or exceeds 5 degrees. To carry out this operation, a continuous measurement system was established by installing an automatic chromaticity monitor.

CONCLUSION

In 2009, an automatic chromaticity monitor was installed at Tsubakidani Water Purification Plant and the new operation (to suspend water intake when raw water chromaticity is over 5 degrees) has been carried out to the present. By this operation, a large increase of DCA concentration has not occurred in tap water and the water supply has not been depleted at times of heavy rain (Fig. 36.4). Therefore it was considered that this countermeasure was generally successful.

Although the countermeasure reported in this research isn't a drastic measure to remove humic substances, it is considered that measuring chromaticity of raw water can provide an effective indicator as a countermeasure against DBPs at a water purification plant with a slow sand filtration system in a mountainous region because the reduction of DBPs is achieved at a minimum expense.

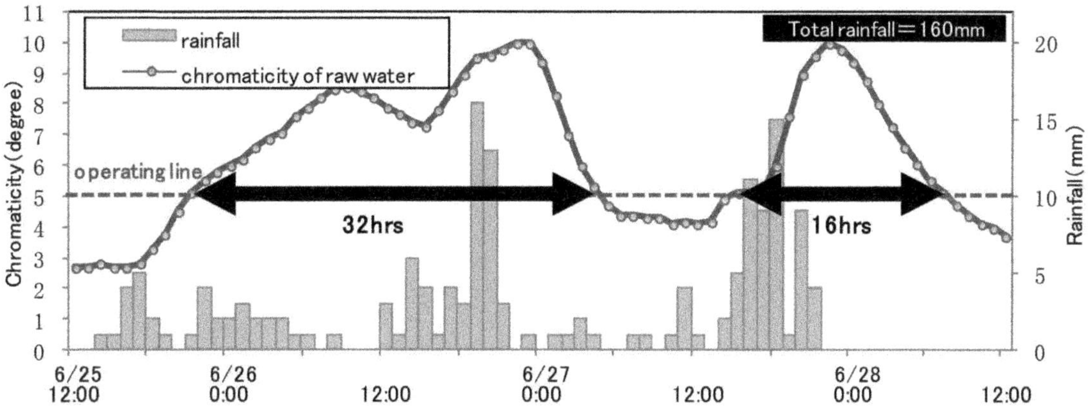

Figure 36.4 A change of the chromaticity of raw water at hard rain in June 2010 (Operating line is the level at which 5 degrees of chromaticity is exceeded and water intake is suspended).

REFERENCE

Japan Water Works Association Standards (2012) Design Criteria for Waterworks Facilities, 240

Chapter 37

On the function of upflow roughing filter for slow sand filter – pilot experiment in Niigata, Japan

Tomoaki ARAKAWA, Morifumi SENO and Nobutada NAKAMOTO
Community Water Supply Support Center of Japan 2-5-17 Nuttarinishi, Chuo-ku, Niigata, 950-0076, Japan (for_you.satoyama@feel.ocn.ne.jp)

ABSTRACT

Pilot plant experiment of upflow roughing filters was done to reduce the load of suspended matter in surface water of rivers. Upflow roughing filter was effective to reduce the load. Function of roughing filter was mechanical trap system and biological function. It was estimated that about one month was necessary to develop the ecosystem. The system function and the maintenance technique were discussed.

Keywords roughing filter; ecosystem; pilot plant

INTRODUCTION

Surface water intake for water purification plant is common in Japan, while subsurface water intake is rare. Turbidity of the surface water of a river is easily influenced by a heavy rain fall. Strong disaster attacked often in Japan by a heavy storm. River bed was easily affected by this disaster. The river administrator hardly accepts a construction of a subsurface intake in a river bed in Japan. As a result, surface intake system for water supply plant is common in Japan. In order to reduce turbid matter in raw water, sedimentation pond is set as a pretreatment for slow sand filter (SSF). And an additional treatment of coagulation process is also strongly recommended in case of heavy turbidity. However, small microorganisms are easily affected by the coagulation reagent. The function of biological treatment is damaged.

Upflow roughing filter system to reduce the turbid water for surface intake has been reported (Wegelin 1996, Nakamoto 2005). The reduction effect of turbid water was also confirmed at Samoan WTP (Alaoa WTP, Malololei WTP and Fuluasou WTP) in South Pacific island country.

The effect of upflow roughing filter (URF) was examined at M city, Niigata in 2008 and 2009 using a pilot plant. Based on our experience and experiment, the ecological function became clear in the process of URF.

BACKGROUND OF PILOT PLANT EXAMINATION

M city (35.51 N, 138.89 E) is located in the center of Niigata plain in Japan. The water treatment plant (A WTP) in M city supplies the drinking water to 56,060 persons for M city and the adjacent N town by the report 2006 of the water supply vision of M city. The design of the capacity of water supply is 62,100 persons and the daily maximum of water supply is 47,000m3/D by A WTP. There are two water sources. One is the surface water of the Kariyata River which runs through the city and the other is the surface water from the Shinano River which is the longest river (367 km) in Japan. The present purification process of A WTP is a rapid sand filtration. Recently, M city was attacked two times by fairly large earthquakes in 2004 and 2007. M city made an evaluation of seismic proof for the A WTP after these earthquakes. The evaluation for A WTP reported the plant should be reinforced and should be repaired.

The city government made a plan to supply more delicious and safer water by SSF. New SSF plant requests a large amount of money to construct. M city tried a SSF experiment in order to get a financial support by Japanese government. Then M city tried to examine the applicability and the performance of slow sand filter. The common knowledge of SSF is weak to muddy water. M city asked about the applicability of SSF for this city to the Community Water Supply Support Center of Japan. Then we examined URF experiment for SSF from April 2008 to March 2010. Three types of URF experiments (5 steps of URF, 3 steps of URF and multi layers URF) were done using a commercial stainless tank of 200 liters.

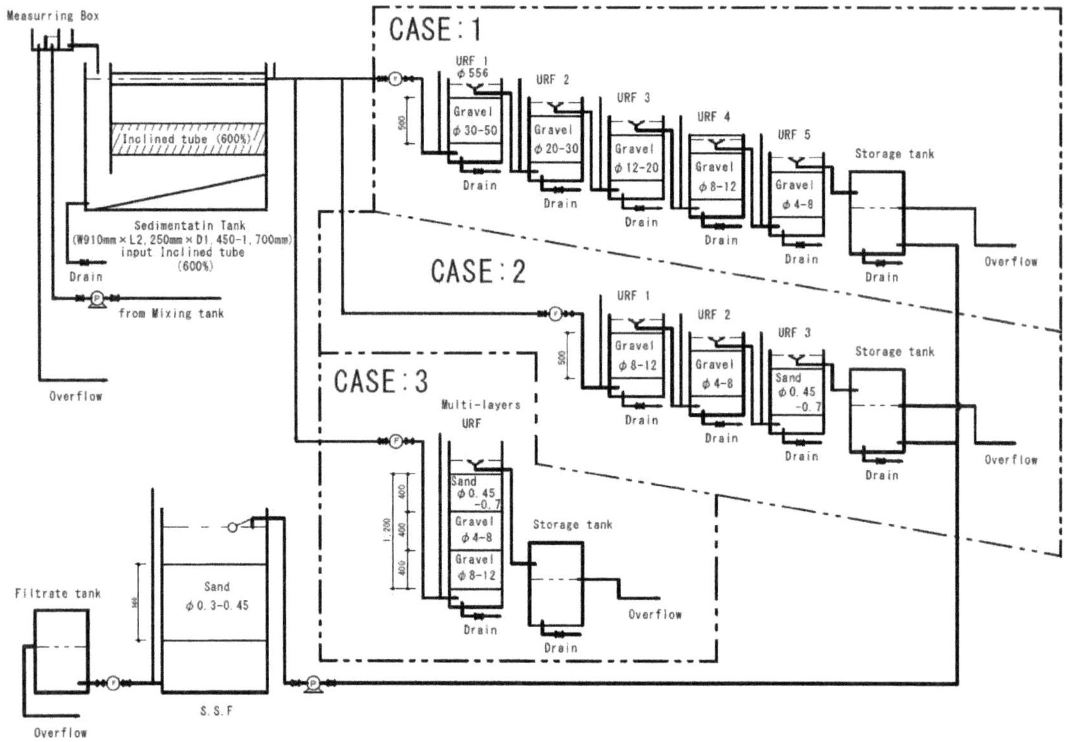

Figure 37.1 Design flow of URF experiment

PILOT PLANT

Figure 37.1 shows a test flow of URF experiments. Figure 37.2 is a photograph of the pilot plant. At first, we checked the sedimentation effect of muddy substance in raw water using a batch test tank. Surface water from rivers flows into the settling tank. Incline plates (honeycomb tube) was input in this tank in order to accelerate the effective sedimentation. The water after the settling tank flew into URF. Case 1 was 5 steps of URF. Case 2 was 3 steps of URF. And case 3 was multi-layers URF.

Figure 37.2 Pilot plant

The size of the batch test tank for sedimentation experiment was φ566 mm x H890 mm. The size of sedimentation tank was L2,250 mm x W910 mm x H1,525 mm (1,400 mm-1,650 mm). The residence time of water was kept to keep 4 hours. The size of URF tank was φ566 mm x H890 mm. The different sizes of gravels were used for 5 steps experiment. The gravel sizes were φ30-50 mm, φ20-30 mm, φ12-20 mm, φ8-12 mm and φ4-8 mm. Depth of gravel layer was 500 mm in each filter tank. The gravel sizes of 3 steps experiment were φ8-12 mm, φ4-8 mm and φ0.45-0.7 mm. The filtration speed in a tank was controlled to 17 m/d. The last tank was covered to avoid sunshine. The size of multi-layers URF was φ556 mm x H1,780 mm. There were 3 layers of different size of gravels. Each depth of 3 layers was 400 mm. The each gravel was used same sizes of gravels as case 2 experiment. The filtration speed in this tank was also controlled to 17 m/d. And the SSF tank was φ900 mm x H1,60 0mm and the sand depth was 500 mm. The sand size was φ0.3-0.45 mm. The filter rate of SSF was kept to 4.8 m/day.

The annual range of water temperature of the plant ranged from 4 C to 28 C. The temperature range during spring experiment (5 steps URF experiment and multi-layers URF experiment) ranged 8 C to 19 C. The temperature range during summer experiment (3 steps experiment) was 24 C to 28 C. The frequency of turbidity from the river during May 2008 to September 2009 was shown in Table 37.1. Water was usually clear but high turbid water was observed sometimes.

Table 37.1 Turbidity distribution of raw water from May 2008 to September 2009.

Turbidity (NTU)	< 10	10-20	20-30	30-40	40-50	50-100	100 <
Days	300	68	22	6	3	8	1

BATCH TEST FOR SEDIMENTATION

When the turbidity is high in source of water, URF is easily blocked without a pretreatment of settling basin. And the back wash of filter is necessary to eliminate the blocked matter. The suitable residence time in a settling basin is evaluated by use of batch test.

The batch test was done two times in snow melting period of March 18 to 19, 2008 (Figure 37.3) and in summer rainy period of July 29 to 30, 2008 (Figure 37.4). Raw water of a plant poured into the batch test tank. Sampling cocks were set at 4 different depths (5cm, 25cm, 45cm and 65cm from the water surface). Water sample of around 200ml was taken from the sampling cocks during 24 hours. Turbidity was measured by HACH 2100Q turbidity meter. Turbid matters were quickly settled down within a short time.

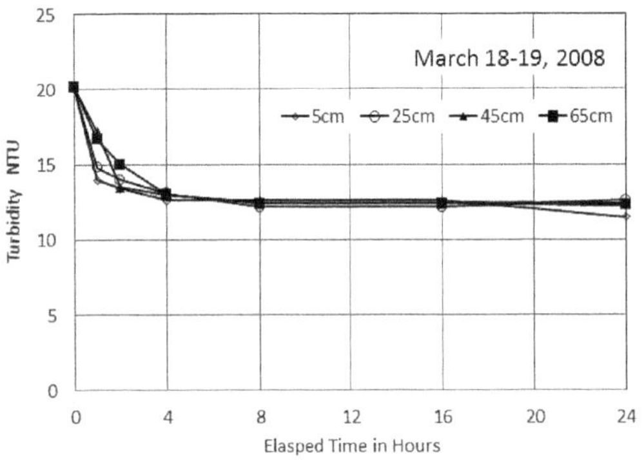

Figure 37.3 Sedimentation experiment in March (snow melting period)

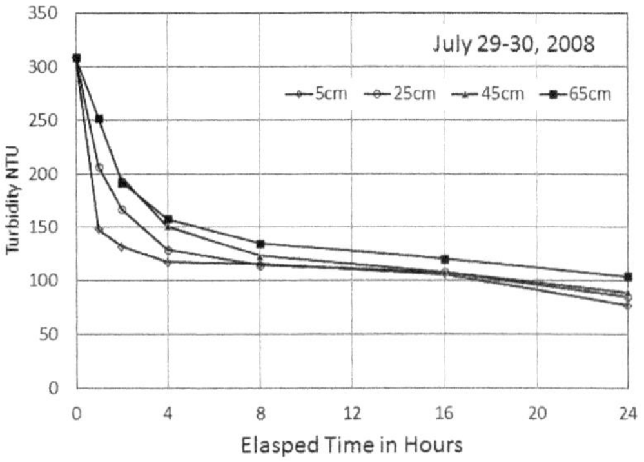

Figure 37.4 Sedimentation experiment in July (turbid water by a summer rain fall)

Major particulate matters in raw water precipitated within 8 hours. Even the major turbid matter of 309 NTU quickly precipitated within 4 hours in July. Remain turbidity may be due to silt or colloidal matters in raw water. In March, 20.2 NTU of raw water became 12.2 NTU (average) in 24 hours. The reducing rate was 60.4 %. In July, 309 NTU of raw water became 88.1 NTU (average) in 24 hours. The reducing rate was 28.5 %. In case of clear water, the ratio of hardly settable substances is large than turbid water. Considering the practical space of a sedimentation tank in a water treatment plant, we assumed 4 hours residence time was acceptable by use of inclined plates (honeycomb tube) in the sedimentation tank. The turbidity after 24 hours settling of 88.1 NTU (average) was serious for SSF. Another pretreatment to reduce the load was necessary to put between sedimentation tank and SSF. URF system was planned. It was gentle treatment without chemical addition. We made our URF experiment referring to Wegelin 1966 and Nakamoto 2005.

UPFLOW ROUGHING FILTER EXPERIMENT

Any type of roughing filter is not only physical reduction. Any filter has a big biological action. It takes time to perform the full of the biological action in URF. Our experiment of Case 1 (5 steps URF) started from April 2, 2008. Turbidity was decreased by step of the series of URF from raw water to SSF. The changes of turbidity are shown in Figure 37.4 during 38 days run (April 2, April 13, May 2 and May 9). Reducing rate of turbidity by URF became high during filter run. From this result we expected there was biological action in this process of URF. However, the development period of biological communities is related with environmental condition such as temperature and nutritional level of water. We expect it takes around one month to develop the suitable biological communities in these URF.

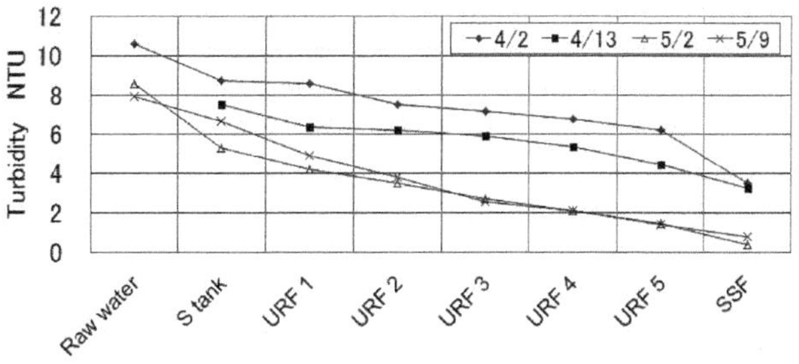

Figure 37.5 Turbidity change from raw water to slow sand filter.

Result of turbidity removal is summarized in Table 37.2. The removal rate was effective in the former 3 URF (No. 1, No. 2 and No.3). The rates were only 9 % in URF No. 4 and 7 % in URF No.5.

Table 37.2 Turbidity removal rate of 5 steps upflow rough filters

Upflow roughing filter and gravel size	No.1 φ30-50mm	No.2 φ20-30mm	No.3 φ12-20mm	No.4 φ8-12mm	No.5 φ4-8mm
Removal rate (%)	32	31	21	9	7
Cumulative rate (%)	32	63	84	93	100

Then, we made next step experiment using 3 steps of URF (Case 2). We chose small size of gravels (φ8-12 mm, φ4-8 mm and φ0.45-0.7 mm). The large size of sand of a standard rapid sand filter was used in the last URF. The filter speed was 17m/day. We had a chance to check the extraordinary loading effect of highly turbid water on August 20, 2008. The maximum turbidity of raw water was 776 NTU.

Table 37.3 Removal effect on the day of the high turbidity (August 20, 2008)

	Raw water	Sedimen-tation tank	URF No.1 φ8-12mm	URF No.2 φ4-8mm	URF No.3 φ0.45-0.7mm	SSFφ0.3-0.45mm
1 hour before peak	621	116	39.0	32.7	12.7	0.98
Peak turbidity	**776**	121	59.1	33.8	12.9	1.02
1 hour after peak	581	116	58.8	29.6	11.8	0.91
Average, NTU	659	118	52.3	32.0	12.5	0.97
Removal rate, %		82.1	55.5	38.7	61.0	92.2
Cumulative rate, %		82.1	92.0	95.1	98.1	99.8

Sedimentation tank was effective to remove the major load in raw water. The extraordinary high turbid water was usually a peak type due to heavy summer storm in Japan. Figure 37.6 shows the turbidity values of this experiment during Aug. 1 to Sept. 25, 2008. There was a peak event on Aug. 20.

Such an extraordinary peak happened by a special large storm. A large amount of heavy particulate matters flew into WTP. The major turbidity was easily reduced in a settling tank. But some amount of turbid matter passed into URF No.1. Finally, we could get a clear filtrate from SSF. The reduction rate between each step of series of URF was clear in the log scale graph (Figure 37.7). The absolute reduction amount was large but its rate was not so large. The first URF was effective than the second URF. However the third URF was large due to smaller gravel of rapid sand filter sand. This high reduction rate in the third URF may contribute the large surface area of gravels. Remarkable increase of head loss was not observed during this experiment. The filtrate of SSF was always low turbidity which was acceptable for Japanese standard level of less than 2 NTU.

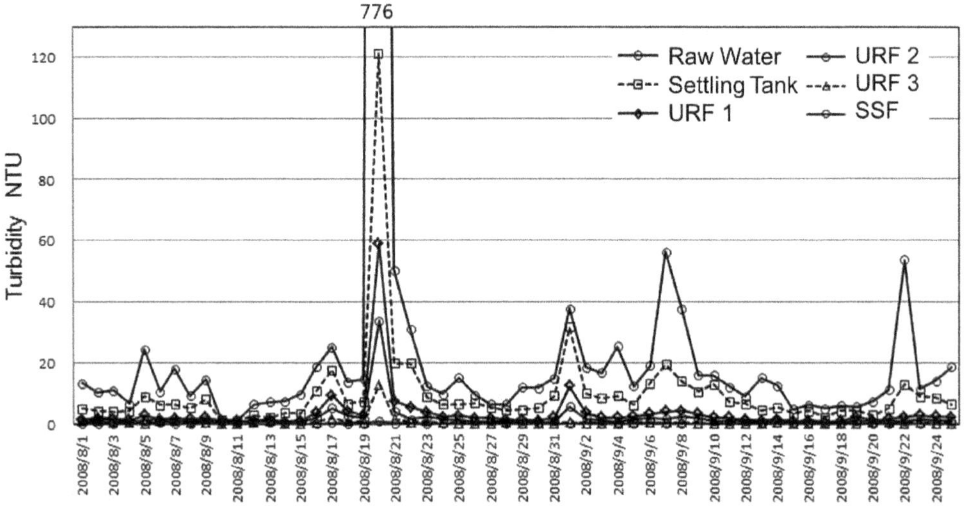

Figure 37.6 Turbidity removal along raw water to slow sand filter from August 1 to September 25, 2008.

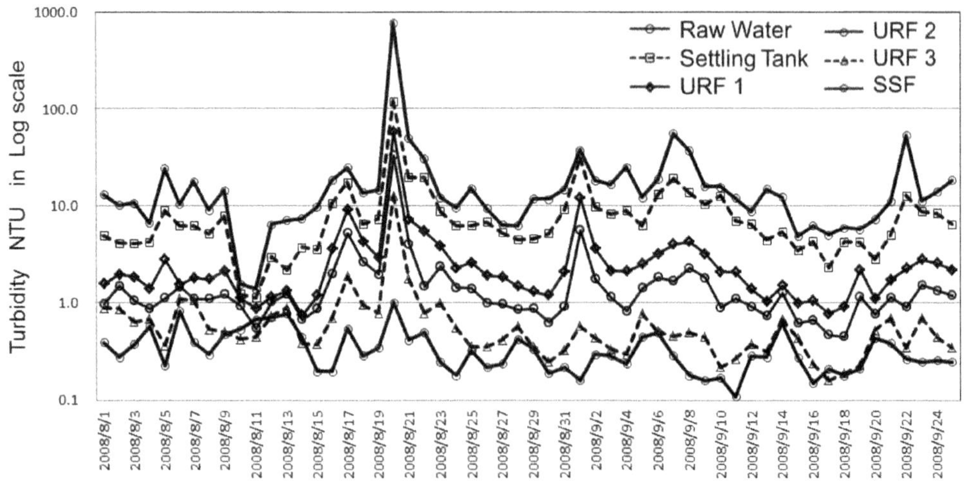

Figure 37.7 Turbidity removal along raw water to slow sand filter (log scale)

The multi-layers URF was also examined (Case 3). The efficiency was around 17 % less reduction in comparison with 3 steps URF. The depth of gravel layers in multi-layers URF was 120 cm. The total depth of 3 steps URF was 150 cm. We could not compare accurately the difference. However, multi-layers URF can reduce to one third of the space of 3 steps URF. But the height of URF becomes high. And the drain work for maintenance URF is easy than 3 steps URF. The multi-layers URF is very attractive, but we need more examinations to make an accurate evaluation.

SLUDGE DRAIN EXPERIMENT

The filter resistant of URF did not increase during a long filter run of Case 1 study. We expected there was a large amount of mud which precipitated at the bottom of the settling tank during a long filter run. And we also expected there was some large amount of trapped mud in each filter. We opened the drain cocks at the bottom tank on July 10. 2008 after 100 days filter run from April 2. The drained water was poured into another spare tank. The drained volume was about 150 liters which was a whole water volume in a URF tank. And the concentration of turbid matter (dry weight per volume) was measured by a Japanese standard test using a glass fiber filter.

The drain water from the settling tank was a condensed soft mud. The concentration of drained muddy water was 215 dry grams per liter. The concentration in the drain water from No. 1 URF was 3.7 dry grams per liter. There was about 60 times difference. Major amount of turbid matter was precipitated in the settling tank within 4 hours. The concentration of drained water of the series of URF decreased step by step (Figure 37.8). We repeated the drain work in case of No. 1 URF and No.2 URF. The major part of trapped particulate matter was drained easily off at one time drain from the bottom cock. We thought that one time drain work for each tank was enough (Figure 37.9). Drained muddy water was quickly settled and the supernatant water became clear soon. This mud was flocculated suspension like an activated sludge. This was confirmed by microscopic observation. Algae and small organisms actively worked in the supernatant water of URF. And many heterotrophic small animals worked among and on the gravels in URF under aerobic condition. They trapped any matter among the gravels of URF.

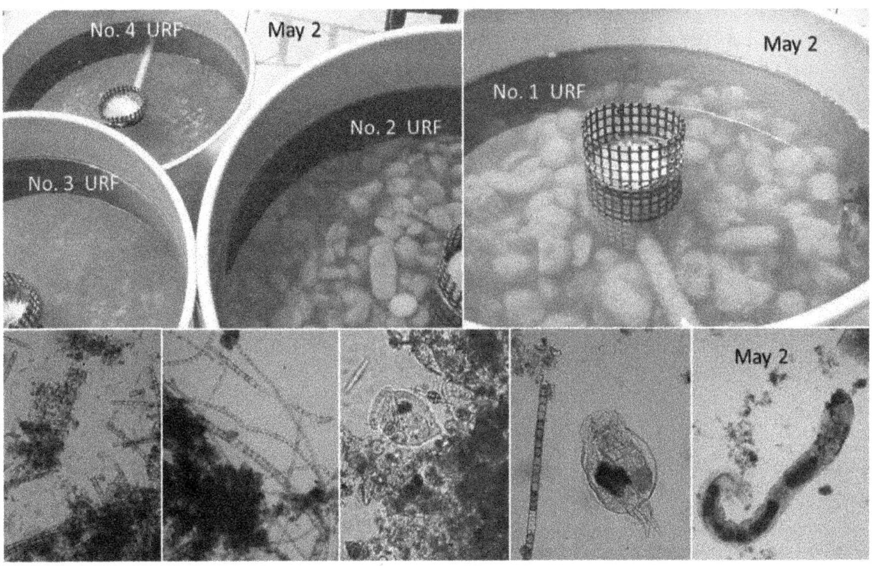

Figure 37.8 Mud and algae on the gravel of URF. Microscopic photographs of surface sample.

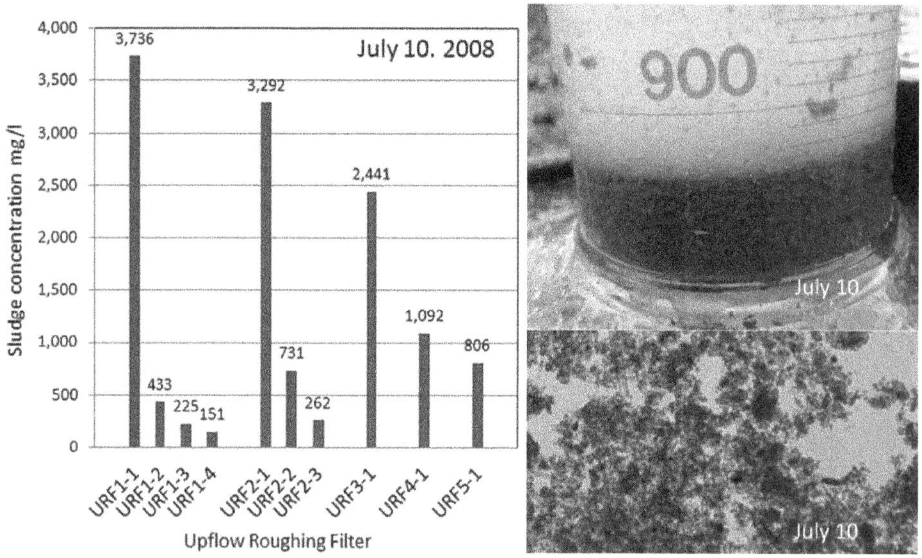

Figure 37.9 Drain test of trapped suspended matter in upflow roughing filter. And microscopic photograph taken on July 10.

The color of drained sludge was brown. It was no smell.

CONCLUSION

SSF is based on Ecological Purification System under aerobic environment where is almost constant downward current. The food for biological community comes from the top. Too much load of inorganic or organic matters is serious for SSF. And the biological function of URF is same as SSF. But the biological action occurs from the bottom to the top among gravels where food for biological communities comes from the bottom. As the size of gravel is larger than sand of SSF, the water passes easily through the gravel layer without resistance. There is enough oxygen supply from the bottom. Inside of URF is always aerobic from the bottom to the top. This system is gentle for aerobic organisms. This is an Ecological Purification System. Biological function is closely related with temperature. This system works well active in the warm region or period.

The drained water is a flocculated suspension. It seems that it is mainly organic debris. Many organisms work in URF. The gravel is fairly large. And there is a large space among gravels. Therefore URF is not so easily blocked with turbid matters. If the filter resistant of URF is increased, one time drain of the blocked matter from the bottom cock is enough. We need not to drain off the mud in URF, so frequently. We conclude that URF is very effective to reduce the load of suspended matters.

REFERENCES

Independent country Samoa (Alaoa water purification plant operation manual).
Nakamoto, N. (2005) How to make delicious water, Tsukiji shokan Publishing Co. ltd., Japan.
Wegelin, M. (1966) Surface Water Treatment by Roughing Filters, SANDEC-SKAT, Switzerland.

Chapter 38

Enhancement of the efficiency of upflow slow sand filtration - effect of combined application with a fiber filter unit

Tatsuo Kato, Ichiro Ozaki, Akihiko Okada
Okada Industry Co. Ltd., Japan

ABSTRACT

The upflow slow sand filtration system was modified by combined application of a fiber filtration unit above the sand bed of the conventional sand filter. The results obtained from pilot scale filtration experiments conducted for removing turbidity contained in river water from a dam reservoir and tracer particles of Cryptosporidium demonstrated clearly that the modification was very effective in alleviating the deterioration of the filtrate in turbidity induced by rainfall and by cleaning of the sand bed. Its effectiveness in lowering the levels of the tracer particles of Cryptosporidium in the filtrate was also confirmed.

Keywords Upflow slow sand filtration; *Cryptosporidium*; fiber filter

INTRODUCTION

As one of the technologies developed for enhancing the efficiency of the slow sand filtration system, the upper-flow sand filtration method has the advantage to alleviate the head loss development, thus enabling extension of the filtration time. In association with this method, several patents were issued. In 1960, Hill et al. discussed in depth about the advantages and disadvantages of this method and since then, experimental investigations of the performance of the upflow flow slow sand filters have been consistently conducted and some improvement methods for the filter equipment are proposed.

In order to improve the efficiency of the slow sand filtration, the authors started to develop the upper flow slow sand filters that could alleviate the faster filter bed clogging and facilitate maintenance in 1960 and so far, the developed sand filters have been introduced for water production in more than 1000 places throughout Japan, particularly in mountainous areas for treatment of stream water with low degree of contamination or sallow groundwater. However, taking into consideration that water pollution by pathogenic microorganisms in some areas is getting worse and that fecal pollution by wildlife is highly likely in some areas where pollution associated with human activities does not exist, further enhancement of the performance of the upflow slow sand filters is required.

For enhancing the treatment efficiency for pathogenic microorganisms, we modified the filters by combined application of fiber filters above the sand bed of the sand filters. The effectiveness of such a modification was studied through pilot scale filtration experiments for removal of turbidity and the homogeneous particles of a well-used *Cryptosporidium* tracer.

METHODOLOGY

The experimental setup of the modified pilot scale upflow slow sand filtration system is shown in Figure 38.1. The filter is consisted of 2 parts: a sand bed and a fiber filter unit. The fiber filer unit, having a pore size of 1.0 μm and a surface area of 0.4 m^2, was installed above the sand bed to treat the water after primary filtration through the sand bed. The sand bed was packed with two layers of sand with different sizes (the effective size for the upper layer was 0.3 mm and that for the lower layer was 0.5 mm); the total bed depth was 65 cm, the sectional area was 1.75 m^2, and the filtration rate was 10 m/d and 5m/d. The experiment consisted of two runs: one was conducted for treating river water stored in a dam reservoir and another one for treating water spiked with a well-used *Cryptosporidium* tracer.

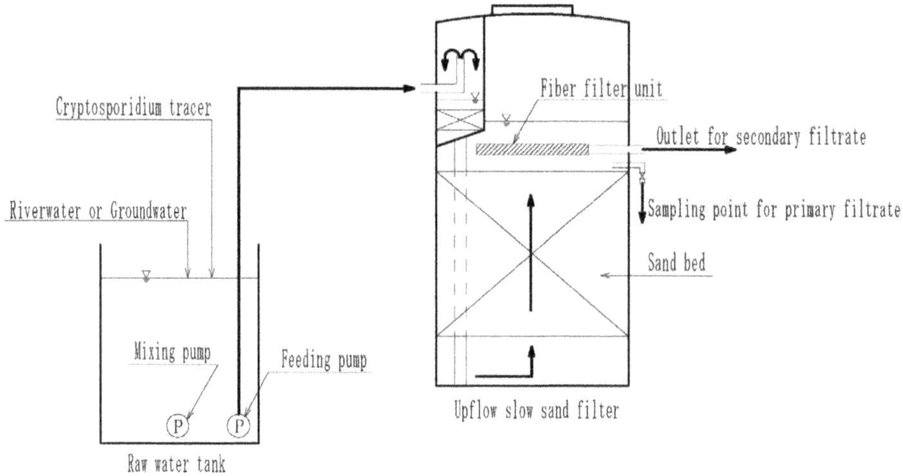

Figure 38.1 Experimental setup of the modified pilot scale slow sand filtration system.

RESULTS AND DISCUSSION

The time profiles of turbidity before and after filtration through the pilot scale experiment run for treating river water from the dam reservoir at the filtration rate of 10m/d are displayed in Figure 38.2. By combined application of the fiber filter unit, the increases in the filtrate turbidity following the cleaning of the sand bed were alleviated obviously; and at the same time, the peak levels of the filtrate turbidity appeared during the sudden raw water quality worsening caused by rainfall were also effectively reduced.

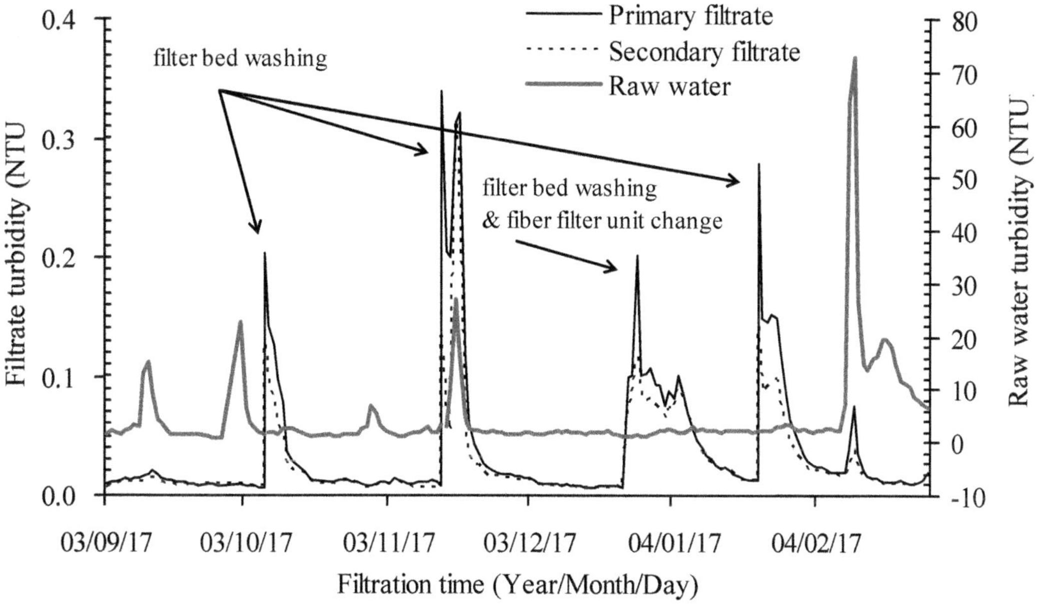

Figure 38.2 Behavior of turbidity during filtration at the rate of 10 m/d. Primary filtrate represents water after filtration through slow sand filter (SSF), and Secondary filtrate represents the filtrate of SSF after further filtration through a fiber filter unit installed inside the SSF.

Table 38.1 shows the results of filtration for removing the spiked homogeneous particles of the *Cryptosporidium* tracer at the san filtration rate of 10m/d. For comparison, the results for filtration at the lower rate of 5m/d are shown in Table 38.2. The combined application of the fiber filter unit enhanced the log removals for the tracer particles under both the sand filtration rate of 10m/d and 5m/d. Its effect in alleviating the bacterial risk associated with detachment occurred inside the sand bed also appeared, as could be seen from the results obtained after the influent water was shifted to water containing no tracer particles after 6 hrs of filtration at either 10 m/d or 5 m/d.

Table 38.1 Behavior of *Cryptosporidium* tracer during filtration at the rate of 10 m/d

Filtration time (hrs)		Raw water number/L	Primary filtrate number/L	Primary filtrate removal	Secondary filtrate number/L	Secondary filtrate removal
0	Raw water added with *Cryptosporidium* tracer was supplied for 6 hours as the influent water	10^6	0	-	0	-
2		10^6	3	5.5	0	>6.0
3		10^6	7	5.2	0	>6.0
4		10^6	8	5.1	2	5.7
5		10^6	17	4.8	2	5.7
6		10^6	7	5.2	0	>6.0
9	Influent water was shifted to raw water that did not contain *Cryptosporidium* tracer after the filtration time of 6 hours	0	6	5.2	3	5.5
12		0	5	5.3	1	6.0
24		0	5	5.3	0	>6.0
48		0	0	>6.0	0	>6.0

removal = $\log_{10}$ (Raw water concentration / Filtrate concentration)

ACKNOWLEDGEMENTS

This study was conducted under a joint research program between enterprises, universities and administrations. The valuable advice from the late professor of Gifu University, Dr. Akira Yuasa, is highly appreciated. We are also very grateful to Kawai Drinking Water Treatment Plant of Kani-city, Gifu Prefecture, Japan, for providing the site for the filtration experiments of this study.

REFERENCES

Akira Yuasa, Akihiko Okada, Akira Nagai, Kiichi Goto (1994) Report on the results of the long-term experiment of upflow sand filtration treating Biwa Lake water, *Water Supply*, 39 (6).

Guidelines of countermeasures for dealing with *Cryptosporidium* in water supply, a notice from the water supply department of the Ministry of Health, Labour and Welfare, Japan, March 30, 2007.

Kiichi Goto, Akihiko Okada, Ikuhiro Komase, Harutoshi Kato (2003) Evaluation of the performance of upflow sand filters by using Fluorescent tracer as the surrogate of *cryptosporidium*, *Water Supply*, 48 (8).

Chapter 39

Surface water treatment plant using ecological purification system in Bangladesh

Md. Mizanur Rahman[a] *and Md. Joynul Abedin Zaman*[b]
[a]*Principle, Technical Training Centre, Nishindara, Bogra, Bangladesh*
[b]*Water Supply Engineer, Asia Arsenic Network, Jessore, Bangladesh*

ABSTRACT

The ecological purification plants for a rural community were constructed in Bangladesh. The surface water treatment plant consists of a sedimentation chamber, upflow gravel filters and slow sand filter. The plant is operated by villager. How to solve the problems to keep suitable maintenance in rural area is described. This is a model example plant for rural water supply in Bangladesh.

Keywords ecological purification system, roughing filter, maintenance, cost effective

INTRODUCTION

Bangladesh is one of the most densely populated and developing countries of the world. Since high level arsenic was detected from tube well in 1993 in Bangladesh, the relevant government sectors and NGOs with development partners constructed numerous safe water devices to overcome the situation. But still 20 million people are using arsenic contaminated water for drinking and cooking. It is urgent matter to supply safe water to those people in Bangladesh. Among safe water devices, deep tube well is the most popular because of low cost, constancy of water temperature and easy maintenance for users. But in some areas it is very difficult to install deep tube well blocked by gravel layer and also not easy to find suitable aquifer due to saline, arsenic and iron contamination. And the groundwater level is going down day by day due to excess extraction of water for irrigation. Nowadays government emphasizes on surface water to use for water supply. Asia Arsenic Network (AAN) has worked in Jessore district for arsenic mitigation from 1997 and made a plan of surface water treatment and pipeline deliver system besides "Baor (oxbow lake)", a big surface water body and invited Prof. Nobutada Nakamoto to provide an appropriate advice. The first ecological purification plant in Bangladesh was completed and started supplying fresh water to 200 house hold in July 2004 in Putkhali Union, Sharsha Upazila, Jessore District. Based on this design, so far AAN constructed 3 surface water treatment plants using ecological purification system (Table 39.1). AAN constructed three piped water supply system as well as also surface water treatment plant. The main reasons are easy maintenance by villager, continuous technical support from AAN and cost effectiveness. So the rural community handle this system very excellent way. This is a remarkable and model example for rural water supply in Bangladesh.

Three surface water treatment plants are almost same in design. But Sl. No 2 and 3 (shown in Table 39.1) are upgrade one than Sl. No 1. For explanation in this paper, we describe only Sl. No 2.

Table 39.1 3 ecological purification systems in Jessore District, Bangladesh

Sl. No	Location	Installation year	House hold (HH)	Water supply (m^3/day)	Monthly service charge (Taka/HH)
1	Viilage: Khalshi, Union: Putkhali, Upazila: Sharsha	2004	180	15	20
2	Viilage: Kustia, Union: Dhuliani, Upazila: Chaugachha	2008	385	25	20
3	Viilage: Punchvulot, Union: Goga, Upazila: Sharsha	2008	300	25	20

DESCRIPTION OF PLANT

For better understanding the detail description of source, design, cost and purification system are describing below. The name of the unit is "Kushitia-Azmatpur Pipe Water Supply System (PWSS)".

Water Source

This area is difficult to find alternative water sources. In dry season surface water dries up and deep tube wells are contaminated with arsenic and salinity. Fortunately the project found some Boars; dead arm of river, called Oxbow lake, full of water throughout the year. Area of baor is 6 km^2 (apx.) and depth is 25 ft.

Design Consideration

Size of raw water tank (RCC: reinforced concrete): 4.2 m x 4.2 m, depth of raw water tank (RCC): 2.8 m & capacity of raw water tank: 48 m^3 .

Capacity of clear water tank and OHT (overhead water tank): 30m^3 & height of the OHT from GL: 12.3 m.

Mode of water supply: Gravitational transmission from OHT through pipeline.

Main pipe length: 8.5 km, Shared tap (Clear water collection point): 101 nos.

Number of pump (1.5 HP): 2 nos, one for to bring water from baor to raw water tank and another one for lifting water from clear water reservoir to OHT.

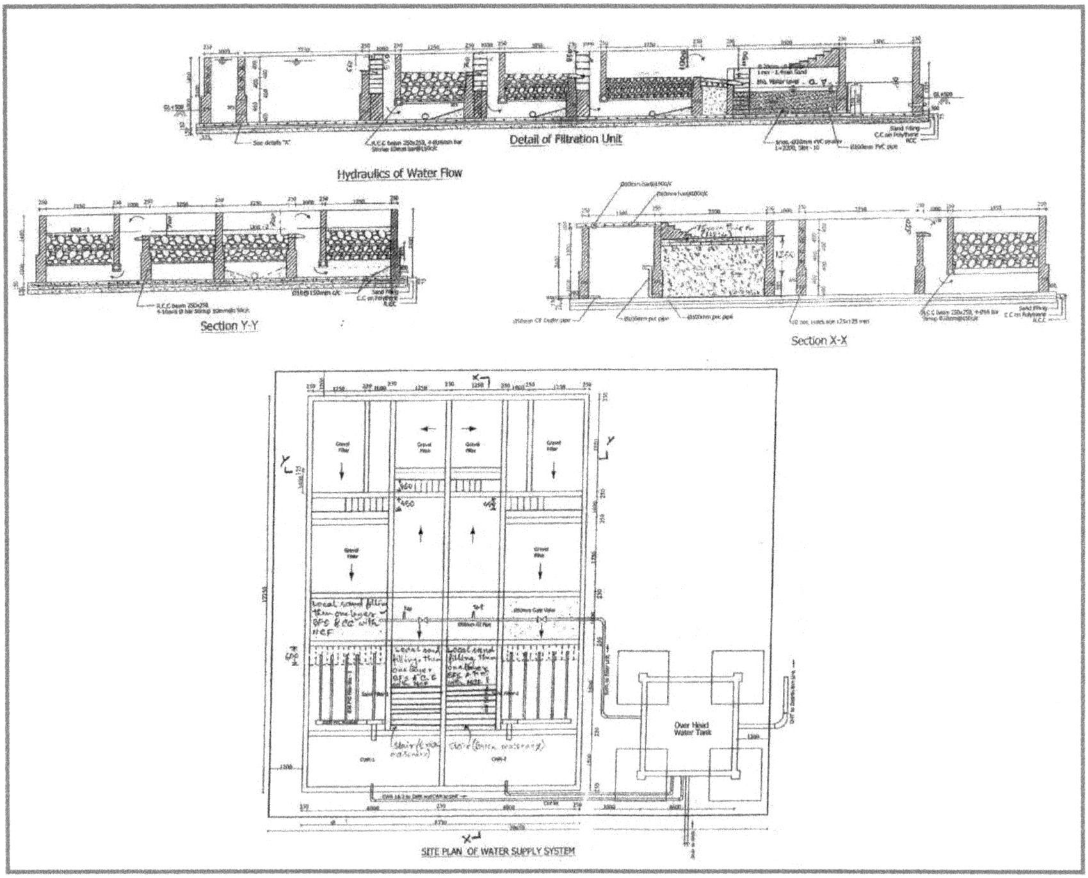

Figure 39.1 Layout of ecological purification system in Bangladesh

Table 39.2 Design of Surface water treatment plant

Sl. No	Description	Unit	Quantity and Specification
1	Filter unit	No	2 (Symmetrical)
2	Size of each unit	m^2	6.5 x 4
3	Height of unit (Depth)	m	2.4
4	Number of gravel filter (Up flow roughing)	No	3 (In each unit)
4	Size of sand bed	m	2.5 x 2.5
5	Flow rates	$m^3/ m^2/ hr$	0.25
6	Material of construction	--	Masonry

Construction Cost

Table 39.3 The construction cost detail are given below (2008)

Sl. No	Description of Item	Amount (Taka)	Amount ($)
01.	Intake Structures	72,400	1,065
02.	Low Lift Pump and High Lift Pump	191,730	2,820
03.	Surface Water Treatment Plant	830,784	12,217
04.	Overhead Water Tank	1,951,722	28,702
05.	Pump House	140,543	2,067
06.	Distribution Pipelines and House Tap Connections in Household	3,041,151	44,723
07.	Sluice valve, Wash out etc	333,000	4,897
08.	Others (Shade, gate and toilet)	248,642	3,656
09.	Material Test (Before and during construction work)	40,000	588
	Total=	6,849,972	100,735

Purification System

At first water is collected from baor by 1.5 HP pump and this water is lifted up in to the raw water tank. Capacity of the tank is 48 m^3. 2 hrs 45 minutes is required to fulfill the tank. After that raw water enter in to treatment plant through gravity through GI pipe. The supply of raw water into a treatment plant is continuous flow (24 hrs/day). So to run the treatment plant, it is necessary to calculate how much water is necessary to supply per day. It was calculated that 500 household will use this water in 2008. Every house hold needs 50 liters water/day for drinking and cooking. For this reason every day 25 m^3 (25,000 lit) treated water is required to fulfill the users demand. Considering this issue 17 lit water need to treat per minutes. A treatment plant has two symmetrical units. So raw water flow rate is 8.5 lit/min for each unit. Thus the two lines of a treatment plant work during 24 hrs. At first raw water falls in inlet chamber then falls in to sedimentation tank. After that this water passing through three consecutive gravel chambers. URF (up flow roughing filtration) system is used in gravel chambers, while water is passing through three gravel chambers, turbidity is reduced. Finally the water passing through sand filter and small animals and microbes remove pathogenic bacteria and water becomes clear and clean to deposit into a reserve tank. After that treated water reserves in an underground tank through gravity. In the underground tank of the tower, some amount of chlorine is added and treated water is lifted up by 1.5 HP pump to the overhead tank. The capacity of this tank is 30 m^3. It takes 1 hr 45 minutes. This water delivered to common taps in the village by gravitational pressure through pipeline. The purification flow diagram is given below.

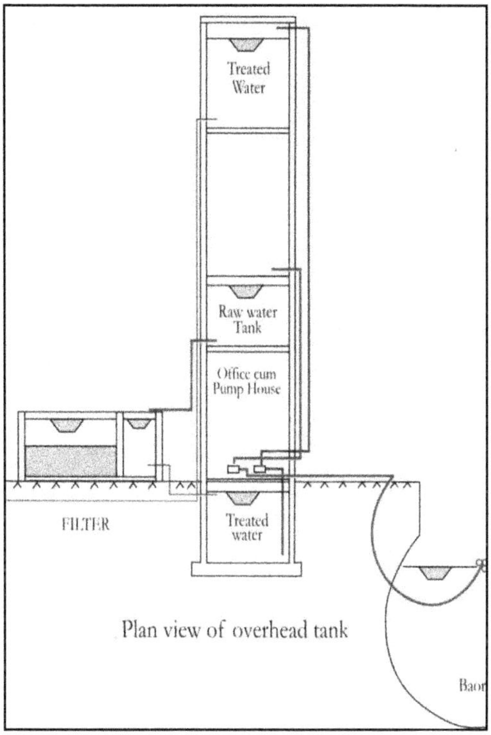

Plan view of overhead tank

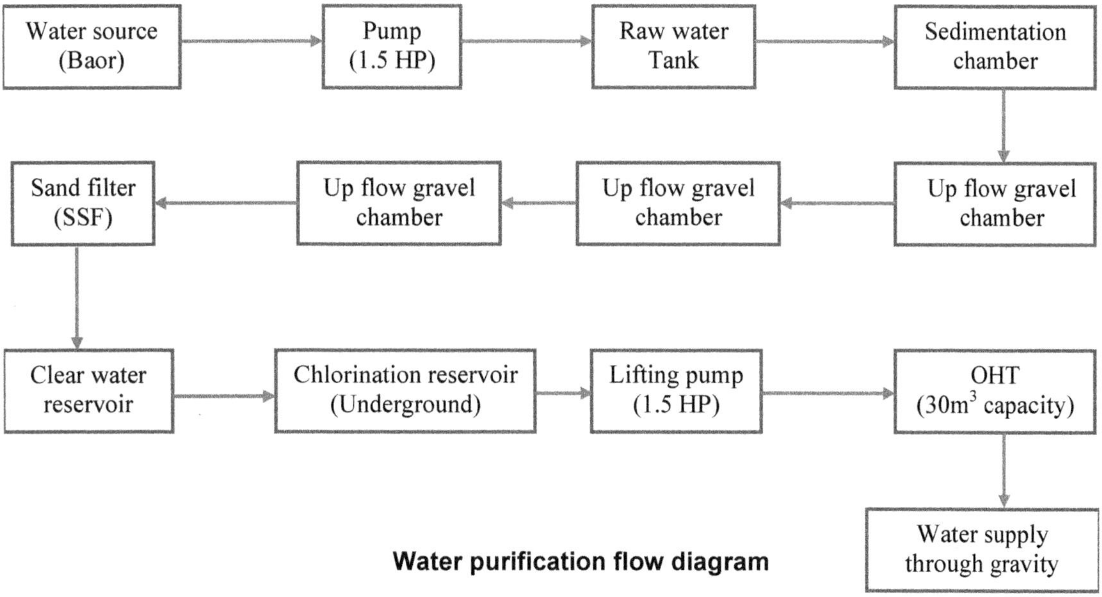

Water purification flow diagram

Water Quality

Table 39.4 The water quality report is given below (Dec-2008)

Parameter	Bangladesh drinking Standard	Baor	Inlet	After Gravel	After SSF
Turbidity (NTU)	10	5	<5	<5	<5
pH	6-8.5	8.19	7.83	8.42	8.23
TDS (mg/l)	1000	138	152	152	166
DO (mg/l)	6	10.14	7.86	7.45	4.96
Hardness (mg/l)	500	144	140		136
AS (μg/l)	50	22	18		<1
Fe (mg/l)	<1	0.18	<.01		<.01
NH4 (mg/l)	0.64	0.14	0.14	0.15	0.16
NO3 (mg/l)	45	1.8	0.3	0.2	0.4
PO4 (mg/l)	6	0.06	0.18		0.06
FC (cfu/100ml)	0	113	113	57	0

OPERATION AND MAINTEANCE

In Bangladesh many treatment plants were installed for piped water supply not only in rural but also municipal area. After long run it was stopped due to lack of proper maintenance. Cost effect is also one of the big factors for smooth running. If the plant or network does not operate proper way, soon water quality also is deteriorated and users lose their interest to use this water. This impact falls on the whole network and finally it is stopped. In 2008 Bangladesh government and JICA surveyed 120 nos of rural pipe water supply systems. According to that survey result 48% were functional, 13% were partial functional and 39% were nonfunctional. There were some reasons for non functional or partial functional. Some of them were water quality (deteriorate), complex operation and maintenance, running cost, good management etc.

Kushitia-Azmatpur PWSS has one management committee. Twenty persons include in this committee. One president, one vice president, one secretary, one treasurer, one vice treasurer and another fifteen persons are general member. Every month they conduct meeting. In this meeting mainly they discuss the present situation of the plant. The main duty of caretaker is, to maintain the plant and collect money ($0.26/HH/month) from house to house. If any house does not pay regularly, committee takes necessary action. The management capacity of this committee is very nice. For this reason the whole system running smoothly which is very important. This is a rear example in rural area of Bangladesh.

Water Supply

The purified water supplies through an overhead tank in circulation with pipe line network by gravitational flow. There are 101 nos stand taps in common place in the village. Every tap point has more or less 5 nos house hold. They collect water from here. Two time water supply in a day. First time is in the morning for 25 minutes (7.00am to 7.25am) and second time at afternoon for 20 minutes (5.00pm to 5.20pm). The water supply hour and amount of water is enough for them.

Maintenance

One caretaker maintains this system. Actually regular maintenance is easy. Every day caretaker takes out dead leaf and other waste materials from the treatment plant and every day he adds chlorine (bleaching powder) into the clear water. Every interval of three or four days, caretaker opens wash out (back wash system) of gravel chamber. When purification flow rate reduces, he scraps sand bed. If require, user committee employs some labors and they wash sand and gravel chamber every year. If any other problem occurs soon the repair by them; if they cannot, they repair it with the support of AAN.

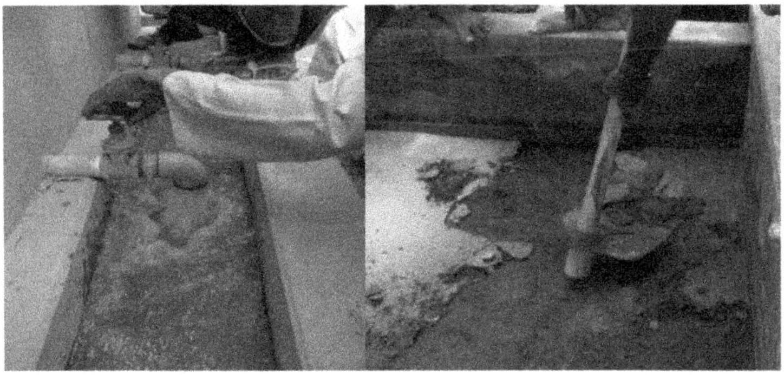

INCOME AND EXPENDITURE

Income

Monthly water tariff $0.26/HH (At present 100% of users are paying monthly tariff). Total income $100.10 ($0.26*385HH)

Expenditure

Salary of caretaker is $64.10/month; Chlorination cost $3.85/month; Electricity bill $20.50/month. Total monthly expenditure is $88.45. Actually the operation and maintenance cost is cheap. So total monthly cash balance is $11.65

At the time of construction users collect $4488 as a user share. User committee has kept this user share in bank for future maintenance since 2008. Above cash balance is also deposited into same bank account.

Cost effect

Calculation says that supply cost per 100 litters is about $0.013 (1 taka). This system is very convenient for rural people.

Constraints

In Bangladesh rural area load shading (electricity off) is very common phenomenon. So some time it hampers the smooth operation. There is no good system for regular water quality check. Sometime the pump becomes out of order. There is no good mechanics or technicians in rural area. So they have to take this pump in town. As they have no extra pump, the water supply stops for one or two days and users suffer a lot. Every year late dry season some people are seasoning their jute in the baor and raw water quality becomes bad. Filter maintenance also becomes difficult.

Another problem was also arose in case of Sharsha pipeline Sl No 1 & 3 (Table 39.1). Sometime saline water came into baor from the sea through river. Then the whole network was also stopped during about two months in the year. But this phenomenon not occurred in every year. In 2011 AAN constructed one gate in middle point of baor. When the saline water comes, users close the gate. So the collection side of raw water is protected from saline water. Again they open the gate, after solving the saline problem. And there is another problem that sometime someusers did not want willingly pay (monthly charge).

Replication

Considering its water quality, maintenance, management and expenditure Bangladesh government also takes it in positive sense. They already replicate this system in Kachua Upazila, Bagerhat District. They have also plan to increase this treatment system in the arsenic contaminated and deep tube well unsuitable area, if there is abundant surface water. Thus the "Ecological Purification System" expands in Bangladesh

CONCLUSION

In Bangladesh rural area many water devices are installed but after a certain period it stops or abundant due to lack of management, maintenance and expenditure. Till now rural people does not habituate to pay money for receiving water. So it is very difficult to run this type of plant. Average monthly tariff of other scheme more or less $1.00/HH. But monthly tariff is $0.26/HH in Kushitia-Azmatpur PWSS cases. Its operation, maintenance and management are simple. During 9 years after launching the first plant, some natural and social problems arose. But still they are running after overcoming these problems. Without any support from government or non-government organization, the management of this plant is appreciative. It is one of the model case and rear example in Bangladesh for rural water supply. It is a learning event for rural water supply. In this connection Bangladesh government is now implement this type of plant.

Chapter 40

Development of ecological purification unit for producing drinking water from organically polluted raw water (BOD~30 mg/L)

Kazuhide. Maruyama and Yoshifumi Hashiguchi**,*

**Main office, Shin Nippon Air Technologies Co. Ltd. 2-31-1 Nihombashi Hama-cho, Chuo-ku Tokyo 103-0007 JAPAN*

***Shin Nippon Lanka (Private) Limited. 309/1 Colombo Road, Welisara, Sri Lanka*

ABSTRACT

In inhabited islands of Maldives, the groundwater pollution is very serious by underground seepage of domestic waste water. Pollution level (BOD) exceeds 30mg/L occasionally. Inhabitants drink rainwater collected on roofs, and use groundwater for daily purpose. But highly populated islands are water-hungry in dry-season. Groundwater consumption exceeds the rainfall and seawater invades and pollutes groundwater. So we propose to recycle living drainage with the water quality control system. The system circulates groundwater charged by rainwater and collects drainage through the perforated PVC pipes. This PVC pipe network will be installed underground beneath the roads and will store water. Collected raw water will be purified into drinking water by the Ecological Purification Unit (EPU) and supplied to each house. We have tested EPU at Shin Nippon Lanka in Sri Lanka, and proved the effectiveness for organically polluted water.

Keywords: ecological; recycle; energy-saving; easy maintenance; low cost; BOD 30 mg/L

INTRODUCTION

Developing countries have problems of river and groundwater pollution by human waste. In 192 inhabited islands of Maldives, inhabitants drink rainwater collect from roofs and use groundwater domestically. But groundwater is polluted by septic tanks and infectious diseases expand. From the reconstruction after the 2004 tsunami, some water treatment plants have been installed in 30 islands supported by developed countries and international organizations. An organization donated RO (Reverse osmosis water purification) units at 40 islands but they have been left unused and damaged because they were too difficult to operate and expensive to maintain. We researched these islands and developed the Ecological Purification Unit, which is safe, low-cost and easy to maintain for inhabitants without extensive plant operation and maintenance knowledge.

Integrated Groundwater Management

In Maldives, the rainfall exceeds 1900mm/year and "lens water (fresh water)" is formed in underground. But rainfall per month decreases to 30-40mm in dry-season. So it needs to keep the balance between consumption and rainfall. For that, it is necessary to return living drainage to underground or use seawater for sanitary, especially in highly populated islands. The perforated PVC pipes network which collects and recycles groundwater will be installed beneath the roads. Recycled water and rainwater on roofs will be collected and then recharged to groundwater with the network. Then it will be purified into drinking water with the Ecological Purification Unit. Especially here, we investigated in Gulhi Island that is highly populated and water-hungry Island. This study was requested by Environment and Energy Ministry of Maldives for case studies.

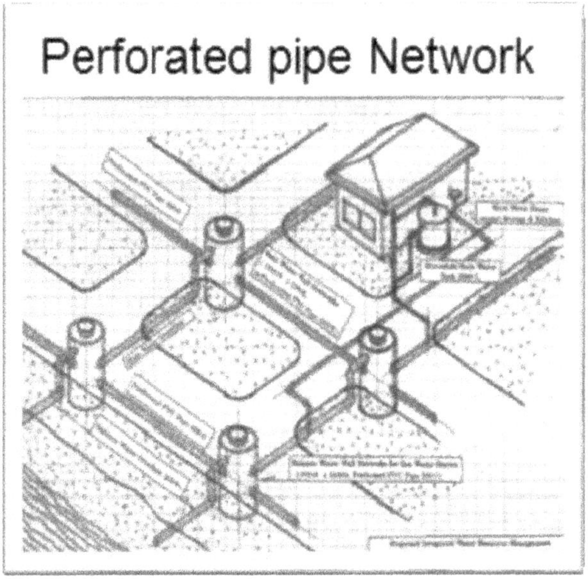

Table 40.1 Annual and monthly rainfall in Maldives (Google information, /mm)

Month	1	2	3	4	5	6	7	8	9	10	11	12	Total
Rainfall	114	38	74	123	219	167	150	176	199	194	231	217	1902

Figure 40.1 Outline of Gulhi Island

- Name of Island: Gulhi Island at South Male Atoll (21.3km from Male)
- Population of Island: 863 persons (124 houses)
- Dimension of Island: Length 400m + Width 225m
- Effective area of the Island: 74,000 m2 (85.7 m2/persons)

Table 40.2 Minimum water requirement per capita islanders in dry season

Description	Consumption per person	Source of Water	Method of Water Treatment
Toilet Flushing	40.0L/day	Sea Water	40.L/day Sewage Treatment
Drinking and Cooking	10.0L/day	Rain Water Tank	Sewage Treatment
Dishwashing	5.0L/day	Rain Water Tank	**Ocean Discharge**
Landry with Soap	5.0L/day	**Purified Water**	**20.L/day/person**
Landry without Soap	20.0L/day	**Purified Water**	Seepage to Underground
Bathing	30.0L/day	**Purified Water**	**Recycling Water**
Wash Basin	10.0L/day	**Purified Water**	**60.L/day/person**
Total	**120.0L/day**	**65.L/day Purified**	**80+40 L/day/person**

Rain water balance in the dry season of Gulhi Island

- Minimum rainfall of February: 38mm/month×50%×74,000m2 = 1,406m3/month = 46 m3/day(50% of the rain water will be evaporates into the atmosphere.)
- Ocean discharge of waste water (rain water) : 20 L/day×863 persons = 17.3m3/day
- Comparison of the minimum rainfall and ocean discharge amount.:
- Minimum rainfall 46 m3/day > Ocean discharge 17.3 m3/day

Integrated water management of Gulhi Island

- 40L/day use of seawater, 15L/day from rainwater tank, 65 L/day from EPU,
- Capacity of EPU 65L/day/person ×863 persons/day = 56,095 L/ day(60 m3/ day)

DEVELOPMENT OF ECOLOGICAL PURIFICATION UNIT

Structure of Ecological Purification Unit (EPU)

- EPU works with "Slanted Soil Chamber Unit" for preprocessing to purify the polluted water with BOD 30mg/L into safe-drinking level. Ecological Purification Unit consists of these three types; "Up-flow Roughing Filter Unit" for middle-processing, and "Slow Sand Filter Unit for end-processing.
- EPU makes easy transportation, construction and maintenance by uniting with FRP (Fiber reinforced plastic).
- The usual capacity of EPU is from 3m3/day to 60m3/day. When it exceeds 60m3/day, multiple EPUs will be installed.

Outline of Slanted Soil Chamber Unit (SSCU)

- Outline of slanted soil chamber unit (SSCU)" was developed by Yonden Consultant Inc (Yonden) as a purification system of kitchen drainage. Yonden and SNK entered into a licenseagreement and developed a purification unit.
- Water process loading per unit area of SSCU is normally $0.2m^3/m^2$/day. SSCUoccupies place more but it can be reduced by multi-stage installation.

- SSCU has 2m wide and divide 1 m of center partition for the easy maintenance. The base plate is corrugated plate to make easy cleaning.
- Sands (3-8mm), anthracites, activated carbon and sponges were adopted for filter media in the test. (In Maldives, coral sands will be applied.)
- We measured removal ratio of nitrate nitrogen by planting Kangkung on the backward tray of SSCU, but we could not measure the quantitative removal ratio. However, organically grown vegetables will be favorable there because the inhabited islands of Maldives have some lack of vegetables.

Outline of Up-flow Roughing Filter Unit (URFU)

- For URFU, we applied several filter media with different sizes. The coarse filter is parted into 3 tiers. The filtration speed is $20m^3$/day.
- Natural aspirators are applied on the supplement side of URFU. It provides the whole-day stability to the dissolved oxygen level in the coarse filter. The sunshade covers are applied on the upper side of roughing filter. They prevent the filters from clogging with algae grown by photosynthesis.
- Sunshade covers were attached on the upper sides of URFU to prevent the growth of algae, they can facilitate filter cleaning.
- Air quantity by aspirators limit is 4-5% of processing water quantity. It provides stability to the saturated-dissolved oxygen in about 80%.

URFU- 6.0m3/day

Outline of Slow Sand Filter Unit (SSFU)

- SSFU was designed by advices from an NPO (Community Water Supply Support Center of Japan). The filtration speed was standardized on 9.6m/day.
- The upper side of SSFU is open for photosynthesis.
- The floating overflow devices are fabricated for SSFU. It minimizes the amount of redundant drainage.

SSFU- 6.0m3/day

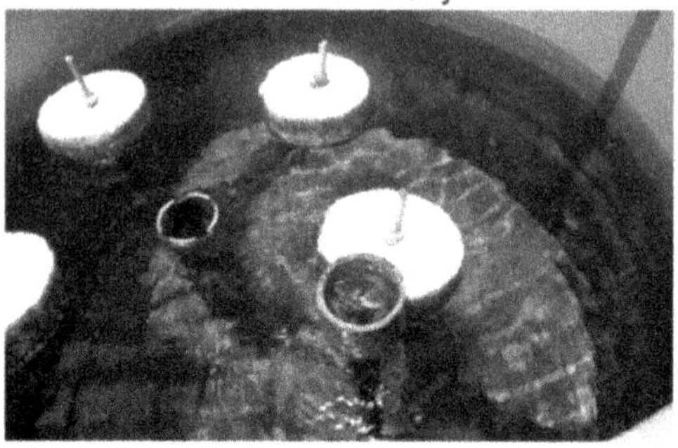

EXPERIMENTAL CONDITION OF THE EPU

Quality of processing raw water

For the testify research in Sri Lanka, we mixed beer, soy milk and ammonia in gutter water to imitate Maldivian groundwater (organically polluted, BOD 30mg/L).

Water quality guidelines for purified water

Minimum guidelines for drinking water quality			
No	Contents of analysis	Unit	WHO or Japanese standard
01	pH	pH	·5.8~8.6 (JWS)
02	Turbidity	NTU	·Below 2.0 (JWS) Target1.0
03	Dissolved Oxygen (DO)	mg/L	·Above 5.0mg/L (JES-3 grade)
04	BOD	mg/L	·3mg/L (JWS -B3grade)
06	Nitrate & Nitrite	mg/L	·Below 10mg/L (JWS)
07	Nitrite (NO2-N)	mg/L	·Below0.05mg/L (JWS)
08	Ammonium (NH3)	mg/L	·Below1.5mg/L (WHO)
09	Total Coli form	pc/mm L	·Below100/mm L (JWS)
10	Fecal Coli form	Pc/mm L	·0pc/mm L (JWS)

JES··Japanese Environmental Standard JWS··Japanese Water Supply standard

VERIFICATION RESULT OF THE EPU

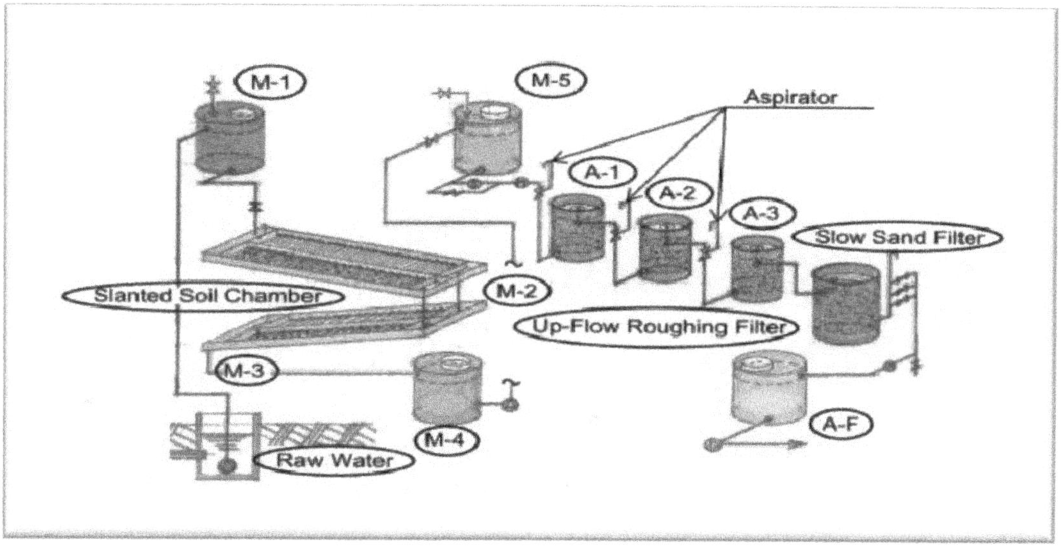

Verification test unit diagram

Summary of the water purification result

Average of a Week Water Quality Tests Data — Period 18thOctober, 2013 to 26th October, 2013

	Measuring Point	Dissolved Oxygen	Turbidity	T. N	Nitrate	Nitrite	Ammonium	COD	BOD	Remarks
		mg/L	NTU	mg/L	mg/L	mg/L	mg/L	mg/L	mg/L	
Slanted Soil Chambers	Canal Water									Added Beer, Ammonia
Slanted Soil Chambers	M-1	2.32	13.48	10.95	3.28	0.010	8.33	10.00	**37.07**	Flow rate: $8m^3$/day
Slanted Soil Chambers	M-2	1.24	0.14	7.00	5.87	0.080	4.33	5.00	**9.87**	Gravel Sands
Slanted Soil Chambers	M-3	3.48	0.01	3.33	3.32	0.000	0.23	4.00	**7.07**	Anthracite with Kangkung
	Water Tank									
Removal Ration			100%	70%		100%	97%	60%	81%	
Up-flow Roughing filter + Slow sand filtration	M-5	5.95	0.00	3.30	3.25	0.000	0.18	4.17	4.67	Flow rate: 6m3/day
Up-flow Roughing filter + Slow sand filtration	A-1	6.61	0.00	3.31	3.30	0.000	0.18	3.33	3.27	Aspirator: ON
Up-flow Roughing filter + Slow sand filtration	A-2	6.66	0.02	3.35	3.32	0.000	0.18	3.33	2.83	Aspirator: ON
Up-flow Roughing filter + Slow sand filtration	A-3	6.57	0.11	2.80	2.80	0.000	0.18	2.50	2.63	Aspirator: ON
Up-flow Roughing filter + Slow sand filtration	A-F	7.11	0.01	3.43	3.42	0.000	0.18	2.33	2.13	
Removal Ration							0%	44%	54%	
Total Removal Ration (M-1 to A-F)			100%	69%		100%	98%	77%	94%	

Purified data chart of organic matter and nitrogenous matter

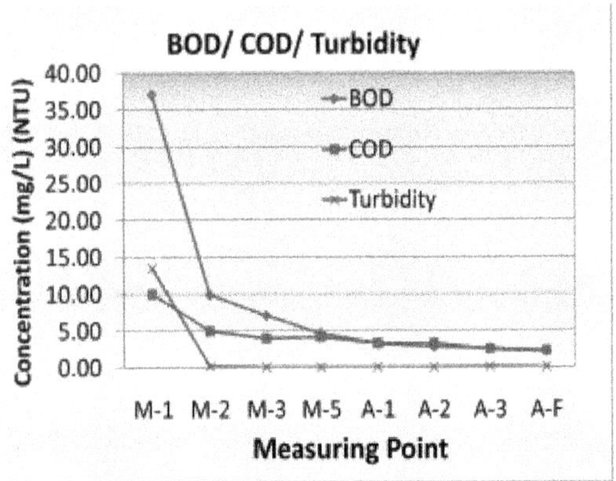

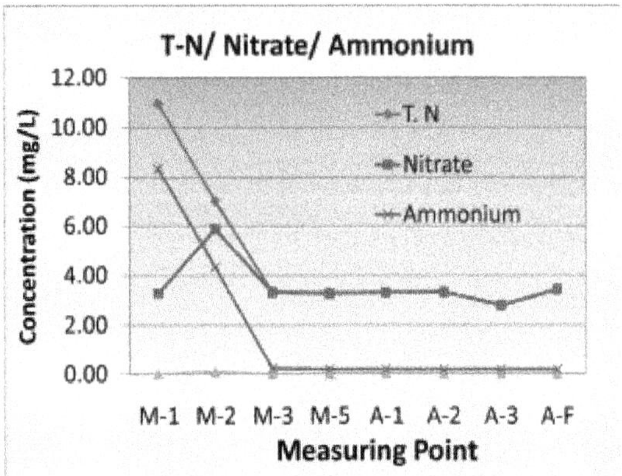

Purified data chart of Fecal Coli form

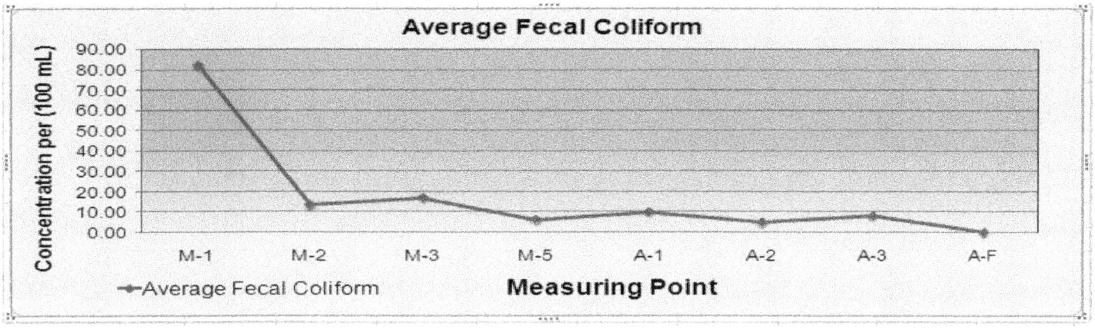

Performance evaluation of each unit

- Turbidity removal rate of SSCU was almost 100% and BOD removal rate was 81%. BOD removal rate of the whole system was 94 %; its ability is very high.
- We measured COD by pack tests. The removal rate of the whole system is a little accurate but it was 77%.
- Oxidation of ammonia and nitrite nitrogen progressed rapidly; it is converted to nitrate nitrogen in the surface of the filter media of SSCU. On the other hand, TN (Total Nitrogen) removal rate by denitrification was 70% reduced rapidly in SSCU filter media at the bottom of the anaerobic state.
- Aspirators for high-level tank were applied on UFRF. Dissolved oxygen level is maintained to 80% by continuous oxygen injection. The sunshade cover installed on the upper side of UFRF prevents filters from clogging by algal growth.
- Fecal Coli form was removed 90% by SSCU and then removed 100% by SSFU bio-film.
- Phosphorus was eliminated from water analysis because the groundwater of inhabited islands of Maldives is low level of it.

OPERATION COST COMPARISION OF EPU WITH RO-PLANT

- Comparision of the operation cost for the Gulhi Island.
- Depreciation and investment interest are not included
- Amount of the product water; 60m3/day×365day=21,900 m3/year

Operation cost for Seawater Desalination Plant (RO)

	Content	Amount /year (US$)
1	Power Consumption RO & Pumps, 609.3kwh/day × 365day × US$0.487/kwh	US$ 108,306.00
2	Spear parts and consumable items RO membrane, Pre filter, Chemicals, Pump spare parts	US$ 8,200.00
3	Man Power Expenses RO, Specialist US$3000 × 12month, Island staff US$1500 × 12month × 2 persons	US$ 72,000.00
4	Indirect Expenses Administration Accommodation, Transportation, etc	US$ 30,000.00
	Total **Unit cost**	**US$ 218,506.00** **(US$ 9.98/m3)**

Operation cost for Ecological Purification Unit

	Content	Amount/year (US$)
1	Power Consumption Pumps, 1.1kw × 4sets × 8.3hr/day × 365days × US$0.487/kwh	US$ 6,492.00
2	Spear parts and consumable items Pump spare parts	US$ 1,000.00
3	Man Power expenses, EPU specialist 5day × 3 times/year × US$200/day + Island staff US$1500 × 12month × 2persons	US$39,000.00
4	Indirect Expenses Administration, Accommodation , Transportation, etc	US$ 16,000.00
	Total **Unit cost**	**US$ 62,492.00** **(US$ 2.85 /m3)**

Operation cost comparison of EPU with RO plant

- Electric consumption of EPU is less than 1/16of RO.
- EPU costs US$ 2.85/m3 although RO costs US$ 9.98/m3. (1: 3.5)
- Electric fee in Maldives costs high because they use small inefficient diesel generator with diesel oil imported from other countries. The imported oil is tanked at the capital Male, transported to each island by small boats then tanked again. Additionally, when the small inefficient RO unit was used with diesel generator, the desalination cost would exceed US$9.98/m3, which was more than three times of Tokyo.

CONCLUSION

- We could testify that the Ecological Purification Unit is able to purify the organically polluted water by our tests in Sri Lanka. SSCU was developed for purification of kitchen drainage but the BOD and turbidity purification ability of BOD is worthy of special mention. It will be a leading unit of moderate-scale purification systems.
- It was said that RO unit is the only way to solve the Maldivian water problem, but RO was too difficult and expensive to maintain. Additionally it needed too much electricity and exhaled CO2 too much. On the other hand, the Ecological Purification Unit is far easier to maintain, and it doesn't need much electricity. It works with solar power and wind-generation. On these accounts, the Ecological Purification Unit is the best system for the inhabited islands of Maldives.
- Many people suffer from water lack in small villages all over the world. This problem has not solved yet because there is no moderate-scale purification system for the organic contamination of groundwater and river water. We aim to popularize EPU globally, being rooted the standard model of the inhabited islands of Maldives.

ACKNOWLEDGEMENT

The authors are acknowledged for helpful advice to Dr. N. Nakamoto and Mr. N. Iwase of NPO, Community Water Supply Support Center of Japan (CWSC) and Mr. Masato Kiji of Yonden Consultants Inc. (Yonden). The authors say a heartfelt thanks to the engineer of air conditioning and building services in Sri Lanka during the development of ecological purification system.

REFERENCES

Ibrahim. S. A (2002) Water resources management in Maldives with an emphasis on desalination, Maldives Water and Sanitation Authority.

Kiji. M (2012) Research on purification by Slanted Soil Chamber Method of BOD, nitrogen, phosphorus contaminated water, Yonden Consultants Inc.

Nakamoto. N, (2005) A manual on slow sand filtration, an ecological water purification system, Shinshu University, Japan.

Chapter 41

Slow sand and slow biochar filtration of raw wastewater

K. Kätzl, M. Lübken*, K. Alfes*, S. Werner**, M. Marschner**, M. Wichern**

**Institute of Urban Water Management and Environmental Engineering, Ruhr-Universität Bochum, Bochum, Germany*

***Department of Soil Science and Soil Ecology, Ruhr-Universität Bochum, Bochum, Germany*

ABSTRACT

Slow sand filtration (SSF) is a well known water treatment. However less research is done in using raw wastewater with a very high total suspended solids (TSS) concentration for SSF. In this study, biochar (Feedstock: Miscanthus, d_{10} = 0.86 mm) was used as an alternative to sand (d_{10} = 0.58 mm) as filter media. Over a period of 70 days, each filter was fed with effluent from the grit chamber of a municipal wastewater treatment plant at a hydraulic loading rate (HLR) of 0.05 m/h. Removal rates for *E. coli* were in the range of 1 to 2 log_{10} units with slightly higher and more constant rates for biochar. Sampling at different filter depths indicated additional back hold of *E. coli* (around 0.5 log_{10} units) in deeper zones for biochar. Overall main removal took place within the schmutzdecke. Biochar achieved also better removal rates, especially at the beginning of the experiment, for main chemical parameters. Increase of supernatant water and therefore filter resistance was similar for all filters. The large grain sizes led to a formation of the schmutzdecke in deeper zones of the filter bed. Adsorption of faecal indicator bacteria and biofilm growth took place over the entire filter bed. This research confirms the suitability of biochar as alternative filter material in SSF of raw wastewater.

Keywords slow sand filtration, biosand filtration, biochar, wastewater reuse, alternative biological filtration, point-of-use.

INTRODUCTION

Water stress or scarcity is an enormous problem all over the world. Currently, around 1.2 billion people live in regions with physical scarcity. Demographic growth and economic development will increase this water stress, especially in arid and semi-arid regions (FAO-Water, 2009). According to Seckler et al. (1999), a third of the population in developing countries live in regions which will be confronted with water scarcity until 2025. Thereby, the main problem is not the daily drinking water requirement of 2 litre per person. Much more, it must be considered that 2,000 to 5,000 litres of water are needed to produce one person's daily food (UN-Water Statistics, 2012). In total, water for irrigation and food production consumes around 70 percent of global freshwater withdrawals (Water, 2012).

Efficient use of water and alternative water sources achieve importance with increasing water scarcity (Langenbach, 2010). In developing countries, approximately 90% of sewage is discharged without treatment and leads to a contamination of surface water and thus potential irrigation water with pathogens (UN, 2002; Qadir et al., 2010). Point-of-use (POU) treatment of water helps to reduce health risks for consumers. Biosand or slow sand filtrations are one of the most suitable available POU treatments to reduce the amount of particular organic substances, pathogens and turbidity (Amy et al., 2006; Drechsel et al., 2008; Sobsey et al., 2008). Main advantages of SSFs are that they can be build out of locally available materials by local trained people, SSFs are cheap in maintenance and a well known treatment method (Jenkins et al., 2011; Tiwari et al., 2009; Young-Rojanschi & Madramootoo, 2014). Reported removal rates of feacal indicator bacteria (FIB) in SSFs differs in literature. Elliott et al. (2008) found an average $\log_{10}$ reduction of *E. coli* of 1.9, Young-Rojanschi & Madramootoo (2014) of 3.71 and Ellis (1987) a *E. coli* reduction of approximately 1.0 (91%). Langenbach (2010) reported a elimination of 2.7-3.4 $\log_{10}$-units for *E. coli.*

The objective of this work was to evaluate the suitability and performance of slow sand filtration treating raw wastewater, and the capability of biochar as an alternative filter material. Biochar can be produced locally out of organic biomass and with its activation potential, biochar is theoretically a most suitable filter material. To our knowledge, there is no current research in the area of using biochar as a filter material for raw wastewater. Hence, further research is necessary to get insights in both potentials and limitations and of biochar filtration.

MATERIAL AND METHODS

The experimental setup consisted of six glass filter columns with a diameter of 50 mm and a total height of 180 cm. Each filter could be divided into two parts. The lower part had a height of 80 cm and five sampling ports and one outlet were installed at the column. The upper part had a height of 100 cm and a emergency overflow was placed on top. Figure 41.1 shows a detailed schematic drawing of the experimental setup. Columns were filled up to 1 cm above the first sampling port. The minimum supernatant water level was adjusted by an outflow weir at 30 cm above the filter bed.

To compare sand and biochar as filter material, three columns were filled with locally purchased sand (Quarzwerke GmbH, Bottrop-Kirchhellen, Germany) and three with biochar produced from Miscanthus (Pyreg GmbH, Doerth, Germany). Before using, the biochar was sieved over 1 mm sieve to separate fine particles. The sand filter media had a porosity of 0.40, maximum diameter of 1.8 mm, effective diameter (d_{10}) of 0.86 mm and a uniformity coefficient (UC) of 1.47. For biochar filter, material with a porosity of 0.59, maximum diameter of >5.6 mm, d_{10} of 0.58 mm and a UC of 3.12 were used. To prevent floating of the light biochar, a mosquito net was filled with approximately 10 cm^3 biochar and weighted with a stainless steel nut. These pillows were set on top of the biochar filter bed.

In both cases a filter bed of 60 cm was supported by two 6 cm gravel layers, out of locally produced gravel (Quarzwerke GmbH, Bottrop-Kirchhellen, Germany). Where the top layer had a diameter range of 3.0-4.0 mm and the coarser bottom layer of 3.0-5.0 mm. After filling the columns, filters were rinsed with tap water for 48 h with a hydraulic loading rate (HLR) of 0.15 m/h to flush filter material. Subsequently, columns were covered with aluminium foil to prevent algal growth in the columns.

The filters were fed by a peristaltic pump (Watson Marlow, Falmouth, United Kingdom) with a constant HLR of 0.05 m/h. Loading rate was controlled ones per week and inlet tubes were flushed with sodium hypochlorite every week to prevent clogging due to biofilm growth. Filter were operated with municipal wastewater from the treatment plant Ölbachtal (Bochum, Germany). For this experiment, effluent from the grit chamber was taken and sieved over a 1 mm curved screen to separate large particles. To reduce the amount of suspended solids by approximately 75%, the raw wastewater was stored for two days in barrel to allow larger particles to settle down. Afterwards, the pre-treated wastewater was stored in a opaque barrel protected from light, cooled to 4-6 °C and continuously stirred with a IKA RW 10 R stirrer (IKA, Staufen, Germany). The reservoir water was renewed every two days. Average effluent temperature of all filters were constant at 20±1°C. Mean surface load of all filter over entire experiment was in the range of 15.3 g_{COD}/m^2h (standard deviation (SD): 5.2) and 4.5 g_{TOC}/m^2h (SD: 1.8).

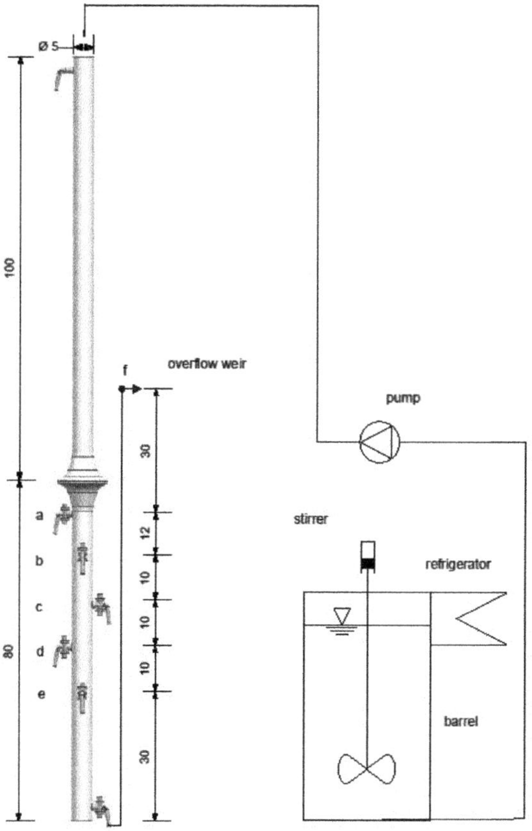

Figure 41.1 Schematic diagram of column design and filter setup.

Before the first sampling occurred, filters were operated with wastewater for six days, to flush out tap water and fine particles. Subsequently, for microbiological analysis and determination of the faecal indicator bacteria *E. coli* and enterococci, influent and effluent samples were taken in 50 ml sterile falcon tubes once per week. Samples were stored at 4 °C and analyzed within 24 h.

Additional samples in 250 ml PE-bottles were taken for physico-chemical analysis (ammonia, TKN, phosphorous, COD and TOC). They were analyzed immediately for temperature, pH, electrical conductivity and turbidity of influent and effluent. Samples for physico-chemical parameter were refrigerated by -20 °C for later analyzes. Influent samples were collected the day before effluent samples to compensate hydraulic retention time (HRT) of the inflow. The theoretical HRT was calculated by

HRT = filter length·porosity/HLR. To develop a concentration profile over filter depth, water samples were drawn on day 50 from sampling ports two (b), three (c), five (e) and effluent (f).

Temperature, pH and electrical conductivity were measured with a Multi 3420 multi-parameter probe (WTW; Weilheim, Germany) and turbidity with a Hach Turbidimeter 2100P Iso (Hach Lange, Düsseldorf, Germany) according to the DIN EN ISO 7027 norm. Chemical oxygen demand (COD) was determined by cuvette tests (Hach Lange, Düsseldorf, Germany). Before analyzing, samples were homogenized with an Ultra Turax T 25 (IKA, Staufen, Germany) for five minutes. COD was measured with a photometer (Dr 5000, Hach Lange, Düsseldorf, Germany). Homogenized samples were also analyzed for total organic carbon (TOC) with a Dimatec 2000 analyzer (Dimatec, Essen, Germany).

Supernatant water levels of each filter were measured every day to control filter resistance and development of the schmutzdecke.

After 70 days, one sand and one biochar filter were removed and samples were taken in sterile falcon tubes from the filter material at 0-5 cm (direct below the schmutzdecke), 10-15 cm, 20-25 cm and 40-45 cm, stored by 4 °C and analyzed within 24 h. For biofilm extraction, approximately two gram of each sample were transferred into a new sterile tube and 10 ml Dehydrated Special Microplates Diluent (DSM) buffer (Birad, Hercules/California, USA) was added. Subsequently, samples were placed in a Ultrasonic Cleaner (VWR, West Chester Pennsylvania, USA) for 10 min, which is a proper treatment for biofilm extraction from sediments (Craig et al. 2002). Afterwards, samples were centrifuged at 650 rpm for 2 minutes with a Heraeus Christ centrifuge (Thermo Fisher Scientific Gmbh, Waltham, Massachusetts, USA) to separate filter material and solution with suspended FIB. The supernatant solution were transferred into a new sterile tube and 1:2 diluted with DSM buffer for analysis of FIB.

Water samples and biofilm were analyzed according to EN ISO 9308-3 for *E. coli* and EN ISO 7899-1 for intestinal Enterococci, using the Most Probable Number (MPN) method. For that, six fold serial dilutions were done with DSM buffer (Biorad, Hercules/California, USA) and eight wells of a microplate (Biorad, Hercules/California, USA) were inoculated with each dilution (200 µl per well). Subsequently, microplates were incubated at 44 °C for 36 to 72 hours. The number and distribution of positive wells of each microplate (Fluorescence under UV light) were used to calculate the amount of *E. coli* and enterococci in each sample. Calculations were done with the MPN calculator of AES Chemunex (AES Chemunex, 2014). The detection limit of this method is in the range of 60 to $6.7 \cdot 10^8$ [MPN/100ml]. Finally, bacteria concentration were $\log_{10}$-transformed for calculations and plots.

RESULTS AND DISCUSSION

The diameter of both material is larger than normally used in slow sand filtration (Bauer et al., 2011; Jenkins et al., 2011; Langenbach, 2010) and more in the range of commonly used media for rapid filtration (Huisman & Wood, 1974).

Because of the high amount of suspended solids in the raw water, large grain sizes were necessary to achieve longer filter lifetimes. Normally, downward flow rate of water in slow sand filtration lies between 0.1 an 0.4 m/h (Huisman & Wood, 1974). To compensate the coarse filter material and the high amount of organic matter, filter were operated with a constant HLR of 0.05 m/h.

Despite different grain size distribution and porosity, the behaviour of the supernatant water layer was similar for both material. The supernatant was constant at adjusted 30 cm for the first 20 days. Subsequently, the water layer rose linear and comparable for both materials with filter life time, until a water overflow occurred after 70 days by one filter. This indicates, that developing of a schmutzdecke and filter clogging is similar for both materials. After 70 days, the upper 10 cm of each filter were removed to reduce the filter resistance through biofilm covering in this area, similar to removal of the schmutzdecke in normal slow sand filtration processes. Contrary to classical SSF, this treatment was not successful and the next filter overflow of both filter types occurred after one week. This indicates that filter ripening takes place in to deeper filter layer than by SSF of river water or groundwater. According to Huisman and Wood (1974) the schmutzdecke will penetrate into greater depths with increasing grain size.

Except turbidity, physical parameters of all filter effluents were similar and approximately in the same range as the inflow. Mean values and standard deviation of filter outflows are represented in Table 41.1. For turbidity removal, a significant difference was found for filter in- and outflow and for sand and biochar columns as summarized in Table 41.1. However, turbidity values of filter inflow differ significantly over the entire experiment.

At the end of the experiment, selected samples were analyzed for the chemical parameters COD, TOC, TKN, NH_4^+-N, PO_4-P which are summarized in Table 41.2.

Declining turbidity removal corresponds with results of Ari and Adin (2006). They found also a decrease in turbidity removal of filter supplied with primary wastewater after 20 days. Their assumption was that high concentrations of biomass in the filtration media due to high nutrient availability are responsible for this effect (Ari & Adin, 2006).

Removal of COD and TOC was in the beginning significantly higher in biochar than in sand columns. For both material, removal decreased during filter lifetime. A similar development is notable for TKN (see Table 41.2), NH_4^+-N and PO_4-P (not shown) removal. For all parameter should be noted that differences in removal rates between sand and biochar filter decreased over the entire experiment. Measured turbidity values in filter outflow indicates better sorption capacity for biochar which can be seen in a later turbidity breakthrough by biochar filtration. This correlates also with significant better removal rates for chemical parameters at the beginning. A possible explanation for higher sorption capacity is the more porous structure of the biochar.

Table 41.1 Mean values and standard deviation of physical parameters of filter in- and outflow

Time [d]	Turbidity [NTU]			Electrical conductivity [µS/cm]			pH		
	Inflow	Sand	Biochar	Inflow	Sand	Biochar	Inflow	Sand	Biochar
6	60	6.5±0.9	6.7±1.6	1226	1187±5	1203±5	7.5	7.8±0.0	8.0±0.1
13	64	20.0±1.0	12.7±1.2	1415	1405±5	1342±7	7.5	7.8±0.0	7.5±0.2
20	152	131.7±18.0	15.5±0.4	1437	1396±11	1358±9	7.6	7.9±0.1	7.4±0.1
27	160	213.7±39.6	219±28.2	967	989±20	1198±57	7.1	7.7±0.0	7.8±0.1
34	950	391.7±30.0	443.3±34.7	1401	1395±2	1428±5	7.6	7.9±0.0	7.9±0.1
41	185	370.7±171.7	307.3±19.4	1376	1403±15	1413±5	7.6	7.1±0.5	6.5±0.1
50	60	22.8±3.9	146.0±42.3	640	660±5	789±44	7.4	7.4±0.1	7.4±0.0
57	16,3	6.6±5.8	105.2±40.0	384	397±11	507±20	7.3	7.4±0.1	7.8±0.1
63	260	156.0±43.0	127.5±32.2	848	880±10	935±11	7.3	7.6±0.0	7.8±0.0

Table 41.2 Mean values and standard deviation of COD and TOC concentration in filter in- and outflow

Time	COD [mg/l]			TOC [mg/l]			TKN [mg/l]		
[d]	Inflow	Sand	Biochar	Inflow	Sand	Biochar	Inflow	Sand	Biochar
6	301	80.3±5.5	35.3±6.4	90.3	45.5±0.6	26.0±2.8	--	--	--
13	385	190.3±47.4	64.6±10.2	124.87	73.1±17.5	36.7±3.1	68.63	57.0±0.8	46.6±1.8
50	375	78.5±2.5	83.5±9.5	51.04	29.0±1.5	27.4±4.5	25.18	21,7±0.2	25.5±2.1
63	160	116.3±4.0	94.3±6.7	--	--	--	--	--	--

Due to problems with microplates for enterococci, only *E. coli* concentration from filter effluent could be determined. Figure 41.2a shows the results for *E. coli* removal over experimental time and Figure 41.2b over filter length at day 50. Inflow concentration of *E. coli* and enterococci varied strong during the entire experiment. Mean values were in the range of $2.86 \cdot 10^6$ [$MPN_{E.coli}$/100ml] (SD: $2.9 \cdot 10^6$) and $7.55 \cdot 10^5$ [$MPN_{enterococcii}$/100ml] (SD: $1.04 \cdot 10^6$). For this, the average values were calculated for the three sand and biochar filters. Removal rates for *E. coli* were in the range of 1 to 2 log units [MPN/100ml], which is comparable to other published results in respect to slow sand filtration (Ellis, 1987). Over the entire experiment, biochar filter showed a more constant and even higher removal efficiency then sand filter. The difference was up to 0.5 log units at the end of the experiment (see Figure 41.2. Unexpected for this experimental setup with large grain size were the relatively high removal rates at the beginning of filtration with clean filter material. With 1 log unit at the beginning the removal rate is in the same range as Bellamy et al., (1985b) found for a new sand bed. The slight decline in removal efficiency after the first sampling could be explained by a decrease of adsorption capacity of the filter material. This correlates also with turbidity removal and chemical parameters, which also decreased compared to the first sampling. Adsorption seems to be an important removal mechanism as also mentioned by Bellamy et al. (1985a). Increasing pathogen removal after the second sampling could be explained through increasing biological activity and development of the schmutzdecke.

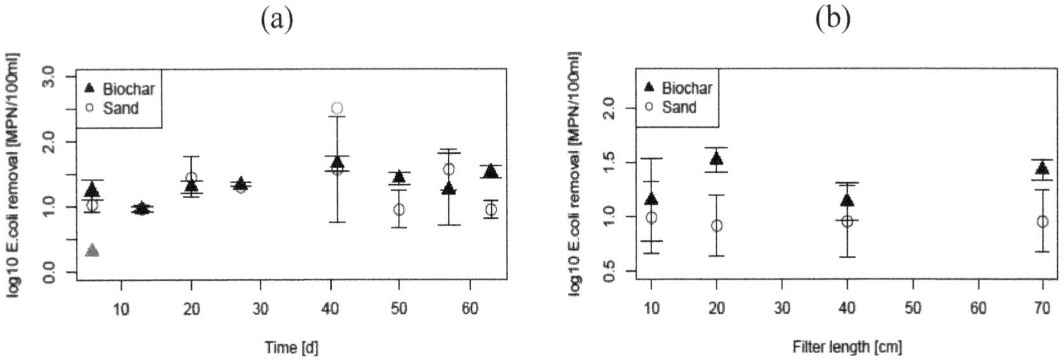

Figure 41.2 Mean $\log_{10}$ removal of *E. coli* through slow sand and biochar filtration over time (a) and filter length (b). Error bars represents standard deviation. Identified gross errors are marked red. Values at the 70 cm mark in Fig 41.2b represent filter effluent.

Sampling at different filter depths indicated that the schmutzdecke is mainly responsible for removal of pathogens (see Figure 41.2) with more than 1 log unit of elimination. This correlates with published results (Bellamy et al., 1985b; Ellis, 1987; Langenbach, 2010), which also identified the schmutzdecke as main removal mechanism. Especially sand filter showed no further removal in deeper filter layers. In contrast, biochar filter seems to be able to remove pathogens in deeper filter zones. Between the first sampling port below the schmutzdecke and the filter outflow an additional removal of nearly 0.5 log units occurred. Extracted biofilm of filter material was analyzed for *E.coli* and enterococci. To standardize results of sand and biochar columns, measured values were based on 1 g dry weight of each filter material. Because of their different density, this method allows only a comparison within each media and a qualitative trend between sand and biochar over filter depth. Results were shown in Figure 41.3a for *E. coli* and 3b for enterococci. The measured concentration was in the range of $7.7 \cdot 10^5$ to $2.6 \cdot 10^3$ and $6.0 \cdot 10^5$ to $1.2 \cdot 10^3$ [MPN/100ml·$g_{\text{Filter material}}$] for *E. coli* and enterococci. Higher values for biochar can be explained through lower weight of this material. A decrease in measured concentration over filter depth applies for both, FIB and filter material. However the decline is slightly stronger for sand than for biochar. Possible explanations for this effect could be a higher sorption capacity and stronger biofilm growth in deeper layer on biochar. The difference between measured FIB concentrations on biochar and sand is higher for enterococci than for *E. coli*. Relatively high concentrations of FIB in deeper layer indicates biofilm growth and adsorption processes over the entire filter bed.

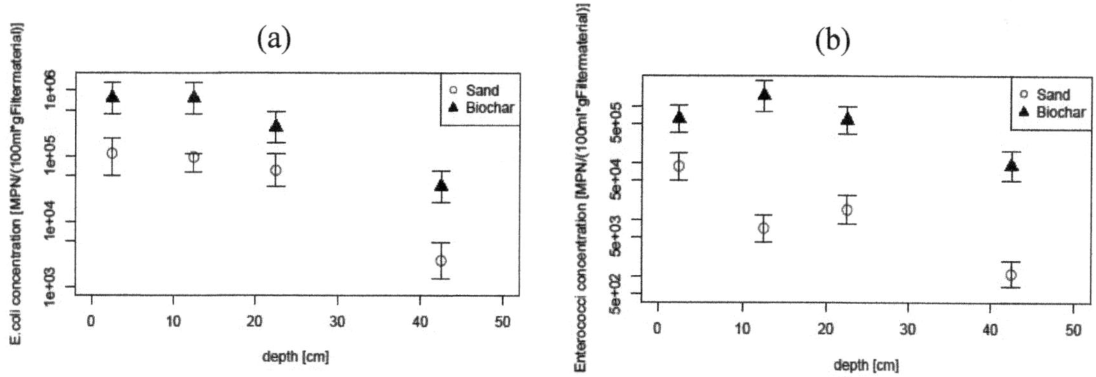

Figure 41.3 Mean $\log_{10}$ concentration of *E. coli* (a) and enterococci (b) in biofilm per gram dried filter material over filter depth. Error bars represents 95% confidence interval.

CONCLUSIONS

Miscanthus-biochar is a suitable material for filtration of raw wastewater and performs better and more constant in removing *E. coli* and chemical parameters than sand. However, the difference between both material is not very high.

With around 1 log unit, main removal of *E. coli* takes place in the schmutzdecke. In contrast to sand columns further removal of 0.5 log units in deeper levels occurred in biochar filter.

FIB are present over entire filter bed, but the concentration of FIB attached on filter material decreases with filter bed depth of around 2.5 log units per gram dried filter material.

Both filter media were able to reduce COD and TOC concentration significant during the entire experiment. Removal of turbidity and chemical parameter decreases over filter life time.

ACKNOWLEDGEMENTS

Funding for this project was provided by the German Federal Ministry of Education and Research (BMBF) through the joint project "UrbanFood^{Plus}", funding number: 031A242B.
Microbiological analyses were done with the assistance of Dr. E. Nettmann and A. Rademacher.

REFERENCES

AES Chemunex. (2014). WATER by AES CHEMUNEX. Retrieved January 22, 2014, from http://water.aeschemunex.com/index.php

Amy, G., Carlson, K., Collins, R. M., Drewes, J., Gruenheid, S., & Jekel, M. (2006). Integrated comparison of biofiltration in engineered versus natural systems. In: Gimble, R. Graham, N.J.D., Collins, M.R.Recent Progress in Slow Sand and Alternative Biofiltration Processes. IWA Publishing (4th ed., Vol. 4). London, UK: IWA Publishing.

Ari, G., & Adin, A. (2006). 23 Down-scaled study of slow sand filtration of secondary effluents. In R. Gimbel, N. J. D. Graham, & R. M. Collins (Eds.), Recent progress in slow sand and alternative biofiltration processes (pp. 194–205). London: IWA Publishing.

Bauer, R., Dizer, H., Graeber, I., Rosenwinkel, K.-H., & López-Pila, J. M. (2011). Removal of bacterial fecal indicators, coliphages and enteric adenoviruses from waters with high fecal pollution by slow sand filtration. Water Research, 45(2), 439–52.

Bellamy, W. D., Hendricks, D. W., & Logsdon, G. S. (1985). Slow sand filtration: influences of selected process variables. Journal - American Water Works Association, 77(12), 62–66.

Bellamy, W. D., Silvermann, G. P., Hendricks, D. W., & Logsdon, G. S. (1985). Removing giardia cysts with slow sand filtration. Journal - American Water Works Association, 77(2), 52–60.

Craig, D. L., Fallowfield, H. J., & Cromar, N. J. (2002). Enumeration of faecal coliforms from recreational coastal sites: evaluation of techniques for the separation of bacteria from sediments. Journal of Applied Microbiology, 93, 557–565.

Drechsel, P., Keraita, B., Amoah, P., Abaidoo, R. C., Raschid-Sally, L., & Bahri, A. (2008). Reducing health risks from wastewater use in urban and peri-urban sub-Saharan Africa: applying the 2006 WHO guidelines. Water Science and Technology, 57(9), 1461–6.

Elliott, M. a, Stauber, C. E., Koksal, F., DiGiano, F. a, & Sobsey, M. D. (2008). Reductions of E. coli, echovirus type 12 and bacteriophages in an intermittently operated household-scale slow sand filter. Water Research, 42(10-11), 2662–70.

Ellis, K. (1987). Slow sand filtration as a technique for the tertiary treatment of municipal sewages. Water Research, 21(4), 403–410.

UN (2002). Factsheet International Year of Freshwater 2003. January 15, 2014, from http://www.un.org/events/water/factsheet.pdf

FAO-Water - Water scarcity. (2009). Retrieved September 28, 2013, from http://www.fao.org/nr/water/topics_scarcity.html

Huisman, L., & Wood, W. (1974). Slow sand filtration. Genua: World Health Organization.

Jenkins, M. W., Tiwari, S. K., & Darby, J. (2011). Bacterial, viral and turbidity removal by intermittent slow sand filtration for household use in developing countries: experimental investigation and modeling. Water Research, 45(18), 6227–39.

Langenbach, K. (2010). Slow sand filtration of secondary effluent for wastewater reuse: Evaluation of performance and modeling of bacteria removal. TU München.

Qadir, M., Wichelns, D., Raschid-Sally, L., McCornick, P. G., Drechsel, P., Bahri, a., & Minhas, P. S. (2010). The challenges of wastewater irrigation in developing countries. Agricultural Water Management, 97(4), 561–568.

Seckler, D., Barker, R., & Amarasinghe, U. (1999). Water Scarcity in the Twenty-first Century. International Journal of Water Resources Development, 15(1-2), 29–42.

Sobsey, M. D., Stauber, C. E., Casanova, L. M., Brown, J. M., & Elliott, M. A. (2008). Point of use household drinking water filtration: a practical, effective solution for providing sustained access to safe drinking water in the developing world. Environmental Science and Technology, 42(3), 4261–4267.

Tiwari, S.-S. K., Schmidt, W.-P., Darby, J., Kariuki, Z. G., & Jenkins, M. W. (2009). Intermittent slow sand filtration for preventing diarrhoea among children in Kenyan households using unimproved water sources: randomized controlled trial. Tropical Medicine & International Health□: TM & IH, 14(11), 1374–82.

UN-Water Statistics: Water, Agriculture and Food Security. (2012). Retrieved September 28, 2013, from http://www.unwater.org/statistics_sec.html

Water, U. (2012). Managing Water under Uncertainty and Risk: The United Nations World Water Development Report 4, 1.

Young-Rojanschi, C., & Madramootoo, C. (2014). Intermittent versus continuous operation of biosand filters. Water Research, 49, 1–10.

Part V
Biosand Filters

Chapter 42

Global Review of the Adoption, Use, and Performance of the Biosand Filter

Dr. T. K.K. Ngai, B.Coff*, D. Baker*, R. Lentz**

**CAWST – Centre for Affordable Water and Sanitation Technology,*
424 Aviation Rd NE, Calgary, Alberta, T2E 8H6, Canada

ABSTRACT

A biosand filter (BSF) is an adaptation of the traditional slow sand filter for intermittent use, and is a popular household water treatment technology. This paper summarizes 32 project reports from 19 countries on the usage, performance, acceptance, and reported problems of the BSF. The reports were either provided by partners of the Center for Affordable Water and Sanitation Technology (CAWST), or were retrieved from published journals. On average, 83% of respondents from the 32 reports used the filter daily, ground water was the primary source, and drinking was the primary use. The filter removed 90% of E. coli bacteria on average and 66% of turbidity from the water. While the BSF is intended to be used as part of a multi-barrier approach, only 30% of users treated water before filtration and 25% after filtration. Eighty-one percent used separate storage containers for treated and untreated water, however 56% of the projects idenitified recontamination during storage as a major problem. User perception of the BSF was high, with 96% of the respondents stating that the filter was easy to use and 94% stating the filter provided enough water for the entire household. The majority of reported problems with use of the BSF stemmed from lack of user education, poor manufacturing quality, and inadequate ongoing maintenance. Only 41% of users knew how to correctly clean the filter, 56% of the filters had correctly installed sand, and 74% had correctly installed diffuser plates. To achieve optimal performance and sustainable use, BSF implementers should implement better quality control processes during manufacturing, provide more household education on proper operation and maintenance procedures, devise better systems for regular follow-up support to communities, and consider the role of the BSF in a multi-barrier approach to water treatment. This will ensure that households are getting the most benefits from the BSF and reaching the end goal of safe and clean water.

Keywords biosand filter (BSF), household water treatment, project planning, project implementation, monitoring and evaluation

INTRODUCTION

According to WHO and UNICEF (2012), of the nearly 800 million people without access to improved drinking water, over 80% of them live in rural, remote regions where conventional large-scale water supply and treatment plants are often technically and economically unfeasible. In addition, many of the water sources that are classified as "improved" are unsafe microbiologically (Onda et al 2012). People without access to safe drinking water are at a greater risk of developing water related diseases such as diarrhea, trachoma, and schistosomiasis (UNDP 2006). Compromised health affects one's ability to attend school or work and increases expenditures on healthcare. Research has demonstrated that improving access to safe water can directly improve one's health, and thus standard of living (Hutton and Haller 2004).

Household water treatment technologies are one of the most promising options to reach vulnerable populations lacking access to sustainable sources of safe water (Clasen 2009). Among these technologies, the BSF (Figure 42.1) has been identified as having high potential to become widely used and sustainable for improving household water quality to reduce waterborne disease and death (Sobsey et al. 2008).

Figure 42.1 Cross sectional image of the biosand filter (CAWST 2012a)

The BSF is a simple, inexpensive ($15 to $40 depending on the country), socially-appropriate technology that is constructed by local technicians using locally available materials including gravel, sand, and cement. When water is poured into the BSF, a biologically active layer develops naturally between the sand and water. Pathogens and suspended particles are removed by predation in the biolayer, or by trapping, adsorption, and natural death in the sand layer. Consequently, ensuring the correct type and size of sand is used, and maintaining the biolayer are important for removing contaminants. Maintenance involves stirring the top layer of water in order to suspend trapped particles, removing the particles by scooping them out (e.g., with a cup), and assuring the sand layer remains level (CAWST 2012b). It requires no external energy or chemicals to operate and maintain, and a properly maintained BSF can last for at least 10-15 years. As of 2012, over 400,000 BSF have been implemented in households by governments, development agencies, non-governmental organizations, community-based organizations, and private businesses, in over 60 countries, serving more than 2.5 million people (CAWST 2012).

This review explores how the BSF is being used globally, compares the performance of filters in different regions, investigates user perception of the filter, highlights common problems with the BSF, and outlines common recommendations resulting from the evaluation studies. The review provides information that will guide existing or future BSF implementers in designing and implementing more effective and sustainable projects, thus, in turn, improving the standard of living for those using the BSF.

METHODOLOGY

Thirty-two evaluation studies of BSF projects were analyzed by CAWST in order to extract common themes on the global adoption, use, and performance of the BSF. The studies were either published papers or reports developed by CAWST's partners and clients, and were published between 2002 and 2012. They were implemented in 19 countries including Afghanistan, Cambodia, Cameroon, Dominican Republic, Ethiopia, Ghana, Haiti, Honduras, Kenya, Lao PDR, Mozambique, Nepal, Nicaragua, Sierra Leone, Tanzania, Uganda, Vietnam, Zambia, and Zimbabwe. The location, year and author of each study is provided at the end of this paper, and all evaluation reports are available in full at http://biosandfilters.info/research.

The studies ranged in their objectives and approach, however, each investigated one or more aspects of BSF usage, performance, acceptance and reported problems. Methods employed in the studies included household inspections, surveys, and water quality tests of filters post-implementation. The number of households visited for each study ranged from 16 to 328, with an average of 87, and the average time of months of use of the filters at the time of evaluation was 22 months post-installation.

Key data and common themes were extracted from each study and grouped into the following four categories: Use and Adoption, Performance, Acceptance, and, Reported Problems and Recommendations for Improved Implementation. The findings for each of the four categories are described throughout the following section.

RESULTS AND DISCUSSION

Use and Adoption of the BSF

The combined studies contained a total of 3139 households, and hence provided significant insight into how BSF are being used around the world. For most studies, one BSF was provided per household, resulting in an overall average of seven people using each filter. Groundwater was the primary source of drinking water in half of the BSF projects. However, surface water and piped water were also important sources (Figure 42.2). Filtered water was mainly used for drinking and food preparation. However, about 29% of respondents also reported using filtered water for washing and for preparing medicines (Figure 42.3).

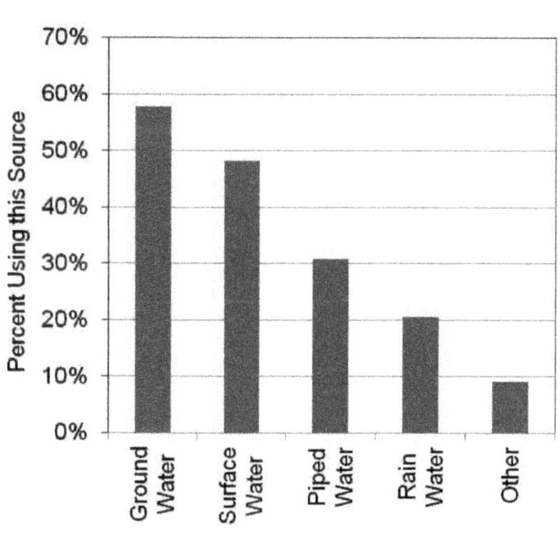

Figure 42.2 Percentage of reported sources of drinking water (summary of 32 studies)

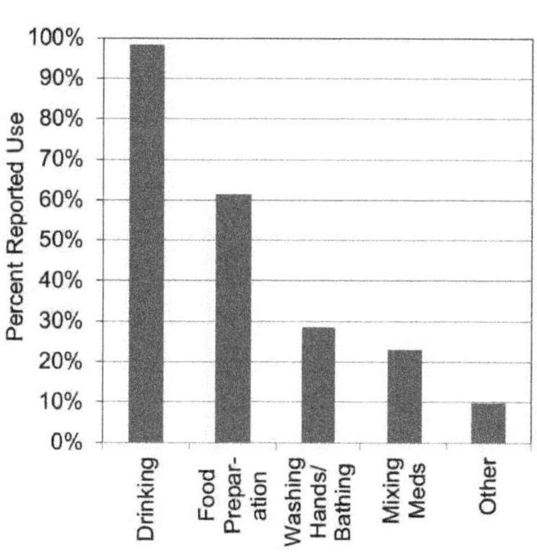

Figure 42.3 Percentage of reported uses for filtered water (summary of 32 studies)

Numerous recent studies (Graf et al. 2010, McGuigan et al. 2011; Enger et al. 2012) have found that health benefits of household water treatment technology can only be realized if the adoption rate is high. Figure 42.4 shows the rate of daily use of the BSF, with studies grouped by the period of time since implementation. Twenty-five of the 32 evaluation studies reported the percentage of households that used their BSF on a daily basis, with an overall average of 83% and a range from 10% to 100%. For BSFs that were implemented between 3 and 6 months before the evaluation, the average percentage of daily use was 97%. For projects with more than 2 years between implementation and the evaluation, the average percentage of daily use was only 56%. However, the figure also shows that relatively low rates of use were found in recently implemented BSF projects (for example Zambia, 2010b), and relatively high rates of use were found for BSFs with longer periods since implementation (for example, Kenya 2004). The most commonly reported reasons for households stopping use of the BSF were lack of follow-up and ongoing support for users and physical breakage of the filters.

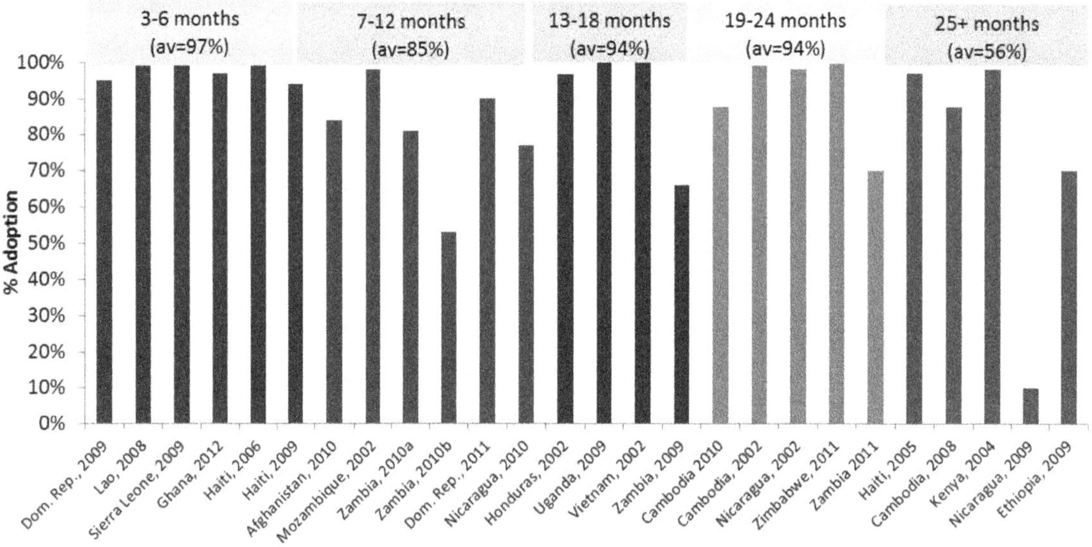

Figure 42.4 Percentage of households using the BSF every day, grouped by the number of months since implementation

Performance

Twenty-two of the 32 studies tested the percentage of E. coli bacteria removal, and nine tested the percentage of turbidity removed by the filters. On average, the filter removed 90% of E. coli bacteria, however, only 60% of samples had levels of E. coli below the WHO recommendation of ten colony-forming units per 100 milliliters (WHO 1997). The average percentage for turbidity removed was 66%, and 94% of samples had turbidity levels below the WHO recommendation of less than five nephelometric turbiduty units (NTU) (WHO 2011). Figure 42.5 shows that there is significant variation in the rate of average E. coli removal between studies.

It is important to note that, no matter how well maintained, the BSF is not designed as a stand-alone technology that will produce microbial-safe water at all times. Local water quality varies dramatically, and no single technology can be equally effective in all situations (Lantagne et al. 2009). The most effective way to secure water safety for both large and small-scale water supply schemes in both the developed and developing world is to apply a multi-barrier approach to water treatment (LeChevallier & Au 2004). A multi-barrier process consists of multiple techniques practised in series to target the various types of contaminants. When applied to household water treatment, the multi-barrier treatment process has five steps; protection of source water, sedimentation, filtration (e.g., use of BSF), disinfection, and safe storage (LeChevallier & Au 2004; CAWST 2012b). The BSF is merely one step in the multi-barrier approach, and it should be complemented by additional steps to ensure water safety (CAWST 2012b).

In the reviewed studies, on average, 30% of respondents reported that they treat the raw water before pouring it into the filters, 25% treat the water after filtration, and 80% store the treated water in a separate container (i.e., not using the same container to fetch raw water and to collect treated water). These indicate that the majority of householders are using the BSF as a stand-alone water treatment technology, rather than as part of a multi-barrier approach, hence further education in the multi barrier approach would be beneficial.

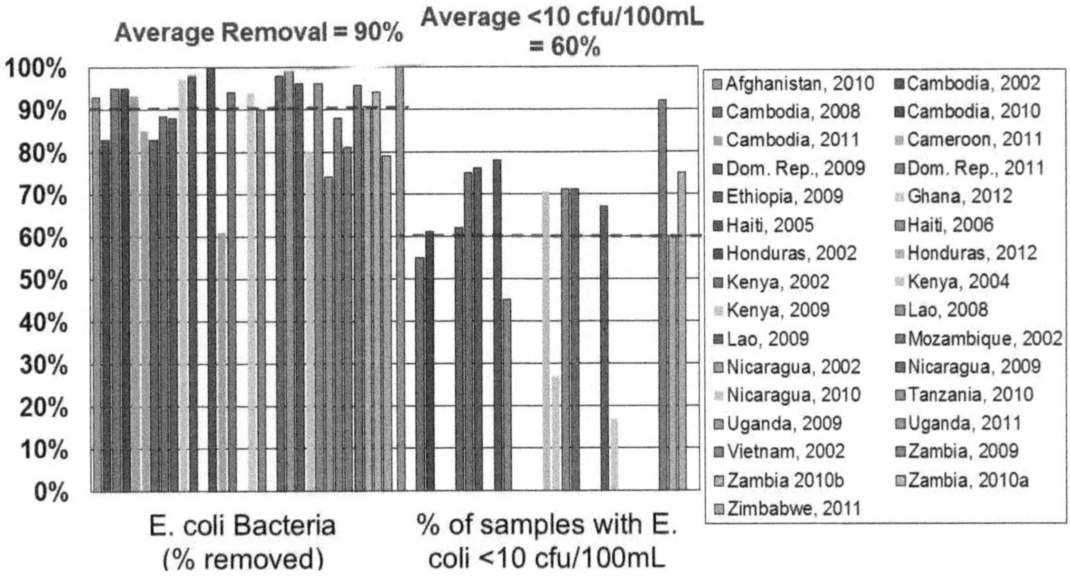

Figure 42.5 The percentage of E. coli bacteria removed and samples with E. coli <10 cfu/100ml average from each study shown

Acceptance

Despite some of the challenges with the performance of the BSF, the filter appeared to be well accepted by households. On average, for respondents who were using the filter regularly, 97% of the respondents indicated they liked the filter, and 96% reported the filter was easy to use. An average of 93%, 93% and 95% of respondents reported that the taste, smell, and appearance of the water improved after filtration, respectively. Ninety-seven percent of respondents said that they liked the filter, although a relatively lesser number (73%) would recommend it to others.

Reported Problems and Recommendations for Improved Implementation

Common challenges and problems associated with the BSF were reported in each of the studies. Figure 42.6 shows the most commonly reported problems from inspection of the filters and survey responses.

Eighteen out of the 32 projects reported that more effective household education and training was needed to properly operate and maintain the filters. An overall average of only 64% of BSF users stated they had attended water, sanitation, and hygiene training, and only 41% of those tested were able to correctly perform a common BSF maintenance task (cleaning of the BSF sand). These findings suggest that user education is a key area requiring improvement, which may have led to other common problems observed with the use and condition of the BSF (Figure 42.6). Hence, there is an opportunity to reduce the rate of problems for future BSF projects through improved implementation.

The following were commonly reported recommendations to improve household education and training from the 32 studies:

- Provide users with adequate training on water, sanitation, and hygiene knowledge and BSF use and maintenance knowledge. Most projects provided initial training, however, several reported that not all BSF users had attended the training, or that it was not comprehensive enough.
- Provision of compensation through payment or provision of tools (bicycle, talktime, maintenance equipment) to improve the ability of community support technititans to undertake their roles.
- Over half of the studies found that follow ups with users were not being conducted regularly, and recommended more resources and more structured processes for scheduling follow-ups.

Other commonly reported problems with the BSFs included:

- *There was recontamination during storage of the filtered water.* Fifty-six percent of the projects identified recontamination during storage as a main problem in the effectiveness of the BSF. Having a separate water storage container that has a well fitting lid and is regularly cleaned would help decrease the E. coli concentrations in the stored water (CAWST 2012b).
- *The sand for the BSF was not properly selected and prepared, resulting in too high of a flow rate, which compromises the filter's performance.* The type and size of sand, and level installation of the sand layer are important for removing contaminants (CAWST 2012b). However, only 52% of the BSFs had correctly installed sand.
- *BSFs were cleaned too often, leading to unnecessary disturbance of the biolayer and reduction of the microbial-removal performance.* There are specific steps to follow on when and how to clean the filter for optimal performance (CAWST 2012b). On average, only 36% understood when to clean the BSF, and 41% how to properly clean the BSF.
- *The filters were not fabricated to an acceptable quality.* Eight of the studies reported that poor filter fabrication quality led to problems with the BSF, including leaks and cracks, resulting in low rates of adoption and sustained use.
- *The diffuser plates were damaged and/or incorrectly placed in the BSF.* The diffuser keeps the water being poured into the filter from rushing into the sand layer and disturbing the bio-layer essential to pathogen removal (CAWST 2012b). Only 74% were correctly installed, which could have also led to ineffective removal of particles and pathogens.

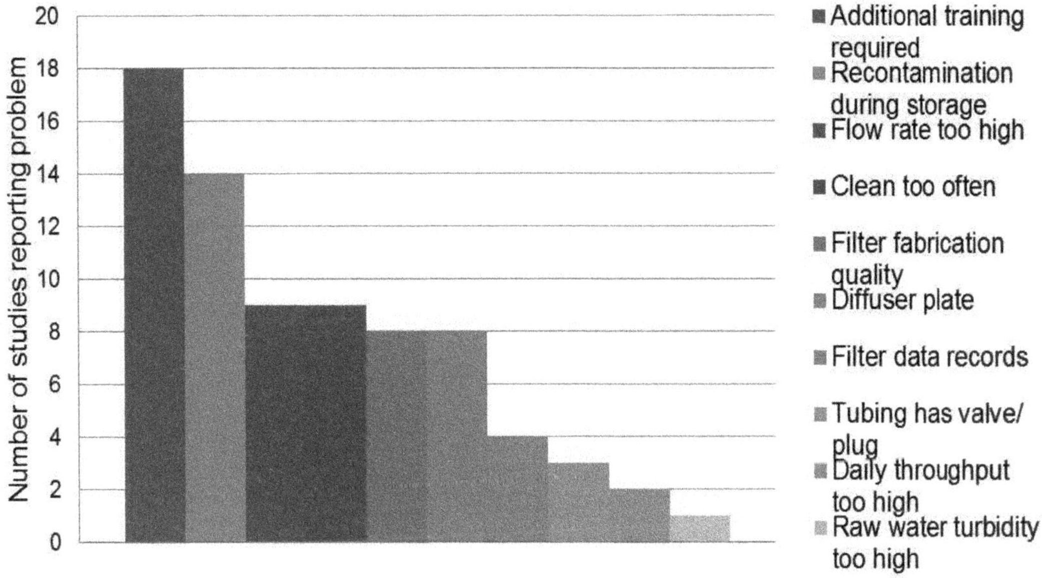

Figure 42.6 Frequent problems identified from project evaluations

CONCLUSIONS

Overall, the BSF was frequently used for a variety of purposes, was found to remove bacteria and turbidity from water effectively, and was well accepted by users. Inspection of the BSF showed common problems related to the use, condition, and maintenance that were affecting BSF performance, such as recontamination of stored water, flow rate too high, and improper cleaning procedures. Many of the observed challenges could be resolved with training and education on proper operation and maintenance procedures for the BSF. Education programs can lead to a behavioral change that will improve users' water treatment techniques and safe storage habits (Quick et al. 2002). This will ensure that households are getting the most benefits from their BSF and reaching the end goal of safe and clean water. Training should ensure implementers are selecting and preparing the sand media according to the recommended sizes. Given the various issues with diffuser plates, implementers may wish to consider making diffuser basins rather than plates, which tend to be easier to maintain (Strand and Sichone 2011). After training, implementers should follow-up with visits to assure that there are no additional problems and the users feel confident in operating and maintaining the BSF independently. Although half of the projects received direct technical support from CAWST, further assistance could be provided to help trainers understand the best techniques in educating BSF users.

REFERENCES

Evaluations Reviewed:

Country	Year	Primary Author
Afghanistan	2010	Tearfund UK
Cambodia	2007	Liang, K.
Cambodia	2008	CAWST
Cambodia	2010	Liang, K.
Cambodia	2010	Murphy, H.
Cambodia	2012	Stauber, E.
Cambodia, Honduras, Kenya, Mozambique, Nicaragua, Vietnam	2002	Kaiser, N.
Cameroon	2011	Klopfstien, L.
Dominican Republic	2009	Stauber, C.E.
Dominican Republic	2011	Aiken, B.
Ethiopia	2009	Earwaker, P.
Ghana	2012	Stauber, C.
Haiti	2005	Duke, W.
Haiti	2006	Frost, D.
Haiti	2009	ACTED
Honduras	2007	Miller, K.
Honduras	2012	Fabiszewski de Aceituno, A.
Kenya	2004	Fewster, E.
Kenya	2009	Tiwari, S.K.
Lao PDR	2008	Nam Saat
Lao PDR	2009	Bhandari, B.
Nepal	2006	Ngai, T.K.K
Nicaragua	2010	Fiore, M.
Nicaragua	2010	Vanderzwaag, J.
Sierra Leone	2009	GOAL
Tanzania	2011	Banda, E.
Uganda	2009	Divine Waters Uganda
Uganda	2011	Mutafya, E.
Zambia	2009	Edwards, E.
Zambia	2010	Mutafya, E.
Zambia	2010	Sichone,V.
Zimbabwe	2011	Banda, E.

Other References:

CAWST, 2012. *Annual Report 2012*. Calgary: Centre for Affordable Water and Sanitation Technology.

CAWST, 2012a. *Biosand Filter* [online]. Calgary: Centre for Affordable Water and Sanitation Technology. Available from: http://www.cawst.org/en/resources/biosand-filter [Accessed 28 September 2012].

CAWST, 2012b. *Biosand Filter For Technicians; A CAWST Participant Manual* [Online]. Calgary: Centre for Affordable Water and Sanitation Technology. Available from: http://www.cawst.org/en/resources/pubs/training-materials/category/34-biosand-filter-for-technicians [Accessed 7 June 2012].

Clasen, T., 2009. *Scaling Up Household Water Treatment Among Low-income Populations. Public Health and Environment: Water, Sanitation, Hygiene and Health.* Geneva, Switzerland: World Health Organization. WHO/HSE/WSH/09.02.

Enger, K.S., Nelson, K.L., Clasen, T., Rose, J.B. and Eisenberg, J., 2012. Linking Quantitative Microbial Risk Assessment and Epidemiological Data: Informing Safe Drinking Water Trials in Developing Countries. *Environmental Science and Technology*, 46 (9), 5160–5167.

Graf, J. et al., 2010. Health Gain from Solar Water Disinfection (SODIS): Evaluation of a Water Quality Intervention in Yaounde, Cameroon. *Journal of Water and Health,* 8(4), 779-796.

Hutton, G. and Haller, L., 2004. *Evaluation of the Costs and Benefits of Water and Sanitation Improvements at the Global Level.* Geneva, Switzerland: World Health Organization. WHO/SDE/WSH/04.04.

LeChevallier, M.W., and Au, K.K., 2004. *Water Treatment and Pathogen Contol: Process Efficiency in Acheiveing Safe Drinking Water.* United Kingdom: World Health Organization.

Lantagne, D.S., et al., 2009. Comment on Point of Use Household Drinking Water Filtration: A Practical, Effective Solution for Providing Sustained Access to Safe Drinking Water in the Developing World". *Environmental Science and Technology*, 43(3), 968-969.

McGuigan, K.G., Samaiyr, P., Du Preez, M. and Conroy, R.M., 2011. High Compliance Randomized Controlled Field Trial of Solar Disinfection of Drinking Water and its Impact on Childhood Diarrhea in Rural Cambodia. *Environmental Science and Technol*ogy, 45, 7862–7867.

Onda, K., LoBuglio, J. and Bartram, J., 2012. Global Access to Safe Water: Accounting for Water Quality and the Resulting Impact on MDG Progress. *International Journal of Environmental Research and Public Health* 9(3), 880-894.

Quick, R., Kimura, A., Thevos, A., Tembo, M., Shamputa, I., Hutwagner, L. and Mintz, E., 2002. Diarrhea Prevention Through Household-level Water Disinfection and Safe Storage in Zambia. *The American Journal of Tropical Medicine and Hygine*, 66(5), 584-589.

Sobsey, M.D., Stauber, C.E., Casanova, L.M., Brown, J.M. and Elliot, M.A., 2008. Point of Use Household Drinking Water Filtration: A Practical, Effective Solution for Providing Sustained Access to Safe Drinking Water in the Developing World. *Environmental Science and Technology*, 42(12), 4261-4267.

Strand, A. and Sichone, A., 2011. *Program for Sustainable Development-Biosand Filter Project Evaluation.* Calgary, Canada: Centre for Affordable Water and Sanitation Technology; Seeds Of Hope International Partnerships; and Africa Manzi Center.

United Nations Development Program, 2006. *Human Development Report 2006: Beyond Scarcity: Power, Pverty and the Global Water Crisis.* New York, USA: UNDP.

World Health Organization (WHO), 1997. Guidelines for Drinking-water Quality, 2nd Edition, Volume 3, *Surveillance and Control of Community Supplies.* Geneva, Switzerland: WHO.

WHO, 2011. *Guidelines for Drinking-water Quality*, 4th Edition. Geneva, Switzerland: WHO

WHO/UNICEF, 2012. *Progress on Drinking Water and Sanitation: 2012 Update.* Geneva, Switzerland: WHO/UNICEF.

Chapter 43

Recent advances in household biosand filter design

T. K. K. Ngai and D. L. Baker**

**CAWST – Centre for Affordable Water and Sanitation Technology, 424 Aviation Rd NE, Calgary, Alberta, T2E 8H6, Canada*

ABSTRACT

The biosand filter is an intermittent-flow adaptation of slow sand filtration technology. Developed over 20 years ago and now with 15+ years' operating experience in households, it has established a reputation for effectiveness, durability, and sustained use. Research, field evaluations, and understanding of the nature of intermittent filter operation have led to advances in the design of the biosand filter as well as the specifications for the hydraulic loading rate, filtration sand, and pause period. Learning has also progressed in virus reduction, the importance of filter ripening, and the maintenance/cleaning of the biosand filter. Finally, different methods of fabricating the filter body and diffuser basin are providing more alternatives for implementing biosand filter projects. As of 2013, 450 organizations have reported implementing biosand filters projects in 55 countries, for a total of over 550,000 filters, impacting more than three million people (CAWST 2013).

Keywords intermittent slow sand filter, household biosand filter (BSF), point of use (POU) water treatment, virus reduction, pause period, hydraulic loading rate, sand specifications, diffuser basin, BSF dimensions, iron-amended BSF

INTRODUCTION

In 1991 Dr. David Manz developed a slow sand filter to be operated intermittently, thus making it more suitable for household applications. By making the highest point of the outflow tubing 5 cm above the sand, standing water covers the sand at all times allowing a biological community of bacteria and other micro-organisms to grow in the top 2 cm of sand. The biosand filter (BSF) stands about 1 m tall and is 0.3 m wide on each side, which makes it a convenient size for use in households. It is filled with layers of specially selected and prepared sand and gravel. The filter container is normally made of concrete or plastic.

The technology is especially relevant to families who do not have piped water and must carry their water into the home from a diverse variety of potentially contaminated sources (up to 50 NTU turbidity). Institutions such as schools, health clinics, community centers, prisons, and places of worship, can also utilize the BSF when piped water is unavailable or frequently contaminated.

Contaminated water is poured into the top of the BSF at least once per day (but not continuously). The water poured into the top of the filter slowly drips through the holes in the diffuser, and flows down through the sand and gravel. Treated water flows out of the outlet tube. No power is required - the filter works by gravity. It takes about one hour to get twelve litres of filtered water. Pathogens and suspended solids are removed through biological and physical processes that take place in the sand. These processes include: mechanical trapping, predation, adsorption, and natural death.

Figure 43.1 Biosand filter and safe storage container (Credit: CAWST)

"The biosand filter has been shown to improve water quality and reduce diarrheal disease in rigorous epidemiological studies, and follow-up studies document sustained, effective performance long after implementation, with filter usage rates remaining high years post-implementation… with very low rates of breakage and disuse over time" Sobsey *et al.* (2008). "Evidently, the biosand filter meets the acceptability criteria of its target users, thus potentially encouraging routine use without extensive intervention to promote behavioural change" (Kubare and Haarhoff 2010).

Many of the recent advances in the design of the BSF pertain to the parameters that make the filter both effective in removing pathogens from the water and acceptable to the household in terms of ease of use and daily quantities of filtered water. These advances were derived from: a) research that has found ways to improve the biosand filter design, b) learning more about how the filter is used in the household, c) understanding the challenges of implementing BSF projects in developing countries, and d) determining the important parameters of the intermittent slow sand filter. Recommendations on how the BSF should be designed constructed and operated are based on this new evidence.

DIMENSION CHANGE

Table 43.1 provides a summary of the biosand filter (BSF) removal effectiveness measured in laboratory and field studies pre-2009, before the design changes discussed here.

Table 43.1 Treatment Efficiency of the Biosand Filter

	Bacteria	Viruses	Protozoa	Helminths	Turbidity	Iron
Laboratory	Up to 96.5%[1,2]	70 to >99%[3]	>99.9%[4]	Up to 100%[5]	95% to <1 NTU[1]	Not available
Field	87.9 to 98.5%[6,7]	Not available	Not available	Up to 100%[5]	85%[7]	90-95%[8]

- Treatment efficiencies provided in the above table require an established biolayer; it takes up to 30 days to establish the biolayer depending on inlet water quality and usage.
- Inlet Water Criteria: Turbidity < 50 NTU (Nephelometric Turbidity Units)

[1]Buzunis (1995) [2]Baumgartner (2007) [3]Elliott *et al.* (2008) [4]Palmateer *et al.* (1997) [5]Not researched. However, as helminths are too large to pass between the grains of sand, up to 100% removal effectiveness is assumed. [6]Earwaker (2006) [7]Duke and Baker (2005) [8]Ngai *et al.* (2004)
From: CAWST (2011) *Biosand Filter Factsheet (detailed)*

The dimensions of the (BSF) include a reservoir volume (the maximum batch or fill that can be poured into the filter at one time) and a pore volume (the volume of the pore spaces within the filtration sand). Research found that bacteria removal by the BSF was significantly improved when the water was retained in the pore spaces of the sand during the pause period, which is when the pore velocity is zero (Baumgartner *et al.* 2007, Elliot *et al.* 2008, Tiwara *et al.* 2008, Jenkins *et al.* 2011).

Earlier designs of the BSF had a pore volume of 8.5 L and a reservoir volume of 18.5 L. This meant that when each batch of water was poured into the filter, the first 8.5 L of filtered water (effluent) was the water that had been retained in the pore spaces of the sand during the pause period and the remaining 10 L of filtered water flowed through the filter without any retention time during the pause period. Bacteria removal efficacy was found to be substantially higher for the first 8.5 L of effluent compared to the remaining 10 L. Research also determined that the initial head in the filter should be reduced to achieve the target 'clean bed filtration rate' (hydraulic loading rate when the filter is first installed) of 400 $L/m^2/hour$ with filtration sand within the recommended ranges for effective size and uniformity coefficient (Baumgartner *et al.* 2007, Tiwara *et al.* 2008). This was necessary to prevent high pore velocity at the beginning of each run.

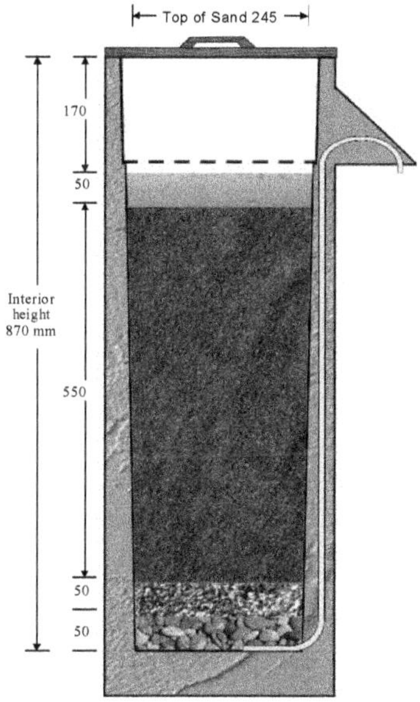

Figure 43.2 Version 10 concrete BSF cross-section (Credit: CAWST) - made with local materials & skills

In response to this research the CAWST recommended standard concrete BSF design was modified in 2009 by increasing the volume of sand in the filter by 22% while keeping the overall dimensions of the BSF the same. This dimension change affected several aspects of the BSF. First, the reservoir volume and the pore volume of the filtration sand were both made equal to 12 L. Tracer tests (Elliot 2008, Bradley 2011) have shown that the BSF operates at a high degree of plug flow. This means that virtually 100% of the filtered water is retained in the pore spaces of the sand during the pause period of the filter cycle. Each batch of water poured into the BSF will not appear at the outlet until the next batch of water is poured in. Second, the head was reduced to 17 cm from 27 cm, decreasing the initial head pressure by 37%, which in turn reduces the maximum pore velocity. Third, the sand bed depth was increased to 55 cm from 45 cm which caused the 'clean bed head loss' in the sand to be 22% greater. This greater volume of sand also provides more sand grain surface area for adsorption and attachment of contaminants. Finally, because the diffuser and outlet spout were moved 10 cm higher to accommodate the increased sand, the height of the outlet tubing spout was raised to 76 cm allowing more space for a safe water storage container below the spout.

This new concrete BSF design was termed CAWST version 10 (v10) (Figure 43.2). Versions 1 to 8 were development designs that Dr. Manz advanced in the 1990s. Version 9 was a thinner-wall design to make the filter lighter to transport and less expensive to construct.

The BSF has been used in developing countries for 15+ years; however, since the v10 filter design is four years old, much of the available research that determined the removal effectiveness of the BSF predates the design change in 2009. Laboratory studies on the v10 are currently underway. Removal effectiveness results, updated for the design change, will be available following the completion of these studies.

FILTRATION RATE AND SAND SPECIFICATIONS

Kubare & Haarhoff (2010) state: "The filtration rate, or hydraulic loading, is the key design parameter for all filtration processes... the heart of BSF design thus lies in the careful and appropriate selection of the media and sizing of the media bed."

While the slow sand filter (SSF) is a constant-rate process, the biosand filter (BSF) is a declining-rate process. For a given water temperature, it is the head loss in the filtration sand bed, combined with the available (declining) head, which determines the filtration rate. The flow control in the BSF is provided by the available head (determined by the design), and the head loss (determined by the sand sizing). Since the filtration rate varies during each run (with declining head) and between runs (with clogging of the sand bed over time), the target filtration rate is defined as the maximum rate that the BSF will encounter. This maximum rate occurs when a filter is first installed in the household, the 'clean bed filtration rate', measured at the start of the run when the filter is filled to the top.

Research, field evaluations, and consideration of the needs of the users in the household have led CAWST to set the recommended target maximum filtration rate at 400 L/hour per square metre of sand surface. Higher filtration rates cause high pore velocities that can scour the sand and cause pathogens to become unattached and re-enter the water stream. Lower filtration rates make the run time unacceptably long to the user, potentially causing them to drink unfiltered water instead. For the CAWST v10 filter, with a surface area of 0.06 m2, this filtration rate provides a flow rate of 0.4 litres/minute. This measurement, made after the filter has just been installed, is a critical quality control parameter for the BSF.

The filtration sand bed provides the resistance to flow, or head loss, necessary to achieve the target filtration rate. The raw sand is processed into filtration sand to attain this target. Sieving the sand with a wire mesh or perforated plate sieve (opening size of 0.7 mm) to remove the oversize sand, then washing the sand with water as many times as necessary to remove sufficient amounts of the finer sand, will achieve the recommended ranges for the effective size and uniformity coefficient of the filtration sand.

A summary of the typical design parameters for the filtration sand and hydraulic loading rate for the BSF and the SSF is shown in Table 43.2. CAWST's recommendations for sand sizes are generally in a tighter range than what others studying the BSF have proposed, and also tighter than what are often used in the SSF. These more restrictive parameters are recommended to make it easier to achieve the target filtration rate. For example, higher uniformity coefficients (greater variety in size) will lead to more packing which will tend to reduce the filtration rate.

Table 43.2 Typical design parameters for the BSF and SSF (Adapted from Kubare and Haarhoff 2010)

Parameter	Biosand Filter (BSF)		Slow Sand Filter (SSF)**
	CAWST Recommendation	Recommendations by Others*	
Filtration Rate/Hydraulic Loading Rate m^3/m^2/hour (or m/hour)	0.4	0.16 – 1.1	0.1 – 0.3
Effective Size, mm (d_{10})	0.15 – 0.2	0.15 – 0.3	0.1 – 0.3
Uniformity Coefficient (d_{60}/d_{10})	1.5 – 2.5	1.5 – 5	< 3
Maximum Grain Size, mm (d_{max})	0.7	-	-
Silt content, %, (d < 0.1mm)	< 4%	-	-
Filtration Sand Bed Depth, m	> 0.5	> 0.4	0.6 – 1.0

*Lukacs (2002), Hillman (2007), Manz *et al.* (1993), Fewster *et al.* (2004), Baumgartner *et al.* (2007), Duke *et al.* (2006) **Fox *et al.* (1994) and Campos *et al.* (2002).

The effective size and uniformity coefficient are the most important sand parameters in determining the filtration rate. A maximum grain size for the filtration sand is recommended by CAWST (dmax < 0.7 mm) to ensure that all sand grains are small enough to contribute surface area to aid in the attachment of pathogens. Maximum silt content (maximum 4% < 0.1 mm) is proposed to avoid situations where the sand meets the criteria for effective size and uniformity coefficient and provides the target filtration rate, but excessive very fine particles become dislodged and create turbidity in the filtered water. Minimum sand depth is needed to provide the necessary pore volume and sufficient total sand grain surface area for effective filtration.

PAUSE PERIOD

A major difference between the slow sand filter (SSF) and the biosand filter (BSF) is the pause period. The intermittent flow of the BSF means that, unlike the SSF, there is a period of time in the filter cycle of the BSF when the water is retained in the pore spaces and the filtration rate/velocity of the water through the pore spaces (pore velocity) is effectively zero. The total retention time (RT) includes the pause period time plus the run time when the water is being filtered through the sand.

Kubare and Haarhoff (2010) state that the pause period is of importance from three competing perspectives:

1. The treatment efficiency is improved because the water stored in the pores of the filtration sand undergoes further quality improvement during this time. Filter effectiveness is generally improved with longer pause periods.

2. The biological layer (biolayer or Schmutzdecke) requires an inflow of nutrients to establish and remain viable. More frequent batches of water (shorter pause periods), will be better from this perspective.
3. Quantity of filtered water produced by the BSF needs to be sufficient for the household daily needs. With shorter pause periods, more filtered water will be available for the family.

It is important to note that the pause period is ultimately at the discretion of the user. CAWST (2011) recommends a pause period of 6 to 12 hours with a minimum of 1 hour and maximum of 48 hours. The maximum pause period is to ensure that in hot arid climates the 5 cm of standing water (supernatant) does not dry out and kill the organisms in the biolayer. Household water requirements for treated water, typically about 40 L/day, mean that the v10 filter (with a 12 L batch capacity) will generally be filled 3 or 4 times per day, equivalent to a 6 to 8 hour pause periods (if the batches are evenly spaced in time).

Figure 43.3 Biosand Filter made in Peru with PVC pipe (Credit: DESEA Peru) Community health worker (CHW, Benita in Kelloccocha) shown measuring filter flow rate. Many of the CHWs don't read numbers well, so colour zones are used to indicate acceptable ranges for flow rates.

To evaluate the impact of user noncompliance regarding filter pause time, Hyde and Lackey (2013) conducted experiments with one BSF that was filled once per day and one BSF filled once every third day. After three months of operation both filters also went through a 'starvation' period when no water was added to either filter for eight days. Coliform reductions were generally (though not consistently) better for the BSF that was filled once every third day over that filled daily, but turbidity measurements showed no significant difference. Both filters demonstrated similar coliform removal effectiveness following the starvation period compared to before.

Definitive evidence on the minimum and maximum pause period that would optimize the operation of the BSF is still lacking, though as yet unpublished research (on the v10 BSF) will help to address this question (Napotnik and Jellison 2011).

CLEANING/MAINTENANCE

The cleaning/maintenance process is significantly different between the biosand filter (BSF) and the slow sand filter (SSF). Both types of filters will clog over time with particles and other materials that accumulate in the top 1 – 2 cm of the filtration sand. In the SSF, the cleaning or maintenance is typically performed when the head loss becomes too great. Due to the very large surface area of the SSF (often <100 m2), the top 1 cm of sand is normally removed from the top of the sand bed using various methods/equipment. Since the top 1 cm of sand is removed entirely with each cleaning, the sand bed depth is reduced by about 1 cm each time it is cleaned. Once the sand bed depth reaches 50 – 60 cm, the removed sand is cleaned and placed back into the SSF to bring the sand bed to the original depth.

An advantage of the BSF is that it is small enough to 'clean in place', using a technique that restores the flow rate without the need to remove the sand. In the BSF, when the flow rate becomes too slow, the user is advised to perform maintenance by 'cleaning'. Fewster *et al.* (2004) suggest this occurs when it takes more than about two hours to filter a batch of water. With the small sand surface area in the BSF, the top 1 – 2 cm of sand can be cleaned manually by filling the reservoir with about 4 L of water, removing the diffuser, swirling the top of the sand with the palm of the hand to stir up the dirt, then scooping out the dirty water with a small container and dumping it into a soak pit (or the bushes). The steps in this 'swirl-and-dump' technique are repeated as many times as necessary to restore the flow rate. Normally it requires 15-20 minutes for cleaning the BSF in this manner.

With both the SSF and the BSF there is concern that the quality of the filtered water is reduced immediately after cleaning because the biolayer is partially removed or impaired. Jenkins *et al.* (2011) found that disturbance of the biolayer by cleaning had no measurable effect on virus removal and only a modest reductive effect on bacterial and turbidity removal, based on measurements taken seven days or more after the disturbance.

DIFFUSER BASIN

The diffuser slows down the water from rushing into the sand layer and disturbing the biological layer essential to pathogen removal (CAWST 2012). This is important because disturbance of the biolayer will allow the influent water to bypass the biolayer and flow through the top portion of the sand without benefiting from the removal mechanisms (or the head loss) that the biolayer provides.

Figure 43.4 Molded plastic diffuser basin (Credit: Clean Water for Haiti & Great Lakes Mold & Manufacturing)

In the slow sand filter (SSF), the water depth above the sand (supernatant) varies with head loss but is typically 30 cm or greater. This water depth helps to expend the kinetic energy of the influent water as it flows into the SSF. Additional methods/equipment are utilized in the SSF to further diffuse the force of the water at the inlet to avoid any disturbance of the biolayer. In the biosand filter (BSF), at the start of the run period when the filter is filled with water from a bucket or other container, the standing water above the sand is just 5 cm deep. This depth of water is insufficient to fully protect the biolayer from disturbance, so a method of diffusing the energy from the influent water is needed.

The original versions of the concrete BSF used diffuser plates rather than basins. They were made with a narrow ledge cast into the concrete on all four interior walls of the BSF concrete body. This ledge, located 2 cm above the level of the water during the pause period, provided the location for the diffuser plate (typically a flat sheet made from plastic or metal) to rest on.

A study by DACAAR (2012) in Afghanistan found diffuser basins having 81 to 144 holes of 1 mm and 2 mm in diameter do not cause any disturbance to the top of the fine sand layer and the biolayer. Jones (2013), of the British Columbia Institute of Technology in Canada, reported similar results – diffusers with too few holes (25), too many holes (225), or too large holes (4 mm) will significantly disturb the fine sand layer and the biolayer. CAWST recommends BSF implementing-organizations use diffuser basins with 64 to 144 holes of 1.5 to 3 mm diameter.

Results of 25 BSF field evaluations in 16 countries found 26% of diffuser plates were damaged and/or incorrectly placed in the BSF (Ngai *et al.* 2012). Diffuser basins use the same size and number of holes as the plates, but differ from plates in that diffuser basins have side walls which extend up to a lip around the top of the basin, so they rest on the top of the filter walls. Diffuser basins prevent any water from escaping around the sides which was a frequent problem with the plates. Basins are more convenient than plates to remove and replace in the filter. Basins also make the fabrication of the concrete BSF easier and less expensive since the ledge is not required if diffuser basins are used. Recently, a diffuser basin for the concrete BSF, made from injection-molded plastic, has been developed (Figure 43.4) which is durable and appears robust enough to last as long as the concrete BSF (15+ years).

RIPENING AND VIRUS REMOVAL

Jenkins *et al.* (2011) found that operating the filter with a long pause period between batches improved performance for bacteria, virus, and turbidity removal versus no pause period. This was especially true of virus removal. It was found that each additional hour of residence time (up to 27 hours) was found to be beneficial. "Pause time may be particularly important for avoiding viral shedding by allowing sufficient time for adsorbed and attached viral particles to decay or be consumed before addition of the next batch."

Microbial challenge studies by Elliot *et al.* (2008) found that reductions of bacteria, viruses, and turbidity by the biosand filter (BSF) could be increased by increasing the retention time of water in the filter. Elliot *et al.* (2011) investigated specifically virus reduction during the pause period. This research, conducted over repeated daily filtration cycles, was to determine the effect of ripening (maturing or aging) of the entire filtration sand bed that involves processes occurring throughout the depth of the sand bed rather than only near the surface in the biolayer. They found that a ripening period of one to two weeks was required before reduction of the viruses began. Over ten weeks the observed rates of virus decline reached 11 – 13% per hour (0.053-0.061 $\log_{10}$/h) of pause period. The cause of virus reduction was found not to be directly from modification of the sand surfaces in the filter bed, but rather by the activity of the microbial community within the depth of the filtration sand bed. The authors state that the most probable mechanisms "are production of microbial exoproducts such as proteolytic enzymes or grazing of bacteria and higher microorganisms on virus particles".

Bradley *et al.* (2011) conducted a long-term (300 days) laboratory study of virus removal in five BSFs: three amended with iron and two conventional (sand-only) BSFs. The unmodified sand-only BSF virus removal varied from zero to 99.9%, averaging about 99% ($2\log_{10}$) removal during the initial 150 days. After 150 days, the sand-only BSF saw a steady increase in virus removal from 99% to over 99.99% ($2\log_{10}$ to $>4\log_{10}$) by 300 days. This efficacy of the BSF to remove more than 99.99% of viruses was attributed to the significant biological 'ripening' leading to ten months of data. This level of virus removal has not been reported in any previous (short-term) research of the BSF and shows the importance of filter ripening (and longer research studies).

IRON-AMENDMENT & ALTERNATE VERSIONS

Addition of zero-valent iron such as common iron nails, iron chips or iron particles either into the diffuser basin or mixed directly in the sand bed is an iron-amended biosand filter (BSF). These have been found to enhance virus, bacteria, and arsenic removal (versus the sand-only BSF). The corrosion of the iron creates continuous formation of new iron oxides (rust). The mechanism for the removal/inactivation of pathogens and arsenic is through electrostatic attraction with positively-charged iron oxide providing adsorption sites for negatively-charged arsenic and virus particles. You *et al.* (2005) found most of the viruses removed were either inactivated or irreversibly adsorbed. Experiments conducted by Bradley *et al.* (2011), using full scale BSFs, indicated that the BSF could potentially be modified to meet the USEPA drinking water standard for virus removal of >99.99% ($4\log_{10}$). Tellen *et al.* (2010) at the University of Buea, Cameroon, found improved removal of total coliform, fecal coliform, and fecal streptococci in iron particles amended BSF compared to the sand only BSF.

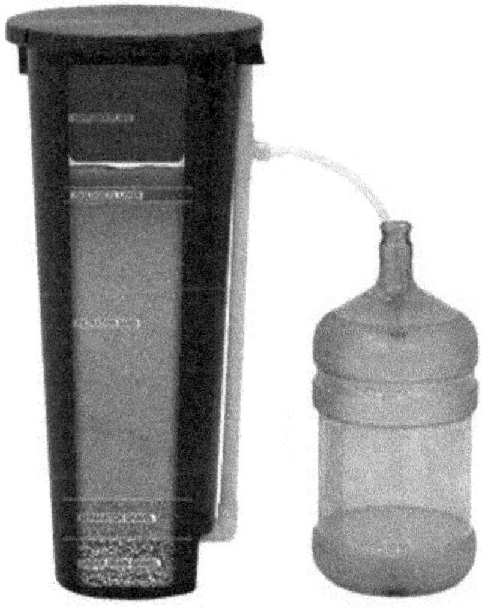

Figure 43.5 Cross Section of Plastic Biosand Filter (Credit: TripleQuest) - available for purchase at various locations in Central America, the Caribbean, and Africa.

In a two-year evaluation of over 1000 Kanchan Arsenic Filters (KAF, BSFs amended with iron nails in the diffuser basin) deployed in rural villages of Nepal, Ngai *et al.* (2007) determined that the KAF typically removes 85–90% arsenic, 90–95% iron, 80–95% turbidity, and 85–99% total coliforms. Similar arsenic removal results have been reported in far western Nepal by the Rural Village Water Resources Management Project (2013) and in central Cambodia by Uy *et al* (2009).

However, the chemistry of the influent water can affect removal. Hard water can cause scaling which prevents rust from forming on the iron. Preferential adsorption by other ions (such as phosphate) can fill up the available adsorption sites on the iron oxide. Studies have examined the options for the added iron (nails, particles or steel wool), and the placement of the iron in the filter or pre-filter (Bradley *et al.* 2011, Noubactep *et al.* 2012).

Numerous situations arise in the 55 countries where the BSF has been employed. Locally available materials and skills often dictate what is feasible. Affordability of the BSF for very poor families is a constant pressure. Challenges in transporting the filter bodies plus the roughly 40 L of sand/gravel per filter are common. Since there are many different situations where safe water is needed, there also need to be many different options.

Several different versions of the BSF have been proposed. The versions suggested are intended to improve the technology in some way such as lighter weight, less expensive, easier to transport, possible to import, or simpler to make. The suggested materials for the BSF body include cast concrete, injection-molded plastic, PVC pipe, sheet metal, concrete pipe, molded plastic bag, ceramic container, large plastic drum, and 5 gallon (20 L) plastic bucket. The shape may be square or round. Some filters are tapered to improve fabrication or transportation.

An example is the sandstorm BSF from Ethiopia (Smith 2013). The filter has a simple constant head device that provides a much lower head (7 cm) which remains constant throughout much of the filtration cycle. The filter body is made with a cylindrical shell of 28 gauge galvanized iron sheet cast into a stable base of concrete.

SUMMARY

(1) The adaptation of slow sand filter (SSF) technology to a point-of-use intermittent slow SSF (the biosand filter (BSF)), has high potential to be an important and significant contribution to water treatment globally. The technology is most applicable to households and institutions without piped water.

(2) The design dimensions of the BSF were changed in 2009 based on compelling research that found that the contaminant removal effectiveness of the BSF could be significantly improved at little or no additional cost. This new design (v10) is strongly recommended to new and existing implementers of BSF projects.

(3) The specifications for filtration rate (maximum hydraulic loading rate) of 400 L/hour per m^2 and for the filtration sand (effective size = 0.15 to 0.20 mm and uniformity coefficient = 1.5 to 2.5) have been carefully selected to balance the reliability of the BSF as an effective water treatment with the needs of the users.

(4) The intermittent flow of the BSF means that there is a pause period between filtration runs. Research has found this pause period to be helpful in further improving the water, with longer pause periods generally better for contaminant removal effectiveness. Further research is needed to quantify the effects and optimal length of the pause period.

(5) The BSF is small enough to allow cleaning by hand, eliminating the need to remove or replace sand. The user is able to operate and maintain the filter herself.

(6) Based on field evaluations, the diffuser basin is preferred over a diffuser plate. The number of holes and their diameter has been the topic of two studies resulting in new learning. An injection-molded diffuser basin has been developed which has substantial advantages over other options.

(7) Recent research has found that virus removal occurs primarily in the entire sand bed rather than in the biolayer (Schmutzdecke). Viruses are reduced by microbial mediation rather than physical/chemical alteration of the sand surfaces. Filter ripening plays an important role in

achieving higher levels of virus removal, with low virus removal initially, increasing to greater than $4\log_{10}$ removal after 6 to 12 months. These findings may be equally applicable to the SSF and the BSF.

(8) When BSFs are amended with iron, the dominant virus removal process at the initial stage is positively-charged iron oxides attracting and adsorbing the negatively-charged viruses. As such, virus removal can be as high as $3\text{-}4\log_{10}$ in iron-amended BSF during the first week of operation, as opposed to less than $1\log_{10}$ for the sand-only BSF. Over time, microbial mediation processes begin to dominate and add to the surface charge processes. Laboratory studies have shown virus removal gradually increasing to 4-6 $\log_{10}$ when ripening reaches 6 to12 months.

CONCLUSION

The simplicity, high user satisfaction, and sustainability of the biosand filter (BSF) have resulted in increasing research interest in BSF technology. Research and development, such as those mentioned above, have led to further performance improvements, making the BSF even more attractive for wider adoption. Considering the majority of the nearly 800 million people lacking access to improved water are rural households (WHO and UNICEF 2013) where conventional, large-scale, drinking water treatment plants are often inappropriate, there is significant potential for the BSF to play a more prominent role in providing safe water. Therefore, further research of the technology should be encouraged.

REFERENCES

Baumgartner, J., Murcott S. and Ezzati M. (2007) Reconsidering 'appropriate technology': the effects of operating conditions on the bacterial removal performance of two household drinking-water filter systems, *Environmental Research Letters*. **2**(2), 024003

Bradley, I., Straub, A., Maraccini, P., Markazi, S. and Nguyen, T. H. (2011) Iron oxide amended biosand filters for virus removal, *Water Research*, **45**(15), 4501 – 4510.

Buzunis, B. (1995) Intermittently operated slow sand filtration: A new water treatment process. Master's Thesis. University of Calgary, Canada

Campos, L. C., Su, M. F. J., Graham, N. J. D. and Smith, S. R. (2002) Biomass development in slow sand filters, *Water Resources*. **36**, 4543–4551.

CAWST (2011) Biosand filter factsheet (detailed) [Online]. Calgary: Centre for Affordable Water and Sanitation Technology. Available from: http://biosandfilters.info/technical/fact-sheet-biosand-filter [Accessed 2 Jan 2013].

CAWST, 2012b. *Biosand Filter For Technicians; A CAWST Participant Manual* [Online]. Calgary: Centre for Affordable Water and Sanitation Technology. Available from: http://www.cawst.org/en/resources/pubs/training-materials/category/34-biosand-filter-for-technicians [Accessed 7 June 2012].

CAWST (2013) Key Performance Indicators, prepared by Ngai, T., and Karia, A. Unpublished data.

DACAAR Water Expertise and Training Center (2012) Action Research on Effects of Numbers and Size of Holes in Diffuser Basin on Sand layer and Biolayer. Unpublished data.

Duke, W. and Baker, D. (2005) The use and performance of the biosand filter in the Artibonite Valley of Haiti: A field study of 107 households, University of Victoria, Canada.

Duke, W, F., Nordin, R. N., Baker, D. and Mazumder, A. (2006) The use and performance of biosand filters in the Artibonite Valley of Haiti: A field study of 107 households, Rural Remote Health, **6**(3), 570.

Earwaker, P. (2006) Evaluation of household biosand filters in Ethiopia, Masters Thesis. Cranfield University, United Kingdom.

Elliot, M. A., Stauber and C. E., Koksal, F. (2008) Reductions of E. coli, echovirus type 12 and bacteriophages in an intermittently operated household-scale slow sand filter. *Water Research* 42, 2662 – 2670.

Fewster, E., Mol, A. and Wiessent-Brandsma, C. (2004) The long term sustainability of household bio-sand filtration, *30th WEDC International Conference*, Vientiane, Lao PDR.

Fox, K. R., Graham, N. J. D. and Collins, M. R. (1994) Slow sand filtration today: An introductory review. In: Slow Sand Filtration. Edited by M.R. Collins and N.J.D Graham. American Water Works Association, Denver, CO, pp. 1–8.

Hillman, A. (2007) Adapting the biosand filter to remove fluoride: An investigation into alternative filter media. MSc Thesis. Cranfield University, UK.

Hyde, M. and Lackey, L. W. (2013) The impact of loading frequency and copper as a biocide on biosand filter performance. *Journal of Water, Sanitation and Hygiene for Development.* 3(3), 355–364.

Jenkins, M. W., Tiwari, S. K. and Darby, J. (2011) Bacterial, viral and turbidity removal by intermittent slow sand filtration for household use in developing countries: Experimental investigation and modeling. *Water Research* 45(18) 6227-6239.

Jones, S. (2013) Biosand filter diffuser basin study, Industry Project, Civil Engineering, British Columbia Institute of Technology (BCIT) unpublished report.

Kubare, M. and Haarhoff, J. (2010) Rational design of domestic biosand filters, *Journal of Water Supply: Research and Technology - AQUA,* 59(1), 1.

Lukacs, H. (2002) From design to implementation: Innovative slow sand filtration for use in developing countries. MEng Thesis. Massachusetts Institute of Technology, Cambridge, USA.

Manz, D.H., Buzunis, B.J. and Morales, C. (1993). Final report on the Nicaragua household water supply and testing project, Division of International Development, University of Calgary, Calgary, AB, Canada.

Napotnik, J., Jellison, K. (2011) Optimizing the Biosand Filter, *WEF Disinfection 2011 Conference*, Cincinnati, Ohio, USA.

Ngai, T., Murcott, S. and R. Shrestha (2004). Kanchan Arsenic Filter (KAF) – Research and Implementation of an Appropriate Drinking Water Solution for Rural Nepal.

Ngai, T. K. K., Shrestha, R. R., Dangol, B., Maharjan, M. and Murcott, S. E. (2007) Design for sustainable development—Household drinking water filter for arsenic and pathogen treatment in Nepal, *Journal of Environmental Science and Health Part A*, 42(12), 1879-88.

Ngai, T.K.K., Lentz, R., Baker, D., Forde, O. (2012) Review of 16 Biosand Filter Projects by 16 Different Implementing Organizations in 14 Countries. *Water Safety Conference*. 13-15, Kampala, Uganda.

Noubactep, C., Temgoua, E. and Rahman, M. A. (2012) Designing Iron-Amended Biosand Filters for Decentralized Safe Drinking Water Provision, *Clean – Soil, Air, Water*, **40**(8), 798-807.

Palmateer, G., Manz, D., Jurkovic, A., McInnis, R., Unger, S., Kwan, K. K. and B. Dudka (1997). Toxicant and Parasite Challenge of Manz Intermittent Slow Sand Filter. *Environmental Toxicology*, **14**, 217- 225.

Sobsey, M. D., Stauber, C. E., Casanova, L. M., Brown, J. M. and Elliot, M. A. (2008) Point of Use Household Drinking Water Filtration: A Practical, Effective Solution for Providing Sustained Access to Safe Drinking Water in the Developing World, *Environmental Science & Technology* **42**(12), 4261-4267.

Smith, A. (2013) Sandstorm: A biosand filter designed for small-scale enterprises. *36th WEDC International Conference*, Nakuru, Kenya. 1-5 July 2013.

Tellen, V., Nkeng, G. and Dentel, S. (2010). Improved filtration technology for pathogen reduction in rural water supplies. *Water*, **2**(2), 285-306.

Tiwara, S. K., Schmidt, W-P., Darby, J., Kairiuki, Z. G. and Jenkins, W. M. (2008) Intermittent slow sand filtration for preventing diarrhoea among children in Kenyan households using unimproved water sources: randomized controlled trial. *Tropical Medicine and International Health* **14**(11), 1–9.

Uy, D., Chea, S., Mao, S., Murcott, S., Mahin T., Adam, M. and Ngai, T. K. K. (2009) Arsenic removal performance of the Kanchan arsenic filter in Cambodia: Findings from a 12 months pilot study, *International Research Colloquium of the Network to Promote Household Water Treatment and Safe Storage* 21-23 September 2009, Royal College of Surgeons in Ireland, Dublin.

World Health Organization & UNICEF (2013) Progress on sanitation and drinking-water 2013 update: Joint Monitoring Programme for Water Supply and Sanitation [Online]. Available from: http://www.who.int/water_sanitation_health/publications/2013/jmp_report/en/ [Accessed 12 Dec 2013].

You, Y. W., Jiehan, Chiu, P. C. and Andyanjin (2005) Removal and Inactivation of Waterborne Viruses Using Zerovalent Iron, *Environmental Science and Technology.* **39**(23), 9263-9269.

Chapter 44

Performance optimisation of household biosand filters

L. C. Campos, Richard Outhwaite***

**Department of Civil, Environmental & Geomatic Engineering, University College London, Gower Street, London, WC1E 6BT, UK*

***Web Earth PTE Ltd, 40 Carpenter Street, Singapore, 059919*

ABSTRACT

This work investigated two methods to optimize the design and performance of the household biosand filter. The first method to optimise the filter was by 'seeding' one filter with 50% of old sand from an existing biosand filter to speed up the maturation of the filter. The second method used was the addition of a tap to the outflow pipe, which would prevent the need for users to store water that had been filtered, thus removing the possibility of recontamination through storage. The seeded filter reached 99% coliform removal after 16 days, 10 days before the unseeded filter, demonstrating the important role of microbial community in water purification by slow sand filtration. Furthermore, the filter with the tap showed that it could still maintain significantly high coliform removal rates (99.2%, $p < 0.01$) even after pausing the flow of water.

Keywords household biosand filter, design, filter performance, maturation

INTRODUCTION

It is estimated that every 21 seconds a child dies from water related disease and that a total of 1.5 million children under five years of age die from diarrhoea each year (UNICEF & WHO, 2009). Developments have been made in the appropriate household water treatment technology in recent years. Developments such as the biosand filter (Manz, 2007; Buzunis 1995) and the Filtron ceramic water purifier (Potters for Peace, 2008) have begun to address this problem, but there still remain a number of challenges.

As new technology develops and becomes available, new technical and socio-technical challenges arise when these technologies are implemented. How people use, and often more importantly misuse, these technologies create new challenges to be solved. Incorrect or unanticipated use of the technology in the field can increase the risk of exposure to contaminants (Baumgartner et al, 2007). This study aimed to investigate whether modifications to the biosand filter can affects its performance in removing water-borne pathogens and avoid possible pathways to misuse that can reduce its efficiency, as well as speeding up the time for the filter to reach its optimum performance.

METHODOLOGY

Filter Construction

Two biosand filters were constructed in accordance to the Biosand Filter Manual (CAWST, 2008). To enable monitoring of the filter during its use, the front of the filters was constructed from Perspex to allow better inspection of the biofilm and schmutzdecke development (Figure 44.1). Coarse and fine gravel were washed and deposited into the base of the each filter as the under-drain in 50mm layers, before the sand layer was installed. The 400mm fine sand layer was installed wet into a shallow layer of water to prevent air bubbles becoming trapped in the filter. The sand used in the study was RH45 sand (effective diameter = 0.182mm and uniformity coefficient = 1.65) from WBB Minerals' quarry in Redhill, Surrey. The sand was previously washed to remove fines, then oven-dried at 105°C before being added to the filter. To identify the effect of old sand (i.e. from an existing biosand filter) on the maturation rate and water quality, the filter with new sand one ("unseeded"), while the other the other ("seeded") had 50% of the new sand (bottom layer) and 50% of old sand (effective diameter = 0.202mm; uniformity coefficient = 1.80). Although the old sand are slightly different from the new one, they are both within typical values of grain size between 0.15mm and 0.30mm and uniformity coefficient of less than 4 for use in slow sand filtration (Huisman & Wood, 1974).

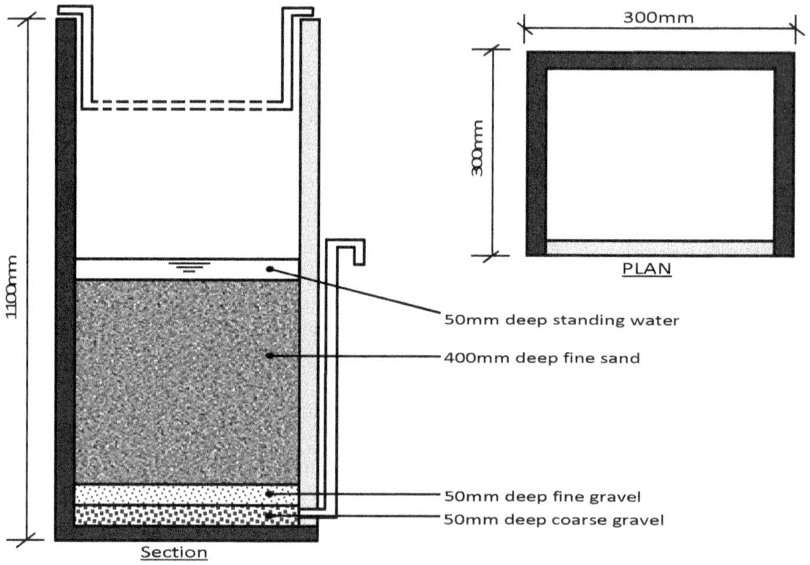

Figure 44.1 Plan and section views of constructed biosand filters.

Raw water

To simulate, as closely as possible, the conditions that the biosand filter is usually under, natural water from Regents Park Lake (London) was used. The water was collected in two 25L containers as the water volume was derived from a minimum basic water requirement of 20L per capita per day (WHO & WEDC, 2011) and the filter reservoir volume of 12L. This represents a basic required amount for drinking, hygiene and cooking, however this may differ depending on the users, physiological, social and cultural norms (Sphere, 2004). The volume of raw water used for each intermitent filtration was 24L and this volume ensured that the filtrate being sampled could be compared directly with the raw water, as verified by intial dye testing and porosity calculations.

Filter maturation study

Following the procedure shown in Table 44.1, the effects of residence time and the effect of seeding the filter to speed up its maturation period could be investigated. This experiment was carried out during 19 days with 3 sets of 5 consecutives filtration days (Monday-Friday) and operational pause of 2 days (Saturday-Sunday). The filtrate was sampled 3 times during the filtration cycle of each water volume (24L), to establish the effect of different residence times within the filter on water quality and to make sure we were collecting the corresponding filtered water to each influent volume.

Table 44.1 Procedure for adding water to the biosand filters and to collect filtrate samples.

Total residence time of the 1st 24L (V_1) to be filtered, which was around 24h					
			Residence time starting point of the 2nd 24L (V_2) and subsequent volumes (V_n) to be filtered		
Time 0 Day 1	Time 45min	Time 90min	Time 0 Day 2	Time 45min	Time 90min
Add 12 L of V_1	Add the remaining 12 L of V_1	Collect Sample 1 (S1) of V_1	Add new 12 L and collect Sample 2 of of V_1	Add the remaining 12 L and collect Sample 3 (S3) of V_1	Collect Sample 1 (S1) of V_2

Filter flow control study

To determine the effect of using a biosand filter with a tap on the quality of the filtrate, one filter was operated with a tap and the other without tap, allowing the entire reservoir to freely pass through the filter. The tap was added to the filter that had been seeded in the previous study. The flow of raw water was paused three times on the filter with the tap, to simulate water being used at different intervals throughout a day when it was required, rather than filtering the entire volume of water in the reservoir each time. The flow on the filter with a tap was paused each time after 8L of raw water had flowed (i.e. after 8L and 16L), then the water was allowed to flowed until all 24L had flowed through and the water reached the 50mm standing water level. The pause period of the water flow was varied between 1, 2 and 3 hours to determine the effect of their pause on bacteria removal. This experiment was carried out during 18 days with 3 sets of 4 consecutives filtration days (Monday-Thursday) and operational pause of 3 days (Friday-Sunday). Again, the filtrate was sampled 3 times during the filtration cycle and the sample procedure for 1 day is shown in Table 44.2.

Table 44.2 Procedure for adding water to the biosand filters and to collect filtrate samples.

Time 0 Day 2	Time 20 min	Time 20min + P1	Time 45min + P1	Time 60min + P2	Time 60min + P1 + P2	Time 90min + P1 + P2
Add 1st 12 L and collect S1	After 8L filtered, close tap and began pause 1 (P1)	Open tap to restart flow	Add the remaining 12L and collect S2	After 8L filtered, close tap and began pause 2 (P2)	Open tap to restart flow	Take Sample S3

Note: P1 = 1h pause; P2 = 2h pause; P3 = 3h pause

Total Coliforms

Total coliform bacteria counts were used as the indicator for bacterial improvement of the filtrate. The membrane filter technique was used to determine the number of bacteria present in the influent and filtrate. A Nalgene filter and 0.45µm cellulose membrane filters were used for all bacteriological tests. Coliforms were counted using a Stuart SC6 colony counter and reported as Colony Forming Units (CFU) per 100mL. Raw water was taken in 10mL samples and diluted to make it up to 100mL, as per regulatory requirements for lake water (APHA, 2005). Filtrate samples were filtered in 100mL quantities. To allow for clear counting of coliforms, this was separated into four 25mL samples as the filter was ripening and reduced accordingly as the coliform counts became lower. Agar plates were produced using a dehydrated culture media. Fluka Modified-Endo medium was used for total coliform growth. Culture media were prepared in batches that were refrigerated and used within one week. The M-Endo media has been noted to produce uncertain results, due to the non-appearance of metallic sheen on colonies that form on the membrane due to the presence of non-coliforms (Burlingame et al., 1983).

RESULTS AND DISCUSSION

Filter maturation study

The unseeded filter initially produced low coliform removal rates (Figure 44.2a). Over the first five days to of operation the filter achieved a maximum of 53% coliform bacteria removal. As the filter and schmutzdecke were still maturing and developing, respectively, in this period, the main removal mechanisms acting are physico -chemical attachment of bacteria to the sand and physical straining of bacteria in the sand bed. The filter matures as the microbiology is established, and the schmutzdecke begins to develop and the biofilm grows on sand grains. The unseeded filter achieved 90% removal after 15 days operation when S2 and S3 were collected with residence time of 24 hours. It took a further 11 days operation to achieve 99% removal rates at 26 days. This rate of bacteria removal is considered to be a mature filter (Manz, 2007). The seeded filter reached higher coliform removal rates much more rapidly. After 7 days of operation it had achieved 90% coliform removal when samples S2 and S3 were collected. (Figure 44.2b). This improved to 99% coliform removal after 16 days of operation. The samples S1taken at the same day they have been filtered (i.e. residence time equal 45min) had consistently lower coliform removal rates in both filters and did not achieve consistently high coliform removal rates. Whilst the removal rate reached 80% removal after 10 days of operation in the seeded filter compared to 18 days in the unseeded filter (Figure 44.3a), after this it showed no further improvement, fluctuating around this mark throughout the remainder of the tests. It can also be seen that the seeded filter started with much higher coliform removal rates (<50%) while the unseeded filter started with lower coliform removal rates (>30%). A one-way analysis of variance was carried out to test the null hypotheses that bacterial removal across with seeded and unseeded filters at each residence time was equal. A p-value of < 0.05 was considered to be significant. For sample S1, seeding did not have a significant effect on the bacteria removal of filtrate ($p = 0.393$). This indicates that after 12L of water have been filtered, there is a decrease in biological activity as the microorganisms within the filter have consumed as much of the pathogens as possible, regardless of whether the filter has been seeded or not.For samples S2 and S3, the seeding had a significant effect on the bacterial quality of the filtrate, with significance factor $p = 0.001$ and 0.026 respectively. Therefore it can be said that when the filter is used as per the standard procedure i.e. using 12L raw water volumes, seeding has a significant effect on the bacterial quality of the filtrate. It does not appear to have significant benefits when volumes of water greater than 12L are filtered. The significance of the sampling point within each filter was compared using independent variable t-tests. For the seeded filter, samples S2 and S3 had significantly higher removal rates than samples S1 ($p = 0.004$ and 0.036 respectively). There was no significant difference between sample s S2 and S3 ($p = 0.280$). The unseeded filter showed no significant difference between samples S1 S2 or S3 ($p= 0.420$, 0.869 and 0.576 respectively). This shows that the effect of increased residence time does not affect performance until the filter is more mature. As a more mature filter will have higher populations of microorganisms, this suggests that it is the process of biological predation that is most prevalent during increased contact time with the filter.

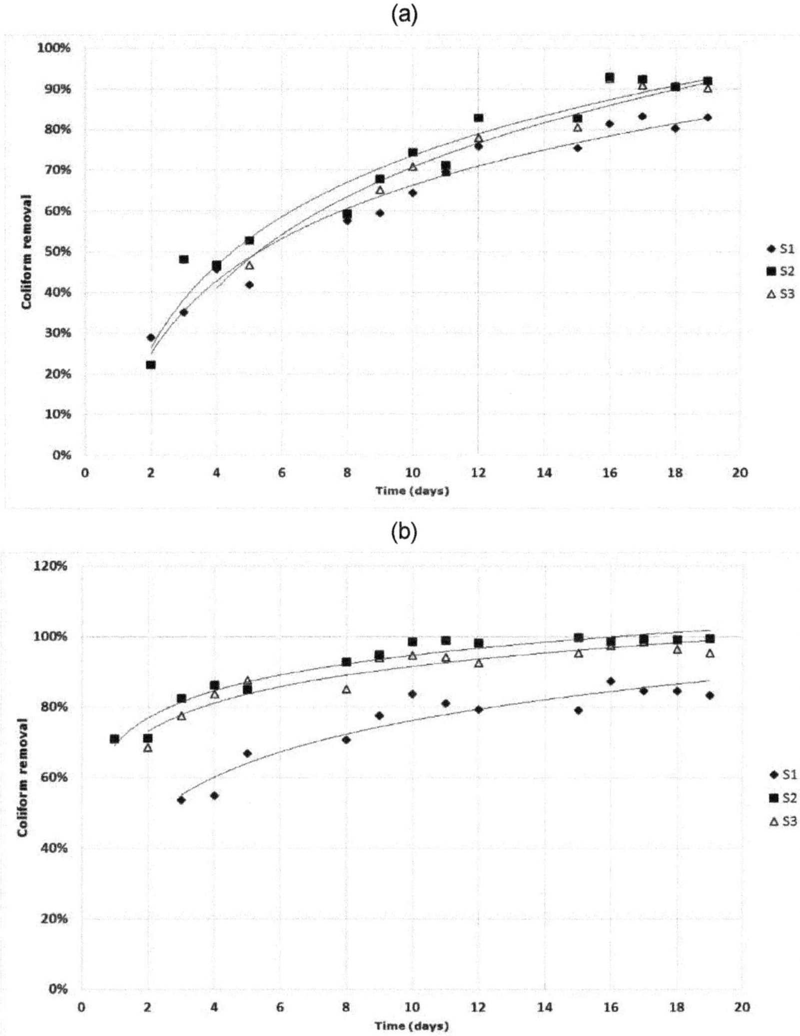

Figure 44.2 Coliform removal rate against time for (a) unseeded filter and (b) seeded filter.

Tap study

The mean coliform removal rate for the filter with tap was above 99% for samples S2 andS3 collected with residence time of 24h, while removal in the filter without tap varied from 80-99%. (Figure 44.3). Even at overflow conditions, samples S1 achieved 97% mean coliform removal rates in the filter with tap. This is contrary to recommendations that the flow should not be stopped during the filtration (Manz, 2007). A one-way analysis of variance (ANOVA) was carried out to assess the effect of the tap on the coliform removal rate at different sampling points. The tap had the most significant effect on sample point 1. The filtrate from the filter with the tap was significantly higher ($p < 0.001\%$) than the filter without. At sample points 2 and 3 tap still provided significantly higher removal rates ($p = 0.042$ and 0.023 respectively). This shows that at the filter with tap consistently provided higher coliform removal rates. The tap increases the residence time and allows smaller volumes of water to be filtered without storage. This follows work by Baumgartner et al. (2007), which identified that smaller dosing volumes of water filtered had lower coliform counts.

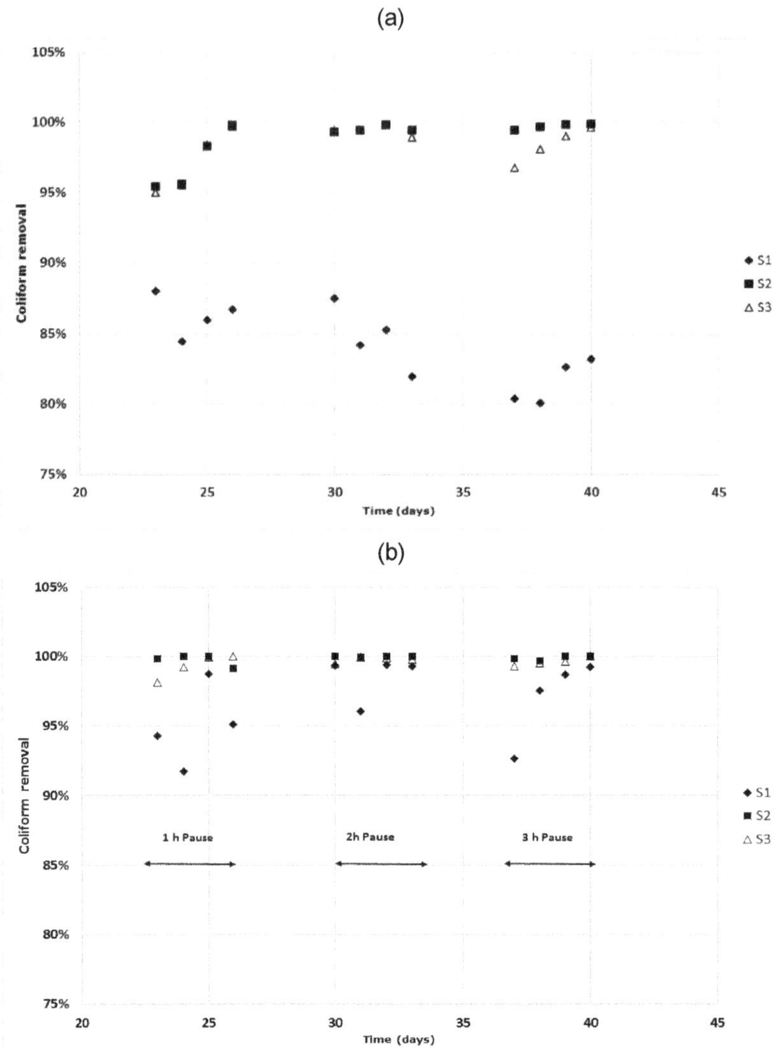

Figure 44.3 Coliform removal rate against time for the (a) filter without tap and (b) filter with tap. Note: Filtration time starts on day 20 following the end of filter maturation experiment on day 19.

There were also significantly higher removal rates between samples at different residence time for filters with and without the tap. Independent t-tests showed that samples S1 had statistically significantly lower coliform removal rates over both samples S2 and S3 for both filters, p= 0.004 and 0.036 for the filter with tap and 0.005 and 0.012 for the filter without tap. This shows that even with the improved performance of the tap, the effect of residence time within the filter is significant. This is in contradiction to current recommendations that pauses in filter flow are detrimental to the filter's performance (Manz, 2007). Clear improvement can be seen in the filtrate of the filter (Figure 44.3b), which was paused using the tap. Even a 1-hour pause increases the performance of the filter. Thus using tap for on-demand supply may be a possible solution to increase performance of the filter and avoid one of the pathways to recontamination. However, the filter was always run down to 50mm pause level after 24L had been filtered during the experiment. The effect of different pause levels on filter performance should be investigated to assess any unfavourable combinations of standing water level and pause time on filter performance.

Headloss development during maturation period and tap tests

Headloss was monitored through the experiment and increased for both filters throughout the maturation period and through the other tests (Figure 44.4). The unseeded filter experienced much higher headlosses throughout the maturation period. The headloss increased 29% from an initial value of 6cm to 8.5cm after 19 days of filtration. This continued to increase until the final test 44 days later, reaching a maximum of 10.8cm, a further increase of 21%. The seeded filter experienced lower headlosses through the tests. Initial headloss was 4.3cm, increasing to 5.9cm after the maturation period (19 days). This is an increase of 27%, a similar increase to the unseeded filter. However, by the final test headlosses in the seeded filter had increased to only 6.3cm, a further 6% increase. This suggests that microbial dynamics therefore play an important role in headloss development in slow sand filtration and the consumption of organic material reduces filter clogging. Campos et al. (2002) through model simulations demonstrated that the schumtzedcke, acting as a porous layer, reduces the input of nutrients and suspended matter into the underlying sand layers causing a decrease in the biomass (i.e. decay of microorganisms) and the resulting bulk specific deposit and headloss within slow sand filters.

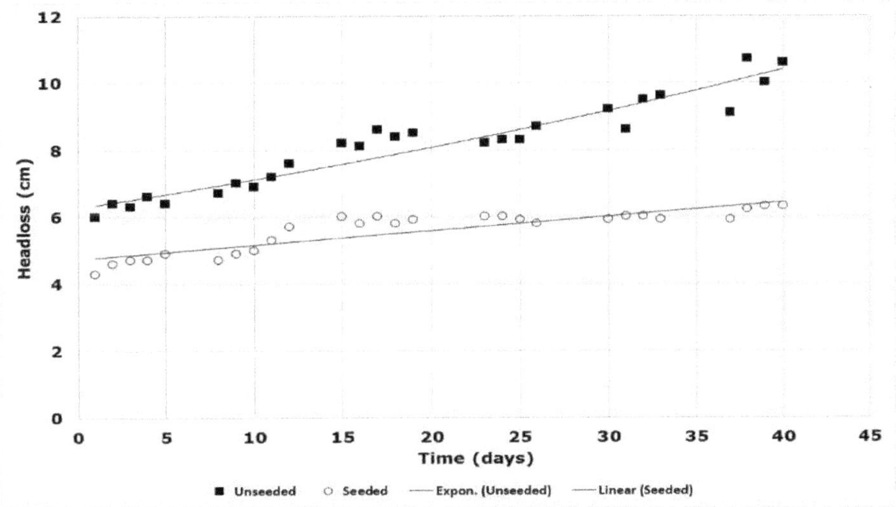

Figure 44.4 Headloss development for seeded and unseeded filters during maturation and tap tests.

CONCLUSIONS

The filter seeded with sand from an existing filter showed greatly improved performance, and identified that the process is greatly speeded up by adding 50% of the sand from another existing filter. The effect of putting smaller volumes of seeded sand should be studied, as this would increase the number of filters that could be seeded using the same volume of sand. The position of where the seeded sand was installed in the new filter could be important an important consideration, as it could affect how the biofilm develops. The tests conducted on the filters with the tap proved that the tap did not adversely affect the performance of the filter to which it was fitted, and the pause in the flow that it introduced did not reduce the bacterial quality of the filtrate. The filter that was seeded in the previous study was the one fitted with a tap; therefore the filter may have been performing better than the other one before the study. It is therefore not possible to definitively say that the performance of the filter with a tap was better than the one without a tap. However, it can be shown that the tap did not have a detrimental effect on the filter it was fitted to. This therefore should help to eliminate a pathway to recontamination of the filtrate, as storage of the water after it has been filtered is no longer required.The provision of the tap may also have an impact on the maturation of the filter, so should be incorporated into a further study of how the filter mature when provided with a tap.

REFERENCES

APHA (2005). Standard Methods for the Examination of Water & Wastewater: 21st ed., American Technical Publishers, New York, USA.

Barrett, J.M., and Silverstein, J. (1988) The effects of high carbon and high coliform feed waters on the performance of slow sand filters under tropical conditions, in Slow sand filtration: recent developments in water treatment technology. Edited by N.J.D. Graham, Ellis Horwood Ltd., Chichester, UK pp 231–252.

Baumgartner, J., Murcott, S. & Ezzati, M. (2007) Reconsidering 'appropriate technology': the effects of operating conditions on the bacterial removal performance of two household drinking-water filter systems. *Environmental Research Letters*, 2 (2), pp 24-33.

Burlingame, G. A., Mcelhaney, J., Bennett, M. & Pipes, W. (1983) Bacterial Interference with Coliform Colony Sheen Production on Membrane Filters, Applied And Environmental Microbiology, Vol 47(1), pp 56-60.

Buzunis, B.J. (1995) Intermittently Operated Slow Sand Filtration: A New Water Treatment Process. University of Calgary, Canada.

Campos, L. C., Smith, S.R., Graham, N.J.D. (2002) Deterministic-Based Model of Slow Sand Filtration. II: Model Application. Journal of Environmental Engineering, ASCE, p.887-894.

Centre for Affordable Water and Sanitation Technologies – CAWST (2008) Biosand Filter Manual Design, Construction, Installation, Operation And Maintenance, Calgary, Canada.

Huisman, L. and Wood, W.E. (1974) *Slow Sand Filtration*, World Health Organisation, Geneva.

Manz, D. (2007) BioSand Water Filter Technology: Household Concrete Design, University of Calgary, Canada.

Potters for Peace (2008) www.pottersforpeace.org, accessed 19 March 2010

UNICEF & WHO (2009) Estimated with data from Diarhhoea: Why children are still dying and what can be done.

The Sphere Project (2004) Humanitarian Charter and Minimum Standards in Disaster Response, The Sphere Project: Geneva, Switzerland. United Nations Children's Fund (2006) Annual Report, New York, US. Accessed at http://www.unicef.org.uk/publications/pdf/ar2006.pdf 5 February 2014.

WHO & WEDC (2011) How much water is needed in emergencies: Technical Note on Drinking Water, Sanitation and Hygiene in Emergencies 9, Geneva, Switzerland.

Chapter 45

Optimizing intermittent slow sand filters (biosand) construction for more consistent microbial removals

K. Sawyer, C. Perron, M.R. Collins
Department of Civil & Environmental Engineering, University of New Hampshire, Durham, NH, USA

ABSTRACT

The intermittent slow sand filter (biosand filter) is a household-scale drinking water treatment system known for its simplicity and ease of use. This research examines several construction factors affecting the flowrate of the biosand filter, volume of water treated, and previous microbial loading to determine their effect on removal of microbes. It is recommended that more flow rate limiting modifications of the influent basket and the underdrain collection system be adopted, the volume of each charge of raw water applied to a biosand filter not exceed 70-80% of the filter's void volume, and that as much time as possible (>8 hrs) separate raw water charges of the biosand filter.

Keywords Slow sand filtration; biofiltration; fabric layers; pre-oxidation; process modelling

INTRODUCTION

The process of slow sand filtration (SSF) has been widely used in the field of water treatment for over 100 years but while important features of the process have been studied in detail, many aspects of the process remain poorly understood. This is partly explained by the greater attention that has given to the more widely applied rapid filtration process, but also because of the inherent complexity of SSF. Much is known of the ecology and fundamental dynamics of the filter process in qualitative terms (eg. common biological species and their interactions), but a comprehensive *quantitative* description of the process remains to be established. The intermittent slow sand filter (ISSF or biosand filter) is a household-scale drinking water treatment system known for its simplicity and ease of use in rural and impoverished areas. It is a modified version of the slow sand filter, a system that is known for its ease of operation (MWH, 2005). The biosand filter was is primarily intended to remove pathogens from drinking water, although modifications can be made to allow it to remove arsenic as well (CAWST, 2008). By reducing the numbers of pathogens in drinking water, biosand filters have demonstrated a significant ability to reduce the risk of disease (Stauber et al, 2008). Their ease of construction and simplicity of operation make them more sustainable than most competing systems (Sobsy et al, 2008).

Figure 45.1 shows the components of the biosand filter. To use a biosand filter, one pours raw water into the reservoir above the diffuser plate and allows gravity to move the water through the filter. The diffuser plate has a number of small holes that slow the flow of the influent water and prevent it from disturbing the biological layer (also known as the schmutzdecke) in the top few centimeters of sand. It is believed that most removals of microbes and particles occur in the upper biological layer (CAWST, 2008; Unger and Collins, 2008) and filter performance improves over time as the filter ripens (Ahammed, 2011; Stauber, 2007). The influent water displaces the water in the filter, which is collected by a drain system at the bottom of the filter and removed via a vertical tube whose outlet is a few centimeters above the surface of the sand. This allows a standing layer of water to remain above the top of the sand, which keeps the media saturated, prevents it from drying out, and preserves the microbial community. After a charge of water has been put into the filter, the filter must remain idle for some time, usually 6 to 12 hours. This allows the microbes to complete the degradation of contaminants (CAWST, 2008).

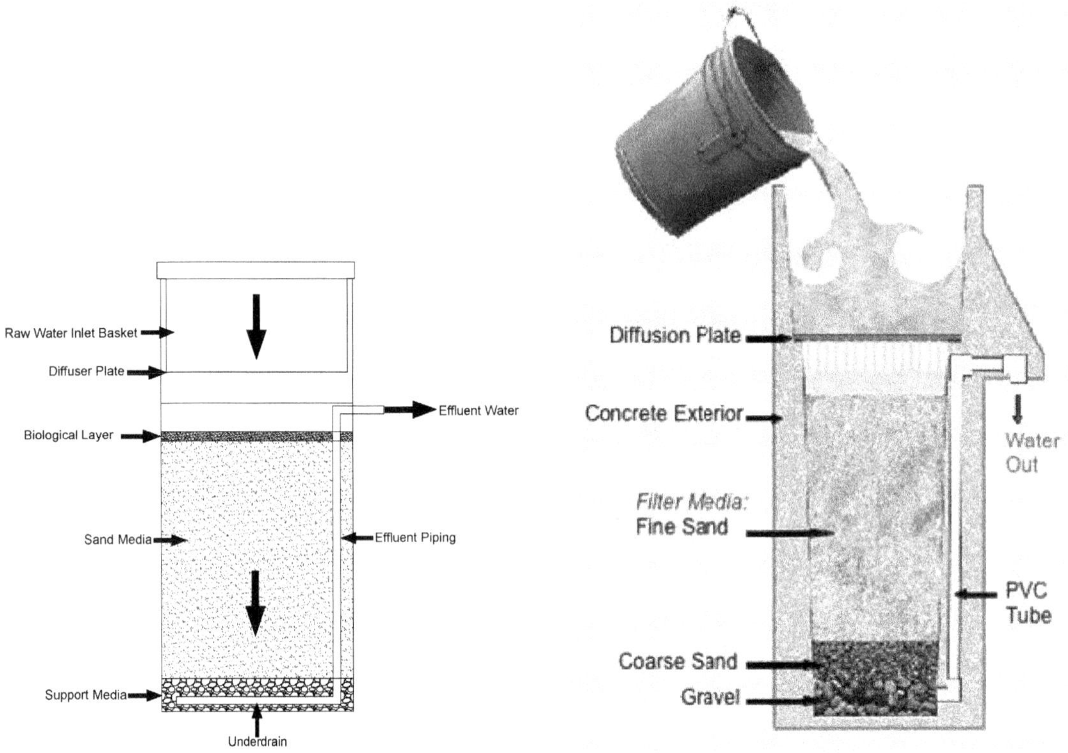

Figure 45.1 Diagrams of an ISSF or biosand filter (CAWST).

A few modified versions of the biosand filter exist. Researchers at the Massachusetts Institute of Technology have added a layer of sand above the diffuser plate, in order to provide a second biolayer (Collin, 2009; Kikkawa, 2008). The TivaWater system, which is being marketed in some African countries, is a lighter and more compact version of the biosand filter that has an integrated storage unit for the filtered water (http://tivawater.com/).

Microbial removals by ISSFs are dependent on several operational conditions to include (i) dosing filters with water quantities less than the void volume, and (ii) utilizing significant resting periods (>6-8 hrs) between dosing. But typical SSFs have an operating range of approximately 0.12 m/hr - 0.30 m/hr hydraulic loading rate (HLR). Since flow rate is one of the master variables in the operation of an SSF, it is important to monitor and control flow rates in ISSF or biosand filter operation to optimize microbial removals too. The principle question this research tried to answer is how does one effectively engineer acceptable flow rates through an ISSF or biosand filter?

METHODOLOGY

The experimental approach taken was to construct a series of 25.4 cm diameter x 61 cm high PVC ISSFs and then assess various design variables including (1) influent flow control devices/baskets, (2) underdrain collector systems, and (3) sand media size distributions. The sands used were considered coarse or fine. The values for the coarse sand were: (Effective Size) d_{10}=0.55 mm and (Uniformity Coefficient) UC=1.75. Analysis of the fine sand revealed d_{10}=0.16 mm and UC=2.13. The underdrains consisted of a conventional single opening (12.7 mm) covered in small pea gravel to prevent plugging by the filter media and one in the shape of a lower-case "t" with one 1.59 mm (0.16 cm) openings per arm for a total of 4 openings. The two influent control schemes consisted of a single influent reservoir basket with either a 2, 3 or 6-1.59 mm drilled openings or two baskets stacked on top of each other with 2 or 3 openings in the 1st basket and 2 offset openings in the 2nd basket. These baskets were sized to hold approximately 7 liters, which was equivalent to 70-75% of the void volume of the ISSF. These reservoirs were charged with solutions containing MS-2 bacteriophage and the bacteria *E. coli*, and the effluents were sampled and monitored every 1-2 liters.

RESULTS AND DISCUSSION

Assessing Influent Flow Control. A series of hydraulic challenges using clean water were made on ISSFs using various influent control devices as depicted in Figure 45.2. Each ISSF, except for the influent flow control system, was filled with coarse sand and utilized an underdrain collection pipe with one 1.27 cm opening as noted in the CAWST biosand construction manual. Frequent hydraulic loading rate (HLR) determinations were made throughout each challenge. No attempt was made in this study to account for biological induced clogging.

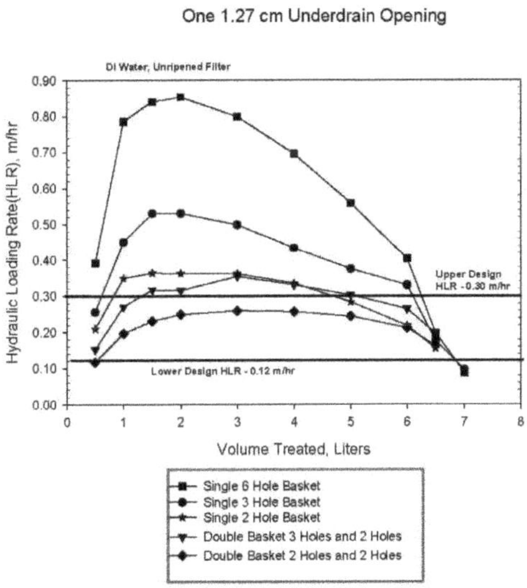

Figure 45.2 Summary of hydraulic loading rates for various inlet basket ISSF designs – coarse sand

A perusal of Figure 45.2 clearly suggest that without adequate flow control, ISSF flow rates can greatly exceed recommended design values by over 3-7x, especially during the initial phases of the charge, when using coarse sand and conventional underdrain collection. The most effective influent control devices were associated with the double basket design. The only influent control structure that was able to consistently achieve conventional SSF HLRs was the double basket configuration with only 2 small openings in each basket.

Assessing Effluent Flow Control. The same series of hydraulic challenges shown in Figure 45.2 were repeated on another set of ISSFs except for the underdrain collection device was changed from one 1.27 mm opening to a 't' with one 0.16 cm drilled opening per arm. The resulting challenge semi-instantaneous HLRs for this more restrictive underdrain collection devive is shown in Figure 45.3. Comparisons between Figures 45.2 and 45.3 highlight the significant influence of the underdrain collection device on ISSF flow rates especially when the total area of the opening(s) was reduced by a factor approaching 16. Not surprisingly, the more restrictive underdrain collection device did not result in a significant reduction in HLRs for the most restrictive influent control system since both flow control systems had an equal amount of perforated opening area.

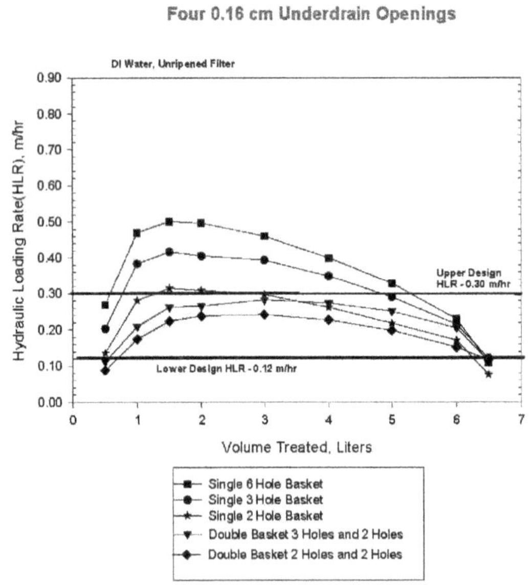

Figure 45.3 Summary of hydraulic loading rates for various inlet basket ISSF designs – coarse sand

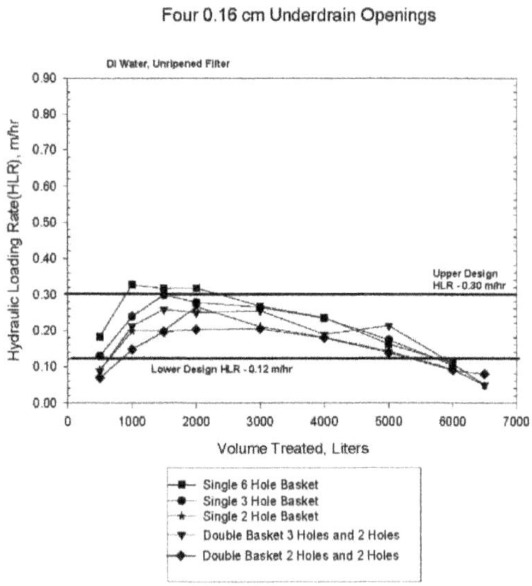

Figure 45.4 Summary of hydraulic loading rates for various inlet basket ISSF designs – fine sand

Assessing the sand media size. The same series of hydraulic challenges shown in Figure 45.3 were repested on another set of ISSFs except the sand media was changed to the finer sand. The resulting HLRs from the various hydraulic challenges on the finer sand media are shown in Figure 45.4. Comparisons between Figures 45.3 and 45.4 depicted the influence of sand size on ISSF flow rates since all other variables were held to be the same. Using the more restrictive underdrain collection system, the finer sand reduced practicably all HLRs into the acceptable SSF flow rate range independent of the influent flow control devices. The use of a very fine sand in ISSFs may be influential enough to make up for more poorly designed influent and effluent flow control systems.

Configuring the most flow constrictive ISSF. The cumulative influence of using multiple flow control systems on semi-instantaneous HLRs from consecutive hydraulic challenges using clean bed ISSFs are summarized in Figures 45.5 and 45.6. Clearly, using coarse sand in an ISSF will result in problematic

HLRs if not properly controlled by enhanced influent and effluent flow control systems. Based on the results of this study, the most flow restrictive for a 25.4 cm diameter ISSF configuration will consist in ascending order of a double influent basket with 2-0.16 cm openings in each basket, a underdrain collection pipe "t" with a 0.16 cm opening in each of the 4 laterals, and very fine sand media (0.16 mm effective size).

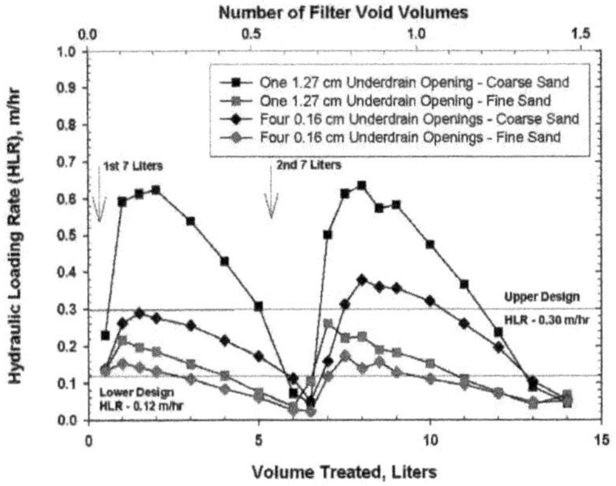

Figure 45.5 Influence of sand size and effluent control device w/ a single 6-Hole basket on HLR for MS2 and *E coli* challenges of ISSF

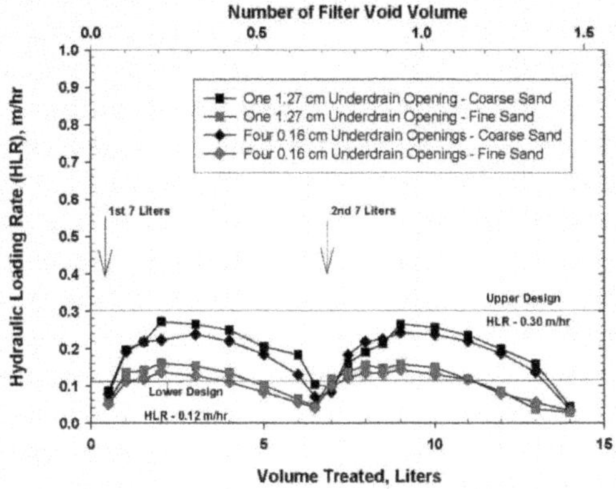

Figure 45.6 Influence of sand size and effluent control device w/ double influent basket with 2-Holes over 2-holes on HLR for MS2 and *E coli* challenges of ISSF

Influence of sand size and underdrain collection design on microbial removals. Four ISSFs were challenged with *E.coli* (feed 10^6 cfu/100ml) and MS2 bacteriophage (feed $4.3x10^4$ pfu/ml) in order to coordinate treatment performance to reductions in filter HLRs. Each ISSF used the double influent basket influent control design with 2 openings in each basket. The four ISSFs were equally divided into fine and coarse sand size and 2 underdrain collection devices with one containing a 1.27 cm opening and the other containing four 0.16 cm openings. The results from the *E coli* and MS2 consecutive challenges are depicted in Figures 45.7 and 45.8, respectively.

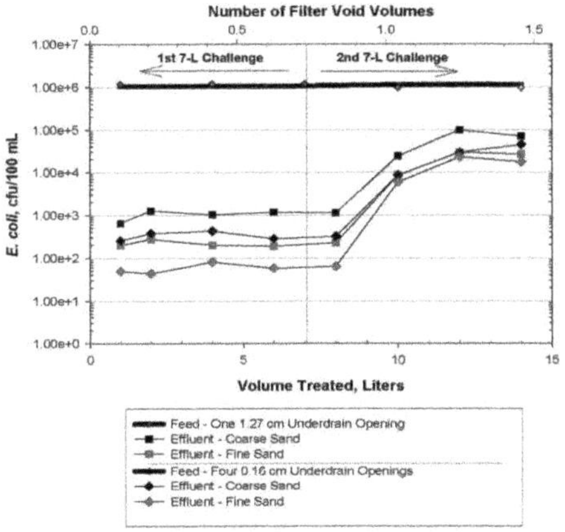

Figure 45.7 Influence of two consecutive 7-L challenges on *E. coli* removals as a function of sand size and effluent control device using a double 2-holes over 2-holes influent basket

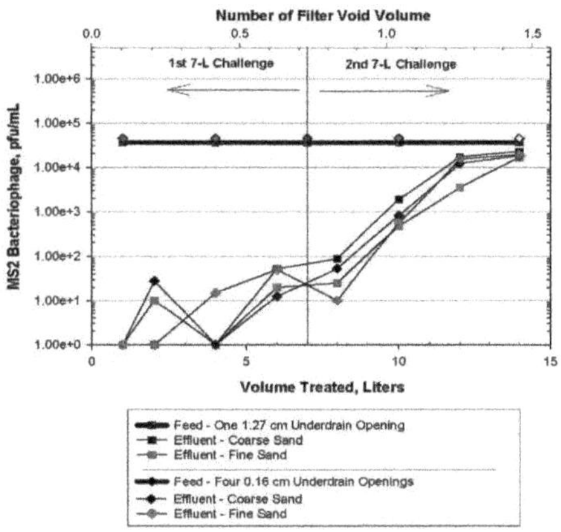

Figure 45.8 Influence of two consecutive 7-L challenges on MS2 bacteriophage removals as a function of sand size and effluent control device using a double 2-holes over 2-holes influent basket

E. coli removals as noted in Figure 45.7 were inversely related to HLR trends depicted in Figure 45.6. Over 4 log removals were observed for the ISSF with the fine sand and most restrictive effluent flow control device while 3 log removals were noted for the ISSF with the coarse sand and least restrictive effluent flow controller. These acceptable removals observed during the first challenge run were all achieved while the HLRs measured throughout the challenge run were within the acceptable SSF HLR range of 0.12-0.30 m/hr. Removals through the second challenge run applied less than 1-hr after the conclusion of the first challenge run were reduced to less than 2 log removals indicating the importance of resting the ISSFs sufficiently between filtered water production runs. Again, the ISSF with the fine sand and most restrictive effluent flow control device achieved higher removals even during this second challenge run.

No discernable removal trend of MS2 bacteriophage as shown in Figure 45.8 were made. Over 2 log removals were observed during the first challenge run but less than 0.5 log removals were observed toward the end of the second 7-liter challenge run. Virsus appeared to be a problematic microorganism for ISSF to remove on a consistent basis. The importance of resting ISSFs between water production runs and charging the ISSFs with less than the filter void volume can not be over emphasized (Elliott et al., 2008).

CONCLUSIONS AND RECOMMENDATIONS

Enhancing microbial removals from ISSFs or biosand filters generally involves the management of 2 major operational components and the implementation of a major design component. The two most influential operational components include (1) the charging of the ISSF with water less than the void volume of the sand media and (2) resting the ISSFs as much as possible between ISSF charges. The major design component that is sometimes overlooked in the construction of ISSFs or biosand filters is the control of the hydraulic loading rate (HLR), which also has a direct bearing on the residence time or empty-bed-contact-time (EBCT) of the source water in the filter sand media.

The focus of this study was to explore various influent and effluent flow control devices and the role sand size has on ISSF flow rates throughout the charging process. There has been a dependency in the ISSF literature for the sand media to be able to restrict the flows sufficiently for effective treatment. This study has shown that the larger sized sand media that also has low uniformity coefficients may be problematic for optimum microbial removals by ISSFS without the incorporation of other mechanisms to reduce the filter flow rates to desired SSF HLRs of 0.12-0.30 m/hr. Based on the results of this study, the most flow restrictive device that could be built into the construction of the ISSF will consist of a double influent basket with 2-0.16 cm openings in each basket and a underdrain collection pipe "t" with a 0.16 cm opening in each of the 4 laterals for a 25.4 cm diameter filter.

It is recommended that a relationship between sand media specification (to include effective size and uniformity coefficient) and following-head flow rate be established to insure acceptable residence or empty-bed-contact times through a typically design ISSF to optimize microbial removals. The sand media specification may need to take the form of surface area per unit volume (m^2/m^3).

ACKNOWLEDGEMENTS

The authors acknowledge the contributions of the following colleagues to aspects of the research described in the paper: Peter Dwyer, Damon Burt and Sinor Ali. The authors also acknowledge the partial support of this project from the USEPA through the Technology Assistance Center network (TACnet). Mention of specific trade names herein does not imply endorsement on the part of the USEPA or the UNH.

REFERENCES

Ahammed, M. M. and K. Davra. (2011). "Performance evaluation of biosand filter modified with iron oxide-coated sand for household treatment of drinking water." Desalination.

CAWST-Centre for Affordable Water and Sanitation Technology. (2008). Biosand Filter Manual. Calgary, Alberta, Centre for Affordable Water and Sanitation Technology.

Collin, C. (2009). Biosand filtration of high turbidity water: modified filter design and safe filtrate storage. Dept. of Civil and Environmental Engineering. Cambridge, MA, Massachusetts Institute of Technology: 154.

Elliott, M. A.; C. E. Stauber; F. Koksal; F. A. DiGiano; M. D. Sobsey. (2008). "Reductions of *E. coli*, echovirus type 12 and bacteriophages in an intermittently operated household-scale slow sand filter." Water Research **42**: 2662-2670.

Kikkawa, I. (2008). Modification of a biosand filter in the northern region of Ghana. Dept. of Civil and Environmental Engineering. Cambridge, MA, Massachusetts Institute of Technology: 128.

MWH (2005). Water Treatment: Principles and Design. 2nd Edition. Hoboken, NJ, John Wiley & Sons, Inc.

Sobsey, M. D.; C. E. Stauber; L. M. Casanova; J. M. Brown; M. A. Elliott. (2008). "Point of Use Household Drinking Water Filtration: A Practical, Effective Solution for Providing Sustained Access of Safe Drinking Water in the Developing World." Environmental Science & Technology **42**: 4261-4267.

Stauber, C. E. (2007). The Microbiological and Health Impact of the Biosand Filter in the Dominican Republic: A Randomized Controlled Trial in Bonao. Department of Environmental Sciences and Engineering. Chapel Hill, NC, University of North Carolina at Chapel Hill: 214.

Stauber, C. E.; G. M. Ortiz; K. R. Liang; B. A. Aiken; M. D. Sobsey. (2008). "The Biosand Filter: A Robust and Sustainable Technology for Household Water Treatment in Developing Countries." Epidemiology **19**(6): S296.

Unger, M. and M. R. Collins. (2008). "Assessing *Escherichia coli* removals in the schmutzdecke of slow-rate biofilters." Journal AWWA **100**(12): 60-73.

Chapter 46

Utilizing structural equation modeling as an evaluation tool for critical parameters of the biosand filter in a pilot study in Para, Brazil

L. E. Voth-Gaeddert, D. W. Divelbiss**, D. B. Oerther**
**Department of Civil, Architectural, and Environmental Engineering, Missouri University of Science and Technology, Rolla, Missouri 65409, United States*
***School of Energy, Environmental, Biological, and Medical Engineering, College of Engineering and Applied Science, University of Cincinnati, Ohio 45221, United States*

ABSTRACT

Biosand filters (BSF) have brought potable water to many communities throughout the world. A significant number of studies have shown the effectiveness of the filter if utilized properly. However, Divelbiss et al. 2013 investigated the BSF from a holistic view of the community. Results showed, while important, the BSF was not among the significant factors for health improvement. This pilot study, as part of a larger case study, investigated the issues behind these findings. Utilizing the SEM model that explained the complex system of interactions within the Guatemalan communities, several Brazilian communities along the Amazon River were studied. While further research is ongoing, it was found that the BSF, again while important, was not the largest factor in the improvement of health within the household. Socio-economic status and improved water source had the largest positive impact on health improvement. It was also found that additional water treatment had the highest negative effect of the usage of the BSF.

Keywords Household Bio-filters, Cleaning, Operation, Economics, Structural Equation Modeling

INTRODUCTION

The biosand filter (BSF) has been utilized within developing regions for over 200 years (CAWST 2012). A strong body of literature has been established on the effectiveness of the biosand filter, in particular the BSF produced by Centre for Affordable Water and Sanitation Technology; CAWST (Stauber, C. et al. 2009; Tiwari, S. et al. 2009; Aiken et al. 2011; Vanderzwaag et al. 2009). Many of these studies show that in a controlled setting, whether in a laboratory or in the field, a removal of at least 1 log is achievable for a majority of harmful contaminants within water. However, this removal is highly dependent on the users understanding of operation and maintenance of the BSF. The BSF can be found in over 60 different countries within 400,000 different households (CAWST 2012). Over this spectrum of communities that are found within these many countries, numerous cultural understandings, education levels, economic statuses, health issues, and other societal issues can be found. Divelbiss et al. 2013 studied one particular set of communities located in northern Guatemala. The studies focus was on the particular intervention, distributing BSFs, and the impact this had on the 'multidimensional' issue of poverty. The primary hypothesis of the BSF having the largest effect on improving community member's health, was not found. Divelbiss reports, "the community is a complex system of interactions which directly and indirectly influence the health of its residents. Policy makers and development practitioners must recognize that single target interventions (e.g., improving water quality) have limited influence on the entire system (Divelbiss et al. 2013)." This study, being limited to northern Guatemala, warranted further investigation.

A secondary study site was selected in the region of Para, Brazil. This site allowed for significant cultural and locational differences. Within this case study, a feasibility and pilot study have been administered by Voth-Gaeddert and colleagues. Voth-Gaeddert et al. 2013 successfully demonstrated the adaptability of the Structural Equation Modeling (SEM) model that described the complex system of interactions within the Guatemalan communities to the Brazilian communities found along the Amazon River. In September of 2013 a pilot study was conducted amongst the communities in the Para region. This region was selected because a Christian organization known as Project Amazon (PAZ) distributed over 10,000 BSFs among rural villages within this area.

Divelbess used the statistical platform SEM to describe the complex system of interactions within the communities. Factors such as filter operation and maintenance (FOM) of the BSFs, household education level (HEL), socio-economic status (SES), diarrheal health burden (DHB), household sanitation, and water consumption habits could be assessed simultaneously. SEM can be either exploratory or confirmatory in its analysis approach and offers a unique combination of features. These include hidden or latent variables (ex. SES in Brazil could be represented by what material the house is made from, ownership of a boat and number of working light bulbs in the house), simultaneous regression analysis, non-summation of errors within aggregated indicators, and a robust graphical interface. Using SEM, assessments of interventions within communities are possible. The significance of understanding what factors have the largest effect within a community is crucial to a successful intervention approach.

METHODOLOGY

Two goals were specified within this pilot project, the first was to confirm what was found by Voth-Gaeddert et al. 2013 showing feasibility in the transition of a region specific survey and SEM model used in Guatemala and the application of these two tools to a region in Brazil. The second goal was to use the revised survey in the first iteration of the SEM model in an exploratory style of analysis. This survey would populate an SEM model that had been revised after the feasibility study performed by Voth-Gaeddert. However, utilizing SEM for exploratory style of analysis requires a full understanding of the techniques involved.

A devotion to a particular set of steps is crucial to producing statistically accurate SEM models. SEM has a broad range of abilities; this study focuses on a single particular ability. The reader is encouraged to study Grace 2006 or Kline 2005. In this particular usage of SEM further care has to be taken due to the use of exploratory techniques and a balance needs to be taken when working with both Exploratory and Confirmatory Factor Analysis (EFA and CFA, respectively) (Grace 2006 and Kline 2005). CFA is, as the name suggests, a confirmatory technique- it is theory driven. Therefore, the planning of the analysis is driven by the theoretical relationships among the observed (indicator) and unobserved (latent) variables. When CFA is used, one hypothesized model (outcome of Voth-Gaeddert et al. 2013) is being compared to a data driven (observed) model using the covariance matrices of both for analysis. A minimization of the difference between matrices is the goal of CFA and is tested through Chi-Square tests as well as others (Schreiber et al. 2006). CFA is used in SEM in the first step of a two-step process. A model of only the latent variables and their indicator variables is created to test for fit within the latent variables; this is called the measurement model (see Figure 46.1). If the measurement model fits according to the fit indices (see Hu, L. et al. 1998; Bentler, P.M. 2007) then the structural model can be assessed.

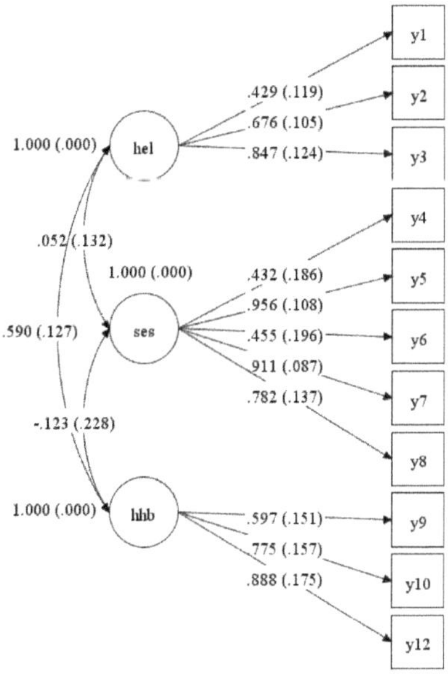

Figure 46.1 Measurement Model with parameter estimates

A structural model describes the relationships between the latent variables as well as other independent observables variables that are hypothesized to have effects within the model. Again, simultaneous regression analysis is used to represent these relationships by way of linear regression (for continuous variables) or log regression (for categorical variables). In Table 46.1 a list of all variables within the model can be found with the type of variables (continuous or categorical) and a brief definition.

EFA is outside the scope of this paper, however it can be an effective tool when relationships between factor and indicator variables are unknown (see Grace 2006).

Once a model has shown 'adequate fit' and a new set of data has been used to populate and confirm adequate fit, the path coefficients can be assessed. Looking at the unstandardized path coefficients, direct and indirect effects can be analyzed. The path coefficients can be interpreted as unit increases between variables. For example in Figure 46.2 a direct path from SES to HHB can be seen with a coefficient of -.115. This means that as SES increases by one unit (an increase in "wealth"), HHB will decrease by .115 units (improved health). However, it is stressed within the literature that causal assumptions are held for the reader to determine but can be discussed within the discussion section of a report (Kline 2005).

Table 46.1 List of variables in SEM model

Variable	Definition	Type
Y1	Reading and Writing Score	Ordinal
Y2	Paternal Education Level	Ordinal
Y3	Maternal Education Level	Ordinal
Y4	Improved Roof	Dictomous
Y5	Owns a TV	Dictomous
Y6	Owns a Shower	Dictomous
Y7	Has Multiple Working Lightbulbs	Dictomous
Y8	Improved Water Storage Material	Dictomous
Y9	Severity of Sickness	Ordinal
Y10	Occurrence of Sickness	Ordinal
Y11	Wait Period of Sickness and Doctor Visit	Ordinal
Y12	Children Under 11	Dictomous
Y13	Filter Operation and Maintenance	Ordinal
Y14	Improved Sanitation	Dictomous
Y15	Improved Water Source	Dictomous
Y16	Improved Water Treatment	Dictomous
Y17	Improved Water Storage	Dictomous

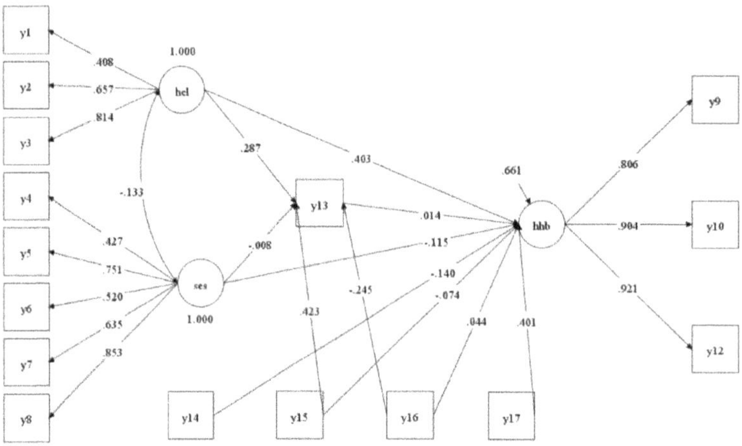

Figure 46.2 Full Model with standardized parameter estimates

A significant part of this style of SEM is testing model fit. A large body of work is available for study on this subject as there are differing opinions on particular tests. However, generally, it is suggest to use Chi-Square (Bentler, P.M. 2007), root mean square error of approximation; RMSEA (Hu, L. et al. 1998), Comparative Fit Index; CFI (Hu, L. et al. 1998), and Tucker Lewis Index; TLI (Muthen & Muthen 2012). The software package MPLUS 7 was used for analysis of data (Muthen & Muthen 2012).

A continued awareness and analysis of cultural differences is crucial when working with multiple global locations. For this reason, changes to the survey and the SEM model from Voth-Gaeddert et al. 2013 were made as a deeper understanding of cultural influences were discovered. These changes occurred mainly within the observable indicator variables used to represent the latent variables in the model. All changes were made either before data collection began or at the time of transition from confirmatory analysis to exploratory analysis. Understanding impacts of BSFs in communities around the world warrants a continued devotion to cultural observations.

RESULTS AND DISCUSSION

The following changes were conducted due to cultural observations made while in the field. Variables are listed and are then followed by an explanation:

- HEL observable variable 'Reading and Writing' test was administered

  The previous SEM model only used two observable variables to describe HEL which in the literature is said to have a weak representation. Three or more observable variables are recommended to help explain the variance within the HEL factor. A reading and writing test does not allow the interviewee to hide illiteracy and helps accurately describe education of that household.
- SES observable variables of ownership of a TV, shower, working light bulbs, material of water storage container, and 'improved' roof, along with others were collected

  A wide spectrum of SES indicators can be found in the literature. This range includes using income level, type of job, net worth of physical objects found in the household, type of objects in household or a combination of these (Jones, E.C. et al 2011; Dressler, W.W. 1998). Further study is needed within this particular subject on strength of indicators. Physical objects found in the household were the main indicators for SES within this study.
- DHB was changed to Household Health Burden (HHB) which included severity and type of illness

  This transition from DHB to HHB was due to a lack of occurrence of diarrhea during the low season of the Amazon River. This could have been due to the location of latrines. During the rainy season, the Amazon River floods the latrines and carries the contents downstream past other villager's households. The top health issues were recorded for each household and these were used to rank the level of the adverse health burden found in the given household.
- Independent variables were changed; 'Soap & Towel Present' to 'Improved Storage' and definition of terms changed

While the importance of soap and towel within a household remains important, it was found that over 90% of the sample population had these items. This term was then incorporated into the 'Improved Sanitation' score partly defined by WHO (WHO 2006). After further investigation the variable 'Improved Storage' was added to test the hypothesis that if a household stores their water properly, the occurrence of diarrhea within the household will go down. Additionally definitions of 'Additional Water Treatment' and 'Adequate Sanitation' were changed to 'Improved Water Treatment' and 'Improved Sanitation' to be consistent with WHO terminology (WHO 2006).

From these changes a new theoretical model was created and data from 41 households were used to populate the altered hypothetical model. A measurement model was first run to assess the fit of observable indicator variables on their respective latent variables. Failures in model identification occurred due to poor explained variance within particular factors. After assessment of raw data and of MPLUS 7 output data (including parameter estimates, residuals, and variances) issues were highlighted and further literature reviews were done. As change in the model should not be data driven alone, alternative models were assessed based off of literature reviews (Guest, G. 2000; Undurraga, E.A. et al. 2010). The process then becomes exploratory and any further analysis should be done with caution, however it can still provide productive insight to the model.

The alternative model that had the best fit (χ^2= 54.5 , p>.077; RMSEA <.171; CFI = .92 ; TLI = .89) is shown in Figure 46.1. The model showed close to adequate fit (p>.05, *<.08*, .90, *.90*, respectively); however it needs to be noted that with a small sample size the fit indices are less accurate. The structural model could then be added to the measurement model to create the full model. Again, adjustments had to be made with relationships inside the model. The alternate model that showed the best fit still fell far short of the goodness of fit tests (χ^2= 54.5 , p>.0012; RMSEA <.01; CFI = .62 ; TLI = .52). However, even with a failed model, some important insights can be gained.

Even with an exploratory style of analysis observations can be gained through both the model failures as well as the final "best fit" model. These observations are;

- HEL needs the addition of indicator variables to improve representative strength, these could include reported reading habits, vocational classes taken or attendance of school for children
- SES was represented the best out of all factors, however, ownership of a shower and TV were problematic, so consideration of the usage of overall income and occupation could improve representation
- SES had a negative effect with HHB (improved health)
- HHB indicator relationship between 'severity of sickness' and 'children under 11' could have been an issue
- HHB may be changed to Activities of Daily Living (ADL) or International Classification of Functioning, Disability and Health (ICF) depending on findings in ongoing research
- FOM had negligible effect on HHB
- 'Improved Water Treatment' has highest negative effect on FOM (worsened use of filter)
- 'Improved Source' had highest negative effect with HHB (improved health the best)
- Poor definitions within the survey of 'Improved Source' may have led to some of the fit issues in the full model

It is important to provide discussion of these observations from a holistic view of the larger body of work between Guatemala and Brazil. Therefore it can be observed that HEL was not well represented within this study. This may have been the main issue in the lack of fit of the full model. The idea that an increase in education will indirectly affect the filters ability to improve overall health within the house was uncertain. Would an increase in education cause them to use alternative methods to filter water; such as using chlorine or a ceramic filter? An improved representation of HEL will allow for further analysis of this question in the next study.

SES was represented well; however, even with "good fit," proper representation (selecting the 'best' indicators) is the goal. SES was shown (in Fig. 46.2) to have the second largest negative effect on HHB and to also have a negative effect on FOM. This could suggest that as "wealth" increases within the household alternative methods are used to acquire clean drinking water; such as buying bottled water or chlorine. This would create a negative effect on FOM but possible improve the overall health within the house.

The "Improvement" variables within the model (Y14-Y17) were all regressed on HHB but only Y15 and Y16 of these four were regressed on FOM. Improved Water Treatment had the largest negative effect on FOM. This suggests that as other treatment methods are utilized within the house, the operation of the BSF in the house decreases. A reexamination of the raw data shows that 20% of household used chlorine and of those users of chlorine, 40% used it incorrectly. The main fallacy that was found was the usage of chlorine before filtration through the BSF. This would compromise the biofilm layer that is an integral part of the contaminant removal process. Improved Sanitation and Improved Source (of water) had a positive impact on health issues within the households.

Finally, FOM had a near negligible effect on HHB. This could have occurred for several reasons. The most likely is poor representation within the latent variable HHB or the cumulative score associated with FOM. Further investigation is needed to help shed light on this issue.

CONCLUSIONS

The second iteration within this study allowed for a significant step to identifying the correct SEM model to represent the environment within households of this region. As outlined in the Results and Discussion section, focus on strengthening representation within latent variables will significantly improve the scores to the goodness of fit tests. The survey used in data collection will be adjusted accordingly. As with all SEM models, a deeper understanding of the scenario that is being modeled will increase the ability of the model to properly represent it. SEM has good potential to serve as a platform for a more standardized approach in analyzing situations within impoverished areas, however further analysis is needed to show what can be accomplished through SEM.

REFERENCES

Aiken, B.A., C. E. Stauber , G. M. Ortiz , and M. D. Sobsey, (2011). An Assessment of Continued Use and Health Impact of the Concrete Biosand Filter in Bonao, Dominican Republic. American Journal of Tropical Medicine and Hygiene, 85(2):309–317

Bentler, P.M. (2007). On tests and indices for evaluating structural models. Personality and Individual Differences 42:825-829.

Centre for Affordable Water and Sanitation Technology (2012). Biosand Filter: History of the Biosand Filter. Cawst.org/en/resources/biosand-filter

Divelbiss, D. et al. (2013). Environmental Health and Household Demographics Impacting Biosand Filter Maintenance and Diarrhea in Guatemala: An Application of Structural Equation Modeling. Environmental Science and Technology 47(3): 1638-1645.

Dressler, W.W., Balieiro, M.C., Santos, E.D. (1998). Culture, Socioeconomic Status, and Physical and Mental Health in Brazil. Medical Anthropology Quarterly. 12(4):424-446

Grace, J.B. (2006). Structural Equation Modeling and Natural Systems

Guest, G. (2000). Using Guttman Scaling to Rank Wealth: Intergrating Quantitative and Qualitative Data. Field Methods. 12(4):346-357.

Hu, L., Bentler, P.M. (1998). Fit Indices in Covariance Structure Modeling: Sensitivity to Underparameterized Model Misspecification. American Psychological Association. (3)4:424-453

Jones, E.C. et al. (2011). Inequality, Socioeconomic Status, and Social Support in Post-Disaster Mental Health in Mexico. Human Organization. 70(1).

Kline, R.B. (2005). Principles and Practice of Structural Equation Modeling.

Schreiber, J.B. et al. (2006). Reporting Structural Equation Modeling and Confirmatory Factor Analysis Results: A Review. The Journal of Educational Research. Heldref Publications.

Muthen, L.K., Muthen, B.O. (1998-2012). MPLUS User's Guide. Seventh Edition. Los Angeles, CA: Muthen & Muthen

Stauber, C.; Ortiz, G.; Loomis, D.; Sobsey, M. A randomized controlled trial of the concrete Biosand filter and its impact on diarrheal disease in Bonao, Dominican Republic. The American Journal of Tropical Medicine and Hygiene 2009, 80 (2), 286–293.

Tiwari, S.; Schmidt, W.; Darby, J.; Kariuki, Z.; Jenkins, M. Intermittent slow sand filtration for preventing diarrhoea among

children in Kenyan households using unimproved water sources: Randomized controlled trial. Tropical Medicine & International Health 2009, 14 (11), 1374–1382.

Vanderzwwaag, J.C., Atwater, J.W., Bartlett, K.H., Baker, D. (2009). Field Evaluation of Long-Term Performance and Use of Biosand Filters in Posoltega, Nicaragua. Water Quality Research Journal of Canada. 44(2):111-121.

Voth-Gaeddert, L.E., Divelbiss, D., Oerther, D.B. (2013). Utilizing Structural Equation Modeling to Correlate Biosand Filter Performance and Occurrence of Diarrhea in Enseado Do Aritapera, Para, Brazil. International Water Association Brazil.

Undurraga, E.A. et al. (2010). Individual Wealth Rank, Community Wealth Inequality, and Self-Reported Adult Poor Health. Medical Anthropology Quarterly. 24(4):522-548.

WHO (2006). Core Questions on Drinking-Water and Sanitation for Household Surveys. United Nations Children's Fund.

Chapter 47
Household filter to remove iron and nitrate for drinking water

Visi Asriningtyas, Luis Villacreces**, Sri Puji Saraswati****
**IWA, Environmental Science Graduate Freelancer, Indonesia*
***Environmental Laboratory, EP Petroecuador, Ecuador*
****Environmental and Sanitation Laboratory, Civil and Environmental Engineering Department, UGM, Jl. Grafika no.2, Yogyakarta, Indonesia*

ABSTRACT
The research of household filtration took place in a dwelling area of Minomartani, Sleman Regency, Special District of Yogyakarta (DIY), Indonesia which has iron excess in groundwater used for drinking water. Nitrate is also prevalent especially in regions where family husbandries surround the wells for drinking water in the dwelling areas. Therefore there are research in this area to overcome this problem for a healthier drinking water to alleviate groundwater quality for human consumption. High Iron is prevalent in Sleman groundwater, DIY (Special District of Yogyakarta), Indonesia, as the geologic materials are olivine, basalt, augite-hypersthene-hornblende andesite at the northern part and augite-hypersthene andesite with hornblende subordinate at the most part of the area. The aquifer in general is also compared with thus of Ecuador.

The experiments using charcoal and sand as the main material for removal of iron and nitrate in drinking water for small communities in Sleman, DIY, Indonesia has been conducted in year 2005 to 2006. Afterwards, the filter is elaborated for more practical use for drinking water which contains Iron and also nitrate from wells. The results of the use of these filters are fine, proved from a house in Minomartani, Sleman, Indonesia, with no more Iron and nitrate excess. Nitrate contains becomes less than 1 mg/l and total Iron does not exceed 0.3 mg/l. The color of water becomes neutral and odorless, if maintained at least once a month by brushing and washing the charcoal (activated carbon) filter and the polypropylene or Cartridge Carbon Filter.

Keywords Iron removal; nitrate removal; groundwater quality; household filter

INTRODUCTION
Sleman regency in DIY (*Daerah Istimewa Yogyakarta* or Special District of Yogyakarta), Indonesia is an area which has excess in Iron in the natural groundwater. The area of research lies between 7°37'30" – 7°47'30" S and 110°24'00" –110°26'30" E. Annual average rainfall during ten years is around 2106 mm to 2526 mm, while the annual average temperature is around 23.34°C to 27.7°C. Groundwater

temperature is 26°C at the time of measurement in February 2013. Merapi Aquifer, Special Province of Yogyakarta (DIY), Indonesia, is one of the best source of potable water in DIY (Asriningtyas and Putra, 2006). Sudarmadji (1994) also stated that the south slope of Merapi active mountain is functioned as an aquifer system, which is the Sleman-Yogyakarta Formation.

Sutikno (1996) mentioned that there are three spring belts and three groundwater regimes viewed by geomorphological and geological characteristics of Merapi volcano landscape. The upper part is the recharge area, the middle part is the storage area, and the lower part is the discharge area.

Compared to Quito aquifer in Ecuador, Merapi Aquifer in Sleman, Indonesia has greater quantity of water with discharge per well of around 30 l/s, yet higher in Iron contain in some part of Sleman area (more than 0.50 mg/l) in water from deep wells. In Ecuador, the total discharge from 30 groundwater operating wells is around 600 l/s (thus around 20 l/s per well) in the Andean regions, which approximately 250 thousands of its inhabitants rely on groundwater for drinking water. This region has average nitrate contain of 3.0 mg/l (Reyno, 2013). Iron is found in an average of 0.43 mg/l and 1.9 mg/l for nitrate in the region. (Narváez, 2010). Very few wells have been drilled in the region of the East due to abundant supply of surface water (Buckalew, *et al.*, 1998). Values of 0.58 mg/l for Iron and 6.5 mg/l for nitrate (Posligua, *et al.*, 2013) are reported in Muisne – Esmeraldas. The example of the drilled well in Riobamba, Ecuador is shown in Figure 47.1 (in the farmost right).

METHODOLOGY

Previous studies have been conducted in year 2005 to 2006 in the research area of Sembung subwatershed, Merapi aquifer, Sleman, Indonesia. There have been 8 dug wells examined for the water quality, including the Iron and nitrate contains. PDAM (*Perusahaan Daerah Air Minum* or the Government Regional Drinking Water Industry) in Merapi aquifer has also examined the Iron and Manganese contain, and has analysed the database together with the Environmental and Sanitation Laboratory, Civil and Environmental Engineering Department, UGM in year 1999 to year 2000. After year 2001, the database were collected and simulations have been done with GMS (Groundwater Modeling System) until 2003 (Asriningtyas and Putra, 2006). The deep well Iron contain was high (some exceeded 0.5 mg/l), and was pre-treated in a drinking water plant in Bedog area, Sleman, with almost the same filtering as the small household filter in this paper, except for the zeolith component which is not used in the household filter. The single data as a sample used in a well in Minomartani was measured in Hydrology and Water Quality Laboratory, UGM, Indonesia for 2 times of measurements, with Standard Method procedure for this paper. This data showed the elimination of Iron contain and reduction of nitrate contain after passing the household filter. The data was taken in February 2013.

Figure 47.1 From left to right: The household filter experiment in 2005-2006, in Sleman; The household filter installed in Sleman, in 2013; a well in Riobamba.

In some other parts of the world, other methode of measurements and analysis have been conducted, as in Bangladesh with Iron concentration measured using a field-based colorimetric kit, then mapped in the area (Merill, *et al.*, 2010). In this short paper, the measurement method has been described in the above paragraph, using accurate Standard Method in the Laboratory, yet in the future research, a kit from EPA (Environmental Protection Agency) for immediate measurement in the field is being purposed for a comparison. The household filter used in Minomartani, Sleman is fabricated by Penguin, Ltd., using mostly polypropylene or Cartridge Carbon Filter. Some have been modified locally to accommodate the elimination of Iron contain, with charcoal and sand paste. The old device is shown in Figure 47.1 on the farmost left, while the fabricated device is shown in Figure 47.1 (in the middle). The result will be discussed in the next section, and further purposed analysis is the behavior of the inlet and the outlet of water passing through the device.

RESULTS AND DISCUSSION

The result in this short paper is the Iron and nitrate reduced by passing water through the household filter, which are nitrate contain become less than 1 mg/l and total Iron contain do not exceed 0.3 mg/l. More complete figures and tables can be seen in the poster section depicted apart from this short paper.

In the previous research in Sleman, Iron contain for drilled wells are some higher than 0.5 mg/l while in dug wells they range only from 0.06 mg/l to 0.2 mg/l. Nitrate contain in drilled wells are not detectable while in dug wells they are mostly high in concentration, ranging from 2.8 to 20.6 mg/l. These high nitrate contains need filtration, and treatment technologies has been made throughout the world, for example The BioSand Filter (Murphy, *et al.*, 2010) and Hom ceramic filter (Omedi, *et al.*, 2010). In this research in Sleman, Indonesia, charcoal-sand-paste and polypropylene household filter are used. Next, more systematic government and public involvement is needed, such as in some other parts of the world (Franks, *et al.*, 2013; Swan and Cooper, 2013) for a better public facility for safe drinking water. Detailed data for long research and simulations will be a further benefit when funds are available (Aubert, *et al.*, 2013; Fleischmann, *et al.*, 2002), while in developing countries where database are mostly scarce, some estimation can be used in some simulations prior to available more complete database (Langergreber, *et al.*, 2008; Asriningtyas, *et al.*, 2005). Results of simulations in the aquifer (Asriningtyas and Putra 2006) showed that there is an urbanization problem to the north (Sleman area); multidisciplinary study, such as in Environmental Engineering program, besides physical and human geography is needed. Better onsite observations and external sources such as the climatic parameters and also population data at certain location and landuse type at certain location are needed, both over a certain time frame together with the water quality data (Iron and nitrate) of the area, to understand better the behavior of Iron and nitrate reduction, for a safer drinking water.

CONCLUSIONS

The conclusions of this short paper are:

(1) Water from shallow wells (dug wells) are usually having quite low Iron contain with high nitrate contain in Sembung subwatershed Sleman area, especially in the dense rural area, where agriculture, family husbandries and dwellings are the most landuse types.

(2) Water from deep wells (drilled wells of more than 15 m in the multilayer Merapi aquifer, Sleman) or wells commonly drilled by the government regional drinking water industry (PDAM: *Perusahaan Daerah Air Minum*) usually have high Iron contain.

(3) The household filter used in this paper, which is easily installed for small community use in each dwelling from its water tower, is good for reducing both Iron and nitrate contain in water, with regular maintenance at least once a month for the sample well in Minomartani, Sleman.

ACKNOWLEDGEMENTS

The authors wish to acknowledge Faisal Buchari for some of the unpublished database, Penguin Ltd. for providing simple household filter for public use, and the team of infield filter maintenance: Munawir and Abdul Aziz.

REFERENCES

Asriningtyas, V., and Putra, D.P.E. (2006) Ten Year Groundwater Simulation in Merapi Aquifer, Sleman, DIY, Indonesia. *Indonesian Journal of Geography*, **38**(1), 1-14.

Asriningtyas, V., Meesukko, C., Gajaseni, N., and Voinov, A. (2005) Seasonal Water Flow Trends in Conjunction with Phytoplankton Biovolume in Petchburi River, Thailand. *J. Sci. Res. Chula Univ.*, **30**(2), 161-178.

Aubert, A. H., Gascuel-Odoux, C., Gruau, G., Akkal, N., Faucheux, M., Fauvell, Y., Grimaldi, C., Hamon, Y., Jaffr´ezic, A., Lecoz-Boutnik, M., Mol´enat, J., Petitjean, P., Ruiz, L., and Merot, P. (2013) Solute transport dynamics in small, shallow groundwater-dominated agricultural catchments: insights from a high-frequency, multisolute 10 yr-long monitoring study. *Hydrol. Earth Syst. Sci.*, **17**, 1379–1391. www.hydrol-earth-syst-sci.net/17/1379/2013/ doi:10.5194/hess-17-1379-2013.

Buckalew, J., James, M., Scott, L., and Reed, P. (1998) Evaluation of Water Resources of Ecuador, Corps of Engineers of the United States of America, Mobile District and Engineering Center Surveying, U.S.A.

Fleischmann, N., Staubmann, K., and Langergraber, G. (2002) Management of sensible water uses with real-time measurements. *Water Science and Technology*, **46**(3), 33–40.

Franks, T., Cleaver, F., Maganga, F., and Hall, K. (2013) Water governance and livelihoods: Outcomes for smallholders on the Usangu plains, Tanzania. *Water Resources and Rural Development*, **1**(2), 1–16.

Langergraber, G., Alex, J., Weissenbacher, N., Woerner, D., Ahnert, M., Frehmann,T., Halft, N., Hobus, I., Plattes, M., Spering , V., and Winkler, S. (2008) Generation of diurnal variation for influent data for dynamic simulation. *Water Science and Technology*, **57**(9), 1483-1486.

Merrill, R.D., Labrique, A.B., Shamim, A. A., Schulze, K., Christian, P., Merrill, R.K., and West K. P. Jr. (2010) Elevated and variable groundwater iron in rural northwestern Bangladesh. *Journal of Water and Health*, **8**(4), 818-825.

Murphy, H. M., McBean, E. A., and Farahbakhsh, K. (2010) Nitrification, denitrification and ammonification in point-of-use biosand filters in rural Cambodia. *Journal of Water and Health*, **8**(4), 803-817.

Narvaez , M. (2010) Design of a water treatment system from groundwater for plant "Los Alamos" Francisco de Orellana. Master's Thesis, School of Chemical Engineering, Faculty of Science, ESPOCH, Riobamba, Ecuador.

Omedi, M. J., and Kipkorir, E. C. (2010) 'Hom': a simple point of use water treatment device. *Journal of Water and Health*, **8**(4), 741-750.

Posligua, P., Gonzalez, A., Delgado, M., Proaño, J., and Banchón, C. (2013) Study of a filtration system with calcium carbonate and natural charcoal for water purification Muisne, Esmeraldas, Faculty Engineering and Agricultural Sciences, University of the Americas (UDLA), Quito, Ecuador.

Reyno, S. (2013) Study of groundwater in the city of Riobamba and their zones of influence : Baseline. Master's Thesis, School of Chemical Engineering, Faculty of Science, ESPOCH, Riobamba, Ecuador.

Sudarmadji. (1994) Some Notes on Groundwater as a Domestic Water Supply of the Yogyakarta Municipality. *Indonesian Journal of Geography*, **26**(68), 1-10.

Sutikno. (1996) Geomorphology of Yogyakarta Area and Its Surrounding Proposed as a Geomorphological Field Laboratory. *Indonesian Journal of Geography*, **28**(71), 1-10.

Swan, A., and Cooper, N. (2013) Innovative funding methods for rural communities and their water pumps. *Water resources and rural development*, **1**(2), 17–26.

Chapter 48

Long-term performance evaluation of biosand filters modified with zero-valent iron

M. Mansoor Ahammed, Chaitali Solanki, Abhilash T. Nair
Civil Engineering Department, SV National Institute of Technology, Surat – 395007, India

ABSTRACT

Performance of two full-scale biosand filters (BSF) modified by introducing zero-valent iron (ZVI) in the form of iron nails and iron filing scrap was compared with that of BSF in a long-duration study spanning three months employing *Eschericia coli* and turbidity spiked water. Results of the tests showed that *E. coli* removal in both the modified filters was significantly better compared to that in BSF. Bacterial removal was always higher in modified filters by at least one-log unit. The *E. coli* removal continuously improved in all the three filters, and it reached a maximum value of about 4 log units in both the modified filters while the maximum removal observed in BSF was around 2.5 log units. All the filters showed good turbidity removal after maturation, and effluent turbidity was around 1 NTU when influent value was 20-25 NTU. Passage of water through the filters resulted in sharp decline in dissolved oxygen. The study thus indicated the potential of ZVI in improving the performance of biosand filters.

Keywords Biosand filter; drinking water quality; household water treatment; microbial removal

INTRODUCTION

Biosand filtration is one of the most promising and accessible technologies for household water treatment (Elliott *et al*., 2011; Jenkins *et al*., 2011). The biosand filter (BSF) is an intermittently operated slow sand filter (SSF) at small scale. It is reported that, as of 2009, over 200,000 BSFs have been implemented in over 70 countries (CAWST, 2009). Sobsey *et al*. (2008) based on performance and sustainability criteria, identified biosand filters as the most effective method among the five different household treatment technologies, and as having the greatest potential to become widely used and sustainable for improving household water quality to reduce waterborne disease and death. Studies reported show BSF can achieve total coliform removal in the range 58.3–99.7% (Baumgartner *et al.*, 2007); *E. coli* removal in the range 0–99.99% (Murphy *et al.,* 2010); greater than 5 $\log_{10}$ removal for *Giardia lamblia* cysts; 99.98% removal for *Cryptosporidium* oocysts (Palmateer *et al.,* 1999); 0 to 0.75 $\log_{10}$ removals of MS2 and PRD-1 bacteriophages; and 1.14 $\log_{10}$ echovirus 12 removal (Elliott *et al.,*

2008). However, reported studies suggest that BSF has its limitations especially in terms of its microbial removal efficiency during the ripening period which typically take 4-6 weeks. Cleaning of the filter also impairs the performance of the filter which persists up to 4 weeks. These findings raise questions about the assurance of safe water provided to users of BSF before ripening and after cleaning (Elliott *et al.*, 2008), and indicates the need for modification of BSF in order to improve the performance of BSF.

Zero-valent iron (ZVI), in the form of either granular or nanoscale particles, has been applied to remediate contaminated groundwater since the mid-1990s. Granular ZVI, typically in the form of scrap iron filings, is commonly used as a reactive material to remove groundwater contaminants. ZVI have been used to remove a broad range of organic and inorganic contaminants including Freons, metals (e.g., Cr(VI), arsenic, and cadmium), radionuclides (e.g., U(VI)), and nutrients (e.g., nitrate) (Chiu, 2013).

It has recently been shown that use of zero-valent iron improves microbial removal capacity (You *et al.*, 2005; Chiew *et al.*, 2009; Tellen *et al.*, 2010; Bradley *et al.*, 2011; Shi *et al.*, 2012). Noubactep *et al.* (2012) suggested the use of metallic iron (Fe^0) to improve the performance of BSF, and conceptually proposed an additional fine sand filter to improve the capture of dissolved iron. No long-duration studies have been reported on the performance of ZVI-modified full-scale BSF at different operating conditions with real source waters found in developing countries. In the present study, performance of full-scale modified BSFs was compared with a BSF in a long-duration study spanning several months during which bacterial removal, turbidity removal, flow rate, physico-chemical characteristics and other relevant parameters were monitored.

METHODOLOGY

The filters used in this study were designed based on the guidelines of CAWST (2010). Plastic containers brought from the local market were used in the study (Figure 48.1). Three biosand filters were used in this study - one conventional BSF (BSF) and two BSFs modified with zero-valent iron (MBSF). The depths of the different layers of media used in the filters are given Table 48.1. Locally available river sand passing through 1.18 mm and retained on 0.150 mm sieve was used. The sand was washed several times using tap water till the wash water became clean. Two types of Zero-valent iron were used in the study. One filter was amended with mild steel nails (20 mm long, 2 mm diameter) obtained from the local market. The other filter was modified by adding iron filing scrap obtained from the workshop of S V National Institute of Technology, Surat, India, where iron materials are made smooth. 7.5 kg of ZVI were used in each filter. The iron nails and scrap were washed several times before putting into the filter. It was mixed uniformly throughout the top 15 cm of the filter media, excluding a 5-cm layer of fine sand as given in Table 48.1. The three filters had a pore volume of 19.1–19.3 L.

Table 48.1 Details of the filter media

Size of media (mm)	Volume of media required (L)	Depth of layer in BSF (cm)	Depth of layer in modified BSF (cm)
4.75-12 (Gravel)	7	5	5
1.18-4.75 (coarse sand)	7	5	5
0.15-1.18 (fine sand)	53	40	20+15+5*

* 20 cm fine sand layer + 15 cm fine sand–ZVI mixture + 5 cm fine sand layer

The three filters, one BSF and two ZVI-modified BSFs (MBSFs), were run in identical conditions for about 3 months. Tap water spiked with *E.coli* to an approximate initial concentration of 10^6 CFU/mL and with turbidity to approximately 25 NTU in the form of natural clay, was used. Sterilised sewage at the rate of 1 mL/10 L of water was also added in order to increase the organic content of the influent. It was thought that addition of sewage would simulate the presence of wastewater in typical drinking water sources of developing countries and accelerate the ripening process (Elliott *et al.*, 2008). A daily charge of 20 L water was chosen based on the typical requirement of a developing country household. Also, it represents about 1 pore volume of the filters. The mean characteristics of the influent during the filter operations were : turbidity 21±3 NTU; pH 8.09 ± 0.18; temperature 28.3 ± 2.3°C; dissolved oxygen 6 ± 0.7 mg/L; *E. coli* $1.16 \times 10^6 \pm 2.34 \times 10^5$ CFU/mL.

Figure 48.1 Photograph of the biosand filter

Turbidity and flow rate were analysed daily while samples for microbial analysis was taken at five days interval. The flow rate reported is the initial flow rate based on the volume of effluent collected during the first minute following the filter charging. For the other parameters the entire effluent for a day was collected and analysed. Enumeration of viable *E. coli* was conducted by plate count test using EMB agar and the results are expressed as CFU/mL. Physico-chemical parameters such as alkalinity, hardness, iron, nitrates, dissolved oxygen and turbidity were analyzed in the laboratory as per Standard Methods (APHA, 1998).

RESULTS AND DISCUSSION

Figure 48.2a shows the changes in filtration rate during the filter run while Figure 48.2b presents the average filtration rates at different phases of the run. Flow rate decreased as the filtration progressed. The average flow rates for all the 3 filters were similar in the beginning. However, flow rate of MBSF(scrap) was significantly less compared to the other two filters during the third month of operation (Fig. 48.2b). Reduction in filtration rate is expected due to particle accumulation and maturation of biofilm layer.

The monthly average effluent turbidity levels for the three filters are presented in Figure 48.3. Influent turbidity was 21 ± 3 NTU. During the first few days of filter operation, effluent turbidity levels of ZVI-modified filters were inferior compared to BSF, which resulted in high monthly averages for those filters for the first month. Though it was thought that this higher turbidity could be due to leaching of iron from ZVI, this possibility was ruled that since the iron concentrations in the effluent were similar in all the filters. Thus the higher turbidity observed in modified filters could be caused by short-circuiting due to higher porosity of the iron scraps/nails or due to improper washing of the iron media used in the filters.

After one month's of operation, all the filters gave comparable performances, and effluent turbidity level was around 1.0 NTU in all the filters which corresponded to greater than 90% turbidity removal. This is similar to the values reported by other investigators (Elliott *et al.*, 2008; Ahammed and Davra, 2011; Kennedy *et al.*, 2012). The results also suggest that MBSFs did not offer any improvement over BSF in terms of turbidity removal.

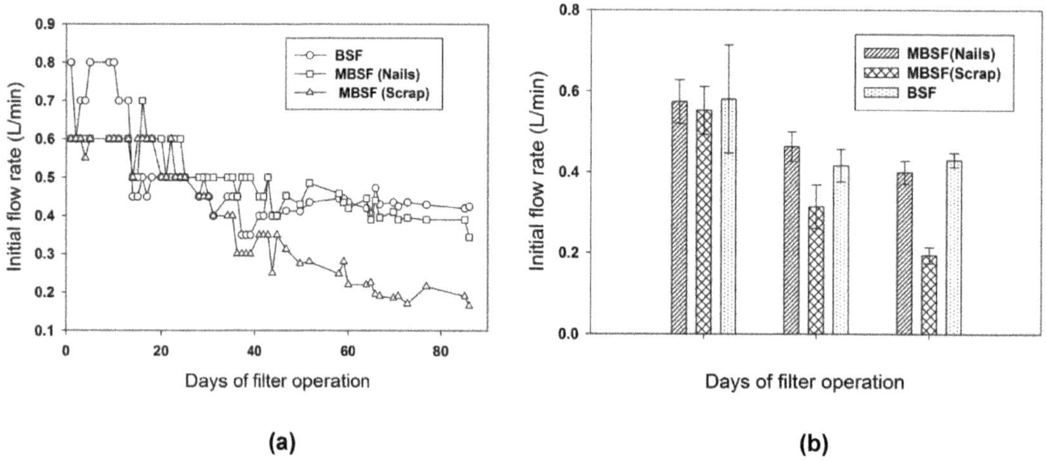

Figure 48.2 Flow rate variation with time (a) daily variation; (b) average monthly variation

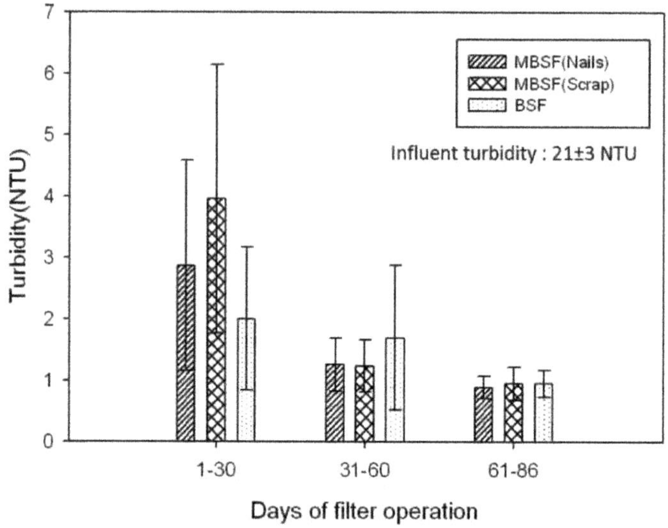

Figure 48.3 Average effluent turbidity for different filters on a monthly basis

Figure 48.4 compares the performance of the three filters over the course of the filter run. All the filters were charged daily with 20 L of *E. coli*-spiked tapwater with an initial concentration of around 10^6 CFU/mL. *E. coli* removal improved in all the three filters over the course of the experiment. However, the removal was always higher in the two MBSFs compared to BSF by at least one log unit. Figure 48.4

also shows that average performance of the two MBSFs was generally similar. The removal efficiency of MBSFs increased from about 1.5 log unit for the first month to about 4 log unit in the third month. Compared to this, BSF gave an *E. coli* reduction of about 0.5 log unit during the first month which improved to about 2.5 log unit in the third month. MBSF(Scrap) outperformed the MBSF(Nail) during the third month. This could be due to the lower filtration rate observed for the MBSF(Scrap) as reported earlier.

Influent and effluent dissolved oxygen (DO) levels were monitored throughout the course of the filter run (data not shown). The passage of water through the filters resulted in reduction in DO in all the filters. The influent DO values were around 6 mg/L. The decline was, in general, higher in MBSFs compared to BSF. For example, during the first month, the average effluent DO was 4.8 mg/L in the case of BSF, while in the two MBSFs these values were 2.5 and 3.5 mg/L. It was seen that the DO fell sharply during the first few days especially in MBSFs. It was also seen that the reduction in DO was less in the subsequent months. This different DO values in the filters at different periods of operation indicates varying removal mechanisms and reactions in the filters. It may be noted that the DO in any of the filters during the filter runs did not fall below 1.5 mg/L indicating that filters were always under aerobic conditions.

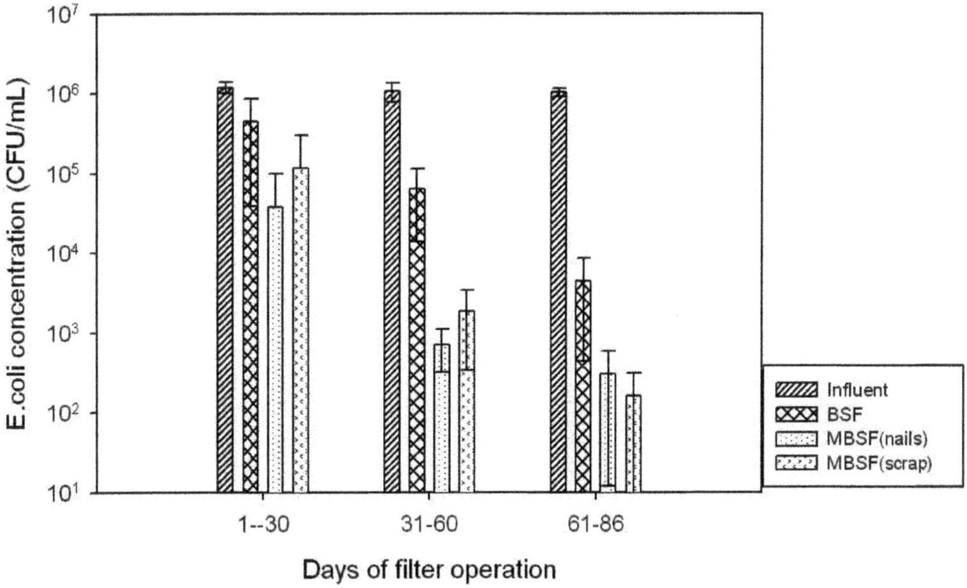

Figure 48.4 Average *E. coli* concentration for different filters on a monthly basis. Monthly data are based on six observations

The variation in pH during the filter run is presented in Figure 48.5. The influent pH was around 8.1 and the effluent pH values in three filters were not significantly different from influent pH values except during the first month's operation. During this period, the effluent pH from MBSF(Scrap) was almost one pH unit higher compared to influent pH. This could be due to various chemical reactions taking place in the MBSF filters.

Little change occurred in other water quality parameters tested such as alkalinity, hardness, nitrates and sulphates during filtration. Influent and effluent irons levels were also not significantly different in both BSF and MBSF.

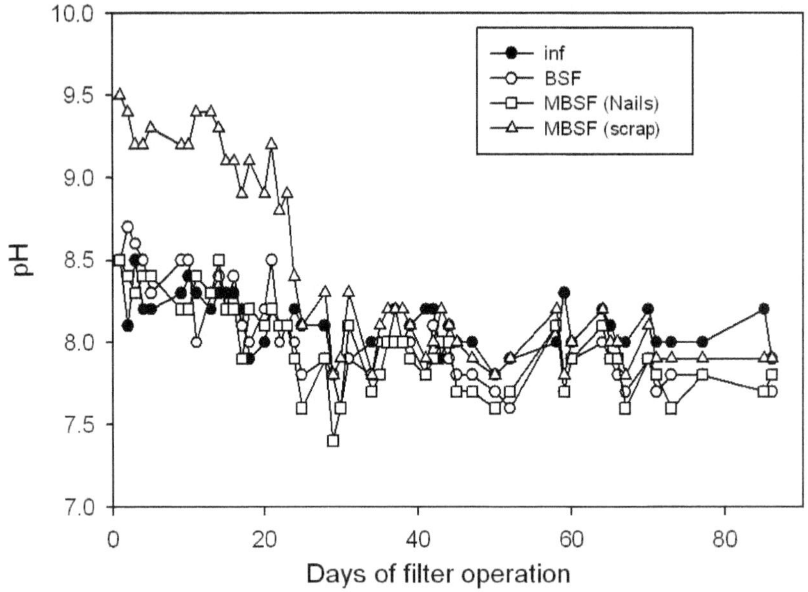

Figure 48.5 Variation in pH values in different filters

Attempts have been made to improve the performance of BSF by modification to its design and operation. These include incorporation of iron oxide-coated sand (Ahammed and Davra, 2011), and reduction of outlet diameter to increase hydraulic retention time (Kennedy *et al.*, 2012) and continuous operation (Young-Rojanschi and Madramootoo, 2014) Such amendments have shown to improve bacterial and virus removal capacities of biosand filter. This study indicates the beneficial effects of incorporating zero-valent iron in BSF. Though ZVI modification did not result in any improvement in aesthetic quality measured in terms of turbidity removal, ZVI-modified BSFs outperformed BSF in terms of microbial removal by at least 1 log unit. This improvement is very important especially during the maturation phase of the filters when microbial removal is generally low.

CONCLUSIONS

The following conclusions can be drawn from the present study.

- Introduction of ZVI in BSF significantly improved the performance of BSF in terms of bacterial removal. ZVI-modified filters showed improved performance by at least one log unit.
- Turbidity was effectively removed by both BSF and MBSFs thus improving the aesthetic quality of water. Effluent turbidity around 1 NTU was achieved in the effluent after maturation.
- Sharp decline in DO was observed during the passage of water through the filters. Decline was different in filters suggesting varying mechanisms of particle removal.
- Filtration through BSF and MBSF did not affect the physico-chemical quality of water including iron.

REFERENCES

Ahammed M. M.and Davra, K. (2011) Performance evaluation of biosand filter modified with iron oxide-coated sand for household treatment of drinking water. *Desalination*, **276**, 287-293.

American Public Health Association (APHA), American Water Works Association (AWWA), and Water Environmental Federation (WEF) 1998. Standard methods for the examination of water and wastewater. 20th Ed., L. Clescerl, A. Greenberg, and A. Eaton, eds., American Public Health Association, Washington, DC.

Baumgartner, J., Murcott, S., and Ezzati, M. (2007) Reconsidering 'appropriate technology': the effects of operating conditions on the bacterial removal performance of two household drinking water filter systems. *Environ. Res. Lett.*, **2,** 1-6.

Bradley I. M., Straub A., Maraccini P., Markazi S and Nguyen T. H. (2011) Iron oxide amended biosand filters for virus removal. *Water Res.*, **45**, 4501-4510.

CAWST (2009) Biosand Filter. Manual for Design, Construction, Installation and Maintenance; Center for Affordable Water and Sanitation Technology, Calgary, Canada. Available online: http://www.cawst.org/assets/File/Biosand_Filter_%20Manual_Version_10_No_Appendices_Sep09.pdf (accessed 10 July 2010).

Chiew H, Sampson M L. , Huch S, Ken S and Bostick B C.(2009) Effect of groundwater iron and phosphate on the efficacy of arsenic removal by iron-amended bioSand filters *Environ. Sci. Technol.*, **43** (16), 6295–6300.

Chiu P C (2013) Applications of zero-valent iron (ZVI) and nanoscale ZVI to municipal and decentralized drinking water systems—a review. *In Novel Solutions to Water Pollution* (Ed) Ahuja, S, American Chemical Society, Washingston, DC, USA.

Elliott, M.A., DiGiano, F.A. and Sobsey, M.D. (2011) Virus attenuation by microbial mechanisms during the idle time of a household slow sand filter. *Water Res.*, **45**, 4092-4102.

Elliott, M.A., Stauber, C.E., Koksal, F., DiGiano, F.A., and Sobsey, M.D. (2008) Reductions of E-coli, echovirus type 12 and bacteriophages in an intermittently operated household-scale slow sand filter. *Water Res.*, **42**, 2662-2670.

Jenkins, M. W., Tiwari, S. K., and Darby, J., (2011) Bacterial, viral and turbidity removal by intermittent slow sand filtration for household use in developing countries: Experimental investigation and modeling. *Water Res.*, **45**, 6227-6239.

Kennedy, T. J., Hernandez, E. A., Morse A. N. and Anderson, T. A. (2012) Hydraulic loading rate effect on removal rates in a biosand filter: A pilot study of three conditions. *Water Air Soil Pollut.*, **223**, 4527-4537.

Murphy, H.M., McBean, E.A., and Farahbakhsh, K., (2010) Nitrification, denitrification and ammonification in point-of-use biosand filters in rural Cambodia. *J. Water Health*, **8**, 804-817.

Noubactep C., Temgoua E. and Rahman M.A. (2012) Designing iron-amended biosand filters for decentralized safe drinking water provision. *Clean-Soil, Air, Water*, **40**(18), 798-807.

Palmateer, G., Manz, D., Jurkovic, A., McInnis, R., Unger, S., Kwan, K.K. and Dutka, B.J. (1999) Toxicant and parasite challenge of Manz intermittent slow sand filter. *Environ. Toxicol.*, **14**, 217-225.

Shi C,Wei J, Jin Y,KnieL K E and Chiu P (2012). Removal of viruses and becteriophages from drinking water using zero-valent iron. *Sep. Purif. Technol.*, **84**, 72-78.

Sobsey, M.D., Stauber, C.E., Casanova, L.M., Brown, J.M. and Elliott, M. A. (2008) Point of use household drinking water filtration: a practical, effective solution for providing sustained access to safe drinking water in the developing world. *Environ. Sci. Technol.*, **42**, 4261-4267.

Tellen, V., Nkeng, G. and Dentel, S. (2010) Improved filtration technology for pathogen reduction in rural water supplies. *Water*, **2**, 285-306.

You, Y., Han, J., Chiu, P. C. and Jin, Y. (2005) Removal and inactivation of waterborne viruses using zerovalent iron. *Environ. Sci. Technol.*, **39** (23), 9263-9269.

Young-Rojanschi, C. and Madramootoo, C. (2014) Intermittent versus continuous operation of biosand filters, *Water Res.*, **49**, 1-10.

Part VI

Alternative Biofiltration – Process Behaviour

Chapter 49

Mechanisms of DOC removal from secondary effluents in a lab biofilter

N. Siebdrath, C. Schaeffer*, W. Uhl**

**Technische Universität Dresden, Institute of Urban Water Management, Chair of Water Supply Engineering, 01062 Dresden, Germany*

ABSTRACT

Biofiltration is an interesting process for pre-treatment of secondary effluents to reduce fouling in membrane filtration processes. Especially with respect to different DOC-fractions a detailed understanding of the removal mechanisms in biofilters is important. In the present study 15% of DOC were removed. For the biopolymer fraction removal was highest (65%) and seemed to continue while for low molecular weight substances removal was limited. The main removal mechanism was shown to be biodegradation. Oxygen consumption in the filter bed correlated well with DOC removal and corresponded to full oxidation of DOC to carbon dioxide. The concentration of bacteria decreased over the filter bed depth in the bulk and on the filter media. Biomass attached to the granular media was shown to be much higher than suspended biomass in the interstitial water. In batch adsorption experiments attachment of organics to the granular media could not be proven.

Keywords biofiltration, mechanisms, adsorption, biodegradation, secondary effluents, ATP

INTRODUCTION

Reclamation of secondary effluents is becoming more and more important, especially in areas with water scarcity. In that, membrane filtration with ultrafiltration or reverse osmosis are important processes. However, organic fouling of the membranes due to attachment of organic substances (Lee et al. 2006; Li et al. 2007) or biofouling, due to the growth of bacteria and production of polysaccharides (Griebe & Flemming 1998), can dramatically impact the performance of the process.

Biofiltration is an interesting low cost tertiary treatment step to further increase the quality of secondary effluents prior to membrane filtration and reduce fouling. Researches on the impact of biofilters onto membrane filtration processes showed that membrane fouling can be reduced considerably (Halle et al. 2009; Hu et al. 2005). For the optimization of such biofilters as well as membrane processes a better understanding of the retention mechanisms of different fractions of organic matter from secondary effluent in the biofilter is important and was striven for in this work.

METHODOLOGY

For the biofiltration experiments a lab biofilter, consisting of 6 glass columns in series with a diameter of 2.8 cm and 40 cm length each and with sampling ports before the first and after each of the six columns was used. The filter bed, consisting of expanded clay with a bed height of 20 cm in each column, was set up with a total bed height of 120 cm. The setup is shown in Figure 49.1. The expanded clay (Filtralite® NC 1.5 … 2.0 mm and inner porosity of the grains of 73 %) was provided by Weber Saint Gobain, Norway.

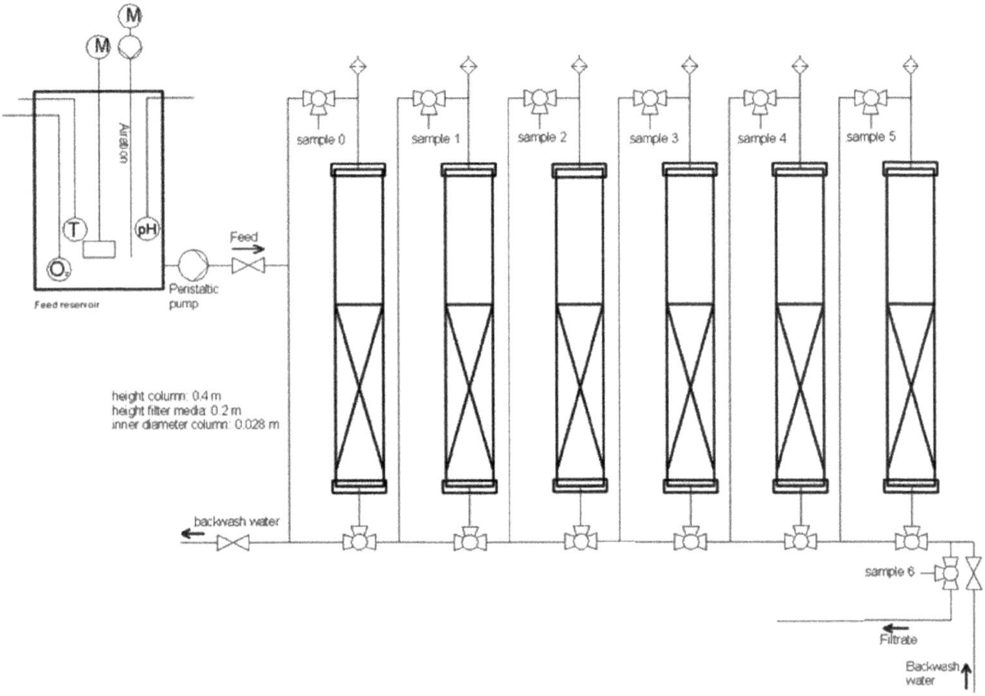

Figure 49.1 Scheme of the biofilter set-up

The filters were continuously fed with prefiltered (20 µm cartridge filter) secondary effluent, obtained from a local waste water treatment plant, and operated in a down flow mode. The initial DOC concentration of the secondary effluent was 7.21 ± 0.44 mg/L, the temperature was 20°C.

The flow rate was adjusted to 0,85 m/h, resulting in a total EBCT of 85 minutes. For biweekly backwashing (after sampling) the flow rate was adjusted such that the filter bed was fluidized but plug-like lifting of the granular media was avoided. Backwash was done for 4 minutes (1st column), 3 minutes (2nd column), 2 minutes (3rd and 4th column) and 1 minute (5th and 6th column).

Samples were taken weekly over a period of 3 months. The feed and the filtrate from the seven sampling points were analyzed for pH, conductivity and oxygen concentration (multi 340i and pH/ oxi 340i WTW, Germany). Dissolved organic carbon (DOC) concentration was determined with a thermal catalytic method (liquiTOC trace, Elementar Analysensysteme, Germany) in duplicate after 0.45 µm filtration.

DOC-fractions were determined by liquid chromatography with organic carbon and organic nitrogen detection (LC-OCD-OND) using a LC-OCD-OND-system (Model 8, DOC Labor, Karlsruhe, Germany). The system automatically filters the samples (0.45 µm) prior to analysis. The software FIFFIKUS, provided with the instrument, was used for analysis of the chromatograms.

Concentrations of suspended bacteria were determined as Total Cell Count (TCC) using flow cytometry (Accuri C6, BD Biosciences, Germany) after staining with SYBR Green I.

As a second parameter for bacteria concentration Adenosine-Triphosphate (ATP) was measured after sample preparation according to the technical bulletin of the BacTiterGlo™ reagent (Promega 2012) with the multi-functional reader INFINITE 200 Pro (Tecan, Germany). Calibration was done using an ATP stock solution (Sigma Aldrich, Germany).

Determination of biomass on the granular media was done using a slightly modified method of Magic-Knezev (Magic-Knezev & van der Kooij 2004). The filter grains were drained and wet weight was determined. 5 g of the wet media were transferred into centrifugal tubes. 50 ml sterile filtered tap water was added and the tubes placed in ultrasonic cleaner (Model USC600TH, output: 45 kHz, 130 W, VWR Germany). Sonication was done five times for 2 minutes, while shaking the samples manually in between to avoid abrasion. The supernatant was analyzed for ATP. ATP measured in the supernatant was converted to ATP per volume of filter bed using the wet bulk density of the filter grains given by the manufacturer (1076 kg m^{-3}).

Adsorption of DOC fractions to the granular filter media was studied with new media (after washing with ultrapure water) as well as with media which had been obtained from the biofilter. For that, 16 g of wet granular media were added to 250 mL of secondary effluent. The samples were shaken on a horizontal shaker over a period of 8 h. In between, samples from the supernatant were taken for LC-OCD analysis. Further details are given in Table 49.1

Table 49.1 Collection of experimental characteristics for adsorption batch experiments.

	Adsorption Batch 1	Adsorption Batch 2	Check plot
Adsorbent	Used expanded clay with biofilm	Purified, unused expanded clay	none
Adsorbent preparation	Pasteurization 70°C, 40 minutes	Autoclaving 121°C, 20 minutes	-
Solute preparation	Secondary effluents: Filtration 2.5 µm, Pasteurization 70°C, 40 minutes	Secondary effluents: Filtration 2.5 µm, Pasteurization 70°C, 40 minutes	Secondary effluents: Filtration 2.5 µm, Pasteurization 70°C, 40 minutes
Amount samples	11	11	11

RESULTS AND DISCUSSION

On average DOC removal in the filter bed of 1.2 m was 15 %. The removal of the fractions of the biggest molecules (biopolymers, > 20.000 Da) and of the smallest molecules (low molecular weight substances, LMW) are shown Figure 49.2. These two fractions are the most interesting ones with respect to potential membrane fouling. While biopolymers are sticky towards the membrane causing organic fouling, low molecular weight compounds are easily biodegradable and may cause biofouling.

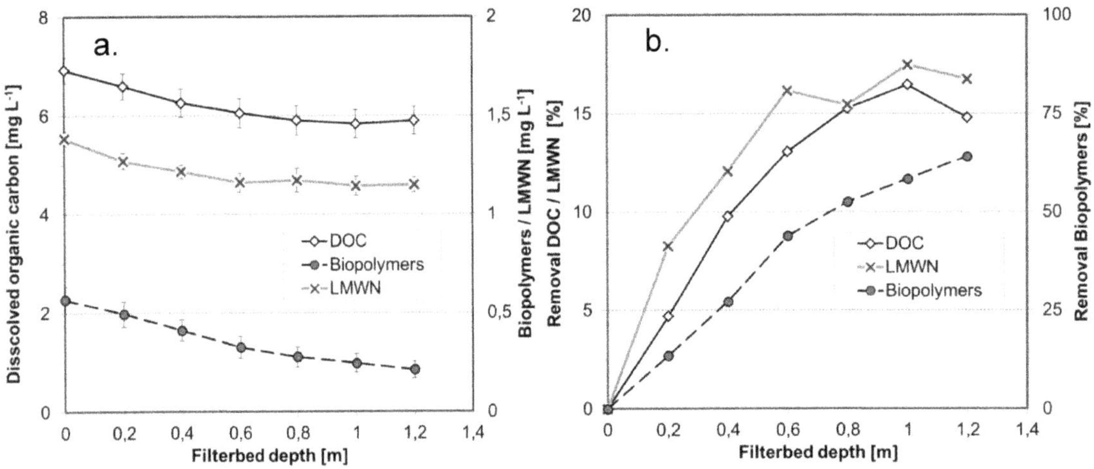

Figure 49.2 Removal of DOC and its fractions over the bed depth of the biofilter. a. Amount of DOC, Biopolymers and Low Molecular Weight Neutrals (LMWN), n=9, α= 33; b. Percentage removal of DOC and its fractions.

The biopolymer fraction was eliminated to the highest extent (65% removal), though the percentage contribution of biopolymers to the total DOC and thus total DOC-removal was the smallest compared to other fractions. For both total DOC and the LMWN fraction the removal did not increase after 0.8 m bed depth, while biopolymers continued to be removed. Thus it seems that the biopolymer fraction might be completely biodegradable while the other fractions are not. Hallé et al. and Zheng et al. report that biopolymers are biodegradable and can be removed via biofiltration processes (Halle et al. 2009; Zheng et al. 2010). However, as the biopolymers are very big and sticky molecules, another possible removal mechanism is simple attachment to the granular media. For the LMW fraction it might be concluded that they are not completely biodegradable.

In order to obtain information whether DOC removal was due to biodegradation or to adsorption, oxygen consumption in the filter was balanced against DOC removal, assuming that all DOC removed was converted to carbon dioxide by biodegradation. It was calculated that 3.20 mg L^{-1} oxygen would be needed for conversion of all the DOC that had been removed. In comparison, on average 3.17 mg L^{-1} oxygen were consumed in the filter. Also, a linear correlation between oxygen consumption and DOC removal (data not shown here) was found. Since nitrification processes can be excluded (no ammonia was present in the influent) it is concluded that all DOC removal in the biofilter was due to biodegradation by heterotrophic microorganisms.

Analysis of the suspended bacteria concentration in the liquid phase as total cell count (TCC) by flow cytometry (Figure 49.3a) shows a decrease over bed depth. This is reasonable as the bacteria can be removed from the liquid by attachment. Also, as they have lower nutrient concentrations with increasing bed depth, they grow slower. The decreasing TCC corresponds to the decreasing ATP concentration in the bulk liquid phase (Figure 49.3b).

Stratification of bacteria attached to the granular media, determined as ng ATP per volume of filter bed, is also shown in Figure 49.3b. For that a volume ratio of 60 % clay grains (including pore water) and 40 % of surrounding liquid was used. It shows that the bed volume concentration of ATP from attached bacteria is on average 760 times higher than that resulting from bacteria suspended in the liquid.

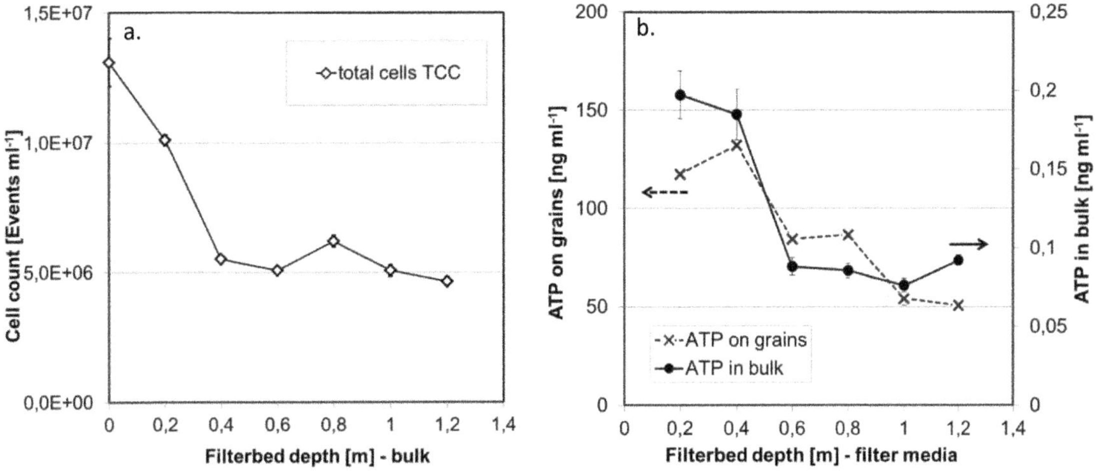

Figure 49.3 (a) Total Cell Count (TCC) over the filter depth in the bulk solution and (b). ATP-concentration of extracted biofilm from the filter material and from the bulk solution over the filter depth.

In order to elucidate the role of adsorption of DOC to the filter media, the results of the adsorption batch tests are displayed in Figure 49.1 which shows the time course of the DOC-concentration when pasteurized secondary effluent was contacted with sterilized virgin or used granular media. No change in DOC in the liquid phase can be observed over time, which means that adsorption most probably does not contribute to DOC-removal in the biofilters. Also, the confidence intervals of the DOC in the liquid in both experiments overlap. Thus, no difference in the adsorption behavior of virgin and used expanded clay can be proven.

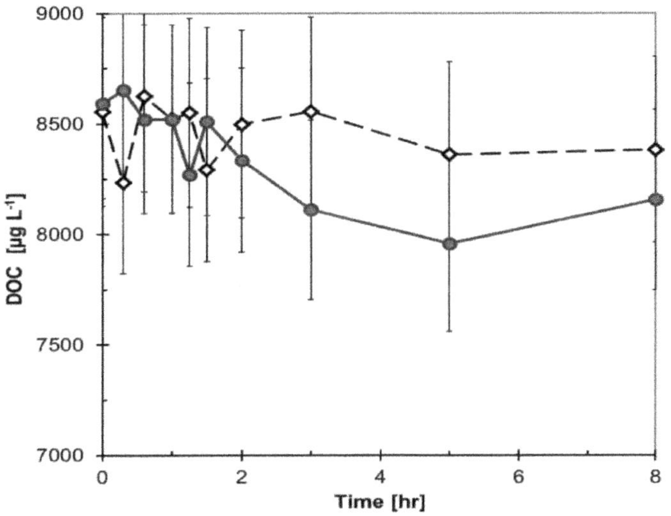

Figure 49.4 Concentration of dissolved organic carbon during adsorption batch experiments: Uncovered, clean expanded clay in secondary effluents and Expanded clay covered with biofilm in secondary effluents. Error bars represent 67 % confidence intervals.

SUMMARY AND CONCLUSIONS

DOC from secondary effluent was removed in the lab biofilter by 15 % on average during 85 min empty bed contact time. The highest removal of 65% was found for the fraction of biopolymers. Low molecular weight substances (LMW) were removed as well, but to a much lesser extent (16%) than biopolymers.

Adsorption to the media or attached biomass was shown not to contribute to DOC removal considerably. Oxygen consumption correlated well to DOC removal and could be balanced for complete oxidation of dissolved organic carbon to carbon dioxide. Thus, biodegradation is the governing DOC-removal from secondary effluents in the biofilter investigated. Measurement of attached and suspended biomass showed that it is much higher in the attached phase, emphasizing the importance of the carrier media for the effectivity of the process.

Generally, as especially biopolymers which are considered as major membrane foulants can be removed very effectively in the biofilter, biofiltration can be considered as a very interesting pre-treatment option for reverse osmosis.

ACKNOWLEDGEMENTS

This study was done in the framework of the joint project "Fouling minimized reclamation of secondary effluents with reverse osmosis". Therefore the authors would like to thank the German Federal Ministry of Education and Research (BMBF) for the financial support (grant no. 02WA1076). Further, Ms. Julia Röhr is acknowledged and thanked for her preliminary studies on the lab biofilter and Mr. Gerit Orzechowski for the determination of bacteria concentrations by Flow Cytometry and providing the template for data-interpretation.

REFERENCES

Griebe, T. and Flemming, H. C. (1998). "Biocide-free antifouling strategy to protect RO membranes from biofouling." Desalination 118(1-3): 153-156A.

Halle, C., Huck, P. M., et al. (2009). "Assessing the Performance of Biological Filtration As Pretreatment to Low Pressure Membranes for Drinking Water." Environmental Science & Technology 43(10): 3878-3884.

Hu, J. Y., Song, L. F., et al. (2005). "Biofiltration pretreatment for reverse osmosis (RO) membrane in a water reclamation system." Chemosphere 59(1): 127-133.

Lee, S., Ang, W. S. and Elimelech, M. (2006). "Fouling of reverse osmosis membranes by hydrophilic organic matter: implications for water reuse." Desalination 187(1-3): 313-321.

Li, Q. L., Xu, Z. H. and Pinnau, I. (2007). "Fouling of reverse osmosis membranes by biopolymers in wastewater secondary effluent: Role of membrane surface properties and initial permeate flux." Journal of Membrane Science 290(1-2): 173-181.

Magic-Knezev, A. and van der Kooij, D. (2004). "Optimisation and significance of ATP analysis for measuring active biomass in granular activated carbon filters used in water treatment." Water Research 38(18): 3971-3979.

Promega (2012). "Technical Bulletin BacTiter-Glo™ Microbial Cell Viability Assay." Promega Corporation; Madison, WI, USA.

Zheng, X., Ernst, M. and Jekel, M. (2010). "Pilot-scale investigation on the removal of organic foulants in secondary effluent by slow sand filtration prior to ultrafiltration." Water Research 44(10): 3203-3213.

Chapter 50

Change of water treatment efficiency in the transition from granular activated carbon to biological activated carbon on an advanced water purification plant

H. Inoue, S. Kodama, M. Ishibashi, Y. Watanabe
Chiba Nogiku-no-sato Purification Plant, Chiba Prefectural Waterworks Bureau, 478-1 Kuriyama, Matsudo, Chiba, Japan

ABSTRACT
Advanced purification treatment by ozone and biological activated carbon (BAC) is effective in removal of NH_4^+-N and organic matters of low molecular weight and periodic replace of activated carbon is considered to be necessary for the appropriate treatment. In this study, we investigated the change of water treatment efficiency in the transition to BAC by comparing the results of water quality analysis about the renewal basins and the control basins. These results suggest that the transition to BAC in the bilogical process was dependent on influent water temperature and quality, but the transition in removal of organic matters was independent of influent water temperature and quality. This knowledge is considered to be applied on the future replace of the activated carbon.

Keywords advanced purification treatment; biological activated carbon; nitrification

INTRODUCTION

Chiba Nogiku-no-sato Purification Plant in Chiba Prefectural Waterworks Bureau intakes raw water from the Edo River and has a production capacity of 60,000 m^3/day. In this plant, NH_3-N, odorants and trihalomethane precursors are removed by advanced purification treatment consisted of ozonation and biological activated carbon (BAC) treatment, and chlorination is usually implemented after BAC treatment. Granular activated carbon (GAC) in two of six GAC adsorption basins was replaced in 2011 for the first time since the operation of this plant started in 2007. In this study, we investigated the change of water treatment efficiency in the transition from GAC to BAC by comparing the results of water quality analysis about two renewal GAC adsorption basins and the control BAC basin.

RESEARCH POINTS AND ITEMS

The structure of advanced water purification faculity of Chiba Nogiku-no-sato Purification Plant were showed in Figure 50.1. This plant has six GAC adsorption basins. In these basins, GAC in Basin 1-1 and 2-3 was replaced and the operation was restarted on 1/9/2007 and 17/11/2007, respectively. For the investigation of change of water treatment efficiency, Basin 1-3 and 2-1 were set as control of 1-1 and 2-3, respectively.

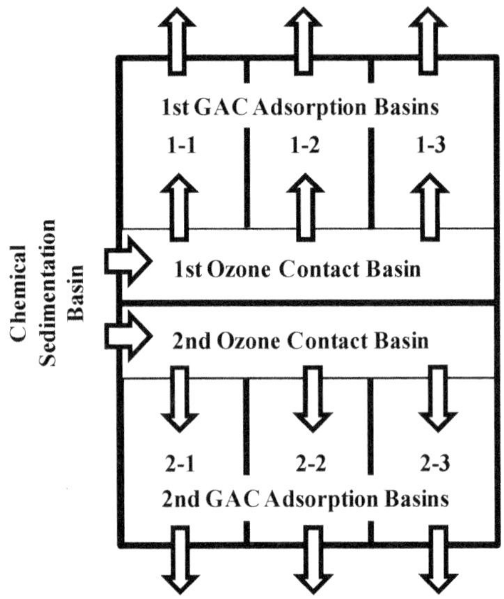

Figure 50.1 Structure of the advanced water purification facility. Arrows show water flow.

To evaluate change of water treatment efficiency during the transition from GAC to BAC, we analyzed the influent and effluent water samples of Basin 1-1, 2-3 and the control basins. The research period was from 1/9/2007 to 28/3/2008 in 1-1, and from 18/11/2007 to 13/6/2008 in 2-3. Analyzed water quality items were described below.

Nitirification process: NH_4^+-N, NO_2^--N, NO_3^--N.

Removal efficiency of organic matters: total organic carbon (TOC), UV absorbance at 260 nm (E260).

Microorganism activity: dissolved oxygen (DO), water temperature.

RESULTS AND DISCUSSION

Nitrification process

At first, NH_4^+-N, NO_2^--N and NO_3^--N were analyzed to investigate the nitrification process by microorganism. In Basin 1-1, concentration of NH_4^+-N in the effluent water was at the same level of the influent water at the beginning of the treatment restart (Figure 50.2a). After 30 days of the restart, NH_4^+-N in the effluent water began to decrease and was completely removed 10 days later. NO_2^--N was not detected in the ozonized influent water at the beginning of the restart, but concentration of NO_2^--N derived from NH_4^+-N increased as NH_4^+-N decreased. After 60 days of the restart, NO_2^--N decreased and became undetectable at all because of further bilogical oxidation to NO_3^--N. Change of NO_3^--N with biological nitirification could not be detected because the concentration of NO_3^--N in the raw water was fifty times or higher than NO_3^--N derived from NH_4^+-N.

On the other hand, NH_4^+-N was not yet removed after 30 days of the restart in 2-3 (Figure 50.2b). Thereafter, the prechlorination at the receiving well of the raw water started for removal of NH_4^+-N because low water temperature caused the leakage of high concentration of NH_4^+-N from the basin. After the prechlorination stopped, control basin 2-1 started to remove NH_4^+-N within one day, but it took about 50 days to remove NH_4^+-N in 2-3. The reason is considered that the first 30 days after the restart was insufficient to establish biological nitirification process because of low water temperature. These results suggest that water temperature is the influential factor for the establishment of biological nitirification process.

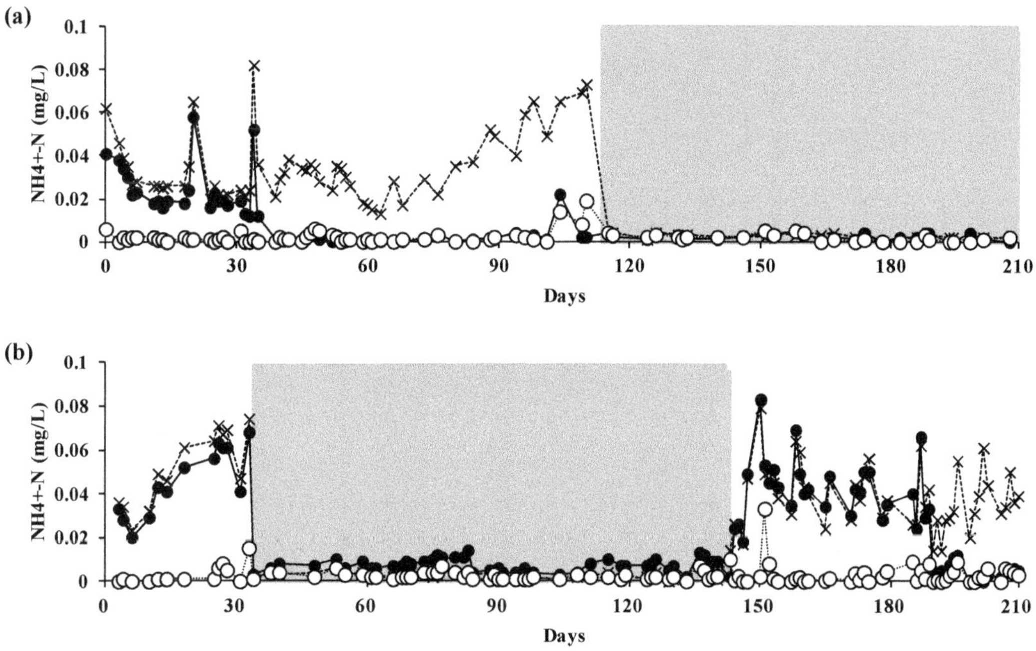

Figure 50.2 Change of NH_4^+-N in the effluent water from 1st (a) and 2nd (b) GAC adsorption basin. X-mark, filled circle and open circle show the influent water, the effluent water from the research basin and the effluent water from the control basin, respectively. Shadow area shows prechlorination period.

Removal efficiency of organic matters

Next, we investigated the removal efficiency of organic matters by analyzing TOC and E260. In both replaced basins, the removal efficiency of TOC and E260 decreased at constant rates regardless of the water temperature or the quality of the raw water during about 100 days after each restart (Figure 50.3). The reason is considered that the adsorption activity still remained and was better than the biological degradation. Thereafter, decrease rates of the removal efficiency of both items became much slower because the adsorption capacity was saturated and organic matters were treated mainly by biological degradation. These results revealed that a time period of 100 days was necessary for the transition to BAC in terms of removal of organic matters.

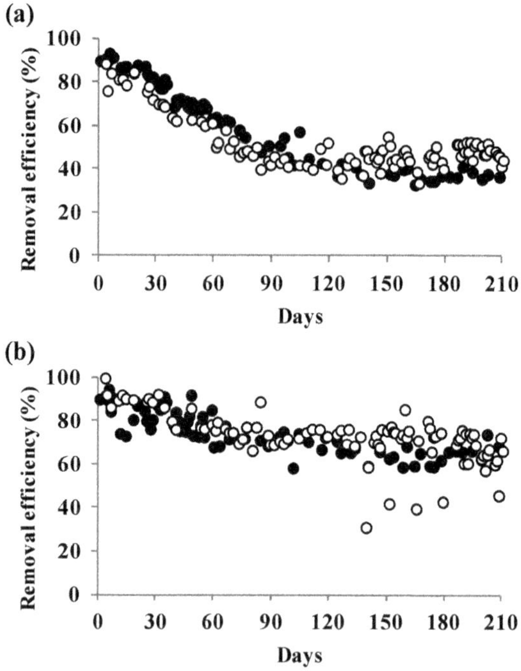

Figure 50.3 Change of removal efficiency of TOC (a) and E260 (b). Filled circle and open circle show the result of 1-1 and 2-3, respectively.

Microorganism activity

Finally, we analyzed DO consumption and pH variation as an index of microorganism activities such as nitrification and oxidative degradation of organic matters. In the control BAC basin, the DO consumption and pH variation significantly changed with dependence on the water temperature. In Basin 1-1, the DO consumption and pH variation were low and did not change as much as the control BAC basin dependent on the water temperature during 40 days after the restart. Thereafter, the DO consumption and pH variation gradually increased and became equal to those of the control BAC basin at 60 days after the restart. On the other hand, the DO consumption and pH variation of 2-3 were not different from those of the control BAC basin probably because of the microorganism hypoactivity by low water temperature.

CONCLUSION

This study revealed that the transition to BAC in the nitrification process and microorganism activity was dependent on water temperature and water quality, and was relatively short-term process. On the other hand, the transition in the removal efficiency of organic matters was independent of water temperature and water quality, and was long-term process. This knowledge is considered to be applied on the future replace of GAC. Further investigation is necessary to elucidate the mechanism of the transition to BAC and perform more efficient operation of BAC treatment.

Chapter 51

Autotrophic growth competition between ammonia-oxidizing archaea and ammonia-oxidizing bacteria in biological activated carbon filter with nitrification potential

I. Kasuga, J. Niu*, F. Kurisu*, H. Furumai*, T. Shigeeda***

**Department of Urban Engineering, School of Engineering, The University of Tokyo, 7-3-1, Hongo, Bunkyo, Tokyo 113-8656, Japan*

***Bureau of Waterworks, Tokyo Metropolitan Government, 2-8-1, Nishi-Shinjuku, Shinjuku, Tokyo 163-8001, Japan*

ABSTRACT

Biological activated carbon (BAC) filtration combined with preceding ozone oxidation has been established as a major advanced drinking water purification process. One of the most important functions of BAC is the biological oxidation of ammonia, which can serve as a precursor for odourous trichloramine. The activity of ammonia-oxidizing microorganisms in the BAC filter is related to the nitrification performance of BAC. In the present study, autotrophic growth competitions between ammonia-oxidizing archaea (AOA) and ammonia-oxidizing bacteria (AOB) associated with BAC were evaluated by DNA stable isotope probing. When low levels of ammonium (0.14 mg N/L) were continuously supplied to BAC collected from a full-scale drinking water purification plant in Tokyo, AOA rather than AOB demonstrated higher autotrophic growth activity following 7 days of incubation. After 28 days, both AOA and AOB showed autotrophic growth. These results indicate that AOA outcompete AOB for the oxidation of low levels of ammonium in BAC filters.

Keywords ammonia-oxidizing archaea; ammonia-oxidizing bacteria; biological activated carbon; DNA-stable isotope probing

INTRODUCTION

Ozone oxidation followed by biological activated carbon (BAC) filtration is intensively introduced to conventional process to meet the demand for high-grade drinking water. The Bureau of Waterworks,

Tokyo Metropolitan Government has advanced a 'Projects for Safe and Better Tasting Tap Water' since 2004. As a part of the project, the introduction of the ozone–BAC process to all major drinking water purification plants that collect raw water from the Tone River system was completed in 2013. In addition to the physicochemical adsorption of dissolved organic matter by activated carbon, it has been previously shown that microorganisms associated with BAC enhance the biological water purification capacity (LeChevallier and McFeters, 1990). One of the important microbiological functions is nitrification because ammonium generates odourous trichloramine produced by the reaction of ammonium with chlorine under some conditions (Hureiki *et al.*, 1994). To ensure odour-free drinking water, it is necessary to maintain the nitrification performance of BAC. However, little is known about the physiological activity of ammonia oxidizers in BAC filters.

For over a century, ammonia-oxidizing bacteria (AOB) have been regarded as the sole players of ammonia oxidation. However, the recent discovery of ammonia-oxidizing archaea (AOA) has shown that AOA are more abundant than AOB in various aquatic and terrestrial environments (Francis *et al.*, 2005; Könneke *et al.*, 2005; Leininger *et al.*, 2006; Wuchter *et al.*, 2006; Di *et al.*, 2010; Liu *et al.*, 2011). Some studies have suggested that AOA are more active than AOB (Beman *et al.*, 2008; Zhang *et al.*, 2010), while a greater involvement of AOB rather than AOA in ammonia oxidation has also been reported (Jia and Conrad, 2009; Mertens *et al.*, 2009; Mußmann et al., 2011; Xia *et al.*, 2011). Thus, the competition between AOA and AOB remains a controversial issue.

AOA have been detected in groundwater treatment plants and water distribution systems in the Netherlands (van der Wielen *et al.*, 2009) and in BAC used in full-scale drinking water plants in Japan (Kasuga *et al.*, 2010a, 2010b). However, little evidence is available regarding the contributions of AOA versus AOB in the ammonia oxidation aspect of water purification. Niu *et al.* (2013) were the first to apply DNA stable isotope probing (DNA-SIP) in the evaluation of the chemoautotrophic growth activity of ammonia oxidizers associated with BAC used for drinking water purification. The principle of DNA-SIP is based on the assimilation of a ^{13}C-substrate by active microorganisms followed by the separation of heavy ^{13}C-DNA and light ^{12}C-DNA. When active ammonia oxidizers assimilate ^{13}C-inorganic carbon during ammonia oxidation, newly synthesized biomass including DNA is labelled by ^{13}C. Buoyant density fractionation of DNA then allows us to easily distinguish between active ammonia oxidizers from dormant ones. This previous study indicated that only AOA had active chemoautotrophic growth activity in response to 0.14 mg NH_4^+-N/L after 14 days of incubation. However, both AOA and AOB could grow at a higher ammonium concentration (1.4 mg NH_4^+-N/L). It is suggested that AOA have a higher affinity for ammonium than AOB. In fact, the first AOA isolate, *Nitrosopumilus maritimus* SCM1, showed a very low half saturation constant (K_m = 133 nM total ammonium) (Martens-Habbena *et al.*, 2009). In the present study, we also focused on the influence of incubation period to further evaluate chemoautotrophic growth competition between AOA and AOB associated with BAC. DNA-SIP was applied to identify active ammonia oxidizers from BAC incubated at a low ammonium concentration for a short (7 days) or long (28 days) time period.

METHODOLOGY

The present study followed the same experimental procedure as described by Niu *et al.* (2013) to evaluate the influence of short (7 days) and long (28 days) incubation times on the chemoautotrophic growth activity of AOA and AOB associated with BAC, which corresponds to GAC-A1 in the previous study. The BAC had been in use for 15 months and was collected from a full-scale drinking water purification plant in Tokyo in June 2012. The plant, which has a production capacity of 1,700,000 m^3/d, uses raw water from the Tone River system. The ranges of water temperature and ammonium concentration in the raw water during 2012 were 3.5–28.1°C (average: 15.0°C) and ND–0.46 mg N/L (average: 0.09 mg N/L), respectively. The configuration process consisted of coagulation/sedimentation followed by primary rapid sand filtration, ozonation, BAC filtration, secondary rapid sand filtration. Chlorine disinfection was applied between the BAC and secondary rapid sand filtration steps. The retention time in the BAC filter was controlled at approximately 20 min. BAC was backwashed by air and finished water with residual chlorine every 5 days.

The ammonium removal potential of BAC was evaluated immediately after 2 h of sampling from the plant. Triplicate batch experiments were conducted by incubating 50 g wet BAC in 300 mL inorganic medium at 20°C with constant agitation (50 rpm). Ammonium concentration of the medium was adjusted to 1.4 mg N/L. Sampling of 10 mL supernatants and their filtration through 0.2-µm-pore size polytetrafluoroethylene membranes (Advantec, Japan) was performed regularly for 24 h. The ammonium concentration was analysed by the indophenol blue colorimetric method using a spectrophotometer (U2010; HITACHI, Japan). Nitrite and nitrate concentrations were determined using an ion chromatograph (861 Advanced Compact IC; Metrohm, Switzerland).

For SIP experiments, reactors filled with 50 g wet BAC were used to continuously expose BAC to low levels of ammonium. Preincubation of BAC in inorganic medium without ammonium or carbon source was performed in the SIP microcosm. After 3 days of preincubation, inorganic medium containing 0.14 mg N/L of ammonium and 12 mg C/L of bicarbonate was supplied. Bicarbonate was labelled with ^{13}C or ^{12}C. The detailed incubation condition has been described elsewhere (Niu *et al.*, 2013). After 7 and 28 days, BAC was collected for DNA extraction and SIP analysis.

DNA was extracted from BAC samples using the FastDNA spin kit for soil (MP biomedicals, USA) following the manufacturer's instructions. DNA concentration was measured using a NanoDrop 1000 (Thermo Scientific, USA). Approximately 3.5 µg of extracted DNA was used for ultracentrifugation in CsCl gradients (Neufeld et al., 2007; Nicol and Prosser, 2011). The initial mixture solution of DNA and CsCl was adjusted so that the buoyant density was 1.700 g/mL. The ultracentrifugation condition was 45,000 rpm (178,000 ×*g*) in a NVT65.2 Beckman rotor (Beckman, USA) at 20°C for 72 h. After ultracentrifugation, gradient fractionation was performed to collect 19 fractions. The buoyant density of each fraction was measured using an AR200 digital refractometer (Reichert, USA). DNA was purified by ethanol precipitation and dissolved in 30 µL TE buffer.

AOA and AOB *amo*A genes that encode a subunit of ammonia monooxygenase in each fraction were quantified by real-time PCR using LightCycler 480 II (Roche, Switzerland) according to a previous study (Kasuga *et al.*, 2010a). The primer sets for the AOA and AOB *amo*A genes were Arch-amoAF/Arch-amoAR (Francis *et al.*, 2005) and amoA1F*/amoA2R (Rotthauwe *et al.*, 1997; Stephen *et al.*, 1999), respectively. The diversity of AOA and AOB *amo*A genes was evaluated by terminal restriction fragment length polymorphism (T-RFLP) analysis. The same primer sets used for the quantification of AOA and AOB *amo*A genes in the real-time PCR experiments were employed. The 5′ ends of Arch-amoAF and amoA-1F* were modified with 6-carboxyfluorescein (6-FAM). PCR products were initially checked on a 1.5% agarose gel by electrophoresis. After purification using a QIAquick PCR purification kit (QIAGEN, Germany), the purified PCR products (approximately 150 ng) were digested by restriction enzymes. AOA *amo*A genes were digested with 10 U *Hha*I at 37°C for 6 h (TAKARA BIO, Japan), while AOB *amo*A genes were digested with 10 U *Taq*I at 65°C for 6 h (TAKARA BIO). The 6-FAM-labelled fragments were separated using an ABI PRISM 3100 Avant Genetic Analyzer (Applied Biosystems, USA). The size standard used was GeneScan™ 1200LIZ (Applied Biosystems). Fragment analysis was performed using GeneMapper version 4.0 (Applied Biosystems).

RESULTS AND DISCUSSION

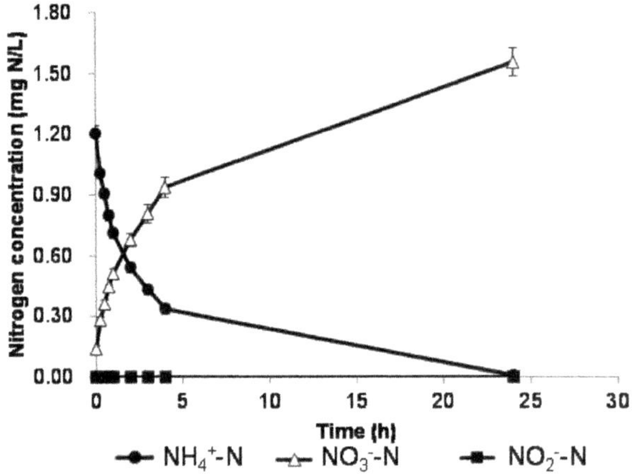

Figure 51.1 Ammonium removal potential of BAC collected from drinking water purification plants.

The result of ammonium removal potential of the original BAC is shown in Fig. 51.1. During 24 h of incubation, 1.4 mg N/L of ammonium was completely removed, while nitrate production was observed simultaneously. Nitrite was not detected throughout the incubation period, suggesting the rapid conversion of ammonium to nitrate by BAC. Excess amount of nitrate was observed, which was probably due to its release from BAC. From the linear regression analysis of changes in ammonium concentration during the first 1 h ($R^2 = 0.971$), the ammonium removal potential of BAC was estimated at 0.028 mg N/L/h/g dry.

The abundance and diversity of AOA and AOB associated with the collected BAC was characterized by analysing the *amo*A genes. AOA *amo*A genes were 4.2×10^8 copies/g dry, while AOB *amo*A genes were 2.8×10^7 copies/g dry. T-RFLP analysis showed that AOA *amo*A genes affiliated with aquatic AOA species comprised two operational taxonomic units (A-OTU1 and A-OTU2). Similarly, AOB *amo*A genes comprised three OTUs (B-OTU1, B-OTU2 and B-OTU3). B-OTU2 and B-OTU3 are closely related to *Nitrosomonas ureae* (Niu *et al.*, 2013).

Approximately half of the ammonium was oxidized to nitrate during the microcosm incubation, while ammonium concentration in the effluent was maintained at approximately 0.07–0.08 mg N/L. Figure 51.2 shows the SIP profiles of AOA and AOB *amo*A genes after 7 and 28 days of incubation. The SIP profiles generated by gradient ultracentrifugation indicate the relative amounts of AOA and AOB *amo*A genes in different buoyant density fractions. After 7 days, a clear shift of AOA *amo*A gene abundance to heavier fractions was observed (Fig. 51.2a), while no difference was found between ^{12}C and ^{13}C cases for AOB (Fig. 51.2b). This demonstrates that the autotrophic growth activity of AOA is superior to that of AOB under this condition. Because it was considered that only AOA were involved in nitrification during 7 days of incubation, the cell-specific ammonia oxidation rate of AOA was estimated from the total ammonium oxidized in the microcosm and the amount of newly produced AOA. This rate was calculated to be 0.087 fmol NH_3 oxidized/cell/h, which was close to the previous estimate based on 14 days incubation (0.11 fmol NH_3 oxidized/cell/h) (Niu *et al.*, 2013). Furthermore, it was similar to the value of AOA culture (0.08–0.53 fmol NH_3 oxidized/cell/h) (Könneke et al., 2005; de la Torre *et al.*, 2008). If all AOA in BAC filtration were active and exhibited the cell-specific ammonia oxidation rate of 0.087 fmol NH_3 oxidized/cell/h, a full-scale BAC filtration could have the potential to treat approximately 0.1 mg N/L of ammonium under the actual operational condition.

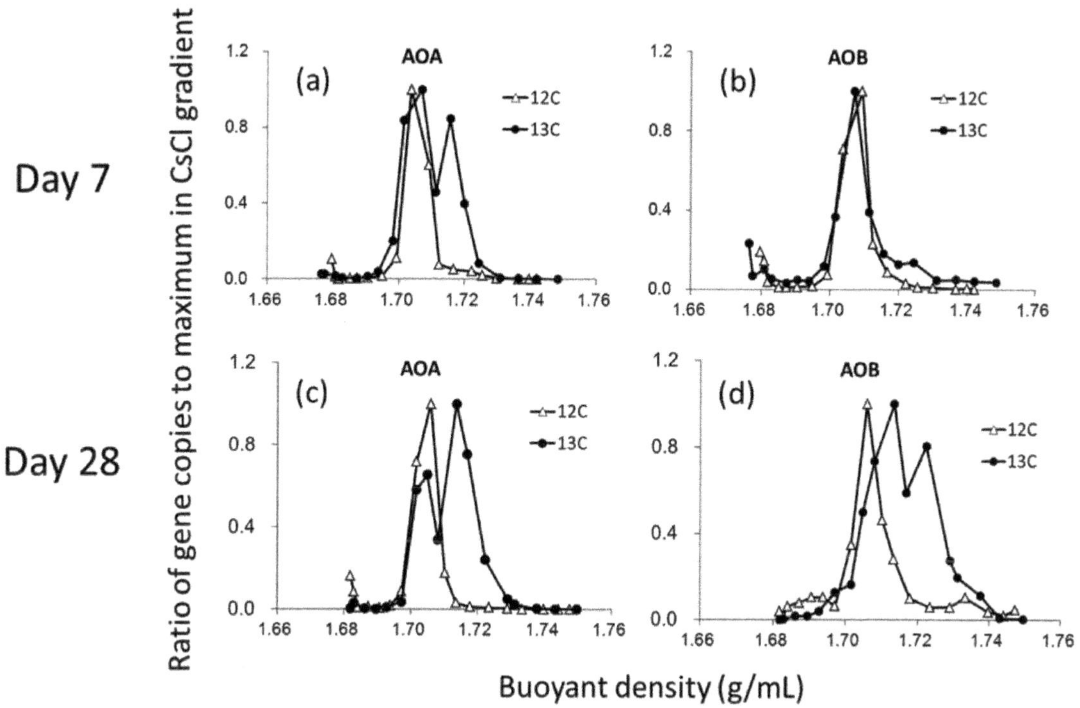

Figure 51.2 SIP profiles of AOA and AOB *amo*A genes of BAC supplied with 0.14 mgN/L of ammonium after 7 and 28 days of incubation. The SIP profiles of (a) AOA *amo*A genes after 7 days, (b) AOB *amo*A genes after 7 days, (c) AOA *amo*A genes after 28 days and (d) AOB *amo*A genes after 28 days.

After 28 days, the profiles of both AOA and AOB *amo*A genes revealed a shift to heavier fractions, and two clear peaks were observed (Fig. 51.2c and 51.2d). This indicates that both AOA and AOB assimilated bicarbonate through ammonia oxidation. A comparison of the SIP profiles of AOA after 7 and 28 days shows that the ratio of the additional peak to the original one was larger after 28 days. This demonstrates that more AOA proliferated after longer incubation. In a previous study (Niu *et al.*, 2013), DNA-SIP revealed that only AOA responded to 0.14 mg N/L under the same condition after 14 days of incubation. The present study clearly showed that the growth activity of AOB is much slower than that of AOA, which may explain the predominance of AOA in BAC.

T-RFLP analysis was applied to the heavy fractions in order to investigate the active members of AOA and AOB (Fig. 51.3). Figure 51.3a shows the diversity of *amo*A genes derived from active AOA after 7 days. The T-RFLP profile comprising A-OTU1 and A-OTU2 was identical to that of the original BAC (Niu *et al.*, 2013). After 28 days, the same members were also detected from the heavy fraction (Fig. 51.3b). Moreover, these T-RFs were previously detected from heavy fractions after 14 days (Niu *et al.*, 2013). Taken together, AOA corresponding to A-OTU1 and A-OTU2 associated with BAC demonstrated high growth activity throughout the incubation period. The diversity of AOB *amo*A genes in the heavy fraction is shown in Fig. 51.3c. B-OTU1, B-OTU2 and B-OTU3 were detected. The identical T-RFs were previously detected from the original BAC (Niu *et al.*, 2013). While B-OTU3 was predominant in the original BAC, B-OTU1 shows the largest relative abundance in the heavy fraction in Fig. 51.3c. This suggests that AOB corresponding to B-OTU1 exhibited the highest growth activity under the incubation condition in the present study.

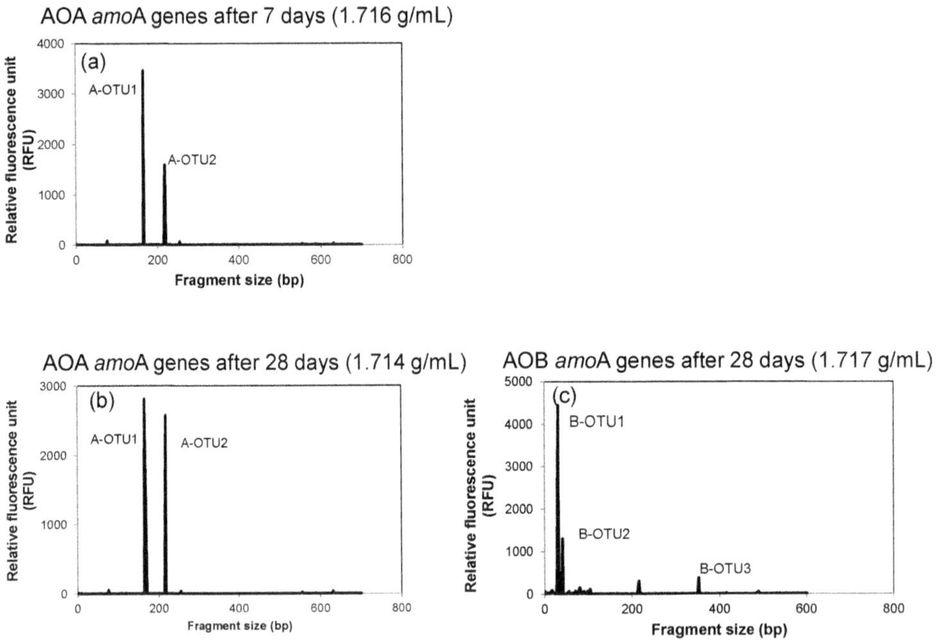

Figure 51.3 T-RFLP analysis of AOA *amo*A genes (a and b) and AOB *amo*A genes (c) in the heavy buoyant density fraction after 7 and 28 days of incubation.

CONCLUSIONS

The main conclusions of the present study are as follows:

(1) DNA-SIP analysis was applied to BAC with nitrification potential to identify active ammonia oxidizers. Only AOA showed chemoautotrophic growth responding to 0.14 mg N/L of ammonium after 7 days of incubation, while both AOA and AOB showed growth after 28 days. Because of the higher growth activity of AOA, it is likely that AOA play a more important role in ammonium oxidation.

(2) The cell-specific ammonia oxidation rate of AOA was estimated to be 0.087 fmol NH_3 oxidized/cell/h, which is consistent with the results of previous studies.

(3) T-RFLP analysis of heavy fractions revealed that all AOA associated with the original BAC had growth potential, while a minor member exhibited a better growth response among the original AOB community under the experimental condition.

REFERENCES

Beman, J.M., Popp, B.N. and Francis, C.A. (2008) Molecular and biogeochemical evidence for ammonia oxidation by marine Crenarchaeota in the Gulf of California. *ISME J.*, **2**(4), 429–441.

de la Torre, J.R., Walker, C.B., Ingalls, A.E., Könneke, M. and Stahl, D.A. (2008) Cultivation of a thermophilic ammonia oxidizing archaeon synthesizing crenarchaeol. *Environ. Microbiol.*, **10**(1), 810–818.

Di, H.J., Cameron, K.C., Shen, J.P., Winefield, C.S., O'Callaghan, M., Bowatte, S. and He, J.Z. (2010) Ammonia-oxidizing bacteria and archaea grow under contrasting soil nitrogen conditions. *FEMS Microbiol. Ecol.*, **72**(3), 386–394.

Francis, C.A., Roberts, K.J., Beman, J.M., Santoro, A.E. and Oakley, B.B. (2005) Ubiquity and diversity of ammonia-oxidizing archaea in water columns and sediments of the ocean. *PNAS*, **102**(41), 14683–14688.

Hureiki, L., Croué, J.P. and Legube, B. (1994) Chorine studies of free and combined amino acids. *Water Res.*, **28**(12), 2521–2531.

Jia, Z. and Conrad, R. (2009). Bacteria rather than Archaea dominate microbial ammonia oxidation in an agricultural soil. *Environ. Microbiol.*, **11**(7), 1658–1671.

Kasuga, I., Nakagaki, H., Kurisu, F. and Furumai, H. (2010a) Predominance of ammonia-oxidizing archaea on granular activated carbon used in a full-scale advanced drinking water treatment plant. *Water Res.*, **44**(17), 5039–5049.

Kasuga, I., Nakagaki, H., Kurisu, F. and Furumai, H. (2010b) Abundance and diversity of ammonia-oxidizing archaea and bacteria on biological activated carbon in a pilot-scale drinking water treatment plant with different treatment processes. *Water Sci. Technol.*, **61**(12), 3070–3077.

Könneke, M., Bernhard, A.E., de la Torre, J.R., Walker, C.B., Waterbury, J.B. and Stahl, D.A. (2005) Isolation of an autotrophic ammonia-oxidizing marine archaeon. *Nature*, **437**(7058), 543–546.

LeChevallier, M.W. and McFeters, G.A. (1990) Microbiology of activated carbon. In Drinking water microbiology, McFeters G.A. (Ed.), Springer, USA, 104–119.

Leininger, S., Urich, T., Schloter, M., Schwark, L., Qi, J., Nicol, G.W., Prosser, J.I., Schuster, S.C. and Schleper, C. (2006) Archaea predominate among ammonia-oxidizing prokaryotes in soils. *Nature*, **442**(7104), 806–809.

Liu, Z., Huang, S., Sun, G., Xu, Z. and Xu, M. (2011) Diversity and abundance of ammonia-oxidizing archaea in the Dongjiang River, China. *Microbiol. Res.*, **166**(5), 337–345.

Martens-Habbena, W., Berube, P.M., Urakawa, H., de la Torre, J.R. and Stahl, D.A. (2009) Ammonia oxidation kinetics determine niche separation of nitrifying Archaea and Bacteria. *Nature*, **461**(7266), 976–979.

Mertens, J., Broos, K., Wakelin, S.A., Kowalchuk, G.A., Springael, D. and Smolders, E. (2009) Bacteria, not archaea, restore nitrification in a zinc-contaminated soil. *ISME J.*, **3**(8), 916–923.

Mußmann, M., Brito, I., Pitcher, A., Sinninghe Damsté, J.S., Hatzenpichler, R., Richter, A., Nielsen, J.L., Nielsen, P.H., Müller, A. Daimes, H., Wagner, M. and Head, I.M. (2011) Thaumarchaeotes abundant in refinery nitrifying sludges express *amo*A but are not obligate autotrophic ammonia oxidizers. *Proc. Natl. Acad. Sci. USA*, **108**(40), 16771–16776.

Neufeld, J.D., Vohra, J., Dumont, M.G., Lueders, T., Manefield, M., Friedrich, M.W. and Murrell, J.C. (2007) DNA stable-isotope probing. *Nature protocol*, **2**(4), 860–866.

Nicol, G.W. and Prosser J. I. (2011) Strategies to determine diversity, growth, and activity of ammonia-oxidizing archaea in soil. *Methods in Enzymology*, **496**, 3–33.

Niu, J., Kasuga, I., Kurisu, F., Furumai, H. and Shigeeda., T. (2013) Evaluation of autotrophic growth of ammonia-oxidizers associated with granular activated carbon used for drinking water purification by DNA-stable isotope probing. *Water Res.*, **47**(19), 7053–7065.

Rotthauwe, J.H., Witzel, K.P. and Liesack, W. (1997) Efficiency of various water treatment processes in the removal of biodegradable and refractory organic matter. *Water Res.*, **31**(3), 639-649.

Stephen, J.R., Chang, Y.J., Macnaughton, S.J., Kowalchuk, G.A., Leung, K.T., Flemming, C.A. and White, D.C. (1999) Effect of toxic metals on indigenous soil β-subgroup proteobacterium ammonia oxidizer community structure and protection against toxicity by inoculated metal-resistant bacteria. *Appl. Environ. Microbiol.*, **65**(1), 95–101.

van der Wielen, P.W.J.J., Voost, S. and van der Kooij, D. (2009) Ammonia-oxidizing bacteria and archaea in groundwater treatment and drinking water distribution systems. *Appl. Environ. Microbiol.*, **75**(14), 4687–4695.

Wuchter, C., Abbas, B., Coolen, M.J.L., Herfort, L., van Bleijswijk, J., Timmers, P., Strous, M., Teira, E., Herndl, G.J., Middelburg, J.J., Schouten, S. and Sinninghe Damsté, J.S. (2006) Archaeal nitrification in the ocean. *Proc. Natl. Acad. Sci. USA*, **103**(33), 12317–12322.

Xia, W., Zhang, C., Zeng, X., Feng, Y., Weng, J., Lin, X., Zhu, J., Xiong, Z., Xu, J., Cai, Z. and Jia, Z. (2011) Autotrophic growth of nitrifying community in an agricultural soil. *ISME J.*, **5**(7), 1226–1236.

Zhang, L.M., Offre, P.R., He, J.Z., Verhamme, D.T., Nicol, G.W. and Prosser, J.I. (2010) Autotrophic ammonia oxidation by soil thaumarchaea. *Proc. Natl. Acad. Sci. USA*, **107**(40), 17240–17245.

Chapter 52

Influence of preozonation on the performance of subsequent bio-filter in water treatment under low temperature condition: Biomass and community analysis

J. Yang, J. Ma, D. Song, Y. Shi, X. Zhai*

State Key Laboratory of Urban Water Resource and Environment, Harbin Institute of Technology, PR China Address: 73 Huanghe Rd. Harbin 150090, China. majun@hit.edu.cn

ABSTRACT

Biomass and community are essential for the performance of drinking water biofilters. In the present study, two biological granular activated carbon (BGAC) filters, feeding with the water after coagulation-sedimentation with or without pre-ozonation treatment, were investigated over a period of 5 months under lower temperature condition. To explore the mechanism of treatment, BGAC particles were sampled from four different heights of the filter (10, 40, 70 and 110 cm), the active biomass and microbial community in the filter, based on 16S rRNA gene-DGGE, were monitored at the mature stage of two BGAC columns. The importance and role of preozonation has been demonstrated.

Key words Drinking water treatment; lower temperature; bio-filter; preozonation; microbial community structure

INTRODUCTION

Filtration by BGAC is proved to be an efficient method for the removal of undesirable dissolved organic carbon (DOC) and ammonia (NH_4^+-N) from water (Boon et al., 2011). Due to the possible presence of chemical contaminants (i.e. benzene, polychlorinated biphenyls (PCBs) and ammonia) (Grabow, 1996) in surface water in some developing areas and the demands on the available supply of high quality drinking water, more advanced technologies or upgrading process of the existing units are employed widely in water utilities (Xu et al., 2007). Therefore, single GAC/BGAC filter or in conjunction with ozonation is implemented in drinking water treatment to remove the organic material (Boon et al., 2011) and ammonia. During ozonation some DOM are oxidized to low molecular weight oxygen-containing organic carbon molecules, resulting in the conversion of some recalcitrant DOC molecules into biodegradable fraction of DOC (von Gunten, 2003; Hammes et al., 2006); thus, ozonation can increase microbial nutrient concentrations and stimulate bio-film formation on the filter media (Hammes et al., 2008). Subsequent removal of this biodegradable fraction is essential and contributes to the biological stability of the water, thereby reducing the likelihood of undesirable bacterial re-growth in the water distribution system (Van der Kooij et al., 1989). Since the major part of DOC removal during biofiltration can be attributed to biological process, this whole process can be of particular interest from a bioengineering perspective (Magic-knezev et al., 2004; Chien et al., 2008; Magic-knezev et al., 2009). Therefore, it is essential to quantify and characterize the microbial biomass, biological activity and to reveal the microbial community structure responsible for this process, especially the mechanism for the removal of organic carbon, ammonia, and nitrite et al. existing in drinking water sources by BGAC/O_3-BGAC filters under lower temperature condition (below 10°C).

In the present study, two parallel biofilters were investigated comparably over a period of 5 months under lower temperature condition. The objective of this study was to explore the influencing mechanism and to evaluate the extent of ozonation pretreatment on microbial community structure and biomass density in biofilters for poor nutritional drinking water treatment under lower temperature condition. These microbial data were compared to the organic carbon and ammonia removal in order to evaluate the extent of biofilters performance enhanced by preozonation.

MATERIALS AND METHODS

The raw water used throughout the current study was taken from the original water pipe of the water work directly and the water for general BGAC filter was treated through coagulation-sedimentation at the water works, the water for O_3-BGAC biofilter was treated after pre-ozonation. The values for water quality are shown in Table 52.1.

The BGAC filters are part of a drinking water treatment pilot-scale plant situated at the City H No.1 Waterworks. The BGAC media occupied 1.2m of the 2m column which had a diameter of 140mm. The setup consisted of two columns constructed with series of sampling ports for both water and media. The two reactors were continuously operated in up-flow mode with an average empty bed contact time (EBCT) in the columns being 30min more than 5 months.

In the pre-ozonation process, ozone was introduced into the ozone contact reaction tower directly from the ozone generator (CF-G-3-010g, produced by GUOLIN, China) with the dosage of 1mg/L. Ozone flow rate was 15mg/min when oxygen flow rate was controlled at 500mL/min.

All the GAC media were obtained from the Huaqing Company, Shanxi province, China. They were used directly without any pretreatment. The carbon is a wood-based granular activated carbon with a very open pore structure type produced by chemical activation using phosphoric acid. Information concerning the GAC media is given in Table 52.2.

Table 52.1 Summary of water quality during the period of winter in the plant

Parameters	Raw water	general BGAC filter		O_3-BGAC filter	
		Influent	Effluent	Influent	Effluent
Temperature (°C)	5.3~8.0	4.5~7.7	3.6~7.0	4.5~7.5	3.4~7.0
pH	7.53~7.69	7.36~7.48	7.41~7.56	7.32~7.45	7.45~7.61
DO (mg/L)	9.45~10.08	8.95~9.89	2.80~4.20	13. 61~16.69	0.48~2.09
Turbidity (NTU)	4.84~14.2	2.59~4.75	0.16~0.51	1.69~4.10	0.11~0.41
Ammonia (mg/L)	1.53~2.74	1.41~2.68	0.19~0.28	1.41~2.65	0.32~0.51
Nitrite (mg/L)	0.11~0.16	0.12~0.13	0.01~0.04	0.01~0.01	0.00~0.00
Nitrate (mg/L)	3.79~5.45	4.16~5.55	5.09~7.21	4.44~5.90	5.09~6.35
DOC (mg/L)	6.08~10.54	4.16~5.84	2.27~3.28	4.72~6.18	1.54~2.14
UV_{254}(cm-1)	0.079~0.092	0.042~0.051	0.018~0.024	0.030~0.036	0.009~0.014

Table 52.2 Characteristics of the GAC used in this study

Characteristics	parameters
Origin	Peach pit
Type of activation	Physical
Mechanical strength (%)	≥92
Iodine value (mg/g)	≥800
Effective size (mm)	1.2-2.4
Specific surface (m^2/g)	700-750
Size of max pores (μm)	<0.1
Bulk density (g/cm^3)	0.48
Moisture (%)	≤10
Ash content (%)	≤5

BGAC samples were collected with a special sampler inserted about 70mm into the biofilter through the purpose-build sampling ports in the filter bed. During the period, BGAC samples were taken every three weeks and the water quality was analyzed more than 3 times for average value prior to every BGAC media analysis.

The direct measurement of ATP contained on cells adhering to BGAC was performed based on a modified method used by Yu et al. (2002). Briefly, the BGAC particles were pretreated as described previously and about 2.0000 g (wet weight) was transferred to sterile iodine flask with 19mL mixed extraction systems (chloroform, methanol and water with volume ratio of 1:2:0.8) and kept eddying for 10min, then stand for 12h, added 5mL chloroform and water respectively and then stand for another 12h. The organic phase was obtained for detecting ATP. The resulting luminescence was converted to ATP concentrations using a calibration curve. BGAC DW was obtained as described previously.

In order to indentify the microbial community structure, PCR-DGGE analysis was employed. DNA was extracted directly from BGAC samples with Powersoil DNA kit. The genomic DNA extracted above was used as a template for PCR amplification. Small subunits of rDNA fragments were amplified with the most Bacteria and the archaea 16S rRNA gene V3 universal primers BSF338- GC(5'-CGCCCGCCGCGCGCGGCGGGCGGGGCGGGGGCACGGGGGGACTCCTACGG GAGGCAGCAG-3') and BAC805R-GACTACCAGGGTATCTAATCC.

RESULTS AND DISCUSSION

The pilot plant used in the present study was fed with actual surface water and therefore subject to natural changes in the influent. DOC, ammonia, natural bioactivity and other parameters in the raw water changed with the time. In the O_3-BGAC filter system, an ozonation step preceded the coagulation and sedimentation. DOC concentrations in influent of O_3-BGAC filter is a little higher than the general coagulation and sedimentation, mainly due to oxidation by hydroxyl radicals produced by decomposition of ozone and some recalcitrant and high molecular weight TOC molecules were oxidized to low molecular weight, resulting in an increase of DOC (von Gunten, 2003; Hammes et al., 2006). Some research work showed that ozonation severely damages algal cells, causing the release/formation of AOC (Müller et al., 2003; Hammes et al., 2007). At the same time preozonation can enhanced the efficiency of coagulation and sedimentation.

At this phase, DOC in inflow water was increased with the decrease of natural temperature as described in Fig. 52.1. The average DOC concentrations in the influent water of the general BGAC filter and O_3-filter were 5.1, 6.25, 7.4 mg/L and 5.2, 6.3, 7.6 mg/L respectively. Under the condition of this study, the average DOC removal efficiencies of the O_3-BGAC filter decreased from 65.4% to 61.8% and in general BGAC filter was from 52.9% to 47.3%, 13.8 percentage point more than the general BGAC filter on average. This removal is much higher to previous studies, which suggest that in surface waters anything between 5 and 49% of the total DOC can be present as BDOC and can be removed biologically (Servais et al., 1991; Volk and Le Chevallier, 2000). So in both the parallel biofilters in the present study, the higher removal efficiency of DOC may be attributed to two main reasons (i.e. the adsorption of the GAC and the sufficient bio-contact oxidation).

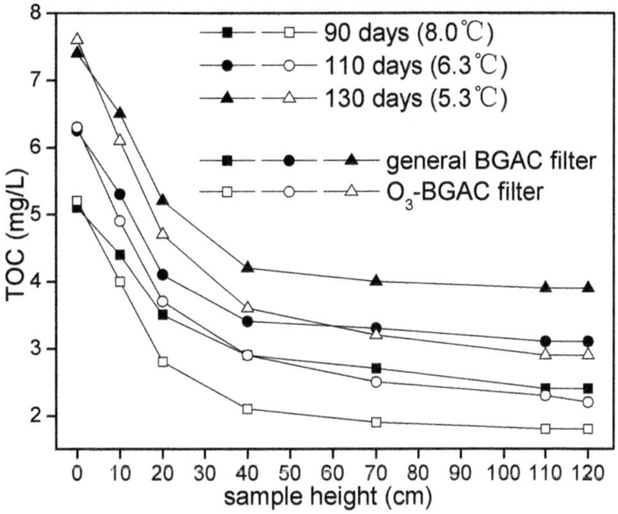

Figure 52.1 DOC filter profiles for three sampling days at steady period. Data in are the means of triplicate independent measurements with a standard deviation always below 5% for three replicate measurements every time.

Higher ammonia removal efficiency was observed at the first several sampling port of the O_3-BGAC filter. While the removal efficiency showed an opposite phenomenon at the upper part of the both biofilters, and the ammonia in the effluent of the O_3-BGAC was a little higher eventually. However, direct effect of pre-ozonation on nitrifier species was not detected using molecular technology set forth in the next section.

Although concentration of dissolved oxygen (DO) in the influent to O_3-BGAC filter was higher than in the general BGAC filter, Fig. 52.1 shows DOC concentration in the effluent of O_3-BGAC filter was lower than in the general BGAC. Table 52.1 shows more DO was consumed which was observed in the O_3-BGAC filter at the same relatively sufficient EBCT. This result is consistent with the removal efficiency of DOC in the biofilter, which may be attributed to the increased biodegradable organic loads by ozone oxidation and higher heterotrophic biomass in the system. Previously research showed that more DO was consumed by outer heterotrophic layer utilizing organic carbon (2~3mg O_2/mg C) inside the reactor. With increasing height, O_3-BGAC filter was gradually losing its advantage of higher DO enhanced by pretreatment. Depth-wise stratification of nitrifiers and heterotrophs was observed at all filter bed depths in previous study (van den et al., 2011). The mass transfer was also restricted by the high biofilm polysaccharide produced by increased heterotrophs. The chemolithotrophic genera e.g. Nitrosomonas sp., Nitrosospira sp., Nitrosococcus, and Nitrosovibrio sp. which play an important role at oxidizing ammonia to nitrite are all aerobic and distributed in inner layer of the biofilm. Previous studies which had profiled oxygen within nitrifying biofilms using microelectrodes have found these depths for autotrophs to be oxygen limited (Okabe et al., 1999). All these subsequently impeded the filter's ability of the upper part to maintain nitrification. The diversity of ammonia removal efficiency in both filters as shown above in Fig. 52.2 interestingly was consistent with the theory.

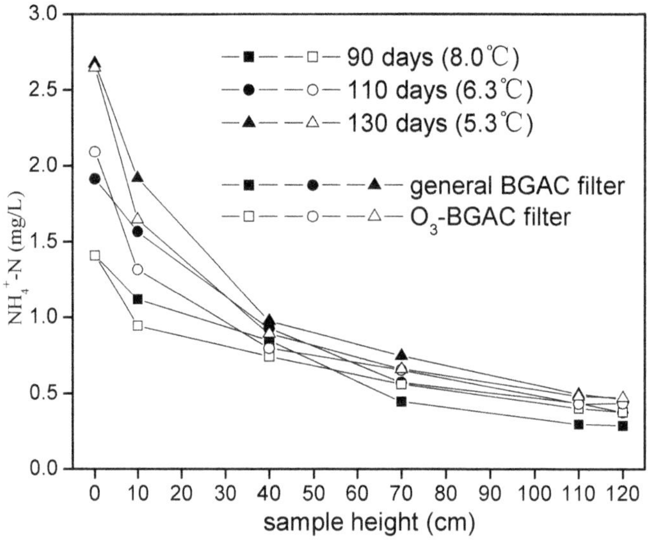

Figure 52.2 Ammonia distributed in the filters for three sampling days at steady state. All data are the means of triplicate independent measurements with a standard deviation always below 5% for three replicate measurements every time.

The average biomass concentration (n=3) for each level during the steady state period was shown in Fig. 52.3. The highest biomass concentration at the general BGAC filter and the O_3-BGAC one (243.63 nmolATP/g (dw BGAC) and 430.16nmolATP/g (dw BGAC), respectively.) were established both at the first sampling point (BGAC1, 10cm from the entry) and decreased abruptly thereafter by average factors of 3.8 and 4.0 respectively to the top of the biofilter.

The previous study showed that the adsorption in the bottom layers would create a nutrient-rich micro-environment on the GAC, which is favorable to maintain the microorganism growth (Herzberg et al., 2003). Regarding the decreasing profile of biomass concentration along the reactors (Urfer and Huck, 2001; Velten et al., 2011), a considerable amount of biomass had established evidently in the O3-BGAC filter and the decreased factor is relatively higher than the general one by 0.2. This just explained the high microorganism growth at the whole O_3-BGAC filter with the higher biodegradable nutrient and DO as the previous study (Urfer and Huck, 2001), especially at entry point; thus, ozonation can stimulate bio-film formation on the front filter media (Hammes et al., 2008). For most resolvable materials, biodegradation acts as the essential mechanism in the contaminant removal by BGAC filters, rather than physical filtration (Kim et al., 1997; Lin et al., 2001), which was consistent with the lower DOC removal efficiency in the upper layer with small biomass and lower bioactivity as shown previously.

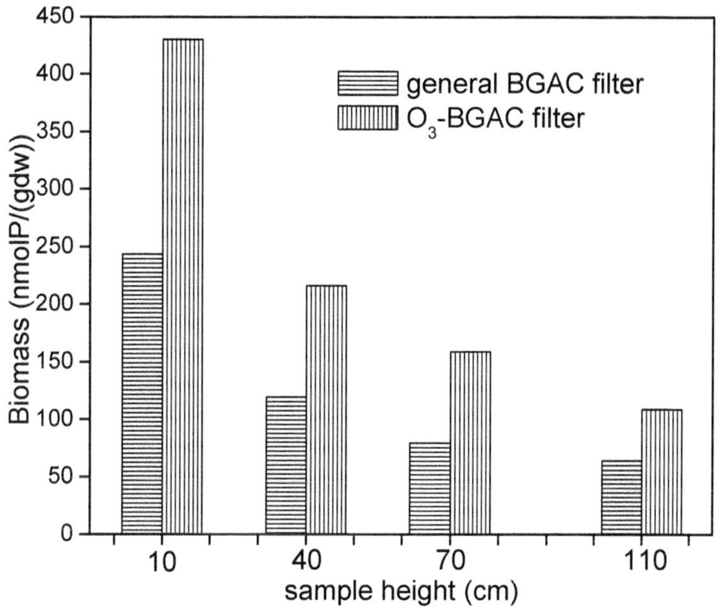

Figure 52.3 Biomass concentration filter profiles for three sampling days at steady period, All data are the means of the three different point in time, with a standard deviation always below 5% for three replicate measurements every time.

PCR-DGGE analysis was employed to study the microbial populations in this work. By amplifying 16SrDNA of bacteria and separating the PCR products with DGGE, DNA fragments were separated, excised and sequenced to substantiate the presence of dominant bacteria in the both bio-filters. A total of 21 bands were excised and yielded unanimous sequences for further analyses. By BLAST search algorithm, high similarity values were found after DNA sequencing, and these results were confirmed by Phylogenetic analysis.

The results revealed a more complex heterotrophic microbial community structure distributed throughout the O_3-BGAC filter. Actinobacterium was dominant at the lower layer of the both biofilters. It belongs to Actinobacteria which is playing an important role in decomposition of organic materials, such as cellulose and chitin, and thereby playing a vital part in organic matter turnover and carbon cycle. However, in the upper part of the general BGAC filter no Actinobacterium was detected, while it still was the dominant one in O_3-BGAC filter. Previous studies showed that the bacteria from the phylum Proteobacteria often dominate in freshwater ecosystems (Zwart et al., 2002); including chlorinated drinking water (Eichler et al., 2006). Beta-Proteobacteria and Alpha-proteobacteria bacterium also played an important role at the lower part of the both filters, especially Alphaproteobacteria bacterium in the whole O_3-BGAC filter.

The performance of the O_3-BGAC system in water treatment was improved through the whole filtration. Nitrospira and Nitrosomonas species detected in O_3-BGAC filter was richer. In the O_3-BGAC filter one strain (Uncultured Thiohalomonas sp. (GQ420958.1)) was identified as a Thiohalomonas denitrificans-like Gammaproteobacteria species, with a similarity of 94%. They grew anaerobically as complete denitrifiers, and aerobically under micro-oxic conditions. This indicated that the ozone preoxidation in the present study did not show any significant impact on the diversity of bacteria in the biofilters.

CONCLUSIONS

From the experiment results explained in the former parts, the following conclusions can be drawn.

(1) The determination of Biomass and PCR-DGGE has been successfully applied to describe the biological filtration system and can be a valuable tool for the characterization of biological filters.

(2) Ozone pre-oxidation had a positive effect on the biomass of microorganism in biofilters for poor nutritional drinking water treatment under lower temperature condition.

(3) The biodegradability of the organic matters in the influent can be promoted by ozone pre-oxidation while the higher biomass of heterotrophic organism in O_3-BGAC filters had a negative effect on the ammonia removal efficiency by depriving the ammonia oxidation bacteria (AOB) of DO and nutrient material.

(4) The ozone preoxidation in the present condition did not show any significant impact on the diversity of bacteria in the biofilters.

ACKNOWLEDGEMENTS

The study is supported by the National Science and Technology Major Project of Water Pollution and Control (index: 2009ZX07424-006) in China and College of Environment and Resources Zhejiang University. We appreciate the assistance from the laboratory of Environmental Microbiology and special thanks to Ph.D Yajuan Xing who shared her experiences about PCR-DGGE analyses. Many thanks also to the Capital Water Co., Ltd.

REFERENCES

Boon, N., Pycke, B.F.G., Marzorati, M. and Hammes, F. (2011) Nutrient gradients in a granular activated carbon biofilter drives bacterial community organization and dynamics. Water Research 45(19), 6355-6361.

Chien, C.C., Kao, C.M., Chen, C.W., Dong, C.D., Wu, C.Y. (2008) Application of biofiltration system on AOC removal: column and field studies. Chemosphere 71, 1786-1793.

Eichler, S., Christen, R., Holtje, C., Westphal, P., Botel, J., Brettar, I., Mehling, A. and Hofle, M.G. (2006) Composition and dynamics of bacterial communities of a drinking water supply system as assessed by RNA- and DNA-based 16S rRNA gene fingerprinting. Appl Environ Microbiol., 72(3), 1858-1872.

Grabow, W. (1996) Waterborne diseases: update on water quality assessment and control. Water South Africa, 22 (2), 193-202.

Hammes, F., Salhi, E., Köster, O., Kaiser, H.-P., Egli, T. and von Gunten, U. (2006) Mechanistic and kinetic evaluation of organic disinfection by-product and assimilable organic carbon (AOC) formation during the ozonation of drinking water. Water Research, 40(12), 2275-2286.

Hammes, F., Meylan, S., Salhi, E., Koester, O., Egli, T., von Gunten, U. (2007) Formation of assimilable organic carbon (AOC) and specific natural organic matter (NOM) fractions during ozonation of phytoplankton. Water Res., 41 (7), 1447-1454.

Hammes, F., Berney, M., Wang, Y., Vital, M., Köster, O. and Egli, T. (2008) Flow-cytometric total Bacterial cell counts as a descriptive microbiological parameter for drinking water treatment processes. Water Research, 42(1-2), 269-277.

Herzberg, M., Dosoretz, C.G., Tarre, S. and Green, M. (2003) Patchy biofilm coverage can explain the potential advantage of BGAC reactors. Environ. Sci. Technol., 37(18), 4274-4280.

Kim, W.H., Nishijima, W., Shoto, E. and Okada, M. (1997) Pilot plant study on ozonation and biological activated carbon process for drinking water treatment. Water Science and Technology, 35(8), 21-28.

Lin, C.-K., Tsai, T.-Y., Liu, J.-C. and Chen, M.-C. (2001) Enhanced biodegradation of petrochemical wastewater using ozonation and BGAC advanced treatment system. Water Research, 35(3), 699-704.

Magic-Knezev, A. and van der Kooij, D. (2004) Optimisation and significance of ATP analysis for measuring active biomass in granular activated carbon filters used in water treatment. Water Research, 38(18), 3971-3979.

Magic-Knezev, A., Wullings, B., Van der Kooij, D. (2009) Polaromonas and Hydrogenophaga species are the predominant Bacteria cultured from granular activated carbon filters in water treatment. J. Appl. Microbiol., 107, 1457-1467.

Müller, K.C., Forster, R., Gammeter, S., Hambsch, B. (2003) Influence of ozonated cyanobacteria on bacterial growth in rapid sand filters. J. Water SRT-Aqua, 52 (5), 333-340.

Okabe, S., Satoh, H. and Watanabe, Y. (1999) In situ analysis of nitrifying biofilms as determined by in situ hybridizaion and the use of microelectrodes. Applied and Environmental Microbiology, 65(7), 3182-3191.

Servais, P., Billen, G., Ventresque, C., Bablon, G.P. (1991) Microbial activity in GAC filters at the Choisy-Le-Roi treatment-plant. J. Am.Water Works Assoc., 83 (2), 62-68.

Urfer, D. and Huck, P.M. (2001) Measurement of biomass activity in drinking water biofilters using a respirometric method. Water Research, 35(6), 1469-1477.

van den Akker, B., Holmes, M., Pearce, P., Cromar, N.J. and Fallowfield, H.J. (2011) Structure of nitrifying biofilms in a high-rate trickling filter designed for potable water pre-treatment. Water Research, 45(11), 3489-3498.

van der Kooij, D., Hijnen, W.A.M. and kruithof, J.C. (1989) The effects of ozonation ,biological filtration and distribution on the concentration of easily assimible organic-carbon (AOC) in drinking-water. Ozone Sci. & Eng., 11(3), 297-311.

Velten, S., Boller, M., Köster, O., Helbing, J., Weilenmann, H.-U. and Hammes, F. (2011) Development of biomass in a drinking water granular active carbon (GAC) filter. Water Research, 45(19), 6347-6354.

Volk, C.J., LeChevallier, M.W. (2000) Assessing biodegradable organic matter. J. Am.Water Works Assoc., 92 (5), 64-76.

von Gunten, U. (2003) Ozonation of drinking water: Part II. Disinfection and by-product formation in presence of bromide, iodide or chlorine. Water Research, 37(7), 1469-1487.

Xu B., Gao N. Y., Sun X. F., Xia S. J., Simonnot M. O., Causserand C., et al. (2007) Characteristics of organic material in Huangpu River and treatability with the O3-BGAC process. Separation and Purification Technology, 57 (2): 348-355.

Yu, X., Zhang, X.J. and Wang, Z.S. (2002) Biomass examination by Lipid-P method for drinking water bio-treatment. Water & Wastewater Engineering, 28(5), 1-5.

Zwart, G., Crump, B.C., Kamst-van Agterveld, M.P., Hagen, F. and Han, S.K. (2002) Typical freshwater bacteria: an analysis of available 16S rRNA gene sequences from plankton of lakes and rivers. Aquatic Microbial Ecology, 28(141-155).

Chapter 53

Deterioration mechanisms of granular activated carbon in fluidized-bed contactor with biological and physical effects

Y. Kobayashi, M. Oya*, Y. Tanaka*, D. Nagashio*, T. Hanamoto**
** Hanshin Water Supply Authority, 3-20-1, Nishi-Okamoto, Higashinada-ku, Kobe, Hyogo, JAPAN*

ABSTRACT

Since 1993, the Hanshin Water Supply Authority (HWSA) has adopted an advanced water treatment system using an ozone contactor and a granular activated carbon (GAC) fluidized-bed contactor. Since the start of operation, 20% of the GAC has been annually replaced with new GAC, maintaining the physical adsorption effect and biological effect. However, ten years after the start of operation, mud balls were detected in the lower layer of the contactor. To deal with this problem, we investigated the deterioration of GAC in different layers. In the upper layer, the grain diameter of GAC was small and its iodine-adsorbing potential decreased. In the lower layer, the grain diameter of GAC was large and the amount of accumulated manganese was also large. This was because of the deterioration and erosion of GAC resulting from the exposure to ozone, and the separated small-diameter GAC grains moved to the upper layer. The formation of the mud balls was probably due to the agglomeration of GAC in the presence of polymers produced by rhizopods.

Keywords: granular actvated carbon; deterioration; fuluidized-bed contactor; mud ball; grain size

INTRODUCTION

The Hanshin Water Supply Authority (HWSA) produces 1,128,000 m3/day of drinking water using the Yodo River as the source of the water and supplies it to 2,500,000 people. Since 1993, HWSA has adopted an advanced water treatment system, as shown in Fig. 53.1. This system incorporates an ozone contactor, a granular activated carbon (GAC) fluidized-bed contactor, and a recoagulation reactor in a conventional rapid filtration system involving coagulation-sedimentation and sand filtration. Since 2001, all the water supplied by HWSA has been processed by the advanced water treatment system. The main aims of this system are to remove substances that cause unusual odor and taste and other chemical substances and to suppress the generation of disinfection by-products (Sasaki T., 1989). In the system, raw water is first treated by coagulation-sedimentation and then subjected to ozonation. Subsequently, adsorption by GAC is performed using an upflow fluidized-bed contactor, which is effective for increasing the effect of the subsequent recoagulation process and suppressing the water head loss before the final sand filtration process. Moreover, the removal of ozone exhaust is not required in the GAC fluidized-bed contactor, enabling the system to have a simple structure. To ensure the high treatment performance of the upflow fluidized-bed contactor, the flow straightener at the bottom should be constantly in good condition to maintain the favorable fluidic state of GAC. Therefore, the system is controlled to keep ozone in the water introduced into the GAC fluidized-bed contactor. In addition, suspended solid (SS) components remaining in the contactor are removed by air washing once a day (SS removal process) (Sasaki T., 1990).

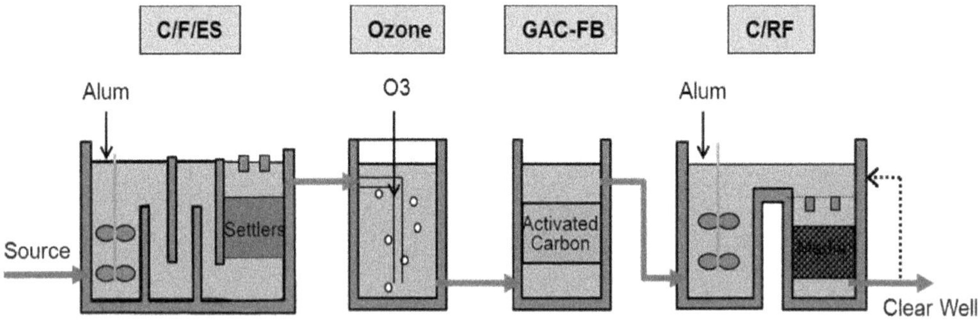

C/F/ES: coagulation / flocculation / enhanced sedimentation,
GAC-FB: granular activated carbon fluidized bed, C/RF: coagulation / rapid filtration

Figure 53.1 Advanced water treatment system of HWSA

GAC placed in the fluidized-bed contactor exhibits a biological effect (as biological activated carbon) with increasing use and thus remove ammonia nitrogen (Omote Y., 1990). Since the start of operation of the GAC fluidized-bed contactor, HWSA has annually replaced 20% of the GAC with new GAC of the same quantity in all the contactors with the aim of achieving a steady removal rate of substrates on the basis of the biological effect and physicochemical adsorption effect. Figure 53.2 shows the cumulative reduction in the trihalomethane formation potential (THMFP) achieved by ozonation and GAC adsorption. The cumulative reduction tends to increase linearly with the number of years of operation. This indicates that the annual replacement of 20% of the GAC results in a constant rate of removal of disinfection by-product precursors. By adopting the advanced water treatment system, the total trihalomethane concentration in the finished water was reduced to about one-third of that before adoption, greatly contributing to the reduction in the concentration of disinfection by-products.

The biological effect in the GAC fluidized-bed contactor decreases as a result of the replacement of GAC. The above high removal rate means that the biological effect is continuously obtained because the proportion of organisms that settle in the contactor is stable as a result of the operation of the fluidized bed. The biological effect in the GAC fluidized-bed contactor reduces the need for chlorine, resulting in a reduction in the dose of chemicals.

The annual replacement of GAC enables the supply of GAC with a predetermined grain diameter distribution and contributes to the stabilization of the properties of the GAC. However, discolored GAC, viscous GAC, and 10-cm-diameter mud balls were detected in a GAC fluidized-bed contactor that had been in operation for more than 10 years since the start of operation (Photograph 53.1). This was probably due to the deterioration of GAC with time. We investigated the properties of GAC and the accumulation of metals in the upper, middle, and lower layers of the GAC fluidized-bed contactor as well as their changes with time and examined the deterioration mechanisms of GAC.

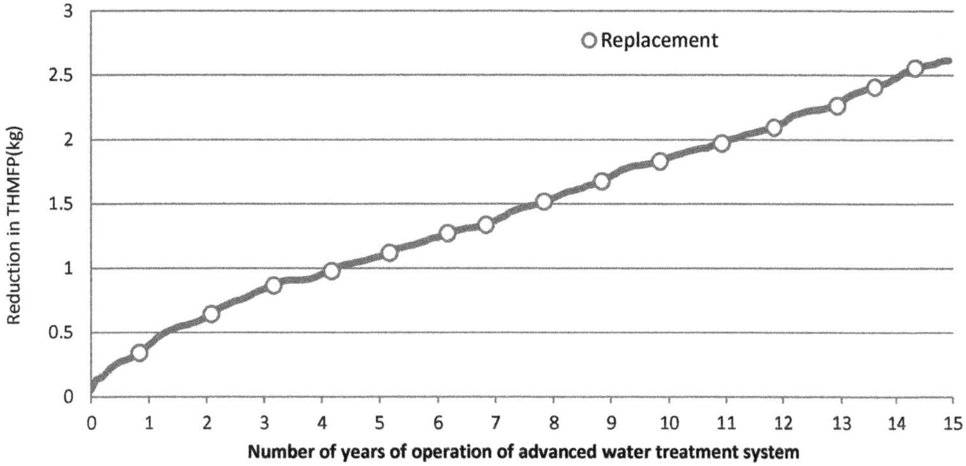

Figure 53.2 Cumulative reduction in THMFP achieved by ozonation and GAC adsorption

Photograph 53.1 Mud balls

METHODS

Sampling

In November 2006, we replaced all the GAC in a GAC fluidized-bed contactor with new GAC with the aim of understanding the state of deterioration of GAC and returning to a normal condition in the contactor. The target contactor was located in the Inagawa Water Treatment Plant of HWSA and had been in continuous operation for 13 years.

Tables 53.1 and 53.2 show the specifications of the target contactor and the as-purchased GAC, respectively. When GAC was removed from the contactor, GAC samples were collected from three layers at different depths (lower layer, 0-700 mm; middle layer, 700-1,400 mm; upper layer, 1,400-2,140 mm) of the GAC fluidized-bed contactor via the system at the bottom. Mud balls found in the lower layer were also sampled. The new GAC used for replacement was used as a control.

The GAC was sampled every three months for three years after its complete replacement. We analyzed the iodine-adsorbing potential, the amounts of accumulated metals (calcium, aluminum, and manganese), and the attached bacteria (heterotrophic and nitrifying bacteria).

Table 53.1 Specifications of GAC fluidized-bed contactor

Type	Upflow fluidized bed
Area for adsorption	47.2m^2
Bed thickness	2.14m

Table 53.2 Specifications of GAC (as purchased)

Material	Coal-based GAC
Hardness	≥85%
Specific surface area	≥950 m^2/g
Pore volume	≥0.25 mL/g
Amount of adsorbed iodine	≥950 mg/g
Grain diameter distribution	Approximately 0.250-

Analysis

The grain size distribution of the GAC samples from each layer was determined by a test in accordance with Japanese Industrial Standards (JIS) K 1474. The GAC samples were crushed so that at least 90% of them passed through a 45-μm-mesh net sieve. The crushed GAC samples were heated and dried at 115 oC for at least 3 h and then cooled in a desiccator to measure the iodine-adsorbing potential and the amounts of accumulated metals.

The measurement of the iodine-adsorbing potential followed Japan Water Works Association (JWWA) A 114. The amounts of accumulated metals were measured as follows. A GAC sample of 0.200 g and 10 mL of hydrochloric acid (1+4) were poured into a beaker and boiled for 10 min using a hot plate. After cooling, the solution was transferred to a 50 mL volumetric flask, into which 1 M hydrochloric acid was added to obtain a 50 mL sample solution. The solution was filtered using a membrane filter of 0.45 μm pore diameter, and the concentrations of metals in the filtrate were measured using an atomic absorption spectrometer. The obtained concentrations were converted to those per gram of the GAC sample to obtain the amounts of accumulated metals.

For the observation of attached bacteria, heat-sterilized purified water was added to the GAC samples obtained by removing moisture, then the samples were subjected to ultrasound at 40 W for 2 min. The supernatant solution was sampled to examine the bacterial counts. The mud balls found in the lower layer were observed by optical microscopy.

RESULTS AND CONDITIONS

Figure 53.3 shows the grain size distributions of GAC in different layers of the GAC fluidized-bed contactor that had been in continuous operation for 13 years. Table 53.3 shows the iodine-adsorbing potential and the amounts of accumulated metals for each layer.

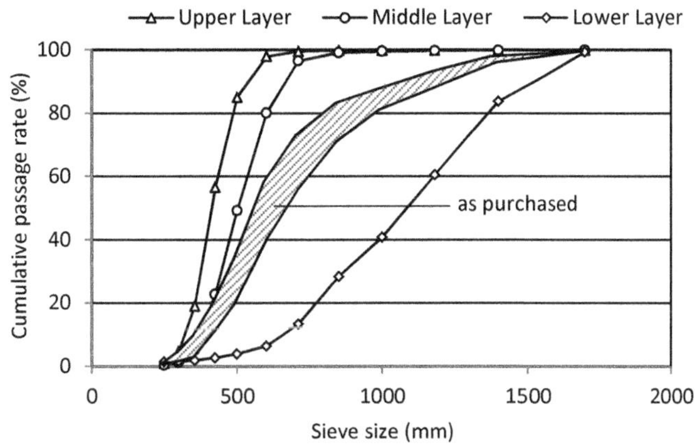

Figure 53.3 Grain diameter distribution for different layers 13 years after start of operation

Table 53.3 Iodine-adsorbing potential and amounts of accumulated metals for different layers

		Upper layer	Middle layer	Lower layer	New carbon
Iodine-adsorbing potential (mg/g)		180	240	460	930
Metals	Calcium (mg/g)	21	15	9.2	0.5
	Aluminum (mg/g)	24	15	17	4.2
	Manganese (mg/g)	1.9	2.1	9.3	0

In the upflow GAC fluidized-bed contactor, large-diameter GAC grains were concentrated in the lower layer, whereas small-diameter GAC grains were distributed in the middle and upper layers. The iodine-adsorbing potential of GAC in all three layers was lower than that of new carbon. Among the three layers, the lower layer exhibited the highest iodine-adsorbing potential, followed by the middle layer, then the upper layer. This was considered to be because the surface of GAC deteriorated and was eroded as a result of the contact between the GAC grains and the exposure to ozone, and the separated small-diameter GAC grains moved to the upper layer. The amounts of accumulated calcium and aluminum were greater in the upper layer than in the lower layer. In contrast, the amount of accumulated manganese was large in the lower layer.

Figure 53.4 shows the change in the iodine-adsorbing potential of GAC with time for the three layers. In the first year after the complete replacement of GAC, the iodine-adsorbing potential decreased, but no marked difference was observed among the three layers. This indicates that the adsorption occurred in all the layers for the first year. After one year, however, a difference in the iodine-adsorbing potential started to develop among the three layers, with the iodine-adsorbing potential tending to be lower in the upper layer and high in the lower layer.

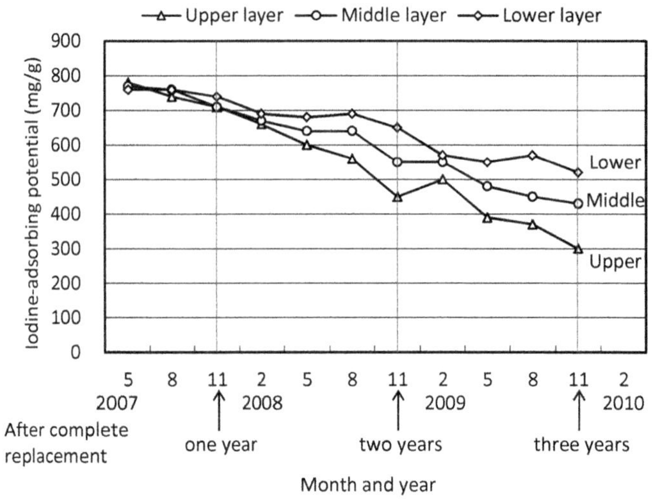

Figure 53.4 Change in iodine-adsorbing potential with time

Figures 53.5 and 53.6 show the changes in the amounts of accumulated calcium and manganese with time, respectively. One year after the complete replacement of GAC, no marked differences in the amounts of these accumulated metals were observed among the three layers. After one year, more calcium had accumulated on GAC in the upper layer, whereas more manganese had accumulated on GAC in the lower layer. In general, the surface area per unit weight of small-diameter GAC is larger than that of large-diameter GAC. Because the amounts of accumulated calcium and aluminum were large in the upper layer, in which small-diameter GAC grains were distributed, these metals were considered to uniformly accumulate on the surface of the GAC particles in all the layers of the contactor. In contrast, manganese that became soluble as a result of ozonation was subjected to the biological effect and physicochemical reaction in the GAC fluidized-bed contactor to become insoluble again and mostly accumulated on the GAC in the lower layer near the inlet.

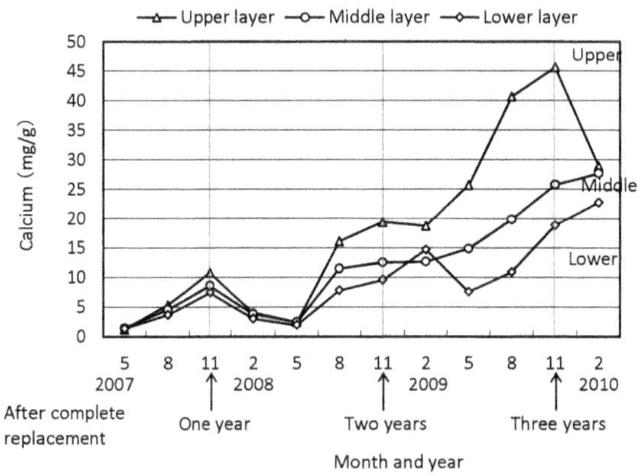

Figure 53.5 Change in amount of accumulated calcium with time

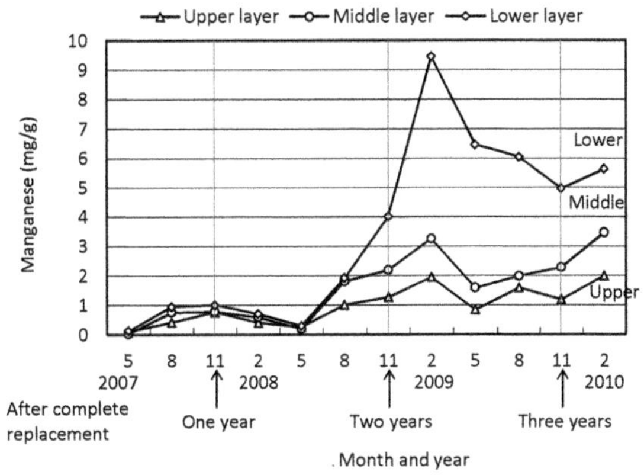

Figure 53.6 Change in amount of accumulated manganese with time

There were no significant differences in the heterotrophic and nitrifying bacterial counts among the three layers.

The mud balls (Photograph 53.1) were black and coreless. Optical microscopy revealed that many rhizopods called *Euglypha* were present in the mud balls (Photograph 53.2). Therefore, it is considered that the GAC agglomerated in the presence of polymers produced by the organisms to form the mud balls. The mud balls consisted of small-diameter GAC at the center and larger-diameter GAC outside. This means that GAC grains came in contact with each other owing to the friction during the operation of the fluidized bed and washing, and the separated GAC grains moved to the upper layer to form a core, which grew and moved to the lower layers, where they settled.

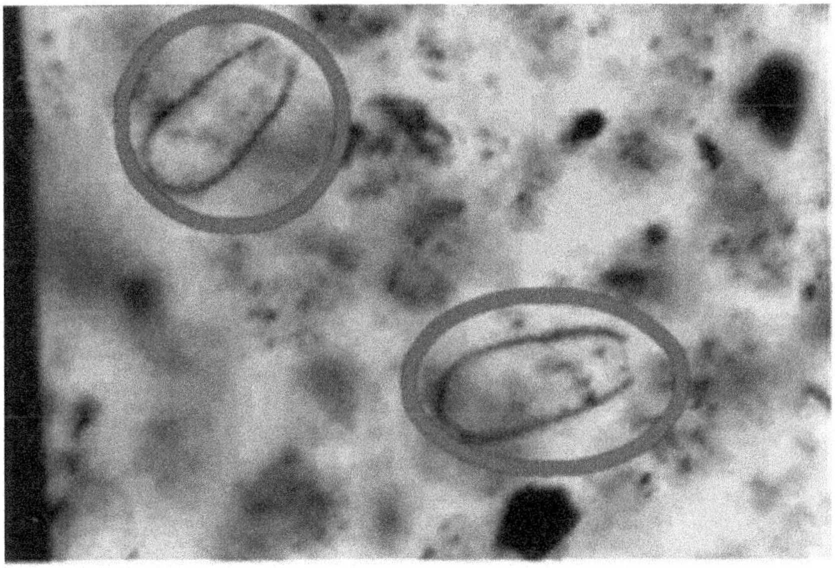

Photograph 53.2 *Euglypha*

CONCLUSIONS

In this study, we investigated the properties of GAC and the accumulated metals and attached bacteria in different layers of the GAC fluidized-bed contactor with the aim of clarifying the state of deterioration of GAC in the contactor in terms of its biological and physical effects. The results indicated that the deterioration of the GAC generally progressed from the upper layer to the lower layer. It is speculated that the surface of GAC deteriorated and was eroded as a result of the contact between GAC grains and the exposure to ozone and that the separated small-diameter GAC grains moved to the upper layer. The mud balls found in the lower layer were considered to have formed in the presence of polymers produced by the organisms.

REFERENCES

Omote Y., Suhara T., Momoya S. Hanamoto T. (1990) Experiment on advanced treatment by ozone and activated carbon fluidized bed.3.Study using demonstration plant (in Japanese). 41th JWWA Annual Conference Proceedings, Japan, 272-274.

Ose R., Michishita K., Yamane H., Nagashio D.(2007) Operation and maintenance of ctivated carbon fluidized bed (II). (in Japanese). 58th JWWA Annual Conference Proceedings, Japan, 234-235.

Sasaki T., Kobayashi K., Yagi Y. (1987) Investigation on the introduction of ozone and granular activated carbon process (in Japanese). 38th JWWA Annual Conference Proceedings, Japan, 201-203.

Sasaki T., Hanamoto T., Kobayashi K. (1990) Fluidized bed adsorption equipment using granular actvated carbon (in Japanese). 41th JWWA Annual Conference Proceedings, Japan, 269-271.

Chapter 54

Living-Filter: an in-reservoir biofiltration system for phytoplankton reduction at the abstraction point

A. T. Castro-Castellon, M. J. Chipps**, J. M. R. Hughes***, N. P. Hankins**

**Department of Engineering Science, University of Oxford, Parks Road, Oxford, OX1 3PJ, UK*

***Thames Water Innovation, Thames Water Utilities Ltd., Reading, UK*

****Department for Continuing Education, University of Oxford, Wellington Square, Oxford, OX1 2JA, UK*

ABSTRACT

The Living-Filter (L-F), a hydroponic bed of floating-reeds, was designed by Thames Water and AquaticEngineering and installed in Farmoor II reservoir, Oxfordshire, in July 2012. The structure (1m H x 21m W x 10m L) supports plants of *Phragmites australis*, *Phalaris arundinacea* and *Carex acutiformis*, with a combined system of curtains and baffles. The aim is to reduce phytoplankton loading onto the water treatment works at the abstraction point, consequently reducing chemical dosage and minimizing costs. Weekly surveys were carried out from July – October 2013 to assess the physical, chemical and biological changes likely to be introduced by the L-F. Physicochemical and biological variables were measured at 16 sampling sites. Results show differences between sites in the L-F with a significant accumulation in chlorophyll *a* (Chla) (as a surrogate of phytoplankton biomass) levels in the L-F platform, although significant Chla reduction has yet to be seen at the abstraction point.

Keywords biofiltration; artificial floating reed-beds; phytoplankton control, Living-Filter

INTRODUCTION

Eutrophic drinking water reservoirs are seasonally affected by nuisance phytoplankton blooms, particularly cyanobacteria [1]. The water industry has generally solved the problem by increasing the number of processing stages for the treatment of reservoir water using new chemicals or innovative technology at the treatment plant [2]. The L-F is a novel technology consisting of a hydroponic floating reed-bed structure (210 m^2). Three reed species are rooted in a non-enriched polyamide matrix. The rationale underpinning the design is to create a biofiltration system driven by a combination of biological and artificial components. The L-F plays a dual role. One role is providing refuge for filter-feeder zooplankton, which in the water column increase their grazing pressure on phytoplankton. The other role is carried out *in situ* by the biofilm attached mainly on the roots and matrix. The matrix and the rhizosphere might also increase the contact time before the water enters the WTW, enhance the physical entrapment of suspended particles, provide refuge for zooplankton and increase the surface area for periphyton/biofilm attachment [3, 4, 5]. Furthermore, a system of curtains and baffles aids the water in flowing through the L-F, hence, all the water abstracted at the tower inlet has passed through the L-F. The L-F may therefore be regarded as an in-reservoir pre-filter treatment, which forms the first stage of the water treatment process. This research aims to evaluate the effectiveness of the L-F at reducing phytoplankton biomass onto the WTW each year from 2013 to 2015, and to optimize the design of the LF. This paper specifically looks at the period from July to October 2013. The following research question was posed for the purposes of this paper: Are there chemical and biological differences between the L-F platform and the inflow and outflows? The following were specific objectives: a. Identify and quantify changes in the physico-chemical variables between the sampling sites. b. Identify and quantify changes in the biological variables between the sampling sites.

METHODOLOGY

To determine changes in the physico-chemical and biological characteristics of the flow pathways through the L-F to the abstraction point, surveys were carried out on a weekly basis since July 2013. A total of 16 sampling sites were identified in Farmoor II reservoir and the L-F structure in Stages 1 to 4 (Figure 54.1).

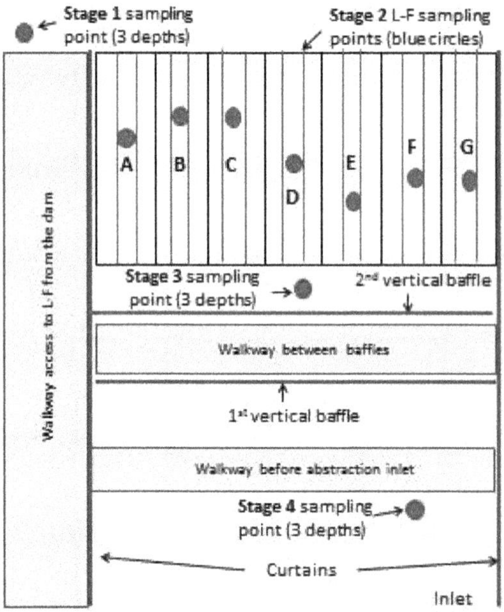

Figure 54.1 A bird's view schematic of the L-F with the location of the 16 sampling sites in Stage 1 to 4. Red circles: (Stage 1, 3, 4) sampling sites at 3 depths.
Blue circles (Stage 2): sampling sites within the L-F platform.
The location of the walkway access, curtains and baffles (red lines) are also shown. (Not to scale).

Sampling point Stage 1 is 2 m to the left of the L-F curtain. The water enters the L-F at Stage 2, the plant-bed structure, which consists of seven rafts, each of 3 x 1m W x 10 m L modules, and here seven sampling sites were identified. From left to right rafts A to C contain primarily *Phragmites australis* (common reed), whilst rafts D and E contain *Phalaris arundinacea*, and rafts F and G contain *Carex acutiformis*. The stage 3 sampling point occurs immediately after water has passed through the L-F and Stage 4 is 10 m from the abstraction inlet. Depth samples were taken at Stage 1, 2 and 3. The selected depths (TOP, MID and BOT) were based on the clinograde observed from the historical and current surveys. The thermocline was generally weak (1°C -1.5°C difference in the water column) from top (0 m) to bottom (11 m) and variable, the reservoir is exposed to wind action, which is evidenced in its popularity for sailing activities. Dissolved oxygen (DO) and temperature (°C) depth profiles were taken with a Hach Intellical LDO rugged sensor. The stratification in the reservoir comprises: epilimnion (TOP: 0 to 4m), metalimnion (MID: 4 to 7 m) and hypolimnion (BOT: 8 to 11m). Samples for abiotic and biotic analyses were collected with a Van Dorn sampler (2L capacity), between 09:00 and 13:00 hours. Samples were pooled and transferred to a plastic container (~5.5L). The Van Dorn is deployed at least three times per depth measurement (Stages 1, 3 and 4). Samples from Stage 2 (A, B, C, D, E, F and G) are taken with a Phil sampler, which consists of a rod of 1.5m length with a 500ml removable bottle for sample collection. This sampler is refilled 10 times per sampling point to fill up the 5.5L containers. The sampling sites were constructed by cutting a square of 20cm at the top of the cage mesh (inside the cage is where the roots and the matrix are contained). A short piece (50 –70cm length) of pipe (15cm diameter) was then introduced and held in place with cable ties. The sample is obtained by plunging the Phil sampler into the cylinder, only one sample depth (~90cm) can be taken due to the restrictions of the physical barrier imposed by the bottom of the cage. Samples are kept in a shaded environment until they are transferred to specific bottles for analysis, following the Standard Operational Procedure of TW laboratory, a United Kingdom Accreditation Service (UKAS) laboratory. To address the question and meet the objectives, subsamples from the containers were taken for pH analysis, conductivity, chloride, total suspended solids, turbidity, alkalinity, total oxidised nitrogen ammonia, silica, biological and chemical oxygen demand, total phosphorus (TP) and soluble reactive phosphorus, total organic, particulate (POC) and dissolved organic carbon and chlorophyll *a* (Chla) (as surrogate of phytoplankton biomass)

RESULTS AND DISCUSSION

Results for the abiotic variables did not show significant differences between samples in Stage 2 compared to Stages 1, 3 & 4, except for TP. TP increased steadily at Stage 2 from Week 7 (100µg L^{-1}), whilst increasing at Stages 1, 3 & 4 from Week 12 (Figure 54.2 Bottom). Maximum TP levels were found at Stage 2 sampling point E (200µgL^{-1}) on one occasion and at Stages 3 & 4 (160-170µgL^{-1}) respectively. Results of Chla concentration for the seven sampling sites in the L-F were always higher than the Chla concentration at Stages 1, 3 & 4 (Figure 54.2 Top). These results suggested that both the TP and Chla were being accumulated by the matrix, roots or both. Furthermore, the estimated population of zooplankton was initially higher in the L-F (Stage 2) than at Stages 1, 3 & 4 indicating the refuge provided by the rhizosphere and matrix were exploited by the zooplankton community. It was expected the zooplankton grazing pressure on the phytoplankton would have reflected some Chla reduction. However, Chla did not show a significant decrease in the water column at Stages 3 or 4. An explanation could be that the existing density of the zooplankton was not high enough for controlling phytoplankton populations, as zooplankton numbers started decreasing from Week 11 at all Stages. Another possible explanation is the flow rate through the L-F might have been too high for effective grazing/trapping by the zooplankton and periphyton. The estimated L-F figures for the filtration rate (9.5–15.9m^3h^{-1}) and flow velocity (2.9–5.5mm s^{-1}) compared to the slow sand filtration rate: 0.2 – 0.6m^3h^{-1} seem too high. The removal efficiency (RE%) of the L-F was assessed for the variables Chla and POC at Stage 1, 3 & 4. Results for RE% of Chla were between 30% and 45% until Week 11 (3 Sep). After this date RE% dropped to zero. Similar results were obtained for POC. The survey effort during July to October 2013 was not focused on biofilm. The Chla levels at Stage 2 could indicate that Chla was trapped by the biofilm colonizing the matrix and roots, but was not consumed at a rate fast enough to decrease levels, being affected by the existing flow conditions. Hence, its role in biological interactions with the phytoplankton could be as relevant as the free-swimming zooplankton. Future research work includes the metagenomic identification of the biofilm, which will provide valuable information of the microorganisms trapping capabilities in the L-F, an experimental design will help to quantify and interpret such processes by regulating flows, matrix and root density.

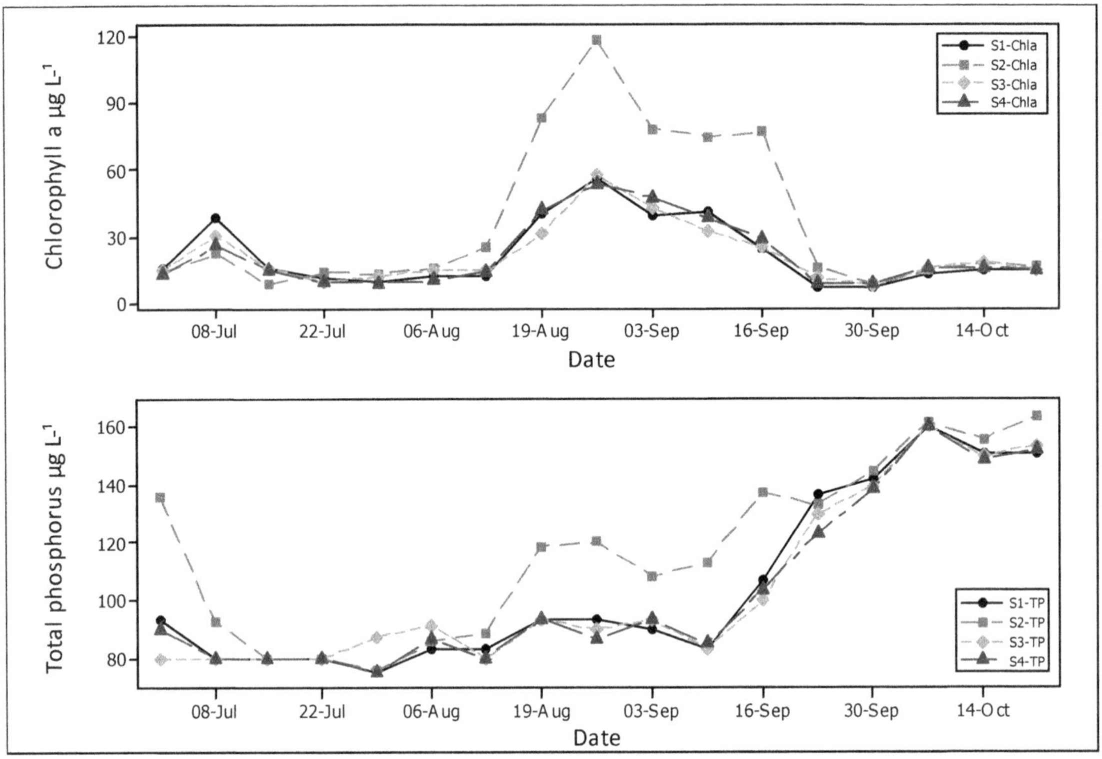

Figure 54.2 Top: weekly average for chlorophyll a (Chla µg L^{-1}). Bottom: weekly average for total phosphorus (TP µg L^{-1}). Stage 1 (black), 3 (green) & 4 (blue) (n=3) and Stage 2 (red) (n=7) (n is the number of samples taken at a particular stage per week). Samples were taken from July to October 2013 (total 17 weeks).

ACKNOWLEDGEMENTS

The authors wish to acknowledge the contribution of members of the Innovation Thames Water (Clear Water) team, field scientists and rangers at Farmoor WTW.

REFERENCES

[1] O'Neil J.M., Davis T.W., Burford M.A., Gobler C.J. (2012) The rise of harmful cyanboacteria blooms: The potential roles of eutrophication and climate change. Harmful Algae, 14, 313-334.

[2] Purcell, D., Parson S.A., Jefferson B. (2013) The influence of ultrasound frequency and power, on the algal species *Microcystis aeruginosa, Aphanizomenon flos-aquae*, *Secenesdmus subpicatus* and *Melosira* sp. Environmental Technology, 34, 1-14.

[3] Tanner, C.C & Headley, T.R. (2011) Components of floating emergent macrophyte treatment wet-lands influencing removal of stormwater pollutants. *Ecological Engineering* 37:474-86.

[4] Moss B., Stansfield J. & Irvine K. (1991) Development of daphnid communities in diatom-dominated and cyanophyte dominated lakes and their relevance to lake restoration by manipulation. Journal of Applied Ecology, 208 (2), 586-602.

[5] Li X.N., Songa H.L., Li W., Lua X.W.Nishimurab O. (2010) An integrated ecological floating-bed employing plant, freshwater clam and biofilm carrier for purification of eutrophic water. *Ecological Engineering*, 36 (4), 382-390.

Part VII

Alternative Biofiltration – Treatment Performance

Chapter 55

Removal of easily and more complex biodegradable NOM by full-scale BAC filters to produce biological stable drinking water

W. A. M. Hijnen, R. Schurer**, B. Martijn***, J. A. Bahlman, W. Hoogenboezem****, P. W. J. J. van der Wielen**

**KWR Watercycle Research Institute, PO Box 1072, 3430BB Nieuwegein, NL*

***Evides Water Company, PO Box 4472, 3006 AL Rotterdam, NL*

****PWN Water Company, Rijksweg 501, 1991 AS Velserbroek, NL*

*****Het Waterlaboratorium, PO Box 734, 2003 RS Haarlem, NL*

ABSTRACT

Distribution of drinking water without chlorine requires low microbial growth potential of the water, which can be assessed with methods detecting biodegradable compounds at low concentrations. Recent insights showed that besides the easily assimilable organic carbon AOC measured with P17/Nox more complex biodegradable compounds or biopolymers are also of significance for the biological stability of water. Therefore, the impact of full-scale Biological Activated Carbon (BAC) filtration processes on the biological stability of water was determined at four full-scale water treatment plants using newly developed methods. The results indicate the importance of BAC filtration for the concentration of more complex biodegradable AOC and particulate organic matter in the water. The removal rate of biodegradable compounds in BAC filtration is affected by influent quality, temperature, and results also indicate an effect of reactivation frequency of the activated carbon.

Keywords BAC filtration; biological stability of drinking water; bioassays; AOC; biopolymers

INTRODUCTION

Drinking Water is distributed without residual disinfectants in the Netherlands. Undesired bacterial regrowth during storage and in the distribution system is controlled by producing biological stable drinking water. In the legislation in the Netherlands (Anonymous, 2001) prevalence of heterotrophic plate counts (HPC) and *Aeromonas* plate counts in drinking water have been introduced as technical parameters and indicators for regrowth. The quantitative standards for these parameters (a yearly geometric average HPC < 100 colony forming units (cfu)/ml; maximum value of 1000 cfu/100 ml for *Aeromonas*) are introduced as signal levels to initiate actions which will minimize potential regrowth problems (potential multiplication of opportunistic pathogens and esthetic problems).

Biological granular activated carbon filtration (BAC) is a frequently used post treatment step in drinking water production from surface water. The process removes (biodegradable) natural organic matter (NOM) to improve taste and odor, color, and the biological stability of the drinking water. Drinking water companies Evides and PWN observe periodically *Aeromonas* regrowth in parts of their distribution systems supplied with BAC filtered drinking water, but not at all production facilities (Table 55.1). Pre-oxidation enhances NOM removal in BAC filtration (Van der Aa *et al.,* 2012). Due to the increased knowledge on the broad efficacy of UV disinfection for water safety (Hijnen *et al.,* 2006) there is a growing interest to apply UV disinfection globally. The full-scale plants (Table 55.1) differ in the use of oxidative steps prior to BAC filtration (Table 55.2).

Table 55.1 *Aeromonas* regrowth and the biological stability of the drinking water, determined with AOC and BFR, after BAC and post-disinfection (<0.1 mg/l ClO_2) and reservoir storage (DWR) (taken from van der Kooij *et al.*, 2010; 2011).

Facility	*Aeromonas* regrowth	Biological stability assessment of the drinking water			
		AOC P17/Nox (µg ac.-C eq/l)		Biofilm formation rate (BFR; (pg ATP/cm^2.d)	
		BAC	DWR	BAC	DWR
#1	Yes	8.6±0.3	15±1	1.1±0.4	26±5
#2	No	3.1±0.8	23±2	3.1±0.8	23±2
#3	Yes	22±1	28±1	3.1±0.3	79±13
#4	Yes	15±3	19±3	1.4±0.2	(15±5)

Assessment of the biological stability of the produced drinking waters, determined by measuring the easily assimilable organic carbon (AOC) and the biofilm formation rate (BFR) using a biofilm monitor, revealed elevated values after reservoir storage. However, the biological stability was not correlated with observed differences in *Aeromonas* regrowth during distribution for the different treatment plants (Table 55.1). An explanation for the lack of correlation between biological stability and *Aeromonas* might be that not all biodegradable NOM is determined with AOC and BFR. The traditional AOC method, with bacterial strains *Pseudomonas fluorescens* P17 and *Spirillum* spp Nox was introduced as a parameter to determine the concentration of easily assimilable organic carbon (Van der Kooij, 2002). These compounds are also largely responsible for the BFR in the biofilm monitor. The biodegradable NOM that might not be detected by AOC and BFR are parts of the more complex biodegradable NOM or biopolymers (i.e. proteins, polysaccharides). Recently, methods have been developed to determine these compounds at the µg-per-liter-level (Van der Kooij and Veenendaal, 2013; Sack *et al.,* 2010; 2011), but their role in biological stable drinking water remains largely unexplored.

The objective of our study is to determine the effect of BAC filtration processes, under variable operational conditions at four full-scale drinking water production plants (Table 55.1), on the removal of easily and more complex biodegradable NOM that can be detected with additional methods for more complex biodegradable NOM. During a period of one to two years the biological stability before and after the BAC filters was assessed with traditional and the newly developed methods to investigate the impact of BAC filtration on the biological stability of water and the significance of the more complex biodegradable compounds.

MATERIALS AND METHODS

Drinking water facilities and processes. The source water and treatment conditions of the four facilities of this study are summarized in Table 55.2. Facilities #1, #2 and #3 use river Meuse water after extended reservoir storage for drinking water production and facility #4 river Rhine water after storage in a local reservoir. The source water quality of the first three facilities was similar. The source water at facility #4 contained more organic matter, chlorophyll-A and ammonium compared to the other facilities. The pre-treatment at the plants differ in the applied disinfection and oxidation processes and their sequence with the rapid sand filtration. The BAC filtration processes operated as final step of the treatment have different process conditions with respect to contact time, back wash and reactivation intervals.

Table 55.2 Source water type and quality, pre-treatment and BAC process conditions.

Facility, river water	#1, Meuse	#2, Meuse	#3, Meuse	#4, Rhine
Extended reservoirs	Yes	Yes	Yes	No
DOC (mg/l)	4.0±0.2	3.9±0.2	4.0±0.4	7.3±0.8
Chlorophyll-A (μg/l)	3.8±4.3	3.3±4.9	2.7±2.6	18.0±11.9
NH_4 (mg/l)	0.04±0.03	0.05±0.02	nd	0.2±0.1
Pre-treatment[a]	COA/SED	COA/FLOT	COA/SED	COA/SED
	RSF- UV	Ozone - RSF	Ozone - RSF	RSF -UV/H_2O_2
BAC contact time (min.)	20-30	40-55	15-20	23-28
Back wash interval (day)	8-16	7	23	56
Reactivation int. (year)	2	6	3	1.5-2

[a] Coagulation/sedimentation or flotation (COA/SED, FLOT), rapid sand filtration (RSF), disinfection with UV, ozone or UV/H_2O_2

Biological stability assessment. The bioassay methods applied to assess the microbial growth potential are the traditional AOC test (Van der Kooij, 2002), the extended AOC test (Sack *et al.*, 2010; 2011) and the biomass production potential (BPP) test (Van der Kooij *et al.*, 2013). The traditional AOC method determines growth maximum of *Pseudomonas fluorescens* P17 and *Spirillum* sp. Nox in 600 ml pasteurized water sample (duplicates); the AOC is expressed in μg acetate-C equivalent/l derived from the yield factors of both bacteria on acetate. Recently, an extended AOC A3 method has been developed to quantify biopolymers to the μg-liter-level using *Flavobacterium Johnsoniae* strain A3 and is expressed as μg biopolymer-C eq./l.

In addition, the BPP was applied. The growth of the autochthonous bacterial population is measured with ATP in 600 ml water samples (duplicate) at 25°C. The parameter derived from the ATP results of a BPP sample is the cumulative biomass production in 14 days (BPC14), which is the integrated amount of ATP measured during 14 days of incubation.

Biomass parameters. The active biomass concentrations in the influent and effluent of the BAC filters was determined with ATP. The total cell count with flow cytometry (FCM) and invertebates biomass were measured in parallel.

Particulate organic matter and particulate carbohydrates. The concentration of particulate organic carbon (POC) and particulate carbohydrate (P-CHC) was determined after hemoflow concentration (hollow fiber ultrafiltration MWCO 30 kDalton). The concentration factor ranged from 20 – 250 depending on the sample (highest factor in the BAC filtrates). Subsequently the TOC, using Shimadzu TOC-VCPH, and carbohydrates, using the method of Dubois *et al.*, (1956) concentrations were measured in the produced concentrates (0.5 – 1.0 l). The POC (μg C/l) was calculated from the TOC concentration corrected for the DOC in the sampled water and the concentration factor. The P-CHC concentration (μg glucose-C eq./l) was calculated from CH concentration and the concentration factor multiplied by 0.4 for the carbon content of glucose.

Data collection and analysis. During one or two years the influent and effluent of a set of BAC filters at the different facilities were sampled for the microbial growth potential (AOC and BPP) and the biomass concentrations (direct ATP, FCM and after concentration). Additional quality parameters such as invertebrates and oxygen consumption were derived from the routine quality monitoring program at the treatment plants. Due to differences in the monitoring programs at the four facilities (sampling points, frequency, timing) the collected data sets were not always complete with different numbers of data. This hampers extended statistical analysis.

RESULTS AND DISCUSSION

Biological stability: bioassays. The microbial growth potential of the water measured with the different bioassays before and after the BAC filtration is presented in Table 55.3.

Table 55.3 Biological stability of water before and after BAC filtration assessed with the AOC methods (AOC P17/Nox and AOC A3) and the BPP test (expressed as BPC14) in different months of the year (average values ± SD; n=4 per year).

Facility		AOC P17/Nox (μg acetate-C eq/l)		AOC A3 (μg biopolymer-C eq/l)		BPC14 (d.ng/l)	
Pretreatment	Year	In	Out	In	Out	In	Out
#1 RSF –UV	2012	14±3[a]	10±3	17±1	11±1	312±31	145±60
	2013	nd[b]	nd	27±22	13±11	379±185	151±116
#2 O_3 –RSF	2012	19±0.0[a]	20±9	4.4±0.1	3.9±1.1	161±11	137±37
	2013	nd	nd	6.2±4.9	3.4±1.2	116±38	104±33
#3 O_3 –RSF	2013	nd	nd	4.0±1.4	6.1±7.2	184±29	122±15
#4 RSF –UV/H_2O_2	2012	45±39	11±2	5.5±0.6	2.5±0.8	356±183	134±58

[a] SD of duplicates of one sampling; [b] nd = not determined

Similar to the results presented in Table 55.1, the AOC concentration, measured with strains P17/Nox, in the influent water was higher at facilities with pre-oxidation processes (O_3 and UV/H_2O_2) than at the facility without these processes (#1). AOC P17/Nox was also higher in the BAC treated water at facility #2 than at facility #1. However, the AOC concentration, measured with strain A3,in the BAC influent and effluent was higher at facility #1 than at facilities #2 and #4. The sequence in AOC A3 concentration in the filtrates was #1 > #3 > #2 > #4. The results of the BPP test, expressed as BPC14, indicated a higher microbial growth potential in the BAC filtrate at #1 compared to the other facilities. The high variability (SD), however, indicates that more data are required for statistically significant conclusions.

Micro-organisms and other suspended biological material in the water. The ATP concentration, cell numbers and invertebrates biomass in the influent and effluent of the BAC filters at facilities #1, #2 and #4 were highest in the effluent of BAC filtration at facility #1 (Fig. 55.1). At facility #1 and #4 a reduction in ATP and cell numbers was observed, whereas at facility #2 both parameters showed an increasing trend. The ATP concentrations and cell numbers in the filtrates correlated with those in the influent. Furthermore, the higher ATP concentrations and cell numbers in the water at facilities #1 compared to water at the other facilities, correspond with the higher biopolymer concentrations (AOC A3) in water at facility #1 (Table 55.2) which demonstrates the correlation of this parameter with bacterial biomass. Clear correlations with the process conditions (Tabel 55.1) were in general not observed, but the increasing trend of ATP and cell numbers at facility #2 might be correlated with the low reactivation frequency of 6 years at this plant. However, additional studies are necessary to demonstrate a causal relationship between these parameters.

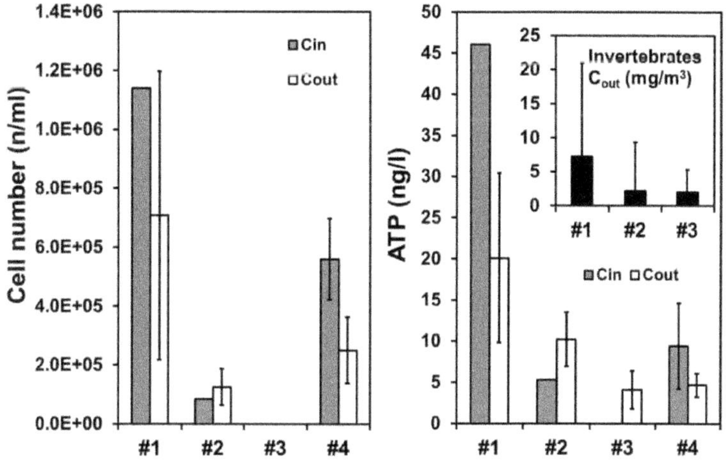

Figure 55.1 Average cell number (n=5-8), ATP concentrations (n=8-52) in the influent and effluent of the BAC filters at the four facilities and invertebrates biomass (n=112-152; error bars SD) in the BAC filtered water at facilities #1, #2 and #3 (#3 no FCM and ATP influent data).

The concentration POC and P-CHC as indicative parameters for biological material in the water are presented in Table 55.4. The ratio of P-CHC/POC indicates a high and steady fraction of biological material (e.g. exo-polymeric substances) in the POC of more than 50% at all facilities (ratios for #1, #2, #3 and #4 were respectively 0.52±0.07, 0.59±0.23, 0.50±0.05 and 0.57±0.05). Thus, 50-60% of the particulate organic matter consists of particulate high molecular poly-carbohydrates, which are common constituents of biomass. The sequence in average POC/P-CHC concentrations in the influent of the BAC filters was #1 > #4 > #3 > #2. This sequence correspond with the sequence for ATP and cell number data, which also demonstrates higher biomass concentrations in the influent water at the facilities #1 and #4 compared to facilities #2 and #3. In the filtrates this sequence changed to #1 = #4 = #3 > #2, indicating that POC and PCHC removal by BAC was different between the facilities.

Reduction of microbial growth potential and suspended biological organic material. The removal capacity of the BAC filtration processes for AOC and BPC14 has been compared (Figure 55.2).

Table 55.4 The average concentrations particulate organic carbon (POC) and carbohydrates (P-CHC) in the water before and after BAC filtration in different periods of the year (n=4 - 7 per year).

Facility Pretreatment	Year	POC (μg C/l) In	POC (μg C/l) Out	P-CHC (μg glucose-C eq./l) In	P-CHC (μg glucose-C eq./l) Out
#1 RSF –UV	2012	189±97	77±32	110±68	34±12
	2013	188±38	118±52	98±29	58±28
#2 O_3 –RSF	2012	66±26	55±15	37±16	30±12
	2013	41±6.1	32±5.6	33±13	17±2.8
#3 O_3 –RSF	2013	94±18	75±17	47±8.5	38±4.8
#4 RSF -UV/H_2O_2	2012	104±17	79±20	59±10	46±13

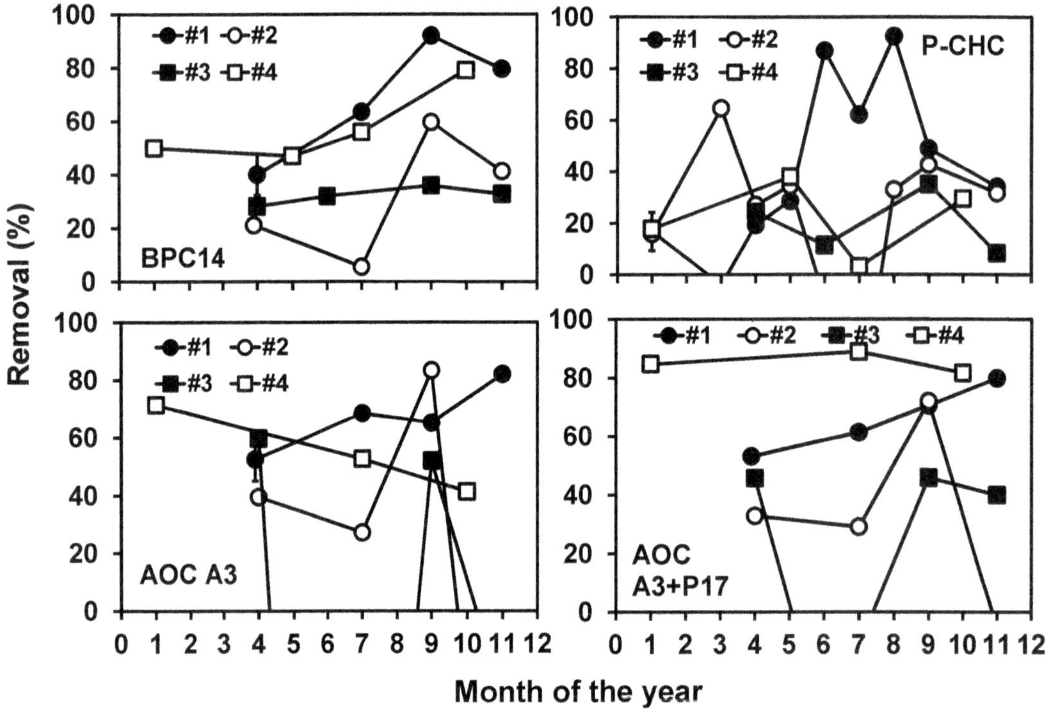

Figure 55.2 Removal (%) of the biopolymers (AOC A3), biopolymers plus low molecular acids (AOC A3+P17), all biodegradable compounds (BPC14) and PCHC throughout the year.

The percentage of removal of biodegradable compounds was calculated for AOC A3 (biopolymers), AOC A3+P17 and BPC14. The removal of AOC A3, as a measure for biopolymer concentration, at facilities #1 and #4 varied between 60-80% and showed an increasing and decreasing trend for facility #1 and #4, respectively. In contrast, the removal of these compounds was generally lower (<50%) and more variable at facilities #2 and #3. Negative removal rates were observed at facilities #2 and #3, indicating breakthrough of produced or accumulated biopolymers in the BAC filter bed. This breakthrough may be correlated with the higher BAC reactivation interval at facilities #2 and #3 compared to facilities #1 and #4 (Table 55.2). The removal of the biodegradable compounds determined with the BPP test (BPC14) were to some degree similar to removal rates for biopolymers (AOC A3). Furthermore, at all facilities removal percentages of BPC14 increased during summer and autumn.

The removal of low molecular compounds and biopolymers combined (A3+P17) showed a high and stable removal (>80%) at facility #4 due to the high AOC P17 concentration in the influent (directly after oxidation with UV/H2O2; Table 55.3). The removal of these compounds at the other facilities showed similar trends as observed for the removal of biopolymers solely, indicating that the contribution of the low molecular compounds to the overall AOC removal at the other facilities was less pronounced than at facility #4.

The removal of P-CHC at the different facilities was generally lower than the removal of AOC or BPC14. Removal percentages was <40% and variable at facilities #2, #3 and #4, sometimes showing negative values (biomass breakthrough) at facilities #2 and #3, which corresponds with AOC A3 results. This might be correlated with the lower reactivation frequency at those two facilities (Table 55.2). In summer high removal of P-CHC was observed at facility #1 indicating more removal at more intensive biological activity.

Oxygen consumption in the filter beds. Oxygen consumption in BAC processes also reflects the biological activity in BAC filter beds. Based on the oxygen concentrations in the filter bed influent and effluent at facility #1, #2 and #3 the specific oxygen consumption was calculated (Figure 55.3).

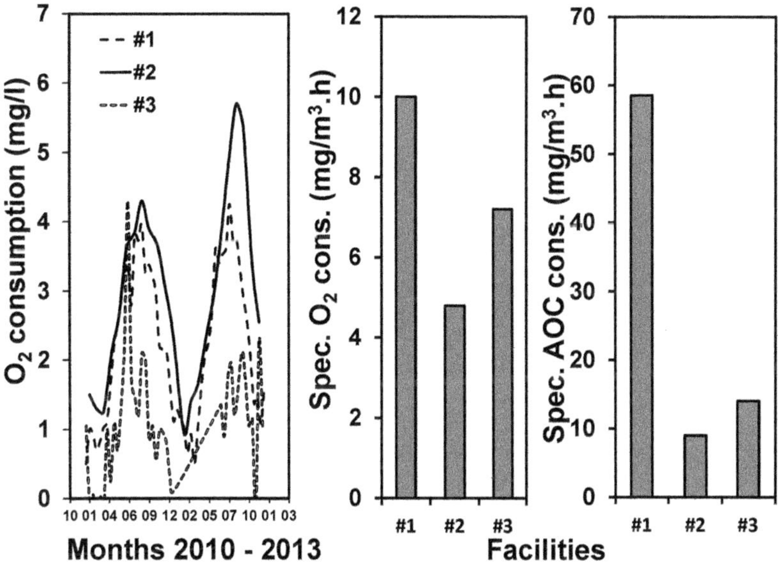

Figure 55.3 Oxygen consumption in the BAC filters at facilities #1, #2 and #3 over a two year period and the calculated specific oxygen and AOC consumption.

The absolute consumption at facility #1 and #2 ranged from less than 1 to 4-5 mg/l, whereas at facility #3 the oxygen consumption ranged from 0 to 2-4 mg/l. Based on the average oxygen consumption during 2010 till 2013, the average AOC reduction and the average volume load of the processes calculated from the specific operational conditions (bed height and filtration rate) the specific oxygen consumption (mg

$O_2/m^3.h$) and the specific AOC (A3+P17) consumption (mg $AOC/m^3.h$) in the filterbeds were calculated (Figure 55.3). The results revealed the highest and lowest values for both parameters at respectively facility #1 and #2.

CONCLUSIONS

The results of this study show that the more complex biodegradable NOM or biopolymers, measured with the new AOC A3 and BPP method, have a significant influence on the biological stability of BAC-treated water. Furthermore, these parameters correlate with the presence of biomass and particulate organic matter in the water. Generally, BAC filter beds improve the biological stability of the water by removing the easily and more complex biodegradable NOM. The results indicate a higher removal rate for these compounds at higher influent concentration, temperature, BAC reactivation frequency and specific oxygen consumption in the BAC filter bed. More intensive monitoring under well defined conditions is required to confirm these observations and to identify more accurately the significant process conditions for BAC filtration to produce biological stable drinking water.

ACKNOWLEDGEMENTS

This study was part of the collaborative BTO Research Program of the Dutch Drinking Water Companies. The authors wish to acknowledge the important contribution of H.A.M. Ketelaars (Evides) and the laboratory of KWR Ton Braat, Lonneke Hensen, Ronald Italiaander, Meindert de Graaf.

REFERENCES

Anonymous (2001) Drinking Water Decree 2001 (Waterleidingbesluit 2001) Staatsblad, The Hague, The Netherlands.

Hijnen, W.A.M., Beerendonk, E.F. and Medema, G.J. (2006) Inactivation credit of UV radiation for viruses, bacteria and protozoan (oo)cysts in water: A review. *Water Res.* **40**, 3-22.

Sack, E.L.W., van der Wielen, P.W.J.J. and van der Kooij, D. (2010) Utilization of oligo- and polysaccharides at microgram-per-litre levels in freshwater by *Flavobacterium johnsoniae*. *J. Appl. Microbiol.* 108: 1430.

Sack, E.L.W., van der Wielen, P.W.J.J. and van der Kooij, D. (2011) *Flavobacterium johnsoniae* as a Model Organism for Characterizing Biopolymer Utilization in Oligotrophic Freshwater Environments. *Appl. Environ. Microbiol.* **77**(19), 6931.

Van der Aa, L.T.J, Kolpa, R.J., Rietveld, L.C. and van Dijk, J.C. (2012) Improved removal of pesticides in biological granular activated carbon filters by pre-oxidation of natural organic matter. *J. Wat. Suppl. Res. And Techn. – AQUA*, **61**(3), 152

Van der Kooij, D. van der Wielen, P.W.J.J., Veenendaal, H.R. and Suylen, G.M.H. (2010) Influence of ClO_2 dosages on the biological stability of the drinking water and the regrowth of *Aeromonas*. KWR report 2009.069, Nieuwegein, NL.

Van der Kooij, D., Veenendaal, H.R. and van der Wielen, P.W.J.J. (2011) Judgement on the biological stability of the drinking water at pumping station Andijk. KWR report 2010.107, Nieuwegein, NL.

Van der Kooij, D. (2002). Assimilable organic carbon in treated water: determination and significance. Encyclopedia of Environmental Microbiology. John Wiley & Sons, New York, US.

Van der Kooij, D. and Hijnen, W.A.M. (1988) Nutritional Versatility and Growth Kinetics of an *Aeromonas hydrophila* Strain Isolated from Drinking Water. *Appl. Environ. Microbiol.* **54**(11), 2842.

Van der Kooij, D. and Veenendaal, H.R. (2013) Regrowth problems and biological stability assessment in the Netherlands. In 'Microbial growth in drinking-water supplies', D. van der Kooij and P.W.J.J van der Wielen eds., IWA publishing, London, UK.

Chapter 56

Renewal and selection of granular activated carbon in advanced drinking water purification in Tokyo

T. Iwamoto, T. Tasaki*, T. Kanami**

**Purification division, Bureau of waterworks, Tokyo Metropolitan Government, Tokyo, Japan*

ABSTRACT

One of the most problematic taste and odor is musty odor such as 2-methylisoborneol (2-MIB) and geosmin because they are rarely removed by conventional drinking water treatment. Since 1974, 2-MIB had been detected in Edo River, one of the water sources of Tokyo, and had caused musty odor in purified water from Kanamachi purification plant (PP). Therefore, Tokyo Waterworks decided to introduce advanced drinking water purification (ADWP), which consists of ozonation and biological activated carbon (BAC) filtration, to Kanamachi PP to remove 2-MIB efficiently and stably. It is necessary to replace GAC with a proper period to keep its function. In order to determine the renewal period, Tokyo Waterworks had conducted follow-up research on the performance of BAC filter after the launch of ADWP. In this paper, renewal period of GAC based on both removal performance and cost-effectiveness is described. Also, a revised renewal period based on the additional investigation on TOC removal is discussed. Moreover, the results of investigation on relationship between GAC performance and base material are introduced.

Keywords Biological activated carbon; adsorption; renewal period

INTRODUCTION

The problem of taste and odor in drinking water is one of a large issue for water utilities around the world; because aesthetic quality of drinking water is usually the only measure available to most consumers to judge the water safe (McGuire 1995). One of the most common and problematic taste and odor is musty odor such as 2-methylisoborneol (2-MIB) and geosmin which is produced by some cyanobacteria and fungi (Izaguirre G 2004, Krishnani KK 2008, Zaitlin B, 2006); because these substances are rarely removed by conventional drinking water treatment such as coagulation, sedimentation, rapid sand filtration and chlorination (Srinivasan R. 2011, Elhadi SL 2004).

In order to remove musty order substances, treatment with ozonation and absorption by granular activated carbon (GAC) is found to be effective (Jung SW. 2004). GAC is originally used for removing organic matters by physicochemical adsorption. However, indigenous microorganisms will accumulate on the surface of GAC layer when not residual chlorine but biodegradable substances are contained in influent water. These microorganisms can also contribute to drinking water purification through biological function. Such biologically activated carbon is called as biological activated carbon (BAC) (Seredyńska-Sobecka et al. 2006; Simpson, 2008). This treatment prolongs the life-span of GAC, because microorganisms on GAC can remove biodegradable substances even after the adsorptive capacity has been exhausted. Moreover biological removal of 2-MIB and geosmin has also been reported in GAC filters (Yagi et al., 1998; Elhadi SLN, 2006, Zhou B, 2011). Therefore ozonation followed by BAC filtration could remove 2-MIB and geosmin more effectively and stably.

Tokyo metropolis is located in the southern part of the Kanto plain which lies near the center of the Japanese archipelago. Tokyo waterworks is one of the biggest water suppliers in Japan with average distribution amount of 4.2 million m^3/day and population served of about 13 million people who live in almost all area of the Tokyo metropolis. Most water resources in Tokyo come from rivers: 80 % of the water resources depend on the Tone River system which runs through the Kanto plain. The Edogawa River, which belongs to the Tone River system, has watersheds which were developed rapidly as commuter towns. Delayed introduction of sewage systems in these areas had caused severe water pollution; especially 2-MIB had been detected since 1974 and they had caused unpleasant odors in purified water. Therefore, Tokyo Waterworks decided to introduce advanced drinking water purification (ADWP), which consists of ozonation and BAC filtration, to the Kanamachi PP, which takes raw water from the Edo River, in order to remove 2-MIB efficiently and stably (Murano 1994; Muramoto 1995).

It is necessary to replace GAC after a certain period to keep its function because aged deterioration and decline of removal capacity are inevitable; while at the same time, frequent replace of GAC requires unreasonable cost against the improvement of removal performance. Moreover it is necessary to consider on GAC regeneration because GAC could recover part of its function through regeneration. Therefore it is important to decide whether GAC should be replaced with new GAC or with regenerated one. In order to determine the renewal period of GAC, Tokyo Waterworks had operated follow-up research on the performance of BAC filter after the launch of ADWP to Kanamachi PP. Furthermore, Tokyo Waterworks has promoted "the Project for Safe, Better Tasting Tap Water" since 2004 in response to rise in demand for safe and high-quality drinking water. This project aims to provide high-quality drinking water with which most customers are satisfied and original water quality targets were set as indicators of odor, taste and appearance. Total organic carbon (TOC) was selected as an indicator of taste and its target value in purified water was set to be under 1.0 mg/L. Therefore, we have conducted further investigation on the GAC function in the aspects of TOC removal as an additional indicator.

Tokyo Waterworks has used GAC made of coal which showed superior performance in water treatment in our previous survey in PPs in Tokyo. However, it has been revealed that GAC made of palm shell charcoal showed an almost equal ability with GAC made of coal recently. Therefore, GAC made of palm shell charcoal was introduced to some actual BAC filters to investigate the ability on trial.

In this paper, renewal period of GAC which was decided based on both removal performance and cost-effectiveness is described. Also, a revised GAC renewal period based on the additional investigation on TOC removal is discussed. Moreover, the results of investigation on relationship between GAC performance and base material are introduced.

METHODOLOGY

Tokyo Waterworks has installed ADWP to five PPs. Specifications of each PP and ADWP are shown in Table 56.1. ADWP in Tokyo consists of ozonation and BAC filtration. In Kanamachi, Misato and Higashimurayama PPs, the ADWP were installed after coagulation-sedimentation process, and BAC treated water were filtrated by rapid sand filter. In Asaka and Misono PPs, additional rapid sand filtration was also introduced between sedimentation and ozonation. Disinfection by sodium hypochlorite is carried out before rapid sand filtration.

Each effective surface area of BAC filter differs in each PP (98-110 m^2), while depth of the GAC layer is 2.5 m in common. The water flows downwards through GAC layer at the linear velocity of 250-300 m/day. Empty bed contact time of BAC filtration is 12-15 min on average. Backwashing are carried out every 3-5 days by using finished water containing residual chlorine.

Table 56.1 Advanced drinking water purification plant (Feb. 1, 2013).

Water resource	Drinking water purification plant	Capacity (m^3/day)	Year of ADWP installation	Number of BAC filter
Edo river	Kanamachi	1,500,000	1992, 1996, 2013	54
Edo river	Misato	1,100,000	1999, 2013	44
Ara river	Asaka	1,700,000	2004	40
Ara river	Misono	300,000	2004	40
Ara river, Tama river	Higashimurayama	1,265,000	2010	32

General properties of GAC which used in actual ADWP are presented in Table 56.2. In BAC filtration, GAC made of coal is mainly used. We introduce GAC made of palm shell charcoal to several test filters to compare their treatment efficiency derived from base material. GAC made of coal are classified into two types based on their effective size and uniformity. PPs without pre rapid sand filtration have GAC with effective size of 1.2 mm and uniformity of 1.3; while PPs with pre rapid sand filtration have GAC with effective size of 0.7 mm and uniformity of 1.7 respectively.

GAC made of coal is a type of crushed activated carbon while GAC made of palm shell charcoal is spherically granulated. The effective size and uniformity of both GAC are the same.

The first study on BAC filtration was started at Kanamachi PP in 1992. The same investigation was operated both in Misato and Asaka PP after the launch of ADWP. In this study, UV260, which is UV adsorption at 260nm, was chosen as an indicator of dissolved organic substances. In addition, NH_3-N, methylene blue active surfactant (MBAS), 2-MIB and trihalomethane formation potential (THMFPs) were measured. Water samples were collected along the process more than once a month. UV260 were measured by spectrophotometer. NH_3-N was analyzed by 1-naphthol method. MBAS was measured by high performance liquid chromatography. 2-MIB and THMFPs were analyzed by purge and trap gas chromatograph method.

Table 56.2 Properties of GAC (Mar. 31, 2012)

	GAC for the actual BAC filter		**GAC for the trial BAC filter**
Purification plant	**Kanamachi, Misato Higashimurayama**	**Asaka, Misono**	**Higashimurayama**
Base material	Coal	Coal	Palm shell charcoal
Effective size (mm)	1.2 (± 0.05)	0.7 (± 0.05)	1.2 (± 0.05)
Uniformity coefficient	<1.3	<1.7	<1.3
Iodine number (mg/g)	>1,000	>1,000	>1,100
Surface area (m^2/g)	>1000	>1,000	>1,100

The second study, TOC was chosen as an indicator instead of UV260. And TOC was measured by the combustion oxidation method using TOC analyzer.

The follow up study on the aged and regenerated GAC had operated with GAC collected from a particular BAC basin in Kanamachi PP. GAC samples collected in different depth (5cm, 0.5m, 1 m, and 2 m from the top layer) were mixed into a composite sample and used for the following tests and steam activation with a rotary kiln. Hardness was measured by the Japanese Industrial Standards K1474 method. Iodine number (IN), methylene blue (MB) adsorption, phenol number (PN), and ABS value were measured by Japan Water Work Association K113 method. Apparent BET surface area (S_{BET}) was determined using adsorption of N_2 under the relative pressure of 0.01 to 0.15 and applying the conventional BET multilayer model.

RESULTS AND DISCUSSION

The transitions of UV260 removal ratio in the first study are shown in Figure 56.1. It is showed that the removal capacities decreased with time in all line including regenerated GAC, and removal capacity of BAC-A became 0 % after 86 months.

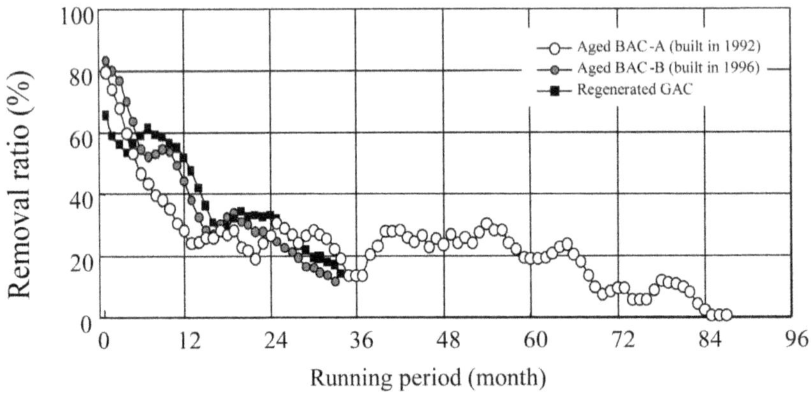

Figure 56.1 UV260 removal ratios through BAC filtration (moving average) in first study

Degradations of removal performance in each water quality indicator in the first study were presented in Table 56.3. With the water temperature decrease, NH_3-N leaks from BAC filters had increased in aged BAC, especially after using it for 6 years: because the removal capacity of aged BAC was unable to catch up with the dynamic change of NH_3-N inflow. This leakage trend had continued after 90 months of use. Therefore, proper renewal period should be 7-8 years in the aspects of NH_3-N removal. Furthermore, the leak frequency of MBAS increased especially after 7 years. Based on these results, the renewal period of GAC in Tokyo was determined to be 7-8 years in 2000.

Table 56.3 Degradation of removal performance of BAC filter in the first study.

Indicator	Degradation of removal	Proper renewal period
UV260	Become 0% after 86 months	7 - 8 years
NH_3-N	Leakage occurred in low water temperature after 6 years	7 - 8 years
2-MIB	Maintain about 100 % through investigation period	-
MBAS	Frequent leakage observed after 7 years	7 years
THMFP	Maintain about 30 % through investigation period	-

Under the consideration of renewal cost, the use of regenerated GAC would be cost-effective when the yield rate and some physical and functional properties are applicable. Therefore several properties of aged and regenerated GAC were surveyed for proper period of regeneration. The indicators mainly used for the evaluation of aged GAC were hardness, IN, and MB adsorption. Hardness tended to decrease cause of abrasion between grains and backwash shear stress especially after 62 month. Indicators for adsorptive capacity decreased with time after 84 months (IN; 1,023 to 210 mg/g, MB adsorption; 190 to 70 ml/g). The indicators mainly used for evaluation of regenerated GAC were S_{BET}, IN and yield ratio. The indicators for adsorption capacity of regenerated GAC decreased with running time, and it revealed that those capacities never make a recovery to the initial potential through regeneration (Fig. 56.2). Furthermore, it is important to find out the transition of yield rate to consider the regeneration cost. The yield rate decreased with time and become ~77 % after using it 5 years. Consequently, it was determined that the regeneration should be done within 5 years based on the recovery of IN and the yield rate.

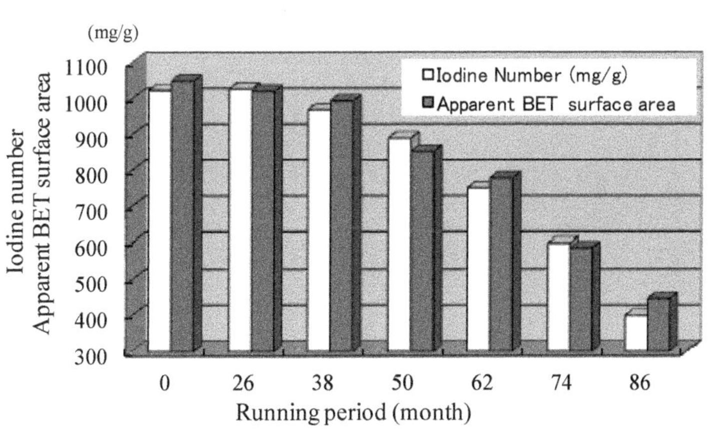

Figure 56.2 Transition of iodine number and apparent BET surface area of regenerated GAC

After that, GAC renewal period was revised in response to the requests for high-quality drinking water in 2009. In Japan, it is legally required to secure 0.1 mg/L of residual chlorine, which causes uncomfortable chlorine odors, in taps. TOC causes the increase of chlorine demand in distribution systems, therefore it is impossible to decrease the chlorine dose at PPs when the concentration of TOC is rather high. Consequently, the reduction of TOC leads to reduce residual chlorine in taps and improve customer satisfactions. Therefore the second survey focused on TOC removal had conducted.

The results showed that the removal ratio of TOC decreased over time and became constant 20-30 % after using it for 5 years (Fig 56.3). The initial drop of removal ratio was assumed to be due to the consumption of sorption capacity, and the persistent TOC removal was likely due to biodegradation. Consequently GAC renewal period was revised to be about 4 years to increase the TOC removal.

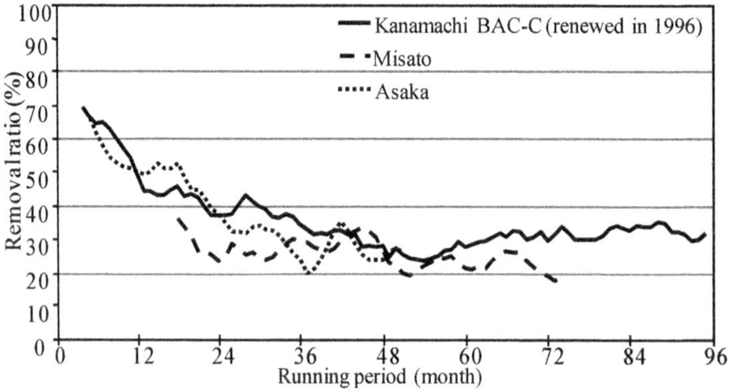

Figure 56.3 Transition of TOC removal ratios through BAC filtration (moving average)

From these results, it could be possible to increase TOC removal capacity by shortening the period, while it also leads to increase the total costs of operation and maintenance of PPs. Therefore cost-effectiveness was estimated by creating a simple model based on TOC removal in the second survey. In this estimation, the result of Misato PP was used because it had the longest investigation period. The changes of TOC removal was simplified using assumptions that TOC removal differed in three periods. The assumptions were described in Table 56.5 and compared with the results of Misato PP in Figure 56.4.

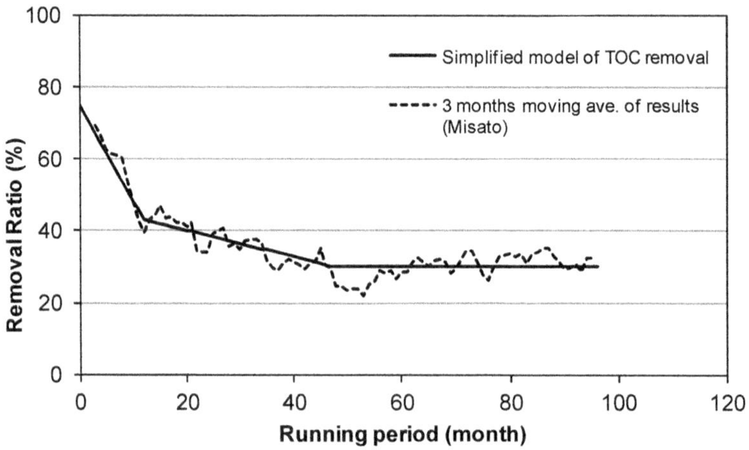

Figure 56.4 Comparison of simplified model of TOC removal with the results of Misato PP.

Table 56.5 Assumptions of TOC removal in the estimation

Running Period (month)	Assumption of TOC removal
From 0 to 12	Removal decreases from 75% to 43%
From 13 to 48	Removal decreases from 43% to 30%
Later than 48	Removal is stable at 30%

Then, costs and improvement of TOC removal of each period were estimated. Upon the estimation, model BAC facilities, which have 24, 25 or 28 filters, was used. Then, renewal month of each filters was determined in accordance with operation rate of the model BAC facility was the best on the assumption that renewal operation of BAC took a month. Then, each running time was allowed to increase gradually until average of running period of all filters became a steady state. Then, TOC removal of each filter was calculated from the model shown in Table 56.6 and Figure 56.4 in the steady state, and the average TOC removal of each renewal period was calculated. In addition, renewal costs were calculated by setting the cost of 1 year was equal to 1. Marginal benefits (B) and marginal costs (C) were calculated by comparing the average TOC removal and costs successive renewal period, respectively. Finally, benefit-costs ratio (B/C) of each renewal period was compared. The results are shown in Table 56.6. The results shows that B/Cs of from 4 to 7 years of renewal period are almost same and B/Cs of shorter than 3 years are smaller than 4 years. Therefore, it was considered 4 years of renewal period is the most cost effective to improve TOC removal sufficiently. Consequently, the renewal period of BAC was shortened from 7 or 8 years to approximately 4 years. The increase of the adsorption capacity ensures the safety of purified water against potential non-biodegradable contaminants such as pesticides and some disinfection byproduct precursors.

Table 56.6 Estimation of removal of TOC in each renewal period and results of cost-benefit analysis

		Renewal period (year)							
		1	2	3	4	5	6	7	8
Average running period (month)		5.5	11.5	17.5	23.5	29.5	35.5	41.5	47.5
Average rate of operation (%)		91.7	95.8	97.2	97.9	98.3	98.6	98.8	99.0
Average TOC removal TOC (%)		59.0	49.6	45.2	41.9	39.5	37.9	36.7	35.9
Marginal benefit (improvement of TOC removal: %)	B	9.4	4.4	3.3	2.4	1.6	1.1	0.9	
Renewal costs (1 year =1)		1.00	0.50	0.33	0.25	0.20	0.17	0.14	0.13
Marginal cost	C	0.50	0.17	0.08	0.05	0.03	0.02	0.02	
Benefit-cost ratio	B/C	19	27	39	48	48	48	48	

In recent years, the use of GAC made of palm shell charcoal has been attracting rising attention in the aspects of environmental load reduction. Furthermore it was showed that the GAC made of palm shell charcoal had almost equal ability to GAC made of coal in our pilot plant. Therefore GAC made of palm shell charcoal was introduced to one actual BAC basin in Higashimurayama PP in 2010 on trial. Thus Tokyo waterworks compared water purification performance of two types of GAC. The removal ratio of NH_3-N in GAC made of palm shell charcoal reached 100 % about 3 months after using it. It was about 2 months faster than that of GAC made of coal. However, decrease in NH_3-N removal ratio was observed in both GAC during the period of low water temperature. Furthermore removal ratios of organic matter, TOC and UV260, decreased to 30 to 40 % against the initial rate after about 3 years in both GAC. Therefore, GAC made of palm shell charcoal was introduced to other PPs in order to further investigation.

CONCLUSIONS

The main conclusions of this work are:

(1) The GAC renewal period in Tokyo was determined to be 7-8 years initially. Thereafter the period was shortened to approximately 4 years under the consideration of adsorption capacity of TOC.
(2) GAC regeneration should be done within 5 years based on the recovery of the IN and the yield rate.
(3) In BAC filtration, GAC made of coal is mainly used while GAC made of palm shell charcoal was introduced to some actual GAC basins to investigate its ability on trial.

REFERENCES

McGuire, M.J. (1995) Off-flavor as the consumer's measure of drinking water safety. *Water Sci. Technol.* 31, 1-8
Izaguirre, G., Taylor, W.D. (2004) A guide to geosmin- and MIB- producing cyanobacteria in the United States, *Water Sci Technol.* **49**(9) 19-24 review
Krishnani K.K., Ravichandran P., and Ayyappan S. (2008) Microbially derived off-flavor from geosmin and 2-methylisoborneol; sources and remediation. Rev Environ Contam Toxicol 194:1-27
Zaitlin B., Watson S.B. (2006) Actinomycetes in relation to taste and odor in drinking water: myths, tenets and truths. *Water Res. 2006* May;**40**(9) 1741-53
Elhadi S.L, Huck P.M, and Slawson R.M. (2004) Removal of geosmin and 2-methylisoborneol by biological filtration. *Water Sc. Technol.* 2004; **49**(9), 273-80
Elhadi S.L.N, Huck P.M., and Slawson RM. (2006) Factors affecting the removal of geosmin and MIB in drinking water biofilters. *Journal of the American Water Works Association.* 98,108-119
Srinivasan R, and Sorial G.A. (2011) Treatment of taste and odor causing compounds 2-methyl isoborneol and geosmin in drinking water: a critical review. *J Environ .Sci.* (China), **23** (1), 1-13
Jung SW, Beak KH, Yu MJ. (2004) Treatment of taste and odor material by oxidation and adsorption. *Water Sci Technol* .2004; **49** (9), 289-95.
Seredyńska-Sobecka B, Tomaszewska M, Janus M, Morawski AW. (2006) Biological activation of carbon filters. *Water Research* **40**(2), 355-363
Simpson DR (2008) Biofilm processes in biologically active carbon water purification. *Water Research* **42** (12), 2839-2838
Yagi M, Nakashima S, Muramoto S. (1998) Biological degradation of musty odor compounds, 2-methylisoborneol and geosmin, in a bio-activated carbon filter. *Water Sci. Technol.* 20, 255-260
Zhou B, Yuan R, Shi C, Yu L, Gu J, Zhang C. (2011) Biodegradation of geosmin in drinking water by novel bacteria isolated from biologically activated carbon. *J Environ. Sci.* 23, 816-823
Ho L, Hoefel D, Bock F, Saint CP, Newcombe G. (2007) Biodegradation rates of 2-methylisoborneol (MIB) and geosmin through sand filters and in bioreactors. *Chemosphere* **66**, 2210-2218
Murano S, Nishino J. (1994) An advanced purification system combining ozonation and BAC, developed by The Bureau of Waterworks, Tokyo Metropolitan Government. Desalination, 98, 207-215
Muramoto S, Udagawa T, Okumura T. (1995) Effective removal of musty odor in the Kanamachi Purification Plant. *Water Sci. Technol.* **31**(11), 219-222

Chapter 57

Biofilm accumulation characteristics of heterotrophic bacteria in tap water produced by biological activated carbon process

S. Kasahara*, M. Ishikawa*

*Department of Environmental Engineering, Osaka Institute of Technology, 5-16-1, Omiya, Asahi-ku, Osaka 535-8585, Japan

ABSTRACT

In order to evaluate the biological stability in a tap water produced by biological activated carbon process, a bench-scale reactor system was operated for about 7 months. The laboratory tap water was used as a test water. Based on material balances on heterotrophic plate count (HPC) and assimilable organic carbon (AOC), the retention time of biofilm cells attached on the surface of polyvinil chloride (PVC) wall (t_r) and the growth yield of heterotrophic bacteria with consuming AOC ($Y_{X/C}$) were estimated. Results indicated that accumulations of biofilm cells and planktonic cells in reactors were dependent on water temperature. Retention times of HPC bacteria for Period 1 (average water temperature was 26 °C) were about 10 times longer than those for Period 2 (average water temperature was 12 °C). In addition, growth yields of HPC bacteria for Period 1 were about 1 order of magnitude higher than those for Period 2. In conclusion, it was suggested that heterotrophic bacteria living in the laboratory tap water, which was produced by ozonation and biological activated carbon filtration processes, would form strong biofilms on PVC wall at the condition of high water temperature.

Keywords Heterotrophic bacteria; distribution system; biological activated carbon

INTRODUCTION

Ozonation and biological activated carbon filtration processes are often applied for assuring drinking water safety. These advanced processes are effective for controlling many chemical risks in tap water. However, it is possible that leak of microorganisms and insufficient removal of biodegradable compounds increase a risk of biofilm formation on pipe walls in distribution systems.

In this study, for evaluating seasonal change in biofilm formation potential in tap water produced by ozonation and biological activated carbon filtration processes, a long term operation of bench-scale reactor system was conducted.

METHODOLOGY

Tap water piped to our laboratory (Osaka City water) was used as a test water. This water has been taken from the downstream of Yodo river which is major water resource of Osaka metropolitan area in Japan. The Yodo river water has high content of effluent from many sewage water treatment plants and contained about 80-600 µg/L of assimilable organic carbon (AOC) (Kasahara *et al.*, 2002). This river water was treated by coagulation, flocculation, sedimentation, ozonation, rapid sand filtration, post-ozonation, granular activated carbon filtration and post-chlorination processes in water purification plant, and then distributed to our laboratory over a distance of about 5 km.

A schematic diagram of a reactor is shown in **Fig. 57.1**. The reactor system was made of polyvinil chloride (PVC) and was placed at dark space for preventing algae growth. In a reactor, a solid PVC drum with a diameter of 14 cm was installed and spun with a rotational speed of 10, 40, 80 and 120 rpm, respectively, for giving various shear stresses to the drum surface. On the solid PVC drum, 24 detachable PVC coupons with a size of 1×15 cm were equipped for collecting biofilm samples. The laboratory tap water was pumped into the reactor with the hydraulic retention time of 0.1 hr. Chlorine residual in the reactor was removed by the addition of sodium thiosulphate.

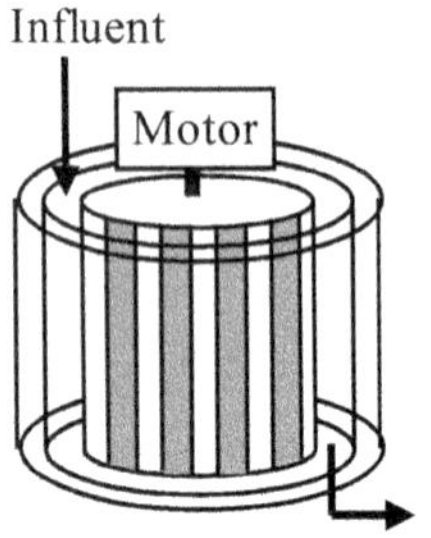

Liquid volume	1.4 L
Surface area	2,300 cm^2
Diameter of drum	14 cm
Size of coupon	1 × 15 cm

Figure 57.1 Schematic diagram of reactor.

Planktonic cell concentrations entering reactors, planktonic cell concentrations in reactors and biofilm cell densities on coupons of each reactor were enumerated with the heterotrophic plate count (HPC) method and the direct total microbial count (DTMC) method. At the elapsed time of 62-111days (Period 1; high water temperature season) and 172-194 days (Period 2; low water temperature season), material balances on HPC and AOC around each reactor were determined, and then retention times of biofilm cells on the drum surface (t_r) and growth yields of heterotrophic bacteria with consuming AOC ($Y_{X/C}$) were estimated by equation (57.1)-(57.3).

$$Q(X_p - X_{in}) = \gamma_d X_b A \tag{57.1}$$

$$t_r = \frac{1}{\gamma_d} \tag{57.2}$$

$$Y_{X/C} = \frac{X_p - X_{in}}{C_{in} - C_{out}} \times 10^3 \tag{57.3}$$

where Q = volumetric flow rate through the reactor (mL/hr); A = surface area for biofilm accumulation in the reactor (cm^2); X_p = planktonic cell concentration in the reactor (CFU/mL); X_{in} = planktonic cell concentration entering the reactor (CFU/mL); X_b = biofilm cell density (CFU/cm^2); γ_d = specific biofilm detachment rate (hr^{-1}); t_r = retention time of biofilm cells on drum surface (hr); $Y_{X/C}$ =

growth yield (CFU/μg of AOC); C_{in} = AOC concentration entering the reactor (μg C/L); C_{out} = AOC concentration in the effluent from the reactor (μg C/L).

RESULTS AND DISCUSSION

Change in temperature of laboratory tap water is shown in **Fig. 57.2**. The temperature of laboratory tap water was over 25°C at the beginning of the operation, and has gradually declined after 70 days has elapsed. Average levels of HPC, DTMC and AOC in laboratory tap water are listed in **Table 57.1**. The level of HPC was 10^2 CFU/mL for Period 1, whereas it was not determined for Reriod 2. The level of DTMC for Period 1 was comparable with that for Period 2. The average concentration of AOC for Period 1 was 50 % lower than that for Period 2. Thus, it was suggested that viability of heterotrophic bacteria in tap water reduced in case of low water temperature, whereas total cell concentration is not dependent on the water temperature.

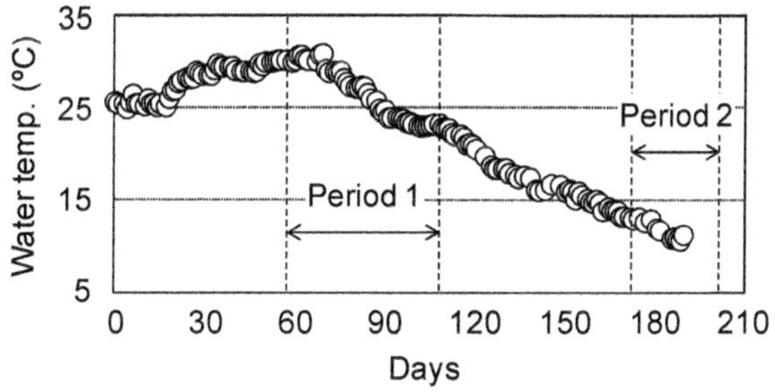

Figure 57.2 Change in temperature of laboratory tap water.

Table 57.1 Average water qualities in laboratory tap water.

Parameter	Period 1	Period 2
Water temperature (°C)	26 (23-31)	12 (11-13)
Free chlorine residual (mg/L)	0.35	0.32
Heterotrophic plate counts (CFU/mL)	5.7x10^2	N.D.
Direct total microbial counts (cells/mL)	6.7x10^5	1.3x10^5
Assimilable organic carbon (μg acetate C eq/L)	24	48

Average numbers of influent cells, biofilm cells and planktonic cells in reactors for Period 1 and Period 2 are shown in **Fig. 57.3**. Obvious differences of cell balances around reactor among the rotational speed were not found, whereas accumulations of biofilm cells and planktonic cells were accelerated at high water temperature. Both numbers of HPC and DTMC determined on the drum surface (biofilm cells) for Period 1 were about 2 orders of magnitude larger than those for Period 2. Those numbers determined in the bulk water (planktonic cells) for Period 1 were 1 order of magnitude larger than those for Period 2. Consequently, as shown in **Table 57.2**, retention times of HPC bacteria on drum surface (t_r) for Period 1 were about 10 times longer than those for Period 2. In addition, growth yields of HPC bacteria ($Y_{X/C}$) for Period 1 were about 1 order of magnitude higher than those for Period 2. Therefore, it is considered that the acceleration of biofilm accumulation with water temperature would be caused by not only rise in growth rate of heterotrophic bacteria but also change in microbial community in biofilms on the sueface of wall.

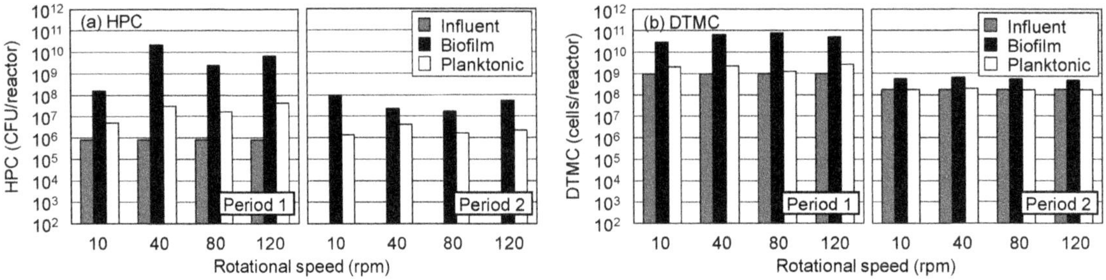

Figure 57.3 Cell balances on (a) HPC and (b) DTMC around reactor.

Table 57.2 Estimation results of retention time on drum surface (t_r) and growth yield of heterotrophic bacteria ($Y_{X/S}$).

Rotational speed of drum (rpm)	t_r (h)		$Y_{X/C}$ (CFU per µg of AOC)	
	Period 1	Period 2	Period 1	Period 2
10	3.9	7.0	2.5×10^5	4.7×10^4
40	70.7	0.6	3.0×10^6	1.7×10^5
80	15.4	1.0	1.1×10^6	7.8×10^4
120	15.4	2.5	3.8×10^6	7.8×10^4

CONCLUSIONS

When the reactor system was operated with the laboratory tap water, both retention time of biofilm cells on the drum surface and growth yield of heterotrophic bacteria with consuming AOC were dependent on the water temperature. Experimental results indicated that heterotrophic bacteria living in the laboratory tap water, which was produced by ozonation and biological activated carbon filtration processes, would form strong biofilms on PVC wall at the condition of high water temperature.

ACKNOWLEDGEMENTS

The authors are grateful to Mr. K. Oyama, S. Taketani and R. Shakunaga, who were students of Osaka institute of Technology, for their excellent assistance in this experiment.

REFERENCE

Kasahara, S. and Ishikawa, M. (2002) Evaluation of assimilable organic carbon in the Yodo river and the effect of coagulation on its control, *Wat. Sci. and Tech: Wat. Suppl.*, **2**(5-6), pp.465-472, 2002.

Chapter 58

Manganese removal in BAC facilities without ozonation in water purification plants

T. Yahata, E. Kawase**, E. Takahashi*, M. Yoshii**

**Water Quality Management Division, Niigata Water Works Bureau, 1-1 Aoyama-suido, Nishi-ku, Niigata City, Niigata 950-2005, Japan*

***General Affairs Division, Niigata Water Works Bureau, 1-3-3 Sekiya-shimokawara-cho, Chuo-ku, Niigata City, Niigata 951-8560, Japan*

ABSTRACT

In the purification of river water, dissolved manganese (D-Mn) is removed by the contact oxidation method, which uses manganese sand in rapid sand filters (RSFs). We surveyed the manganese removal process by a biological activated carbon (BAC) treatment without ozonation. After one BAC bed was replaced, the D-Mn concentration was monitored at various sampling points. The survey results showed that D-Mn was not detected in the water treated by the aged BAC beds, but high concentration of manganese was detected by the new BAC bed. A follow-up survey indicated that a BAC bed needs six months to obtain manganese removal functionality. After six months of operation, the concentration of manganese was almost zero at all sampling points. In the next survey, we measured the quantity of adhered manganese (ad-Mn) in the GAC beds and RSFs. The results from the Aoyama and Shinanogawa purification plants were compared. In the Aoyama plant, which did not adopt BAC treatment, D-Mn was removed by manganese sand formed at the RSF surface. The quantity of ad-Mn in the RSFs was between 5.8 and 13.5 mg/g. Before beginning operation of the Shinanogawa purification plant, we performed a test operation without BAC treatment, and mangaese removal was required during the test period. However, after six years of operation, the quantity of ad-Mn in the RSFs decreased to less than 0.3 mg/g. Based on these facts, we concluded that the quantity of ad-Mn decreased gradually because the concentration of D-Mn in water treated by RSFs was almost zero when the BAC facility obtained manganese removal functionality. We also concluded that BAC beds should be replaced during the summer season because the concentration of manganese is at its lowest level because of the high temperature of the river water.

Keywords Tap water; Biologically activated carbon treatment; manganese removal; BAC bed replacement

INTRODUCTION

Niigata city, which has a population of 800,000, is located in the middle of Japan. There are six water purification plants in the city, and all of their water resources come from river surface water. In 2005, a new purification plant named Shinanogawa was built to replace an older purification plant. BAC treatment equipment was adopted at Shinanogawa to reduce the risk of trihalomethanes (THMs). Figure 58.1 shows the water treatment processes in Shinanogawa. Note that the Shinanogawa plant has a BAC facility without an ozonation process. A BAC facility is composed of six beds. To investigate the effect of BAC treatment, we evaluated the parameters of removable matter related to drinking water quality, such as trihalomethane formation potential (THMFP), ultraviolet absorbance (E260) and pesticides.

Horikawa et al. (2009) found a correlation between E260 and THMFP in water treated by BAC, and calculated the removal rate of organic compounds based on E260 values using the following equation.

Removal rate of organic compounds (%) = A/B × 100

A: Outflow of E260 values (from GAC facility)
B: Inflow of E260 values (into GAC facility)

The results showed that the removal rate decreases to less than 30% after four years of operation. Our investigation revealed that this is due to a reduction in the adsorption effect in BAC treatment. The principal functions of BAC treatment include adsorption and biodegradation. The characteristic results from investigations on BAC treatment are as follows. The adsorption effect in BAC treatment can remove organic compounds such as humic acid and pesticides, but their removable quantity gradually decreases with aging. Biodegradation in BAC treatment can remove biodegradable matter such as ammonium nitrogen and nitrate nitrogen, but the removable rate changes with the season because the biological activity depends on the water temperature.

In order to recover the adsorption effect in BAC treatment, we replaced one BAC bed in 2011. D-Mn in the treated water was detected after this replacement. This does not typically occur because D-Mn is usually removed by the contact oxidation method, which uses manganese sand at the filtration sand surface. To investigate the reason for this detection, we monitored the annual transition of D-Mn at various sampling points and measured the quantity of ad-Mn at RSFs and the granular activated carbon (GAC) facility. This paper considers a manganese removal process by BAC treatment without pre-ozonation and proposes a replacement plan for BAC beds for a sustainable removal of manganese.

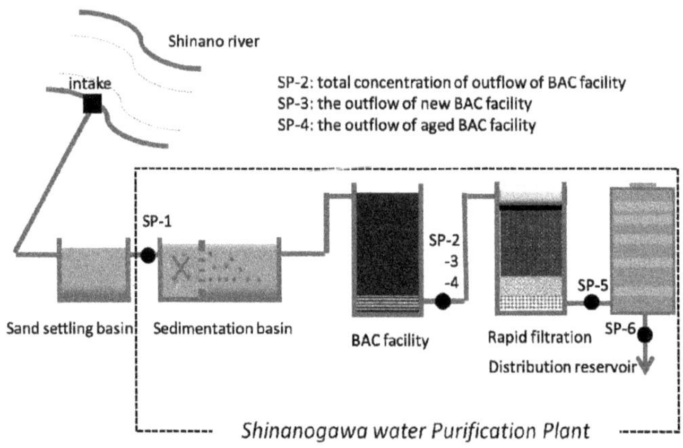

Figure 58.1 Sampling points at the Shinanogawa purification plant, where D-Mn was monitored for several treatment processes.

EXPERIMENTAL METHOD

To monitor D-Mn concentration, the sample water was collected at the outflow from various processes shown in Figure 58.1. The samples were measured twice a month from April 2010 through March 2011. The Japanese government standard method was used to measure D-Mn by ICP-MS.

To measure the quantity of ad-Mn, sample sands were collected at various points giving in Figure 58.2(a). These were extracted by the method shown in Figure 58.2(b), and the dissolved manganese was measured by ICP-MS.

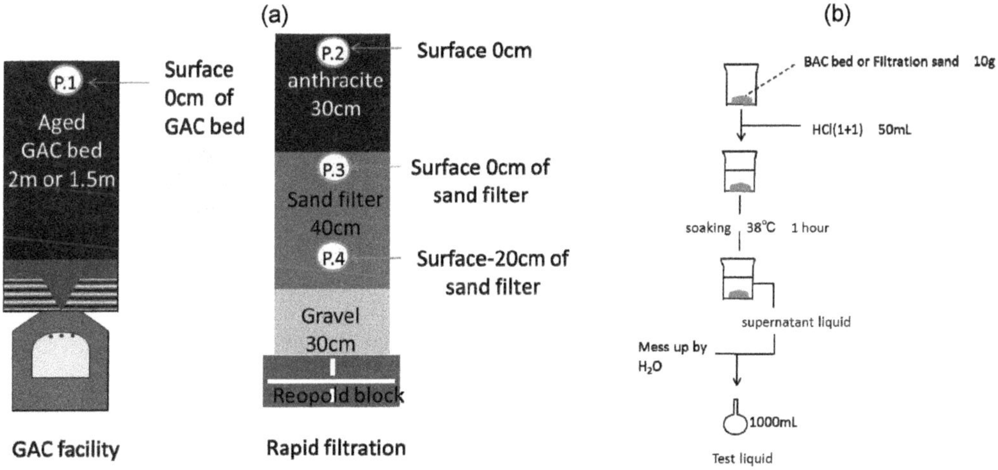

Figure 58.2 (a) Sampling points for the filtration sand at the GAC facility and RSF when ad-Mn was measured. (b) The method for extracting ad-Mn from several sample sands.

RESULTS

Figure 58.3 shows that D-Mn was detected in the treated water sampled from the new BAC facility after GAC replacement. The same concentration was detected in both raw and treated water. This is because GAC cannot remove D-Mn that exists as a D-Mn ion in the aqueous phase. However, we did not detect manganese from the sample collected from the outflow of the aged BAC facility. The results show that D-Mn could be removed by the aged BAC facility.

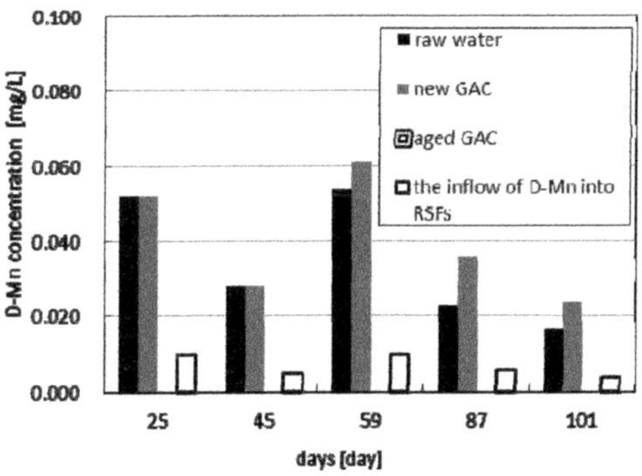

Figure 58.3 Variation in soluble manganese in raw water and the outflow from new and aged BAC facilities when D-Mn was monitored for 100 days at a GAC facility.

Figure 58.4(a) indicates that after 180 days, the manganese concentration leaking from the new BAC bed decreased, as compared to the aged BAC level. It takes six months for a BAC bed to obtain manganese removal functionality. As shown in Figure 58.4(b), after 180 days, the manganese concentration leaking from the new GAC was higher than that from the aged GAC. This shows that the functionality of manganese removal was not yet stable.

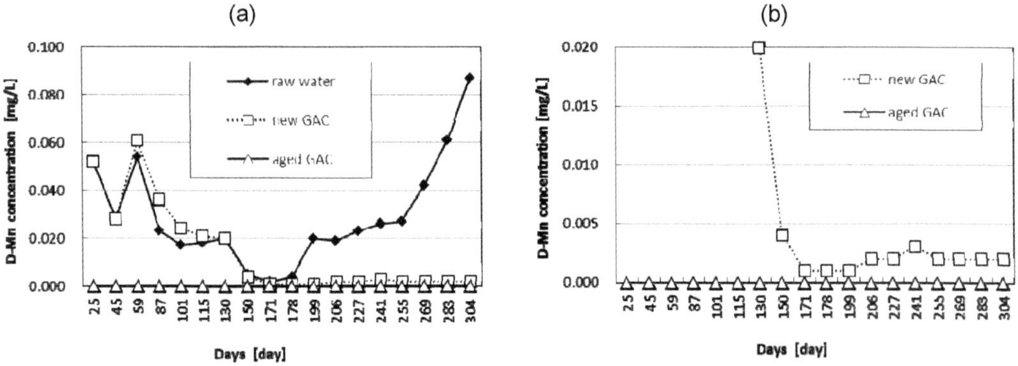

Figure 58.4 Variation in soluble manganese from raw water and the outflow from new and aged BAC facilities when D-Mn was monitored for a year at a GAC facility. (a) The range of high D-Mn concentrations. (b) The range of low D-Mn concentrations.

To determine where the manganese removal processes performed, we measured the quantity of ad-Mn at various sampling points shown in Figure 58.3. In the filter media of p.1 GAC, the quantity of ad-Mn from aged GAC was 3.1 mg/g, which is higher than the standard value specified for manganese sand, i.e., 0.3 mg/g. On the other hand, the ad-Mn in the new BAC was less than 0.3 mg/g. These results suggest that aged D-Mn was oxidized to manganese dioxide at the surface of aged GAC.

We also measured the quantity of ad-Mn at other sampling points. When comparing the filter sand of p.3 and p.4, the results from the Shinanogawa water purification plant were significantly less than those from Aoyama. The results show that the manganese sand is performed during test operation in Shinanogawa lost mangaese removal functionality after six years in operation.

Table 58.1 Concentration of ad-Mn for several filtration sands when the sample sands were collected before and after GAC replacement.

Filter media	Water purification plant	The state of BAC bed	Quantity of adhered manganese [mg/g]
p.1 GAC	Shinanogawa	aged BAC	3.10
		new BAC-replaced BAC-	0.01
p.2 Anthracite	Shinanogawa	aged BAC	0.26
		new BAC	0.35
p.3 Filter sand	Shinanogawa	aged BAC	0.05
Surface 0cm		new BAC	0.06
	Aoyama	Without BAC	13.50
p.4 Filter sand	Shinanogawa	aged BAC	0.06
Surface -30cm		new BAC	0.05
	Aoyama	Without BAC	5.80

DISCUSSION

To show the influence of aging and GAC replacement on the manganese removal process, we indicate the change in the removal rate of D-Mn by biological oxidation or contact oxidation, including the start of the operation and before and after GAC replacement. As shown in Figure 58.5(a), the manganese removal process changes to BAC oxidation from contact oxidation a few years after the operation starts. We summarize the result from Figure 58.5 below.

Regarding Contact Oxidation at RSF:

i. At the start of the operation, all of the D-Mn in the water resource is removed by contact oxidation in RSFs.
ii. The quantity of ad-Mn decreases gradually because D-Mn concentration inlet to RSFs is almost zero when the BAC facility obtains manganese removal functionality.
iii. The removal rate of D-Mn in RSFs decreases to 50% at the time of GAC replacement.

Regarding Biodegradation at BAC Facility:

iv. Biodegradation begin to perform the function of D-Mn removal after few months from the start of the operation
v. Bioderadation loses the function of D-Mn removal when GAC bed is replaced
vi. Six months after GAC replacement, the BAC bed can remove 90% of the D-Mn in raw water. After 18 months, it can remove 100%.

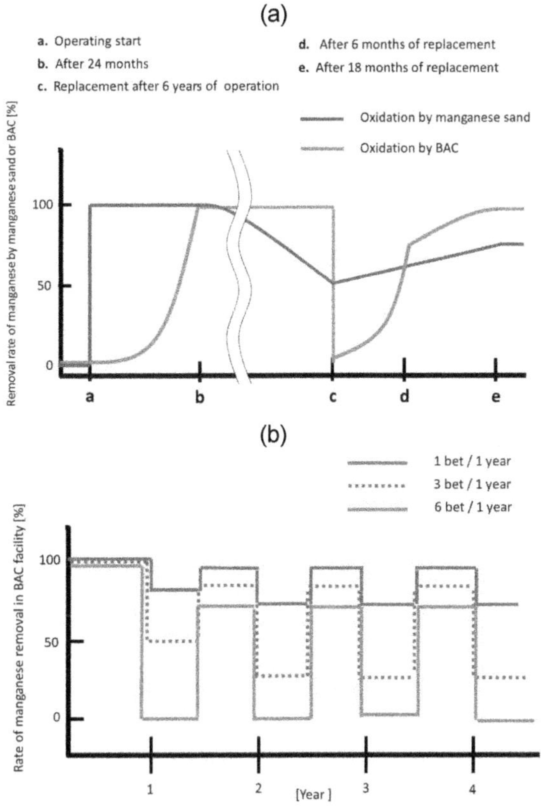

Figure 58.5 (a) Change in D-Mn removal rate by biological oxidation and contact oxidation (above) and (b) removal rate of D-Mn by changing the number of replacement GAC beds in a year (below).

As shown in result 4, we should be able to predict the change in the D-Mn removal rate based on GAC replacement because the GAC bed replacement decreases the removal rate of D-Mn. The prediction results are shown in Figure 58.5(b). When several BAC beds are replaced over time, the removable quantity of manganese decreases significantly over six months. Therefore, if the volume of the replaced GAC bed is large, it has a greater influence on the D-Mn removal process. the way to prevent this case, the manganese sand in RSF is insufficient to remove D-Mn, and that at the rapid filtration needs to be added.

To check the decrease in the D-Mn removal rate because of GAC bed replacement, the seasonal variation in D-Mn in raw water is shown in Figure 58.6. Based on this result, BAC beds should be replaced during the summer season because D-Mn concentration is at its lowest value because of the high temperature of the river water.

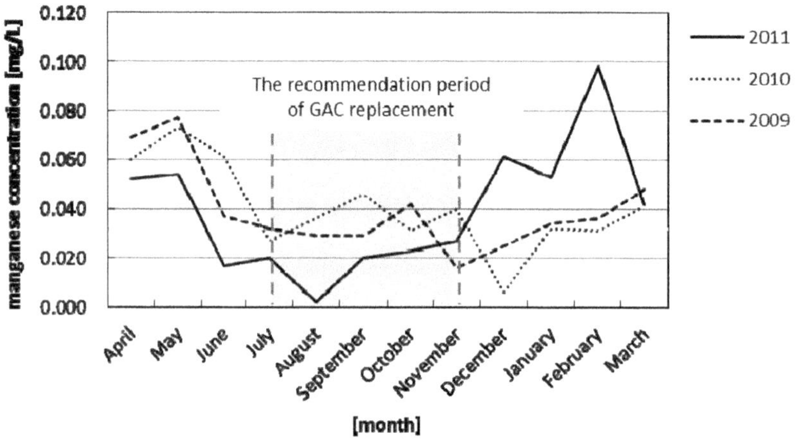

Figure 58.6 Seasonal variation in soluble manganese in raw water from 2009 to 2011.

CONCLUSIONS

The main conclusions of this study are as follows:

(1) BAC facilities without ozonation can still remove manganese.

(2) New BAC facilities can perform manganese removal after six months of operation.

(3) The quantity of ad-Mn in RSFs decreases gradually in the Shinanogawa purification plant. Because D-Mn concentration inlet to RSFs is almost zero, the BAC facility obtains manganese removal functionality.

(4) BAC beds should be replaced during the summer season because D-Mn concentration is the lowest because of the high temperture of the river water.

REFERENCES

W. Horikawa, M. Yoshii, M. Kondo, (2009) Evaluating a performance by BAC treatment and planing a replacement of GAC bed in Shinanogawa purification plant, Proceedings of the 61st National Conference on Water Researchs, Japan.

S. Takai, H. Nakanishi, (1987) Removal of iron and manganese in ground water, *The Industrial Water Institute.*

Y. Sato, K. Fukushi, N. Tanbo, (1993) Evaluation of the bio-chemical function rapid sand filtration revaluating the bio-chemical function of sand filtration(2). *Journal of Japan Water Works Association* Vol.61 No.12, JWWA.

Osaka Water Supply Authority, (2009) The effective means for the problem of leaking a manganese, *Annual report of Water Examination of OWSA.*

Chapter 59

Use of water sensitive urban design systems for biofiltration of urban stormwater: laboratory biodegradation batch studies

*K. Zhang**, **, *X. Tian**, *D. Page****, *A. Deletic**, **, *D. McCarthy**, **

**Monash Water for Liveability, Department of Civil Engineering, Monash University, Clayton, VIC 3800, Australia*

***CRC for Water sensitive Cities, Melbourne Vic 3800, Australia*

****CSIRO Land and Water, Waite Laboratories, Urrbrae SA 5064, Australia*

ABSTRACT

Water Sensitive Urban Design (WSUD) systems are a form of alternative biological filtration that is increasingly being used for stormwater management. To date, there is limited knowledge on the behavior of organic micropollutants in such systems and applicability to water reuse. In this study, biodegradation of a model micropollutant, phenol was investigated in the laboratory from two types of media typically used in stormwater biofiltration systems (also known as biofilters, bioretentions, and rain-gardens). Conditions, including temperature, pH, starting concentrations, moisture contents and redox conditions (aerobic and anaerobic) were varied to cover common environmental ranges. Ultraviolet absorbance at 254 nm (UVA) was also evaluated as potential surrogate for phenol biodegradation. Phenol examined in all types of soils was degraded biologically, with high percentage of removal ($k = 0.125 - 1.26\ h^{-1}$) and short half-lives ($t_{1/2}$=0.550 - 5.54 h). The results indicated that the biodegradation rates were affected by temperature, soil moisture and redox condition. The highest biodegradation rates were achieved based on temperature 30^0C and with soil moisture of 100% under aerobic conditions. pH and initial phenol concentration had little impact on biodegradation of phenol in this study. Analysis of spearman rank correlation indicated that UVA reduction was significantly related to phenol biodegradation in Sand Deep soils under all conditions ($P<0.01$).

Keywords stormwater; biodegradation; surrogates; ultraviolet absorbance (UVA); phenol

INTRODUCTION

Urban stormwater runoff volumes are increasing with urbanization and the increase of impervious surfaces (Goonetilleke et al., 2005). The reuse of urban stormwater has the potential to augment urban non-potable and potable uses. However, stormwater runoff may contain various pollutants, including nutrients, heavy metals, microorganisms and organic micropollutants (Makepeace et al., 1995; Göbel et al., 2007). These may become an obstacle to the reuse of stormwater, especially for potable uses.

Stormwater biofiltration systems (also known as biofilters, bioretention systems or rain garden) have the potential to provide both water quality and quantity benefits in water recycling applications (FAWB, 2009). They are commonly used low energy stormwater treatment method consisting of a planted soil-based filter media, which usually has a high sand content (e.g. loamy sand) underlaid by a sand transition layer. They are implemented under integrated water management approaches such as Water Sensitive Urban design (WSUD) and stormwater Best Management Practices (BMP) (Wong, 2006). Stormwater biofiltration systems are very effective for the removal of sediments, metals and nutrients (Henderson et al., 2007; Blecken et al., 2009), thus are becoming increasingly popular for stormwater harvesting. However, little research has been done with regard to the fate of micropollutants in stormwater biofiltration systems. Numerous studies have documented the removal process of organic micropollutants in similar treatment systems such as wetlands (Matamoros et al., 2008) and aquifers (Pavelic et al., 2006), which indicated that adsorption and biodegradation being the most important processes within these soil-based systems. In order to improve stormwater biofiltration systems, knowledge of micropollutant fate in these systems is desirable.

This study aims to investigate the process of biodegradation of one micropollutant that existing in stormwater. There are various types of micropollutants detected in stormwater, such as polycyclic aromatic hydrocarbons, pesticides, phthalates, phenols and polychlorinated biphenyls (Makepeace et al., 1995; Göbel et al., 2007). They have a wide range of physical-chemical properties, therefore may undergo different removal mechanisms in stormwater biofiltration systems. Phenol is a good micropollutant to explore biodegradation process because researchers have found it to be biodegradable (Broholm and Arvin, 2000), while the others like polychlorinated biphenyls biodegrade very slowly and thus adsorb (Mackay et al., 2006). Several studies have reported biodegradation of phenol in water treatment systems (Broholm and Arvin, 2000; Shibata et al., 2006). In general, the results indicate that a high percentage of phenol (>50%) was removed due to biodegradation. The rate of biodegradation may be affected by many factors such as initial phenol concentrations (Broholm and Arvin, 2000), temperature (Onysko et al., 2000) and pH (El-Naas et al., 2009). Also, the soil moisture level (Shelton and Parkin, 1991) and anaerobic condition (Chang et al., 2004) have been observed to inhibit chemical biodegradation rate.

Moreover, study of surrogate parameters of micropollutant removal is important for monitoring the performance and further validation of these systems (Drewes et al., 2008; DHV, 2013). Drewes et al (2008) have developed a suite of readily measurable surrogates for organic micropollutants during wastewater treatment and recycling systems. Surrogates such as UVA (ultraviolet absorbance), DOC (dissolved organic carbon), ammonia, and nitrate were identified as suitable for monitoring biodegradation. However, to date the suggested surrogates have not been tested for stormwater biofiltration systems.

In this study, biodegradation was hypothesized as a main process contributed to the removal of phenol in a stormwater biofiltration system. Due to the variability of stormwater quality and stormwater treatment system conditions, some influential factors, including initial concentration, temperature, pH, soil moisture and redox conditions were studied. UVA was hypothesized to be a good surrogate for phenol biodegradation. Laboratory batch experiments were conducted to measure biodegradation of phenol in two types of stormwater biofiltration systems. The specific aims of this study were:

1) Study the biodegradation rate and half-lives of phenol in stormwater biofiltration systems;
2) Investigate the main factors influencing phenol biodegradation;
3) Identify the potential of ultraviolet absorbance (UVA) as a good suitable surrogate for phenol biodegradation in stormwater biofiltration systems.

MATERIALS AND METHODS

Soil samples and chemicals

Soil samples were collected from the two field stormwater biofiltration systems located at Monash University, Melbourne, Australia. Surface soils of the biofiltration system usually consist of original filter media, stormwater sediments and humus and therefore contain different organic matter content and also microbe communities in comparison to deep soils. Thus both surface (top ~5 cm) and deep soils (≥10 cm) were collected for study. The characteristics of four soils collected are presented in Table 59.1.

Table 59.1 Characteristics of biofiltration soils used in this study.

	Loamy Sand Deep (LS-D)	Loamy Sand Surface (LS-S)	Sand Deep (S-D)	Sand Surface (S-S)
Bulk Density (g cm^{-3})	1.58	1.59	1.61	1.60
Total Organic Matter (%)	3.6	5.3	0.35	4.2
Total phosphorus (mg kg^{-1})	260	470	30	420
TN (mg kg^{-1})	1,200	2,000	300	1,400
NH_3 (mg kg^{-1})	1	2	< 1	3
NO_x (mg kg^{-1})	2	6	< 1	8
Sand (%)	92.8	91.4	99.4	95.3
Clay (%)	3.1	2.5	0.32	1.0
Silt (%)	4.1	6.1	0.28	3.7

Phenol (ACS) used in this study was purchased from Sigma-Aldrich (Australia) and stock solutions of 1 g L^{-1} were prepared for use. Synthetic stormwater was prepared following the procedure described previously by Blecken et al (2009), by mixing de-chlorinated tap water with sediment collected from a stormwater wetland and then used to prepare phenol solutions of different concentrations.

Biodegradation experiments

Table 59.2 presents the experimental matrix of the batch study. Biodegradation experiments were performed in 1L screw-cap glass jars (Rowe Scientific, Australia) containing 198.8 g of the collected soils and a certain volume of synthetic stormwater within known phenol concentrations to get the target initial concentration and soil moisture (Table 59.2). After the addition, the glass jars were incubated in the dark without shaking for up to 32 hours under different temperatures. Sterile soils were obtained by gamma irradiation at 25kgy of irradiation level and these were used as controls under appropriate conditions. Blank controls (no soil) were also conducted under different temperatures, pH and initial concentrations. All the experiments were prepared in triplicate.

Effect of initial concentration (C_0). 25.00 mL of synthetic stormwater with 2.400 mg L^{-1} and 1.800 mg L^{-1} phenol solutions were added to 198.8g Sand Deep soils separately to obtain initial phenol concentrations. The initial concentration was in agreement with the relatively low phenol concentrations found in stormwater (0.100-167.0 μg L^{-1}) (Dugan and Christakos, 1983;Grynkiewicz et al., 2002).

Effect of temperature. According to our temperature monitoring data of the studied biofiltration system at the depth of 3cm, 15cm and 30 cm from September 2010 to August 2011 (data not shown), soil temperature changed over time, especially for surface soil. The 1st, 5th and 99th percentile temperature values of the recorded temperature of surface soil were 4^0C, 15^0C and 30^0C, respectively. Therefore the impact of temperature on phenol biodegradation rate in these surface soils was studied under these three temperatures. As a point of comparison, the Sand Deep soils were also studied at the highest temperature of 30^0C.

Effect of pH. The impact of stormwater pH on phenol biodegradation was studied in Sand surface and Sand Deep soils, where the pH value of the synthetic stormwater was varied from 5 to 7 (typical stormwater pH value according to Duncan, 1999) by adding 1M HCl or 0.1 M NaOH solutions.

Effect of soil moisture. Stormwater biofiltration systems experience drying cycles between storm events; it can range from 25% during dry periods and nearly saturated (100%) during wet weather events (Daly et al., 2012). 50% soil moisture (half saturation) corresponded to the medium moisture value of field soil moisture monitoring. Hence, the phenol biodegradation rate was studied under three different soil moisture contents: 25%, 50% and 100%, which were achieved by mixing 12.50 mL, 25.00 mL and 50.00 mL synthetic stormwater with 4.800 mg L^{-1}, 2.400 mg L^{-1} and 1.200 mg L^{-1} phenol with 198.8 g soil separately to get the same initial concentration (i.e. 300 μg kg^{-1}).

Effect of redox condition. Stormwater biofiltration systems experience anaerobic conditions in the deep layer of biofiltration systems after long dry periods (FAWB, 2009). Anaerobic biodegradation of phenol in Sand Deep soil was studied by placing the 1.000 L glass jar with loose lid into a 3.500 L anaerobic jar packed with Oxoid™ AnaeroGen™, which reduces oxygen content to <1.0%.

Table 59.2 Experimental matrix of the biodegradation batch study.

Factors	Values	Soil type tested**
Temperature	4°C, 15°C*, 30°C	Loamy Sand Surface, Sand Surface, Sand Deep
pH	5.0, 7.0*	Sand Surface, Sand Deep
Initial concentration (C_0)	200 μg kg^{-1}, 300 μg kg^{-1}	Sand Deep
Redox	aerobic*, anaerobic	Sand Deep
Soil moisture	25%, 50%, 100%	Sand Surface

Note: * 'base-case' value, other influential factors were fixed at the 'base-case' values when studying the effect of the target influential factor; ** Not all the influential factors were tested for each type of soil. The target biofiltration system with sand filter media was designed according to the best Australian practice (FAWB, 2009) and therefore only sand surface/deep soil were tested for most of the factors.

Sampling and analysis

During the incubation, the soil samples were extracted at 0, 0.5, 2, 4, 8, and 24 h (and 30 h if required). Preliminary experimental results indicated that phenol was below the detection limit after 30 hours. The sampling scheme was adjusted according to the change of phenol concentrations during the experiment. 30.00 g soil samples were transferred to 50.00 mL centrifuge tube, to which was added 30.00 mL deionized water and shaken for about 5 minutes. The samples were centrifuged at 4000 rpm for 10 minutes and the supernatant extracted and analysed for phenol concentration and ultraviolet absorbance at 254nm (UVA). Phenol was analysed according to the 8047 4-Aminoantipyrine method Adapted from Standard Methods for the Examination of Water and Wastewater, Method 5910 (APHA et al., 2012). HACH DR 5000 UV-Vis Laboratory Spectrophotometer was used for both phenol and ultraviolet absorbance at 254 nm (UVA) analyses.

Phenol biodegradation were described by $C_t = C_0 \times e^{(-kt)}$, where C_0 denotes to the initial concentration of phenol, C_t is the concentration of phenol at time t and k is the biodegradation rate constant (h^{-1}). The phenol half-lives, $t_{1/2}$ (h) were also estimated by the by the equation $t_{1/2} = \text{In}2/k$.

RESULTS AND DISCUSSION

Biodegradation of phenol in stormwater biofiltration soils

Figure 59.1 shows the change of phenol concentrations in four different soils, while Table 59.3 presents the biodegradation rate constants and half-lives of phenol in four different soils under all studied conditions. Phenol concentrations declined to detection levels (<3.000 μg kg^{-1}) within 30 hours in all tests except controls, regardless of the difference of soil characteristics. The change of phenol concentration in the blank control was within ±1.3%, indicating minimal hydrolysis. The reduction of phenol concentration in sterile controls was probably due to adsorption. The adsorption of phenol onto loamy sand surface, loamy sand deep, and sand surface soils was high (about 30%) due to the high organic matter content in these soils (3.6%-5.3%; Table 59.1), while only around 15% phenol adsorbed onto sand deep soils, which contains only around 0.35% of organic matter. These results suggested that microbial activity is the main process responsible for the phenol degradation. The most rapid removal was observed in Sand Surface soil, followed by Loamy Sand Surface, Loamy Sand Deep and finally Sand Deep. The biodegradation of phenol in Sand Surface soil at stormwater pH=5 had the shortest half-life (0.550±0.085 h), while the same soil with anaerobic conditions had the longest half-life (5.54±1.06 h) (Table 59.3).

Influence of different conditions on phenol biodegradation

Effect of temperature. Biodegradation rates of phenol were observed to vary with different temperatures in Loamy Sand Surface (LS-S) and Sand Surface (S-S) soils (Table 59.3). The biodegradation rate of phenol was the slowest under the lowest temperature (4^0C). These results suggest that temperature could affect the ability of biodegradation in stormwater biofilter. El-Naas *et al.* (2009) and Onysko *et al.* (2000) reported that lower temperatures hindered bacteria activity. In addition, the higher production rate of metabolites at 30^0C is also a key reason contributes to taking 30^0C as the optimum temperature for biodegradation (Sá and Boaventura, 2001).

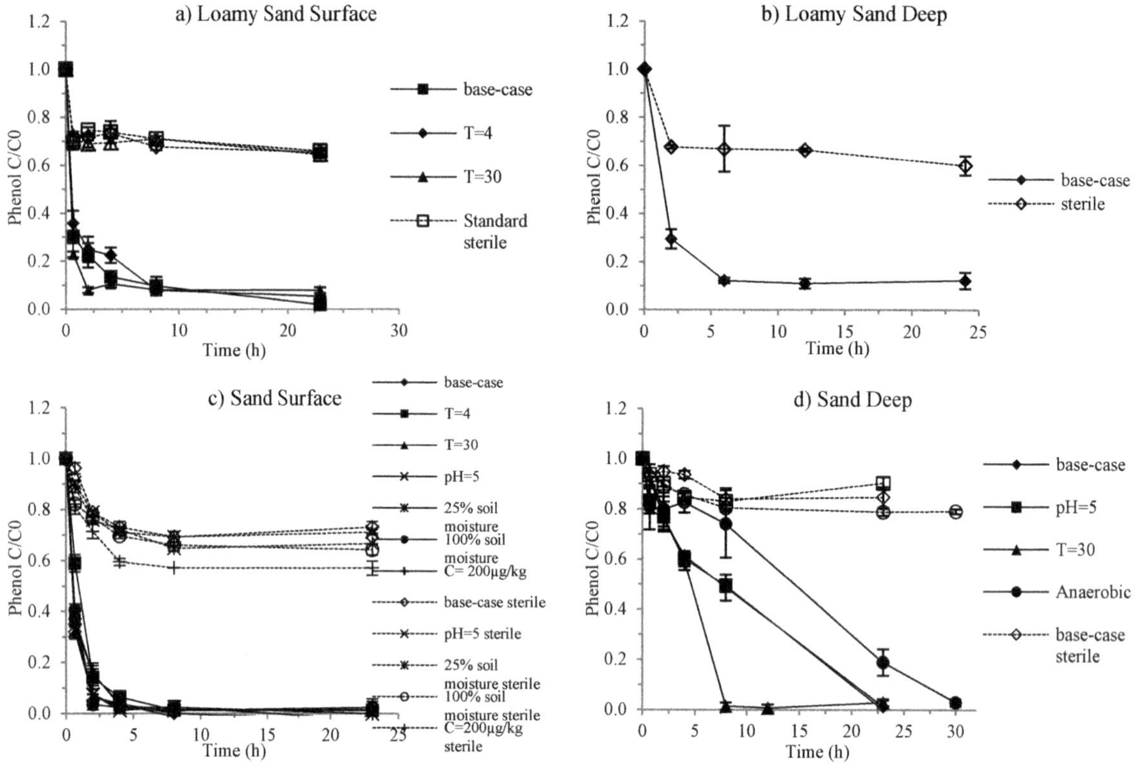

Figure 59.1 Biodegradation of phenol in four different soils. All conditions are 'base-case' (initial concentration C_0 = 300 μg kg^{-1}; T = 15°C; pH = 7; 50% soil moisture; aerobic condition) otherwise stated.

Effect of pH. The pH of stormwater had little impact on biodegradation of phenol, where the degradation rate of phenol at pH = 5 was 0.132 h^{-1} and pH = 7 was 0.165 h^{-1} in Sand Deep soils (Table 59.3). In Sand Surface soil, the higher rate (1.26 h^{-1}) occurred at lower pH (=5), but the difference was small (< 25.0%). Alkalinity of soils served buffer effect of pH, which led to slight changes of soil pH when added a small amount of stormwater in this study.

Effect of initial concentration of phenol. The phenol biodegradation rates in Sand Surface soil with lower initial concentrations (200μg kg^{-1}) had higher rates (1.08 h^{-1}, Table 59.3) as compared to the higher initial concentration (300μg kg^{-1}; 0.947 h^{-1}). This is in agreement with the findings by Broholm and Arvin (2000) who argued that higher phenol concentrations cause inhibitory effects on the activity of the biomass.

Effect of soil moisture. The biodegradation rate of phenol was higher at the highest soil moisture content (100%, Table 59.3). This positive effect of soil moisture content or soil water potential on the activity of microorganisms in soils is well documented (Wilson and Griffin, 1975). The results from this study are similar to those previous reported (Anderson, 1981;Shelton and Parkin, 1991). Anderson (1981) stated that water content indirectly impacts the biodegradation rates due to its influence on both the size and survival of biomass. Shelton and Parkin (1991) reported that drying soils have consistently been observed to retard or inhabit rates of microbial metabolism, which is consistent with the decrease in phenol biodegradation rates in the lower moisture content soils.

Effect of redox condition. Experiments were conducted in Sand Deep soils under both aerobic and anaerobic conditions (Figure 59.1) and the results showed that phenol degradation was observed within 32 h under both scenarios. The result suggests that biodegradation of phenol in Sand Deep soil can occur anaerobically. However, compared to aerobic conditions (half-life 4.20±1.94 h, Table 59.3), anaerobic biodegradation was slower, with a half-life of 5.54±1.06h. Bak and Widdel (1986) reported that anaerobic conditions can have detrimental impacts on the activity of bacteria and hence these conditions may hinder biodegradation of phenol in this study.

Table 59.3 First-order degradation rates, half-lives of phenol and spearman rank coefficient between the biodegradation removal rates of phenol and UVA in studied biofiltration systems

Type of Soil	Condition	Degradation Rate k (h^{-1})**	R^2	$t_{1/2}$ (h)	Spearman coefficient***
Loamy Sand Surface (LS-S)	Base-case*	0.395±0.123	0.824	1.76±0.615	0.50
	T = 4 °C	0.332±0.022	0.900	2.09±0.139	0.32
	T = 30 °C	0.742±0.078	0.810	0.934±0.099	0.27
Loamy Sand Deep (LS-D)	Base-case	0.377±0.036	0.946	1.84±0.177	0.24
Sand Surface (S-S)	Base-case	0.947±0.132	0.903	0.732±0.104	-0.08
	pH = 5	1.26±0.190	0.953	0.550±0.085	0.48
	25% soil moisture	0.881±0.158	0.927	0.787±0.146	0.00
	100% soil moisture	1.22±0.208	0.837	0.568±0.100	0.00
	T = 30 °C	0.974±0.044	0.837	0.712±0.032	-0.18
	T = 4 °C	0.779±0.122	0.940	0.890±0.143	0.24
	C_0 = 200μg kg^{-1}	1.08±0.177	0.909	0.642±0.109	0.19
Sand Deep (S-D)	Base-case	0.165±0.052	0.900	4.20±1.49	**0.74**
	pH = 5	0.132±0.020	0.969	5.26±0.817	**0.8**
	T = 30 °C	0.488±0.138	0.809	1.42±0.442	**0.64**
	Anaerobic	0.125±0.023	0.838	5.54±1.06	**0.7**

*: 'base-case': All other conditions are 'base-case' conditions (initial phenol concentration C_0 = 300 μg kg^{-1} at T = 15 °C, pH=7, in the soils with 50% soil moisture and under aerobic condition) otherwise stated;
** ± standard deviation, based on analysis of the triplicate experiments;
*** Spearman rank coefficient, BOLD values means P value <0.01while the others indicate P>0.05.

Implications for phenol removal in field stormwater biofiltration systems

These results suggest that biodegradation will contribute to the removal of phenol in stormwater biofiltration systems. Half-lives of phenol varied between 0.550~5.54 hours, which is great news for a typical stormwater biofiltration system that has a designed hydraulic residence time of 3~5 hours (FAWB, 2009). During wet events, when the system is nearly saturated (soil moisture ~100%), the phenol biodegradation rate will be high. Also, during dry weather when soil moisture content decreases the biodegradation rate slows down, but this does not matter as only a little phenol is adsorbed on soils and this little amount can be then degraded over dry weather period that may last days to weeks (e.g. 95th percentile of the length of dry period in Melbourne is 19 days, Zhang K.F. et al., 2013). According to our previous study on DO analysis in stormwater biofiltration systems, anaerobic condition was only found in deep zones after a long dry period (>3 weeks), which indicates that the slow biodegradation rate of phenol under anaerobic conditions will not be a big problem because phenol may have already been biodegraded before anaerobic condition occurs in the system. The differences of phenol biodegradation rates observed between different soils indicate that filter media with higher organic matter may support larger microbe communities, which will then result in a higher phenol biodegradation rate. Therefore appropriate filter media types may be a key design parameter of stormwater biofiltration systems for phenol removal. The analysis of temperature influence on phenol biodegradation rate implicates that stormwater biofiltration system will perform better in summer than that in winter. Inflow stormwater quality (i.e. pH &phenol concentration) also affects the phenol biodegradation rates, however, within the range of their values tested in this study, phenol biodegrades quickly in regardless of different pH values and initial concentrations.

Identification of UVA as a surrogate of phenol biodegradation

The spearman rank correlation coefficients between the biodegradation removal rates of phenol and ultraviolet absorbance (UVA) are presented in Table 59.3. UVA reduction (ΔUVA) is significantly related to phenol biodegradation (Δphenol) in Sand Deep soils under all conditions ($P<0.01$). However, non-significant correlations were found between ΔUVA and Δphenol in all of the other soils (including some negative correlations). This is hypothesized to be linked to UVA being a measure of the aromatic component of the water soluble dissolved organic matter (DOM) in soils, and that the levels of total organic contents varied significantly between the four soils. Indeed, the total organic matter content in the three soils which showed poor correlations (Loam sand surface, Loamy sand deep and Sand surface) were very high (3.6%-5.3%,Table 59.1), while the opposite was true for the Sand Deep soil (0.35%). As such, it is proposed that soils with a high level of organic matter will result in higher background UVA readings, meaning that it is more difficult to detect changes in the UVA signal. This suggests that ΔUVA could only serve a good surrogate for phenol biodegradation in the stormwater biofiltration systems using sand as filter media.

CONCLUSIONS

The biodegradation of phenol in stormwater biofiltration systems was studied in the laboratory using soils collected from the field. Phenol equilibrium concentration of up to 300 μg kg^{-1} in soils was biodegraded to non-detectable levels (<3 μg kg^{-1}) within 30 hours. The rates of biodegradation were described using first order rates and varied from 0.125 to 1.26 h^{-1} with corresponding half-lives from 0.550 to 5.54 hours. The highest rates were achieved based on temperature 30^{0}C and with soil moisture of 100%. Phenol was also found to be biodegraded under anaerobic conditions, but with lower rates compared to aerobic condition. pH and initial concentration of phenol had little impact on biodegradation rates. The results suggest that biodegradation will contribute to the removal of phenol in stormwater biofiltration systems. Ultraviolet absorbance (UVA) reduction was found to be a good surrogate for phenol biodegradation in the stormwater biofiltration systems that use sand as filter media.

ACKNOWLEDGEMENTS

This study was funded by Australian Cooperative Research Centres (CRC) program and Chinese Scholarship Council (CSC).

REFERENCES

Anderson, J., 1981. Soil moisture and the rates of biodegradation of diallate and triallate. Soil Biology and Biochemistry 13(2), 155-161.

APHA, AWWA and WEF., 2012. Standard methods for the examination of water and wastewater (22nd Edition), American Public Health Association, Washington, DC.

Bak, F. and Widdel, F., 1986. Anaerobic degradation of phenol and phenol derivatives by Desulfobacterium phenolicum sp. nov. Archives of microbiology 146(2), 177-180.

Blecken, G.-T., Zinger, Y., Deletić, A., Fletcher, T.D. and Viklander, M., 2009. Impact of a submerged zone and a carbon source on heavy metal removal in stormwater biofilters. Ecological Engineering 35(5), 769-778.

Broholm, M.M. and Arvin, E., 2000. Biodegradation of phenols in a sandstone aquifer under aerobic conditions and mixed nitrate and iron reducing conditions. Journal of Contaminant Hydrology 44(3), 239-273.

Chang, B., Yu, C. and Yuan, S., 2004. Degradation of nonylphenol by anaerobic microorganisms from river sediment. Chemosphere 55(4), 493-500.

Daly, E., Deletic, A., Hatt, B.E. and Fletcher, T.D., 2012. Modelling of stormwater biofilters under random hydrologic variability: a case study of a car park at Monash University, Victoria (Australia). Hydrological Processes 26(22), 3416-3424.

DHV, 2013. Guidelines for validating treatment processes for pathogen reduction-Supporting Class A water recycling schemes in Victoria. Department of Health, Victoria.

Drewes, J.E., Sedlak, D., Snyder, S. and Dickenson, E., 2008. Indicator and surrogates to assess removal of wastewater-derived contaminants in wastewater treatment and reclamation. WateReuse Foundation, Alexandria, Virginia.

Dugan, G.L. and Christakos, C.E., 1983. Water Quality of Airport Storm Runoff WRRC technical report 153. Water Resources Research Center, University of Hawaii at Manoa.

Duncan, H., 1999. Urban stormwater quality : a statistical overview. CRC for Catchment Hydrology, Melbourne.

El-Naas, M.H., Al-Muhtaseb, S.A. and Makhlouf, S., 2009. Biodegradation of phenol by Pseudomonas putida immobilized in polyvinyl alcohol (PVA) gel. Journal of Hazardous Materials 164(2–3), 720-725.

FAWB, 2009. Adoption guidelines for Stormwater Biofilter systems. Facility for Advancing Water Biofiltration, Monash University.

Göbel, P., Dierkes, C. and Coldewey, W.G., 2007. Storm water runoff concentration matrix for urban areas. Journal of Contaminant Hydrology 91(1-2), 26-42.

Goonetilleke, A., Thomas, E., Ginn, S. and Gilbert, D., 2005. Understanding the role of land use in urban stormwater quality management. Journal of Environmental Management 74(1), 31-42.

Grynkiewicz, M., Polkowska, Z., Kot-Wasik, A. and Namiesnik, J., 2002. Determination of phenols in runoff. Polish Journal of Environmental Studies 11(1), 85-90.

Henderson, C., Greenway, M. and Phillips, I., 2007. Removal of dissolved nitrogen, phosphorus and carbon from stormwater by biofiltration mesocosms. Water science and technology 55(4), 183.

Mackay, D., Shiu, W.Y., Ma, K.-C. and Lee, S.C., 2006. Handbook of physical-chemical properties and environmental fate for organic chemicals. CRC Press, Taylor &Francis Group.

Makepeace, D.K., Smith, D.W. and Stanley, S.J., 1995. Urban stormwater quality: Summary of contaminant data. Critical Reviews in Environmental Science and Technology 25(2), 93-139.

Matamoros, V., García, J. and Bayona, J.M., 2008. Organic micropollutant removal in a full-scale surface flow constructed wetland fed with secondary effluent. Water Research 42(3), 653-660.

Onysko, K.A., Budman, H.M. and Robinson, C.W., 2000. Effect of temperature on the inhibition kinetics of phenol biodegradation by Pseudomonas putida Q5. Biotechnology and bioengineering 70(3), 291-299.

Pavelic, P., Dillon, P.J. and Nicholson, B.C., 2006. Comparative evaluation of the fate of disinfection byproducts at eight aquifer storage and recovery sites. Environmental Science & Technology 40(2), 501-508.

Sá, C.S.A. and Boaventura, R.A.R., 2001. Biodegradation of phenol by Pseudomonas putida DSM 548 in a trickling bed reactor. Biochemical Engineering Journal 9(3), 211-219.

Shelton, D.R. and Parkin, T.B., 1991. Effect of moisture on sorption and biodegradation of carbofuran in soil. Journal of Agricultural and Food Chemistry 39(11), 2063-2068.

Shibata, A., Inoue, Y. and Katayama, A., 2006. Aerobic and anaerobic biodegradation of phenol derivatives in various paddy soils. Science of The Total Environment 367(2), 979-987.

Wilson, J.M. and Griffin, D.M., 1975. Water potential and the respiration of microorganisms in the soil. Soil Biology and Biochemistry 7(3), 199-204.

Wong, T.H.F., 2006. Australian runoff quality : a guide to water sensitive urban design. Crows Nest, N.S.W. : Engineers Media for Australian Runoff Quality Authorship Team.

Zhang K.F., Aguiar L., Randelovic A., McCarthy D.T. and A., D., 2013. Determination of operational and challenge conditions for validation of stormwater biofilters and wetlands, Gold Coast, Australia.

Part VIII

Alternative Biofiltration – Process Developments

Chapter 60

Applicability of spherical coconut shell activated carbon to biological activated carbon for advanced drinking water treatment

Azusa Okada, Koji Takahashi*, Takahiro Okawa**, Katsuaki Satoh**

**Swing corporation, Chemical Products Business Division, Environmental Treatment Chemical Development Department*
35 Nakasode, Sodegaura City, Chiba 299-0267, Japan
***Swing corporation, Chemical Sales & Marketing Division, Chemical Sales Promotion Department*
7-18, Konan 1-chome, Minato-ku, Tokyo 108-8470, Japan

ABSTRACT

High-performance spherical activated carbon (AC) derived from coconut shell for advanced drinking water treatment was newly developed. It is produced by adding a special binder to crushed coconut shell, carbonizing/activating after kneading/mixing, compressing, and spheroidizing. The mesopore and macropore volumes of spherical coconut shell AC are about three times as much as those of crushed coconut shell AC, and it has almost the same macropore volume as crushed coal-based AC.

The spherical coconut shell AC shows the earliest start of nitrification reaction. This reaction occurs from around the contact volume 2500, and ammonia nitrogen was not observed around the contact volume 3200. It is estimated that well-developed pore volume is related to this fast reaction. Moreover, to evaluate the abundance of Ammonia-Oxidizing Archaea (AOA) in practical waterworks facility, AOA *amo*A genes were quantified. In the result, 1.2×10^7 copies g-dry^{-1} were detected in spherical coconut shell AC. However 4.0×10^6 copies g-dry^{-1} were detected in coal-based AC. It suggests that spherical coconut shell AC has better capacity to settle AOA rather than crushed coal-based AC, which effect on difference of nitrification reaction.

This report describes several our findings regarding the applicability of the spherical coconut shell AC to biological AC filtration for advanced drinking water treatment.

Keywords spherical coconut shell activated carbon; macropore; biological activated carbon; advances drinking water treatment

INTRODUCTION

Under increasing expectations of demand for safe drinking water, the importance of advanced water purification is drawing more attention than ever, not to mention the case of formaldehyde water pollution caused by the spill of hexamethylenetetramine into the Tone River.

At the same time, we are concerned about the progress of global warming and power shortages due to the shutdown of nuclear power plants. Therefore, an effort to reduce environmental burdens and save energy is required.

CURRENT STATUS OF GRANULAR ACTIVATED CARBON

The pore distribution of activated carbon (AC) highly depends on its material because the manufacturing method of granular activated carbons (GAC) for advanced water purification process is limited to the steam activation method[1)] with consideration for health and hygiene. Generally, coconut shell AC has a great developed micropore which range is 2 nm or less in size, but mesopore and macropore which range is 2 nm or more in size is poorer than that of coal-based AC[2)]. For this reason, coconut shell AC is considered to be suitable for the removal of low-molecular musty odor or trihalomethane, but not suitable for the removal of chromaticity from high-molecular-weight organic substances, trihalomethane precursors, etc.

Therefore, coal-based AC has been most commonly used at waterworks facilities in large cities due to the differences of pore structure and treatment effect[3)] as described above.

OVERVIEW OF SPHERICAL COCONUT SHELL ACTIVATED CARBON

In such circumstances, we have developed the spherical AC derived from coconut shell (hereinafter collectively called "spherical coconut shell AC"), as a new GAC for water treatment.

We believe that the spherical coconut shell AC is positioned as a special AC with higher performance, hardness, and environmental-consciousness than that of conventional coal-based AC.

MANUFACTURING METHOD

Figure 60.1 illustrates a typical manufacturing process of AC. Typical coconut shell AC (crushed carbon) is manufactured by merely crushing, and carbonizing/activating.

On the other hand, the spherical coconut shell AC is produced by adding a special binder to pull crushed coconut shell into spherical shape by kneading/mixing, compressing and spheroidizing. This new manufacturing process has enabled us to produce the spherical coconut shell AC, which has a high-performance and high-hardness with an enhanced proportion of mesopore and macropore, although it's made from coconut shell.

Photo 60.1 shows electron micrograph of the spherical coconut shell AC. All data shown below are those of the pH adjusted spherical coconut shell AC.

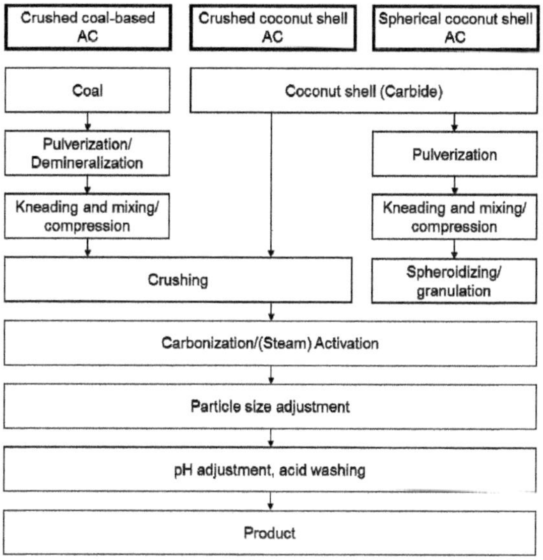

Figure 60.1 Manufacturing process

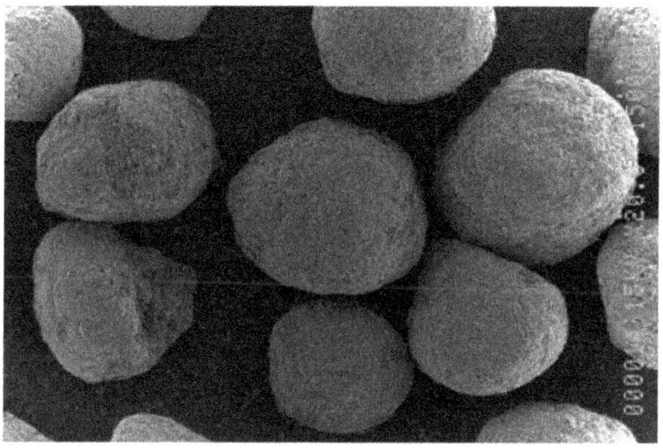

Photo 60.1 Spherical coconut shell AC

PROPERTIES OF EACH ACTIVATD CARBON

The properties and specifications of the spherical coconut shell AC, the properties of typical coconut shell AC (crushed carbon) and coal-based AC (crushed carbon), and requirements of JWWA A114 GAC for water treatment [1)] were shown in Table 60.1.

In terms of representative indicators, iodine adsorption performance and methylene blue adsorption performance, the spherical coconut shell AC greatly exceeds the requirements of JWWA A114; particularly, the methylene blue adsorption performance is significantly higher than that of crushed coal-based AC or crushed coconut shell AC. Further, the spherical coconut shell AC has the largest surface area and highest micropore volume associated with iodine adsorption performance. In addition, the spherical coconut shell AC shows a significantly lower percentage of ignition residue, the indicator of impurity volume, than that of crushed coal-based AC, and has the highest hardness indicated abrasion strength. Moreover, the mesopore and macropore volumes of the spherical coconut shell AC are about three times as much as those of crushed coconut shell AC, and it has almost the same macropore volume as crushed coal-based AC.

Table 60.1 Properties of activated carbon

Properties \ AC		Spherical coconut shell AC	Specifications of spherical coconut shell AC	Crushed coconut shell AC	Crushed coal-based AC	JWWA A114 requirements[1)]
Iodine adsorption performance	mg/g	1260	> 1100	1220	1030	> 900
Methylene blue adsorption performance	mL/g	250	> 200	200	190	> 150
Bulk density	g/mL	0.421	0.4-0.5	0.495	0.482	> 0.4
Ash	%	2.3	< 5	0.9	8.8	< 10
Hardness	%	98	> 95	98	95	> 90
BET (specific) surface area	m^2/g	1330	> 1100	1250	1080	-
Micropore volume	mL/g	0.58	-	0.55	0.48	-
Mesopore volume	mL/g	0.20	-	0.06	0.28	-
Macropore volume	mL/g	0.43	-	0.15	0.41	-

TREATMENT PERFORMANCE

The adsorption isotherms of the following items was shown in Figures 60.2, 60.3, and 60.4 respectively; (1) Chloroform, a representative substance of disinfection by-product, (2) 2-Methylisoborneol (2-MIB), a representative substance of odor (musty odor), (3) Reagent humin assumed natural chromaticity components (humic acid was produced by Wako Pure Chemical Industries. Analyzed as total organic carbon, TOC)

We found the superiority in the spherical coconut shell AC and crushed coconut shell AC with a larger surface area and micropore volume in terms of low-molecule chloroform (molecule weight 119.4) and 2-MIB (molecule weight 168). Moreover, there was superiority in the spherical coconut shell AC and crushed coal-based AC with larger mesopore and macropore volumes in terms of the reagent humin, which consists of organic substances with a wide range of molecule weight distribution.

From the above results, we suggest that the spherical coconut shell AC possesses specific pore properties and the characteristics of both crushed coconut shell and coal-based AC.

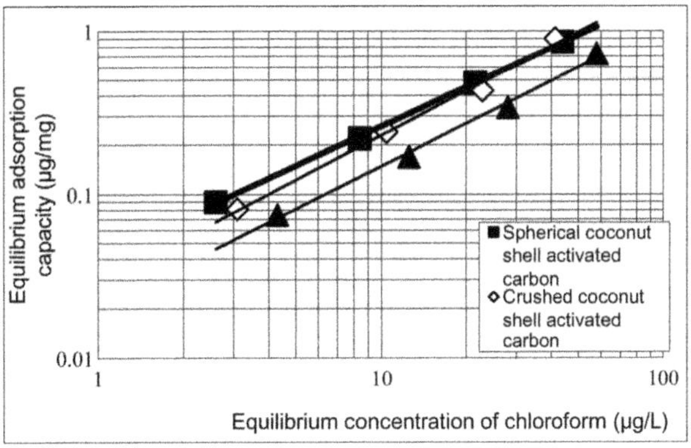

Figure 60.2 Adsorption isotherms of chloroform

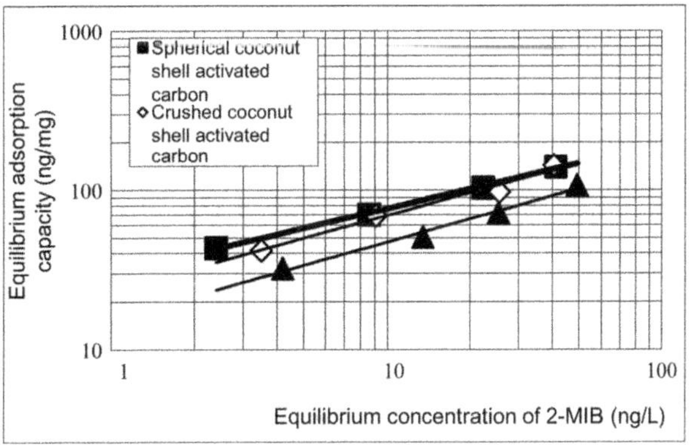

Figure 60.3 Adsorption isotherms of 2-MIB

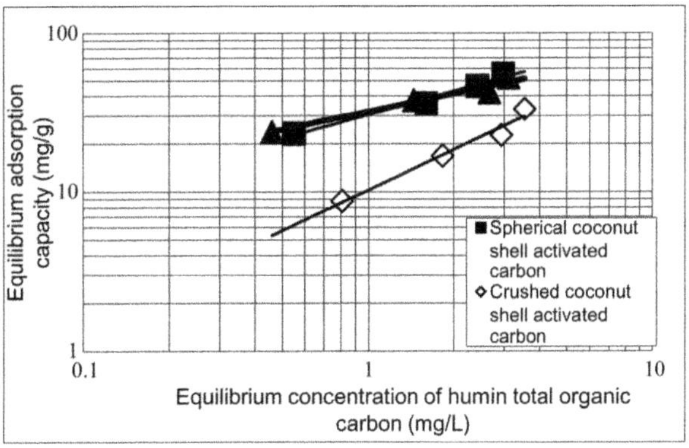

Figure 60.4 Adsorption isotherms of fumin

COLUMN FLOW TESTS (DYMAMIC TESTING)

Flow tests were performed by using ozone treated water in a waterworks facility (Ultraviolet ray absorbance at 260nm, E-260: 0.016 to 0.034cm^{-1}, total trihalomethane forming potential, THMFP: 0.009 to 0.023mg/l) to assess the treatment performance of organic substances, etc. in practical facilities. Figure 60.5 and 60.6 show the following respectively; (1) Relationship between THMFP removal rate and contact volume (total volume of flow divided by the volume of AC), (2) Relationship between E-260 removal rate and contact volume.

For example, as for the THMFP removal rate in a contact volume of approx. 35,000 times, the spherical coconut shell AC demonstrated 33%, while crushed coconut shell AC demonstrated 18%, and crushed coal-based AC demonstrated 30%. It was confirmed that the THMFP removal rate of the spherical coconut shell AC was similar to that of crushed coal-based AC, and higher than that of crushed coconut shell AC. Similar results were obtained for the E-260 removal rate. It is estimated that the properties of spherical coconut shell AC and crushed coal-based AC with a well-developed pore distribution are reflected in this result.

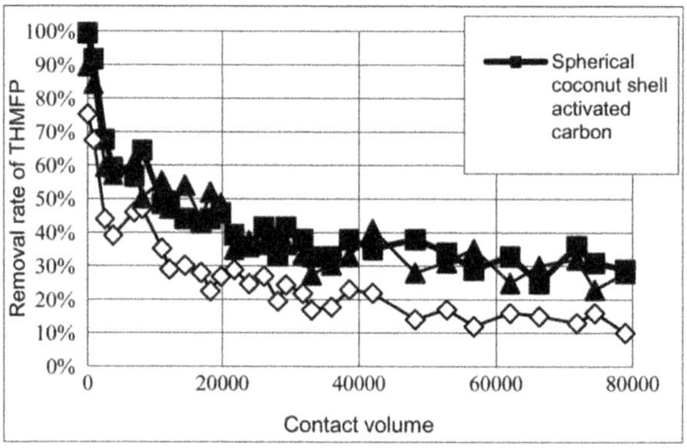

Figure 60.5 Relation between removal rate of THMFP and contact volume

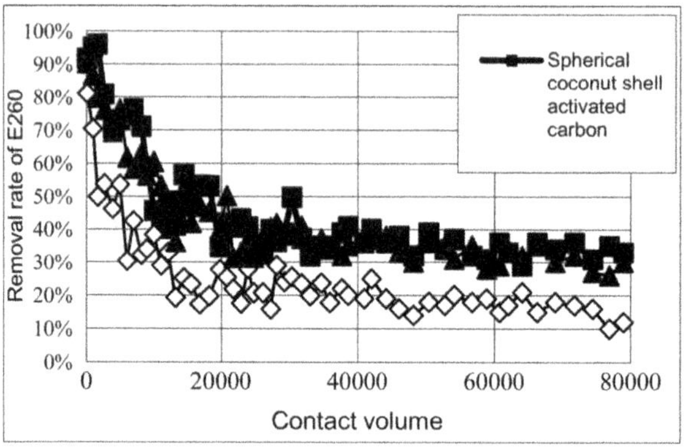

Figure 60.6 Relation between removal rate of E260 and contact volume

TURBIDITY AND WASHING

Washing tests were performed based on JWWA A 103 filtration materials for water treatment. The result was shown in Figure 60.7.

The spherical coconut shell AC and crushed coconut shell AC reduced high turbidity at the initial washing after filling AC compared with crushed coal-based AC. The turbidity decreased about third part of the first run. Further, spherical coconut shell AC could contribute to operation/maintenance improvement and reduction of washing water volume/consumption and electric power consumption because it can reduce particle generation during the steady-state backwash process.

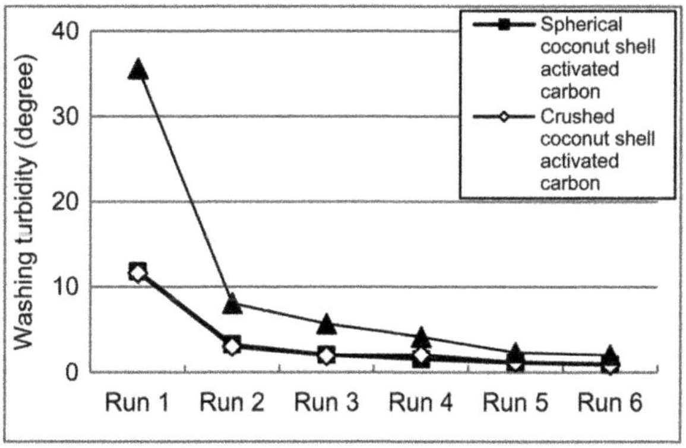

Figure 60.7 Test results of washing turbidity

REGENERATION OF ACTIVATED CARBON

The performance of GAC for water treatment can be recovered through the same steam activation method as in the manufacturing process. A number of water utilities have been adopted regenerated AC (effective use of used AC)[3].

Regeneration tests were performed for the used AC by the column flow tests using ozone treated water in a waterworks facility, with an indirect-rotary kiln regeneration testing apparatus. The test results were shown in Table 60.2.

Table 60.2 Test result of regeneration

Properties	AC	Crushed coal-based AC			Spherical coconut shell AC		
		Unused AC	Used AC	Regenerated AC	Unused AC	Used AC	Regener ated AC
Iodine adsorption performance	mg/g	1030	470	810	1260	650	1080
Methylene blue adsorption performance	mL/g	190	110	150	250	160	220
Bulk density	g/mL	0.482	0.541	0.471	0.421	0.479	0.425
Hardness	%	95	93	91	98	97	96
Yield rate (of reactivation)	%	-	-	88	-	-	93

The results show that spherical coconut shell AC has better recovery of adsorption performance, less regeneration damage (reduced hardness) and a higher yield rate (yield) than crushed coal-based AC. Therefore, the spherical coconut shell AC is regarded as a suitable GAC for regeneration/reuse.

EVALUATION AS BIOLOGICAL ACTIVATED CARBON AMMONIA NITROGEN IN SIMULATION WATER

Flow tests were performed by using simulation water with ammonia nitrogen at approx. 1mg/L to assess the properties as biological AC. Figure 60.8 shows the relationship between decreasing ratio of ammonia nitrogen of AC treated water to AC influent water and contact volume.

The spherical coconut shell AC shows the earliest start of nitrification reaction compared with other two AC. This reaction occurs from around the contact volume 2500, and ammonia nitrogen was not observed in the AC treated water around the contact volume 3200.

Figure 60.9 shows the pore distribution by using the mercury press-in method. Crushed coal-based AC has a peak around 1 μm in the macropore range, while the spherical coconut shell AC has peaks around 0.5 μm and 4 μm. It is estimated that more adequate voids were achieved among spherical coconut shell AC grains compared with crushed ACs, in addition to the well-developed macropore range of spherical coconut shell AC with its high peaks around 0.5μm and 4μm as described above, may give favorable effect on the settlement and growth of biological membrane.

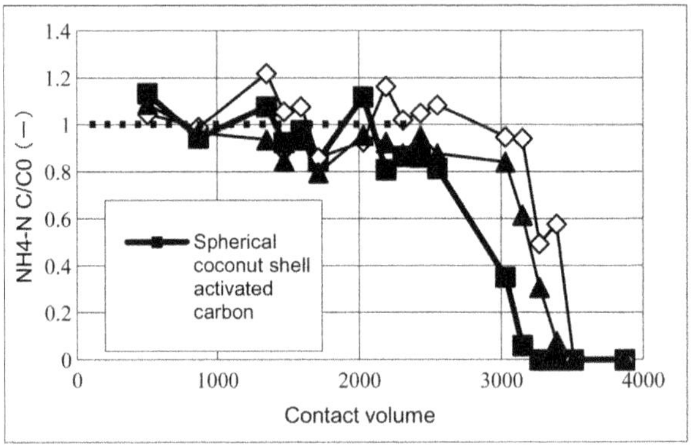

Figure 60.8 Relation between decreasing rate of NH_4-N and contact volume

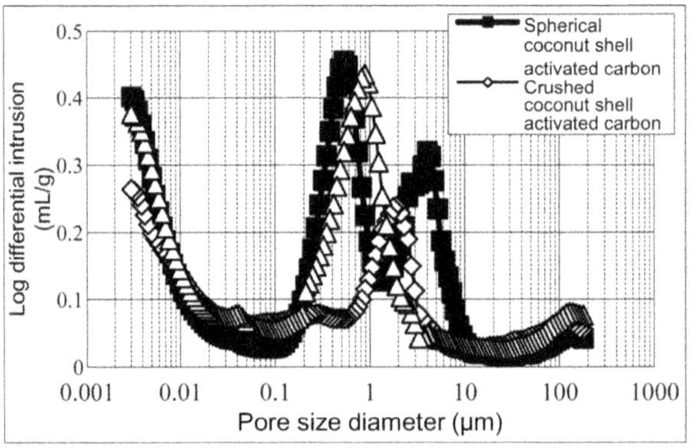

Figure 60.9 Pore distribution

ABUNDANCE OF AMMONIA-OXIDIZING ARCHAEA AND BACTERIA

It has been known that Ammonia-Oxidizing Archaea (AOA) and Bacteria (AOB) are involved in ammonia oxidation[4)]. Quantitative-polymerase chain reaction (Q-PCR) was performed to determine abundance of AOA and AOB[4)5)]. The target functional genes were AOA and AOB *amo*A, which encode a subunit of ammonia monooxygenase of AOA and AOB, respectively. Abundance of AOA and AOB were evaluated for the following four samples. Spherical coconut shell AC (AC1) and, regenerated spherical coconut shell AC (AC2) used for 9 months were collected from waterworks facility A. Crushed coal-based AC (AC3) and spherical coconut shell AC (AC4) served for 11 months were collected from waterworks facility B. The results were shown in Figure 60.10.

AOA *amo*A genes were considerable more abundant than AOB *amo*A genes for all the samples. The abundances of AOA *amo*A genes in AC1, AC2, AC3 and AC4 were 1.4×10^7, 1.6×10^7, 4.0×10^6 and 1.2×10^7 copies g-dry^{-1}, respectively. This result was almost consistent with the previous studies[4)5)]. There was no significant difference of AOA *amo*A gene quantity between AC1 and AC2. On the other hand, AC3 and AC4 differed significantly ($p<0.05$). AC4 demonstrated three times as much AOA *amo*A genes as AC3.

From these results, we suggest that regenerate process has little influence on settlement of AOA. Moreover, AOA is likely to settle in spherical coconut shell AC rather than crashed coal-based AC, which effect on difference of nitrification reaction (Figure 60.8).

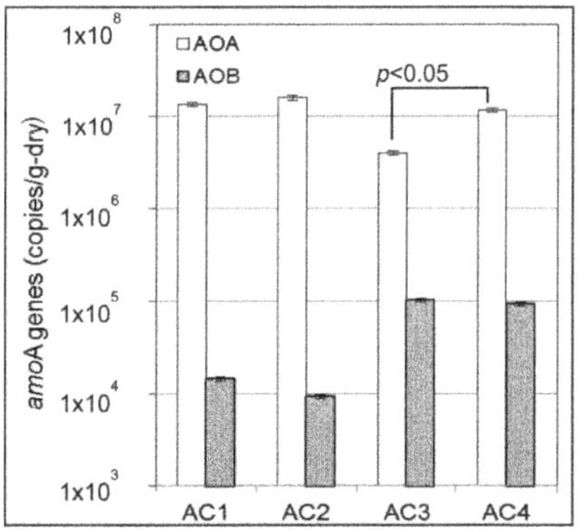

Figure 60.10 Abundance of AOA and AOB *amo*A genes

DIVERSITYIS OF AOA *AOMA* GENES

Figure 60.11 and 60.12 show the T-RFLP profiles of AOA *amo*A genes to determine diversity of AOA for sample AC[4)5)]. The fragment of 217 bp was dominant for AC1 and AC2. In contrast, the T-RFLP profiles of AC3 and AC4 were consisted of 165 bp and 217 bp. This compositional difference in *amo*A genes is probably attributed to raw water quality or water treatment process.

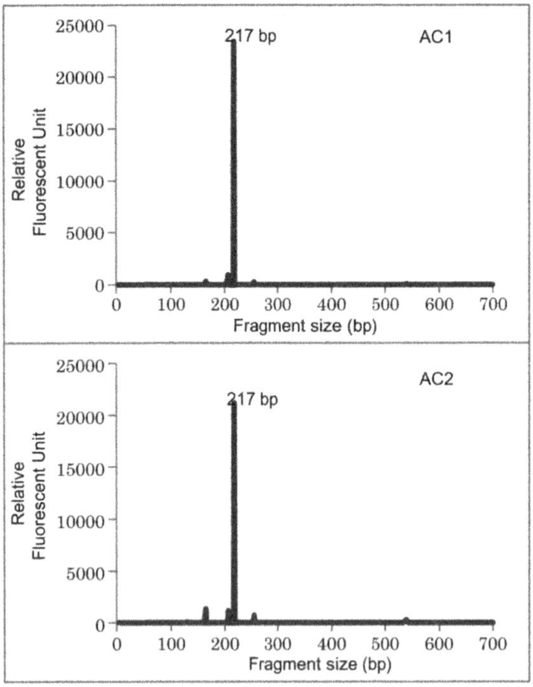

Figure 60.11 Diversity of AOA (AC1 and AC2)

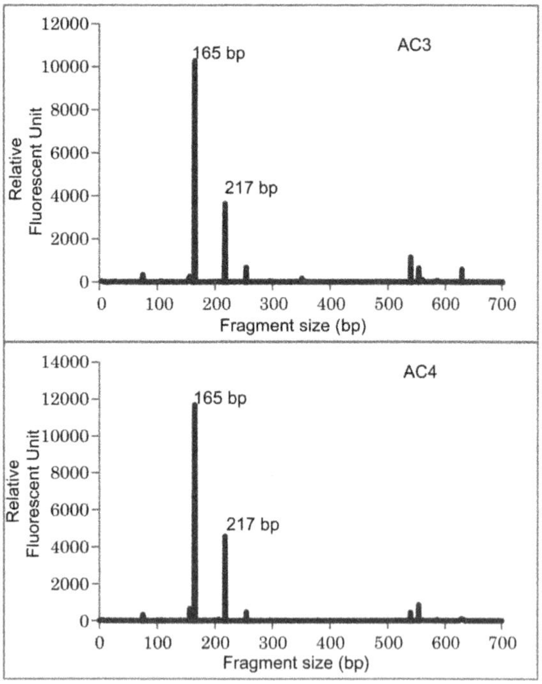

Figure 60.12 Diversity of AOA (AC3 and AC4)

CALCULATION OF CO_2 EMISSION AT MANUFACTUING

Plant-derived coconut shell is carbon neutral. Therefore, carbon dioxide (CO_2) emissions from combustible carbon derived from coconut shells can be regarded as zero. Calculation of CO_2 emission at the time of manufacturing for each AC was shown in Table 60.3.

The CO_2 emission of the spherical coconut shell AC is approx. 3t-CO_2/t-AC higher than that of crushed coconut shell AC due to increased power consumption involved in (spherical coconut shell AC) granulation and combustible carbon resulting from the binder. However, it is around 1/4 of that of crushed coal-based AC made of fossil fuels. Therefore, little under 20t-CO_2/t-AC reduction of CO_2 emissions is expected.

Additionally, the calculation of emission can be further reduced through regeneration/reuse. For instance, if we assume that waterworks facility with a volume of treated water of 1,000,000m^3/day (amount of GAC consumption: 8,000m^3 - SV 5 h^{-1}) replaces a gross quantity of AC, it is estimated that spherical coconut shell AC would reduce CO_2 emission approx. 83,000t through reactivation/reuse, compared with a full replacement of unused crushed coal-based AC (approx. 1/8 of unused crushed coal-based AC). The comparison of CO_2 emissions was shown in Figure 60.13.

Table 60.3 Calculation of CO_2 emission at manufacturing[6)]

Items	CO_2 emission t-CO_2/t-AC		
	Crushed coal-based AC	Crushed coconut shell AC	Spherical coconut shell AC
Fuel (gas oil consumption)	10.16	3.11	3.19
Electricity consumption	2.78	0.82	1.74
Water consumption	0.02	0.01	0.01
Combustible carbon	12.93	0.00	1.90
Total	25.89	3.94	6.84

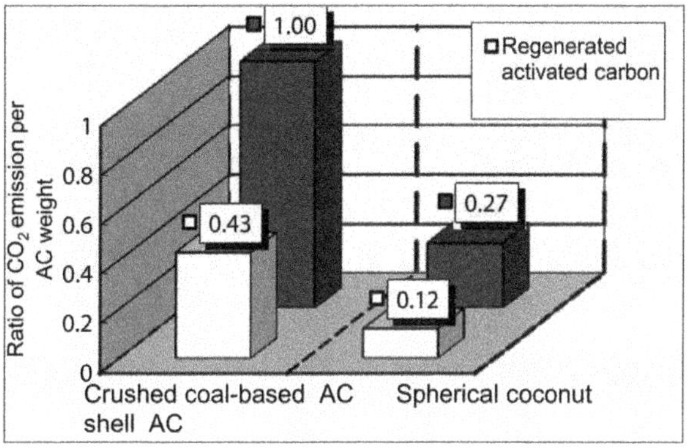

Figure 60.13 Comparison of CO_2

CONCLUSIONS

The applicability to advanced drinking water purification plants and properties as a biological AC of the spherical coconut shell AC was outlined as above. The features of the spherical coconut shell AC are summarized as below;

(1) High capability to treat organic substances, etc;

(2) Excellent biological carrier (prompt nitrification reaction);

(3) High hardness;

(4) Low environmental load (reduction of CO_2 emission);

We earnestly hope spherical coconut shell AC will serve as one of the useful option to provide safe and tasty drinking water in a stable manner.

ACKNOWLEDGEMENTS

We thank Dr. Kasuga, Water Environment Technology Lab, Dept. of Urban Engineering, The Univ. of Tokyo, for technical support in assessment of nitrification potential. We thank some waterworks facilities for cooperation in sample collection.

REFERENCES

1) Japan Water Works Association (JWWA): JWWA A 114 Granular Activated Carbon for water treatment (2006)
2) Japan Water Research Center: Guideline for water purification technology 2010 (2010)
3) Japan Water Research Center: Technical information on the status and consideration of the introduction of advanced water purification into water utilities with commission expenses of Health, Labour and Welfare Ministry
4) Niu, J., Kasuga, I., Kurisu, F., Furumai, H., Shigeeda, T., 2013. Evaluation of autotrophic growth of ammonia-oxidizers associated with granular activated carbon used for drinking water purification by DNA-stable isotope probing. *Water Res.* 47 (19), 7053-7065.
5) Kasuga, I., Nakagaki, H., Kurisu, F., Furumai, H., 2010. Predominance of ammonia-oxidizing archaea on granular activated carbon used in a full-scale advanced drinking water treatment plant. *Water Res.* 44 (17), 5039-5049.
6) Ordinance of Ministry of Environment, "Ministerial ordinance on calculation of the emission of global warming gas involved in business activities of designated emitters" March 2006.

Chapter 61

Development of biological roughing filter as an alternative to coagulation-sedimentation process

T. Okamoto, K. Ogura*, Y. Kobayashi*, M. Oya*, D. Nagashio*, T. Hanamoto**

**Hanshin Water Supply Authority, 3-20-1, Nishi-Okamoto, Higashinada-ku, Kobe, Hyogo, JAPAN*

ABSTRACT

A biological roughing filter (BRF) that can be used as an alternative to the coagulation-sedimentation process was incorporated into a rapid coagulation-sedimentation reactor with slant-pipe to construct a pilot plant, and its performance was experimentally examined. The BRF realizes both biological treatment and solid-liquid separation, and the rapid coagulation-sedimentation reactor operates only when the turbidity of the raw water is 50 degrees or higher. Experimental results showed that the mean turbidity of the BRF-treated water was 0.6 degrees when the injection dose of aluminum sulfate was 15 mg/L, which was less than half that in the case of the coagulation-sedimentation process. Larger amounts of manganese and ammonium nitrogen, indices of the biological effect, were removed by the BRF than by the coagulation-sedimentation process, demonstrating that the developed process can be used as an alternative to the coagulation-sedimentation process.

Keywords Biological roughing filter; alternative to the coagulation-sedimentation; biological effect; ammonium nitrogen; manganese;

INTRODUCTION

Many water utilities in Japan have adopted a rapid filtration system involving coagulation-sedimentation and sand filtration as a means of solid-liquid separation (Japan Water Works Association, 2013). The Hanshin Water Supply Authority (HWSA) has also conventionally adopted a system based on these processes. In 1993, HWSA started to adopt an advanced water treatment system using ozone and granular activated carbon (GAC) with the aim of removing undesirable substances, such as those causing an unusual odor and taste, chemical substances, and disinfection by-products (Sasaki T., 1989).

Surface water including water in dams and river water accounts for approximately 75% of the sources of tap water provided in Japan (Japan Water Works Association, 2013). Recently, the amount of algae in dams, lakes, and ponds has increased as a result of eutrophication, and the turbidity of river water has decreased. In addition, there is concern that the probability of generation of blue-green algae in these sites may increase owing to climate change (Ministry of the Environment, 2008).

Aluminum-based coagulants (e.g., polyaluminum chloride and aluminum sulfate) account for at least 99% of coagulant used by Japanese water utilities (Japan Water Works Association, 2013). In Japan, the injected dose of aluminum-based coagulants has tended to increase in turbidity because of the increased amount of algae in the raw water and the decreased turbidity of river water. HWSA, using the surface water from the Yodo River as a water source has concerned that the concentration of aluminum remaining in the treated water and the amount of soil produced at the water treatment plant (water-treatment-plant soil) may increase, the dewaterability of such soil may deteriorate, and the cost of chemicals used may increase.

With the above background, HWSA has developed an alternative to the coagulation-sedimentation process, that is, a biological roughing filter (BRF) that can reduce the injected dose of coagulant and realize both biological treatment and solid-liquid separation.

METHODOLOGY

Outline of pilot-scale plant

To develop the BRF, we constructed a pilot-scale experimental system in our demonstration plant for advanced water treatment located within the Inagawa Water Treatment Plant of the HWSA. The developed BRF was tested in the pilot plant.

Table 61.1 summarizes the specifications of the BRF reactor, and Photograph 61.1 shows the filter media. The BRF reactor consists of an upper layer filled with a porous hollow polypropylene filter medium with a diameter of 4 mm and a length of 4 mm and the lower layer filled with a porous spherical ceramic filter medium with an effective diameter of 0.83 mm. The hollow polypropylene filter medium has high porosity and a large capacity for trapping turbid substances, and it is expected that flocks will grow and be trapped in the pores. The spherical ceramic filter medium traps small flocks that cannot be trapped in the upper layer. The double-layer structure with the two different filter media is expected to realize biological treatment and help reduce the turbidity. The composition of the filter media is determined so that the turbidity of the treated water is maintained at less than 1 degree for at least 24 h when the turbidity of the raw water is 50 degrees (Hanamoto T., 1996).

Table 61.1 Specifications of BRF filter media

Upper filter medium	Material	Porous hollow polypropylene
	Shape	Hollow and cylindrical
	Dimensions	4mmφ×5mmL
	Layer height	2.0m
Lower filter medium	Material	Porous spherical ceramic
	Shape	Spherical
	Dimensions	Effective diameter 0.83mmφ
	Layer height	0.5m

Figure 61.1 shows a flowchart of the pilot plant. The coagulant is injected into the raw water in a rapid mixing reactor before it reaches the BRF reactor. The turbidity of the surface water from the Yodo River increases to up to several hundred degrees when it heavily rains, although it is 10 degrees or lower under normal conditions. In the pilot plant, the raw water flows directly from the mixing reactor to the

BRF reactor when its turbidity is normal. However, a preliminary examination revealed that the turbidity was insufficiently reduced by the treatment using the BRF alone when the turbidity of the raw water was high. Therefore, when the turbidity of the raw water exceeds 50 degrees, the raw water is introduced into a rapid coagulation-sedimentation reactor with slant-pipe to undergo preliminary treatment before it reaches the BRF reactor. Tables 61.2 and 61.3 summarize the specifications of the rapid coagulation-sedimentation reactor and the BRF reactor, respectively. The results for the BRF-treated water were compared with those of the water treated by coagulation-sedimentation in the demonstration plant for advanced water treatment.

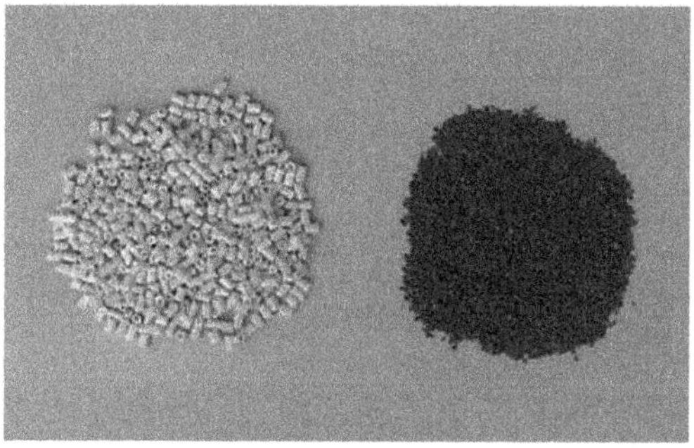

Photograph 61.1 BRF filter media

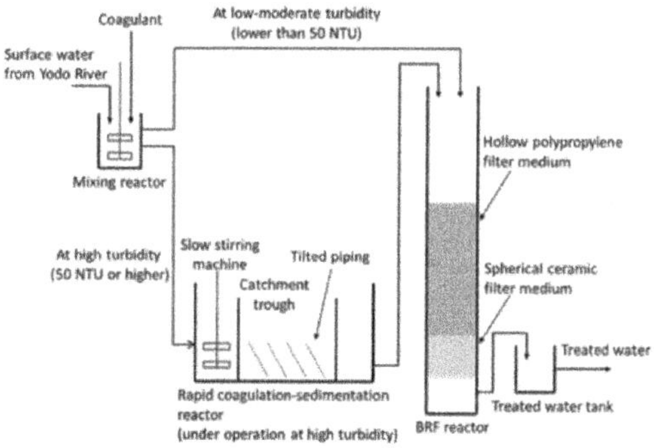

Figure 61.1 Schematic of water treatment in pilot plant

Table 61.2 Specifications of rapid coagulation-sedimentation reactor

Retention time	Slow stirring machine	15 min
	Sedimentation reactor	18 min
Dimensions of tilted piping		W0.7 m, L0.7 m, H1.3 m
Upward flow velocity		425 mm/min
Surface load factor		32 mm/min
Flocculator	Number of revolutions	25 rpm
	G value	64
	GT value	57,000

Table 61.3 Specifications of BRF reactor

Amount of water to be treated		120 m^3/day
LV		240 m/day
Cleaning condition		When head loss reaches 1800 mm
Cleaning method	Air-water flow washing	
	Air-water washing	1.0 m/min, 3 min
	Water-only washing	1.2 m/min, 10 min

Operating Conditions

The surface water from the Yodo River stored in a water tank in the demonstration plant was used as the raw water. Aluminum sulfate (Al2O3 concentration, 8%) was used as the coagulant injected into the rapid mixing reactor. In the BRF treatment, the injected dose of the coagulant was maintained at 15 mg/L (0.64 mg/L of aluminum) when the turbidity of the raw water was less than 50 degrees (low-moderate turbidity). When the turbidity of the raw water was 50 degrees or higher (high turbidity), the injected dose was varied between 30 and 60 mg/L (1.3-2.5 mg/L of aluminum) in accordance with the turbidity of the raw water. In the demonstration plant, the injected dose of the coagulant (aluminum sulfate) during the period of low-moderate turbidity was 40 mg/L (1.7 mg/L of aluminum). Thus, the injected dose in the BRF treatment during the period of low-moderate turbidity was almost half that adopted in the coagulation-sedimentation process.The filtration rate in the BRF reactor was fixed at 240 m/d. When the water head loss reached 1,800 mm, the BRF reactor was cleaned by backwashing involving air/water washing (1.0 m/min, 3 min) and water-only washing (1.2 m/min, 10 min).

In winter, the biological effect in the BRF on, for example, the removal of ammonium nitrogen, decreases because of the decreased water temperature. Therefore, to maintain the biological effect, phosphoric acid was added to the raw water at an injected dose of 1.0 mg/L immediately before the raw water is introduced into the BRF reactor.

Analysis

To evaluate the water treatment performance of the BRF, we measured the concentrations of ammonium nitrogen, iron, manganese, aluminum, and total organic carbon (TOC) as well as UV260 and the potassium permanganate consumption in the raw water, pretreated water, and BRF-treated water once a week. The turbidity of these types of water was continuously monitored using a continuous measurement instrument. As a control for comparison with the BRF-treated water, the water treated by coagulation-sedimentation in the demonstration plant (sedimentation-treated water) was similarly monitored.

RESULTS AND CONDITIONS

Main factors related to water quality

Table 61.4 shows the mean values for the total concentrations of aluminum, iron, manganese, ammonium nitrogen, and TOC along with UV260 and potassium permanganate consumption as well as their mean removal rates in the raw water, sedimentation-treated water, and BRF-treated water. The mean removal rate indicates the mean of the removal rates in all measurements.

The total aluminum concentration was 0.27 mg/L in the raw water and 0.19 mg/L in the BRF-treated water, both of which were lower than that in the sedimentation-treated water (0.39 mg/L). This is considered to be due to the effect resulting from the reduced injected dose of the coagulant in the BRF treatment (15 mg/L) compared with that in coagulation-sedimentation (40 mg/L). The coagulant also affects the dewaterability of the water-treatment-plant soil, which is improved with increasing silica/alumina (SiO2/Al2O3) ratio. The measurement results indicate that the silica/alumina ratio of the water discharged by backwashing the BRF reactor (1.68) was higher than that of sludge produced during coagulation-sedimentation (0.97). The BRF is considered to contribute to improving the dewaterability of the water-treatment-plant soil because the injected dose of the coagulant can be suppressed when the BRF is used.

The total manganese concentration was 0.036 mg/L in the raw water and 0.005 mg/L in the BRF-treated water, which was lower than that in the sedimentation-treated water (0.019 mg/L). The removal rate of manganese by the BRF decreased during a low-water-temperature period, when the biological effect decreases. Therefore, we infer that manganese becomes insoluble owing to the biological effect of the microorganisms in the BRF filter layers and is trapped at these layers.

The ammonium nitrogen concentration was 0.07 mg/L in the raw water and 0.03 mg/L in the BRF-treated water, which was lower than that in the sedimentation-treated water (0.05 mg/L). Similarly to the case of manganese, the removal rate of ammonium nitrogen by the BRF tended to decrease when the water temperature was low in winter probably because of biological oxidation, although it was almost 100% in summer.

Table 61.4 Mean values for main items related to water quality and mean removal rates

Item	Raw water	Sedimentation water	Mean removal rate	BRF–treated water	Mean removal rate
NH_4–N[mg/L]	0.07	0.05	19.7	0.03	58.0
$KMnO_4$ consumption[mg/L]	5.5	3.2	41.9	3.9	30.5
Fe[mg/L]	0.364	0.045	87.6	0.064	82.4
Mn[mg/L]	0.036	0.019	48.5	0.005	85.7
Al[mg/L]	0.27	0.39	–44.6	0.19	30.8
UV260[–]	0.238	0.115	51.7	0.144	39.5
TOC[mg/L]	1.8	1.3	26.1	1.4	19.3

Figure 61.2 shows the removal rate of ammonium nitrogen by the BRF when phosphoric acid was added to the raw water. The removal rate of ammonium nitrogen gradually increased from approximately 40% on January 25th, the day before phosphoric acid was first added, to approximately 90% on March 1st (35 days later). This result confirmed that ammonium nitrogen can be effectively removed by adding phosphoric acid even when the water temperature is low. Moreover, the nitrifying bacteria count per amount of filter media in the upper layer was 3.3×105 N/100 g before the addition of phosphoric acid and 7.9×107 N/100 g after this addition. This indicates that the nitrifying bacteria count is increased by adding phosphoric acid. In this test, the phosphoric acid was first added after the temperature of the raw water decreased to 9 °C. Hence, the metabolic function of microorganisms in the BRF reactor had not been fully activated at the start of addition and it took at least one month to achieve sufficient nitrifying function. If the addition of phosphoric acid is started before the water temperature decreases to maintain the nitrifying function of the microorganisms, the high removal rate of ammonium nitrogen can be maintained even when the water temperature is low in winter.

In terms of the potassium permanganate consumption, UV260, and TOC concentration, which are indices of organic substances, the treatment performance of the BRF represented by the removal rate was inferior to that of coagulation-sedimentation. This is considered to be due to the presence of turbid substances resulting from the lower turbidity-reducing ability of the BRF than that of coagulation-sedimentation because of the lower injected dose of the coagulant. Note that the reductions in the above indices per injected dose of coagulant in the BRF-treated water are greater than those in the sedimentation-treated water.

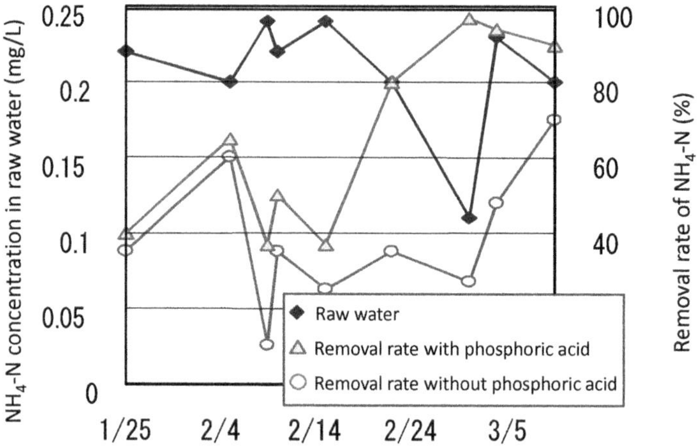

Figure 61.2 Concentration and removal rate of ammonium nitrogen with and without addition of phosphoric acid

Turbidity

Table 61.5 shows the treatment performance of the BRF in the pilot plant for reducing the turbidity. The continuous water-running duration was 5,200 h, during which the rapid coagulation-sedimentation reactor operated (i.e., the turbidity of the raw water was 50 degrees or higher) for 64 h. During the continuous water-running duration, the turbidity of the BRF-treated water was less than 1 degree for 4,542 h (87.3%). The reduction in the mean turbidity was 94.4% and the mean duration of continuous filtration was approximately 100 h. Figure 61.3 shows the turbidity and the head loss of the BRF-treated water during a period of low-moderate turbidity. Even when the turbidity of the raw water was approximately 15-25 degrees, the turbidity of the BRF-treated water was maintained below 1 degree for a continuous filtration duration of at least 80 h. This means that continuous filtration can be performed for approximately 3 days, during which the turbidity of the BRF-treated water remains below 1 degree for 90% of the time provided the pilot plant is operated under suitable conditions. However, the turbidity of the BRF-treated water exceeded 1 degree for the remaining 658 h (12.6%) because of operational factors, such as delays in the switching on of the reactor and the setting of the reactor to increase the injected dose of coagulant. To deal with this problem, the rapid coagulation-sedimentation reactor was constantly operated regardless of the turbidity of the raw water; as a result, the percentage of the duration in which the turbidity of the BRF-treated water exceeded 1 degree was decreased to 0.4%.

The above results revealed that when the BRF is applied to water sources in which the turbidity of the raw water frequently exceeds 50 degrees, the load on the subsequent sand filtration process can be alleviated by constantly operating the rapid coagulation-sedimentation reactor. Recently, membrane filtration technologies have been adopted by many water utilities. If the BRF is applied as a pretreatment before membrane filtration, pretreatment systems such as the rapid coagulation-sedimentation reactor are not required because membrane filtration reduces the turbidity with higher certainty than rapid sand filtration.

Table 61.5 Treatment performance of pilot plant for reducing turbidity

Continuous water-running duration (h)		5200 hour
Operating conditions for rapid coagulation-sedimentation reactor		Turbidity of raw water ≥ 50 degrees
Raw water	Maximum turbidity	443 degree
	Mean turbidity	10.7 degree
	Duration during which turbidity <1 degree	64 hour
BRF-treated water	Mean turbidity	0.6 degree
	Duration during which turbidity <1 degree	4542 hour
	Percentage of hours with turbidity <1 degree	87.3 %

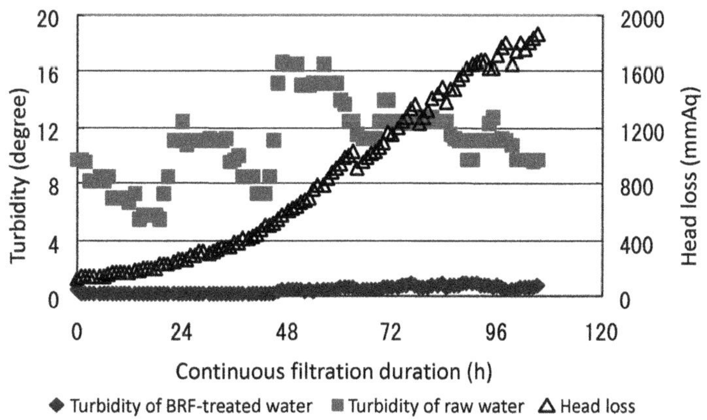

Figure 61.3 Concentration and removal rate of ammonium nitrogen with and without addition of phosphoric acid

CONCLUSIONS

The BRF can reduce the injected dose of coagulant to half that in coagulation-sedimentation and can be used as an alternative to this process by installing a rapid coagulation-sedimentation reactor for pretreatment. Moreover, the BRF can effectively remove ammonium nitrogen and manganese. The removal rate of ammonium nitrogen, which is considered to decrease when the water temperature is low, can be maintained by adding phosphoric acid. When the BRF is used, the concentration of chlorine-consuming substances, such as ammonium nitrogen, is reduced because of the biological effect, enabling a reduction in the use of chlorine-based agents.

The developed BRF can be incorporated into a conventional filtration system as an alternative to the coagulation-sedimentation process without pretreatment when the turbidity of the raw water is less than 50 degrees. A simplified pretreatment system is required before the BRF when the turbidity of the raw water is 50 degrees or higher.

As described above, we developed a BRF that realizes both biological treatment and solid-liquid separation as an alternative to the coagulation-sedimentation process. The BRF is expected to reduce the concentrations of substances that cause membrane fouling during filtration, such as iron and manganese. In the future, we will attempt to apply the BRF as a pretreatment before membrane filtration.

REFERENCES

Hanamoto T., Nagashio D., Sasaki T., Nuno M., Mizusawa H., Mizuno K. (1996) Biological Course Filter: as Alternative to Coagulation/Sedimentation, 47th JWWA Annual Conference Proceedings, Japan, 634-635

Japan Water Works Association. (2013) 2011 Water Statistics (in Japanese), Japan.

Ministry of the Environment, Japan, 2008, Wise Adaptation to Climate Change - Report by the Committee on Climate Change Impacts and Adaptation Research - Part I: Wise Adaptation to Climate Change, Japan, 23.

Sasaki T., Omote Y., Hanamoto T., Kobayashi K. (1989) Examination about new water treatment system (in Japanese). 40th JWWA Annual Conference Proceedings, Japan, 150-152.

Chapter 62

The environmentally-friendly effects of biotreatments in Osaka Water Supply Authority

T. Hoson, Y. Takeda***

**Niwakubo Water Treatment Plant, Osaka Water Supply Authority, 2-30-18, Oba-cho, Moriguchi, Osaka, 570-0009, Japan*

***East Waterworks Office, Osaka Water Supply Authority, 4-1-27, Shimokosaka, Higashiosaka, Osaka, 577-0803, Japan*

ABSTRACT

The Osaka Water Supply Authority (OWSA) has been operating biotreatment in Mishima Water Treatment Plant (WTP) and Niwakubo WTP located in the lower areas of the Yodo River. The biological contact oxidation process has been running in Mishima WTP, and the biological contact filtration process in Niwakubo WTP. The average removal ratio of ammonium nitrogen by the biotreatment in Mishima WTP is 79.8%. However, the performance is affected by water temperature. On the other hand, the average removal ratio in Niwakubo WTP keeps over 95% independently of water temperature. Removal of ammonium nitrogen in biotreatment enables the stable operation of residual chlorine control and the reduction in the dosage of sodium hypochlorite. As for dissolved manganese, the concentration in biotreated water has been below 0.04mg/L in Mishima WTP and 0.01mg/L in Niwakubo WTP. It is also confirmed that turbidity can be reduced by biological contact filtration process in Niwakubo WTP. Therefore, the dosage of the coagulant has been reduced. Biotreatments in OWSA not only exert a beneficial effect on reducing ammonium nitrogen and dissolved manganese, but also contribute to a decrease in the dosage of chemicals.

Keyword Ammonium Nitrogen; Biotreatment; Manganese.

INTRODUCTION

OWSA started the business inherited from Osaka Prefectural Waterworks as of April 1, 2011. To municipal water services, OWSA provides about 70 percent of the tap water consumed in all cities and towns in Osaka Prefecture (except Osaka City), which amounts to 524 million m^3 in FY 2012.

OWSA has three water treatment plants for tap water: Murano WTP, Niwakubo WTP and Mishima WTP. All source water for these plants is taken from the Yodo River. The Yodo River system mainly consists of the Uji River, the Katsura River and the Kizu River and it covers 6 prefectures in Kansai area embracing various cities and industries. Therefore, in the lower areas of the Yodo River, the water quality comes under the influence of miscellaneous drainage and nutrient from the upper areas.

Especially, as Mishima WTP and Niwakubo WTP are located in the lower areas of the Yodo River, both plants had some difficulties in controlling residual chlorine affected by ammonium nitrogen in source water. For this problem, Osaka Prefectural Waterworks introduced the biological contact oxidation process into Mishima WTP, and the biological contact filtration process into Niwakubo WTP.

RESULTS AND DISCUSSION

Figure 62.1 illustrates the schematic of the treatment flow of Mishima WTP. Mishima WTP adopted the biological contact oxidation process using the honeycomb tube, which began operation in 1991. Then, the advanced water treatments (ozonation and granular activated carbon treatment) began operation in 1998.

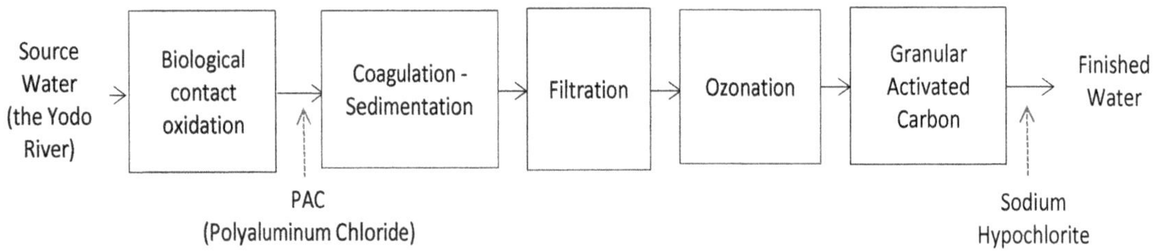

Figure 62.1 The schematic of the treatment flow of Mishima WTP.

The biological contact oxidation process is one of the biotreatment which take advantages of the characteristics of microorganisms. This process requires the media and fresh air. As the media, we use the honeycomb tube of polyvinyl chloride. In the basin, the media (honeycomb tube) and fresh air provide a beneficial environment for aerobic microorganisms to thrive. The colony of microorganisms consumes nutrient such as ammonium nitrogen. Ammonium nitrogen in the water can be reduced through this process. The specification of the biotreatment in Mishima WTP is shown in Table 62.1.

Table 62.1 Specification of the biological contact oxidation basin in Mishima WTP

Treatment capacity	87,800m^3/day/unit x 4 units
1 Unit	3 series of 3 basins
Number of basins	36 basins
Contacting media	Honeycomb tube
Material	PVC (Polyvinyl chloride)
Surface area	180m^2/m^3
Residence time	1.2 -1.5 hours
Air volume	7.5Nm^3/hour/m^2 (maximum volume)
Washing method	Air from diffuser tube

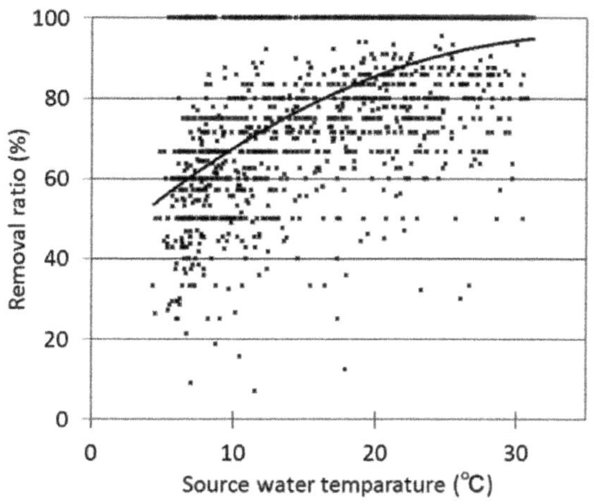

Figure 62.2 Relationship between removal ratio of ammonium nitrogen and water temperature.

The biotreatment was installed in Mishima WTP for the purpose of removing ammonium nitrogen (Hamano *et al*, 1985). Around that time, the ammonium nitrogen concentration in source water was considerably high, and the maximum in FY1990 was 1.12mg/L. However, it keeps low level in recent years. The maximum in 2012 was 0.29mg/L.

In the last 10 years, the average removal ratio of ammonium nitrogen by biotreatment in Mishima WTP is 79.8%, and reaches up to 90.8% in cases when source water temperature is 20 degrees Celcius or higher. If it drops to 10 degrees Celcius or lower, the ratio is 62.3%. Through biotreatment, the concentration of ammonium nitrogen has been kept low enough to control residual chlorine in the later chlorination process. Figure 62.2 implies that a decrease of biological activities in winter causes a decline in the removal ratio.

We have a difficulty in eliminating dissolved manganese completely even by advanced water treatments. As dissolved manganese has a characteristic of getting colored by oxidization in the presence of free chlorine, it could be a major cause for color in treated water(Yokote *et al*, 1997). In the Yodo River, the dissolved manganese concentration sometimes reaches 0.05mg/L, which is the threshold level for imposing some color on treated water. However, the dissolved manganese concentration in biotreated water has been below 0.04mg/L in Mishima WTP in the last 10 years and the average removal ratio is 93.5%. Thus, the biotreatment proves to be effective to reduce dissolved manganese to the satisfactory level. Table 62.2 shows the Removal ratio by the biotreatment in Mishima WTP.

Table 62.2 Removal ratio by the biotreatment in Mishima WTP.

	Ammonium nitrogen (%)	Dissolved manganese (%)
Average removal ratio in Mishima WTP from FY 2003 through FY 2012	79.8	93.5
In case of 20°C or higher in source water temperature	90.8	97.3
In case of 10°C or lower in source water temperature	62.3	89.8

Figure 62.3 illustrates the schematic of the treatment flow of Niwakubo WTP. The biological contact filtration process began operation in 2005 in Niwakubo WTP. In cases of low turbidity of source water, neither coagulant dosing nor coagulation-sedimentation is done, and source water directly flows into the biological contact filtration basin. This is because the coagulation-sedimentation process removes the necessary nutrient (phosphorus) for microorganisms nitrifying ammonium nitrogen. Thus, to keep the biotreatment basin functional, the coagulation-sedimentation process is operated only in cases of high turbidity of source water (Uchihashi *et al*, 2003). After the biological contact filtration, PAC (polyaluminum chloride) is applied in order to form microfloc in the secondary coagulation basin. This microfloc is directly filtered without sedimentation. The filter consists of two layers: sand and anthracite. Then, advanced water treatments follow as the next processes. Niwakubo WTP adopted the biological contact filtration process using the spherical fiber carrier. The spherical fiber carriers of polyester are bedded 2 meters thick in the basin. The water is brought into contact with the colony of micro-organisms attached to the carrier. The specification of the biotreatment in Niwakubo WTP is shown in Table 62.3.

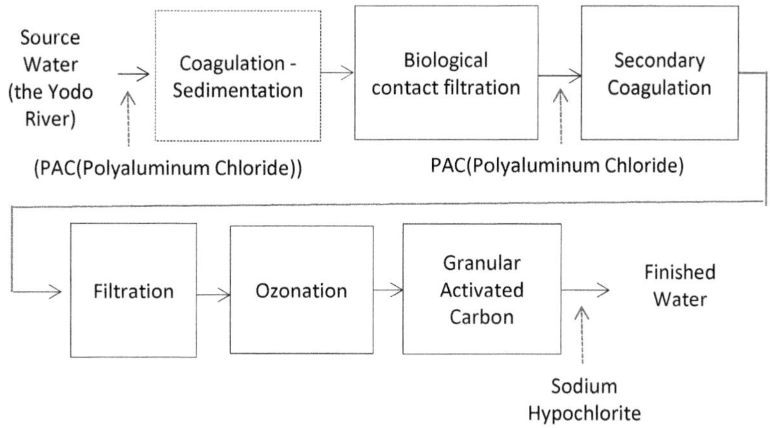

Figure 62.3 The schematic of the treatment flow of Niwakubo WTP

Table 62.3 Specification of the biological contact filtration basin in Niwakubo WTP.

Treatment capacity	216,000m^3/day
System	Greenleef filters of downflow
Filter media	Spherical fiber carrier of polyester (approximately 5 mm in diameter)
Bed height	2m
Filtration area	75m^2/basin
Number of basins	10 (5 in the North Facility, 5 in the South Facility)
Filtration speed (max.)	360m/day
Contact time	8 minutes (in maximum filtration speed)
Washing method	combination of water and air
Washing cycle	every 2-3 weeks
Washing time	5 minutes with water and air, 32 minutes with water

As for ammonium nitrogen, the concentration in the influent of biotreatment has ranged between less than 0.01mg/L and 0.37mg/L since 2005. The concentration in the effluent (biotreated water) has been 0.06mg/L or less. The average removal ratio of ammonium nitrogen by biotreatment in Niwakubo WTP is 97.8%, and keeps over 95% even in cases when source water temperature drops to 10 degrees Celcius or lower. It is confirmed that the removal performance is maintained independently of water temperature.

Since 2005, the dissolved manganese concentration in the influent of biotreatment has ranged between less than 0.01mg/L and 0.07mg/L. The concentration in biotreated water has been 0.01mg/L or less. From all data in this period, it is calculated that the average removal ratio of dissolved manganese by the biotreatment in Niwakubo WTP is 99.1%.

The maximum turbidity in the influent of biotreatment is 16 degrees of kaolin turbidity, and the one in biotreated water is 6 degrees. From all data in this period, it is calculated that the average removal ratio of turbidity by the biotreatment in Niwakubo WTP is 75.7%. Table 62.4 shows the removal ratio by the biotreatment in Niwakubo WTP.

Table 62.4 Removal ratio by the biotreatment in Niwakubo WTP

	Ammonium nitrogen (%)	Dissolved manganese (%)	Turbidity (%)
Average removal ratio in Niwakubo WTP from Dec. 2005 through Mar. 2013	97.8	99.1	75.7
In case of 20℃ or higher in source water temperature	99.1	99.9	80.5
In case of 10℃ or lower in source water temperature	95.3	98.5	68.8

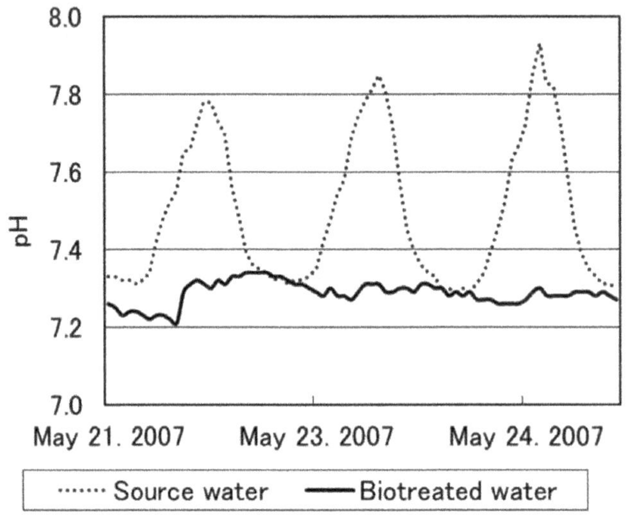

Figure 62.4 Daily variation of pH in Niwakubo WTP.

Figure 62.4 shows the daily variation of pH in source water and biotreated water. Daily variation of pH can be levelled off by biotreatment. This effect contributes to the stable treatment in the following processes. Buffer action against pH is working not only on a daily basis but also on an annual basis. For example in FY2006, pH in source water had ranged between 6.64 and 8.14. On the other hand, pH in biotreated water converged between 6.55 and 7.53. It can be considered that a large amount of microorganisms keep optimal pH to thrive in a beneficial environment and this behavior influences pH in the treated water as buffer action (Kumasawa *et al*, 2008).

Table 62.5 shows the dosage of water treatment chemicals to be compared before and after the introduction of biotreatment.

After the introduction of biotreatment, PAC is only dosed in secondary coagulation to form microfloc unless turbidity gets high in source water. Now that PAC is not always necessary in coagulation-sedimentation basin, it is consumed less than before. The dosage of PAC is reduced by 58%. Removal of ammonium nitrogen in biotreatment enables the stable operation of residual chlorine control and the reduction in the dosage of sodium hypochlorite by 35%.

Table 62.5 The average dosage of water treatment chemicals before and after the introduction of biotreatment.

	Before (FY2002)	After (FY2008)	Reduction ratio
PAC (mL/m^3)	24.7	10.4	58%
Sodium hypochlorite (mg/L)	1.78	1.16	35%

CONCLUSIONS

The main conclusions of this work are:

(1) Ammonium nitrogen :The average removal ratios of ammonium nitrogen by biotreatment in Mishima WTP and Niwakubo WTP were 79.8% (from FY2003 to FY2012) and 97.8% (from Dec.2005 to Mar.2013), respectively. Removal of ammonium nitrogen in the biotreatment enables the stable operation of residual chlorine control and the reduction in the dosage of sodium hypochlorite.

(2) Dissolved Manganese:The average removal ratios of dissolved manganese by biotreatment in Mishima WTP and Niwakubo WTP were 93.5% (from FY2003 to FY2012) and 99.1% (from Dec.2005 to Mar.2013), respectively. To remove color caused by a rise of dissolved manganese, we used to dose additional PAC just before ozonation. Now the biotreatment allow us to suppress color without chemical dosing.

(3) Tubidity :It is confirmed that turbidity can be reduced by biological contact filtration process in Niwakubo WTP. The dosage of PAC was reduced by 58% (in comparison between FY2002 and FY2008). In the event of a sharp rise of turbidity in source water, coagulation-sedimentation process permits a proper treatment without a burden to microorganisms in basins.

(4) pH :pH change in source water can be levelled off by the buffering action of biotreatment. For example, on May23 2007 in Niwakubo WTP, pH in source water ranged between 7.29 and 7.85, and in biotreated water converged between 7.27 and 7.33. This effect contributes to the stable treatment in the following processes.

REFERENCES

Hamano, M., Sato, T., Fujiyoshi, K., Kawai, N. (1985) Experiments of biological treatment in Osaka Prefecture. *Journal of JWWA*, **54**(9), 12-31.

Yokote, M., Nakao, N., Nishikawa, M. (1997) Removal of dissolved manganese by biological contact oxidation using honeycomb. *Water Quality Test and Investigation Report*, Osaka Prefectural Waterworks, **38**, 543-549.

Uchihashi, T., Nakiri, T., Okuno, M. (2003), Study on biological contact filtration using spherical fiber carrier and direct filtration. *Water Quality Test and Investigation Report*, Osaka Prefectural Waterworks, **44**, 435-438.
Kumasawa, M., Takahashi, K., Hirataka, K., Mukai, S., Nakao. N. (2008) Advanced water treatment facilities in Niwakubo Water Treatment Plant(II). *Proceedings of the 59th Annual Conference and Symposium,* JWWA, 228-229.

Chapter 63

Biofilm development in upflow gravel filters in layers with a fabric cover

L. D. Sánchez, E. A. Dominguez*, N. Muñoz*, J. T. Visscher**, L. C. Rietveld****

**Universidad del Valle, Facultad de Ingeniería, Instituto Cinara, Cali, Colombia*
***JTF-Rotterdam BV, Rotterdam, The Netherlands*
****Delft University of Technology, P.O. Box 5084, 2600 GA Delft, The Netherlands*

ABSTRACT

Upflow gravel filters covered with filter fabrics were studied looking at chlorophyll-a and biofilm development. The main biological activity related to chlorophyll-a occurred in the filter fabrics and the top 15 cm of the gravel bed. The daily increase in the thickness of the biofilm on filter fabrics ranged between 0.0182 and 0.031 mmd^{-1}; reaching a maximum thickness between 1.2 -1.35 mm. Biofilm thickness simulated by AQUASIM against chlorophyll-a indicate that a higher concentration of chlorophyll-a coincides with a greater thickness of the biofilm. Chlorophyll-a concentration stabilized at 3.8 gm^{-3} in the first and 4.2 gm^{-3} in the second run. A direct relationship was found between head loss in the filter fabric and the biofilm thickness. The study suggests that removal efficiency of TSS and turbidity for a layer of fabric cover of 0.56 cm may be equivalent to a gravel layer of 0.20 m with a gravel size of 3.2 mm at the top of the filter bed.

Keywords Algae; biofilm; upflow gravel filters; filter fabrics

INTRODUCTION

Upflow gravel filtration (UGF) is a pretreatment technology used in multi-stage filtration (MSF), to reduce the load of suspended solids to facilitate the performance of the final stage of slow sand filtration (SSF). This technology is being applied in rural areas in Latin America and elsewhere to cope with considerable variations in the level of turbidity and total suspended solids in surface water sources. UGF units consist of gravel beds of different sizes with coarse gravel at the bottom, to fine gravel at the top. The treatment process is a combination of physical, physical-chemical and biological processes in the filter layers (Galvis, 1999).

Whereas the biological activity in SSF has been studied in more depth it may be expected that a similar type of biofilm develops on top of the surface layer in the UGF which may be composed of inert particles, organic matter, and a variety of microorganisms (bacteria, algae, protozoa, etc.) (Di Bernardo and Sabogal, 2008). The clarity of the supernatant water on top of the UGF allows penetration of sunlight which reaches the top of the granular medium thus facilitating algae growth. Part of these algae also can penetrate into the filter bed due to the shape, size and mobility (Di Bernardo and Sabogal, 2008).

The use of filter fabrics on top of the UGF can facilitate biological growth on their surface and retain microorganisms in this biofilm which may contribute to the removal of impurities. The biological growth on the filter fabric feeds on the organic matter and inorganic nutrients present in the water, primarily nitrogen and phosphorus which serve as an energy source (Sanchez *et al.* 2006). The filter fabrics are light and relatively easy to remove which allows for efficient surface cleaning. It is anticipated that the filter fabric will enhance the biological activity and thereby, may contribute to a reduction in the necessary filter bed height, which, in turn, has an impact on the reduction in investment costs.

In this paper, the behavior of algae in UGF was studied measuring the chlorophyll-a because this pigment has good correlation with algal biomass, being an indicator of biological or biofilm growth (Di Bernardo and Sabogal, 2008), and determining its influence on treatment efficiency. Biofilm growth was analyzed using the software AQUASIM developed by EAWAG (Reichert, 1998).

MATERIALS AND METHODS

The study was conducted on a pilot scale in which an uncovered circular filter was conditioned with four gravel layers and a filter fabric on top, operating at a filtration velocity of 0.5 mh^{-1}. The height of the supernatant was kept constant (0.09 m above of the filter fabric). Gravel sizes and layer thickness were based on Galvis (1999) with the coarse gravel at the bottom and the smallest gravel at the top (Fig. 63.1). The UGF comprised of four layers of gravel (size 25.4 to 3.2 mm; porosity 35-50%; layer thickness between 25 to 16 cm covered by a set of three filter fabrics with the following characteristics: filter fabric 1: thickness: 1.6 mm, surface area: 15563 m^2/m^3, porosity: 87.9%; filter fabric 2: thickness: 1.8 mm, surface area: 15860 m^2m^{-3}, porosity: 87.7%; filter fabric 3: thickness: 2.2 mm, surface area: 26061 m^2m^{-3}, porosity: 83%; fiber size of 33 microns. The UGF was fed with natural water from the Cauca River, Cali-Colombia. To reduce the entrance of peaks in turbidity and TSS a dynamic roughing filter (DyRF) unit was installed before the UGF. Gravel layers were washed with clean water and grain size analyses were performed to check the grain size of each layer. Furthermore the uniformity coefficient was estimated, U_c ($d_{60}d_{10}^{-1}$) and the porosity (ε_o) of the filter material was determined following the procedure defined by Ives (1990). A piezometer and sampling point was installed to guarantee measurements of head loss (*hf*) and water quality sampling throughout the duration of the filter run.

Filter operation: Two filter runs were analyzed taking into account the following two criteria for suspending the operation of the filter: The maximum total head loss (HT maximum) was limited to 15 cm (Galvis, 1999); and the reduction of the efficiency for TSS or turbidity removal. Once the filter stopped its operation for any of the two criteria, the surface of the unit was washed as well as a bottom backwashing operation was carried out. When it was ensured that *hf* was recovered to the initial conditions a new filtration run was implemented.

Measurements: Daily flow control was carried out in a channel with a triangular weir and the head loss (*hf*) was measured with piezometer tubes located at different heights in the UGF (Fig. 63.1) on a daily basis in each filter layer. Turbidity, TSS, *E-coli*, chlorophyll- a, phosphate and nitrates were measured daily for each of the layers (0.25, 0.50, 0.75, 0.91 and 0.96 m) as these are the main input data for the model that is being used. Turbidity measurements were performed using a Hatch 2100P turbidimeter. The other parameters were measured according to standard methods (APHA, AWWA, WPCF, 2005). Sampling of chlorophyll-a was conducted by placing pieces of fabric cover of 1 cm * 1 cm above the surface and were properly marked for identification and registration. Measurements of chlorophyll-a in the gravel were also done at depths of 10 and 15 cm by small gravel containers with a volume of 10 ml (Fig.63.1) as here algae are also expected to be present due to their mobility. The removal efficiency for TSS, turbidity and *E-coli* of the UGF was calculated based on the mass balance of each gravel layer.

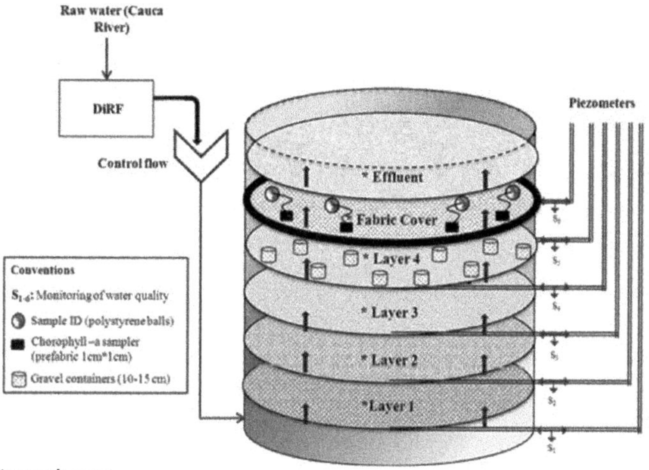

Figure 63.1 Pilot filter scheme.

Biofilm development simulation: The simulation tool AQUASIM (Reichert, 1998) provides models for several aquatic systems based on a compartmentalized reactor. Links between compartments, processes and variables were specified in order to model the experimental UGF with filter fabrics, considering the mathematical model for biofilm thickness explained by Reichert (1998). Variables processes and compartments defined for the simulation of the two filter runs are shown in Table 63.1. The AQUASIM simulation program was used to estimate the thickness of the biofilm adopting a model with one compartment corresponding to the filter fabric. The input data that were used include: the concentration of chlorophyll-a at the top of the filter fabric, and nitrogen and phosphate concentrations below and above the filter fabric. Two main assumptions were made: 1) The biofilm is a uniform layer on the top of the filter fabric formed by filamentous algae (Campos *et al.*, 2006); 2) biofilm growth occurs in a completely homogeneous medium and biofilm thickness is considered uniform throughout filter fabric.

Table 63.1 Simulation variables, process and compartments for biofilm growth.

a. Variables

Variable	Description	Type	Value (Unid)
CDS1	Diffusivity coefficient substrate 1(nitrogen)	FV	0.0001 (m^2d^{-1}) (1)
CDS2	Diffusivity coefficient substrate 2 (phosphates).	FV	0.0001 (m^2d^{-1}) (1)
CO1	Chlorophyll-a concentration 1	SV	Experimental (gm^{-3})
CS1_1	Substrate 1(nitrogen) concentration, filtration run 1	RLV	Experimental (gm^{-3})
CS1_2	Substrate 1(nitrogen) concentration, filtration run 2	RLV	Experimental (gm^{-3})
CS2_1	Substrate 2 (phosphates) concentration, filtration run 1	RLV	Experimental (gm^{-3})
CS2_2	Substrate 2 (phosphates) concentration, filtration run 2	RLV	Experimental (gm^{-3})
DB	Maximum concentration (Chlorophyll-a)	FV	4.6 (gm^{-3})

(Continued)

Table 63.1 Simulation variables, process and compartments for biofilm growth. (*Continued*).

GB	Biofilm thickness	PV	(m)
Q	Flow	FV	Measured (m^3d^{-1})
RO1S1	Algae yield - substrate 1 (nitrogen)	RLV	Measured
RO1S2	Algae yield - substrate 2 (phosphates).	RLV	Measured
TCE01_1	Algae growth rate, filtration run	FV	0.13 (d^{-1}) (2)
TCE01_2	Algae growth rate, filtration run 1	FV	0.2 (d^{-1}) (2)
Time	Filtration run time	PV	(d)
TRO1	Respiration rate for algae (chlorophyll *a*)	SV	(d^{-1})
TSS1	Saturation rate substrate 1 (nitrogen)	FV	2 (gm^{-3})
TSS2	Saturation rate substrate 2 (phosphates)	FV	2 (gm^{-3})
b. Process			
Process	**Description**	**Type**	**Program equation**
CO1S1_1	Algae growth - substrate 1 – Filtration run 1	DP	TCEO1_1*CS1_1 / (TSS1+CS1_1)*CO1
CO1S1_2	Algae growth - substrate 1 – Filtration run 2	DP	TCEO1_2*CS1_2 / (TSS1+CS1_2)*CO1
CO1S2_1	Algae growth - substrate 2 – Filtration run 1	DP	TCEO1_1*CS2_1 / (TSS2+CS2_1)*CO1
CO1S2_2	Algae growth - substrate 2 – Filtration run 2	DP	TCEO1_2*CS2_2 / (TSS2+CS2_2)*CO1
RO1	Algae respiration	DP	TRO1*CO1
c. Compartments			
Compartments	**Description**	**Type**	**Inputs**
Fabric (1-3)_CA_1	Biofilm growth on fabric cover; filtration run 1	BR	CS1_1:CS1_1*Q CS2_1:CS2_1*Q
Fabric (1-3)_CA_2	Biofilm growth on fabric cover; filtration run 2	BR	CS1_2:CS1_1*Q CS2_2:CS2_1*Q

FV: Formula variable; **SV**: State variable; **RLV**: Real list variable; **PV**: Program variable; **PD**: Dynamic process; **BR**: Biofilm reactor. (1) Reichert, (1998); (2) Calculated according to Monod equation, Madigan *et al.* (2003).

RESULTS AND DISCUSSION

Chlorophyll-a behavior: Chlorophyll-a was monitored for the filter fabric and the first 10-15 cm of the top gravel layer to identify biological activity. Sampling showed that the top gravel layer (layer 4, gravel size 3.2 mm) and the set of filter fabrics had the highest concentrations of chlorophyll- a. Fig. 63.2 shows behavior of chlorophyll-a in filter fabric and in the gravel bed at 10 and 15 cm depth which is in line with our expectation it is related to algae that have moved from the surface into the filter. Chlorophyll-a development in the filter fabric shows three phases: 1) the adaptation phase (ripening phase) which took 20 days in the first filter run (4 days in second); 2) the growth phase between day 20 - 60 and 3) the stabilization or equilibrium phase. For the first filtration run the growth rate of chlorophyll-a was lower probably because starting the process needed more time for colonization of organisms in the gravel and the filter fabric, which is in accordance with Di Bernardo (1995) and Sánchez *et al.* (2006). Results of the study indicate that in the UGF with filter fabrics the main biological activity related to chlorophyll-a was in the filter fabric. But even at lower depth some algae are present which confirms their ability to move from the surface against the flow deeper into the filter bed.

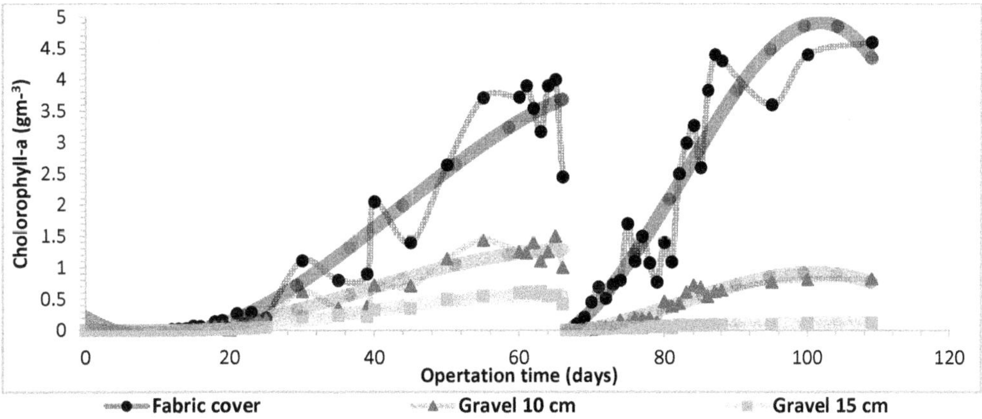

Figure 63.2 Chlorophyll- a behavior with depth.

Biofilm thickness development on the surface of the filter fabric, simulated by AQUASIM 2.0, is presented in Fig. 63.3. A progressive increase of biofilm thickness on the surface of the filter fabric was evident over time for both filtration runs, which conforms to the progressive development of chlorophyll- a biomass presented in Fig. 63.2. The indicated increase in the thickness of the biofilm on the filter fabrics, according to the simulation was 0.0182 mmd^{-1} in the first filtration run; reaching a total thickness of 1.2 mm in 66 days. For the second filtration run thickness grew by 0.031 mmd^{-1}, reaching 1.35 mm in 43 d (Fig. 63.3). For biofilm model a methodology for calibration has not been suggested. We therefore check the reliability of the results of the model by carrying out a sensitivity analysis for nitrogen and phosphorus. This analysis showed low sensitivity in respect to biofilm thickness and a very small error contribution ($2\text{-}7*10^{-8}$), which is possible as they are abundant and no limitation occurs. Chlorophyll-a behavior versus simulated biofilm thickness on the surface of the filter fabric for the first and second filtration run in UGF is shown in Fig. 63.4. Results indicate that there was a direct relationship ($r^2 > 0.90$) between the concentration of chlorophyll- a and the increase in biofilm thickness. It was observed that in a UGF, progressive maturation implied a biofilm growth that could be correlated with a logarithmic model, in which the biofilm reached a relative stabilization of chlorophyll-a concentration of 3.8 gm^{-3} (Fig. 63.4a-filtration run 1) and 4.2 gm^{-3} (Fig. 63.4b-filtration run 2) for a biofilm thickness of 1.2 and 1.35 mm respectively. This relative stabilization can be explained in terms of space availability and competition for nutrients, as reported by Di Bernardo (1995).

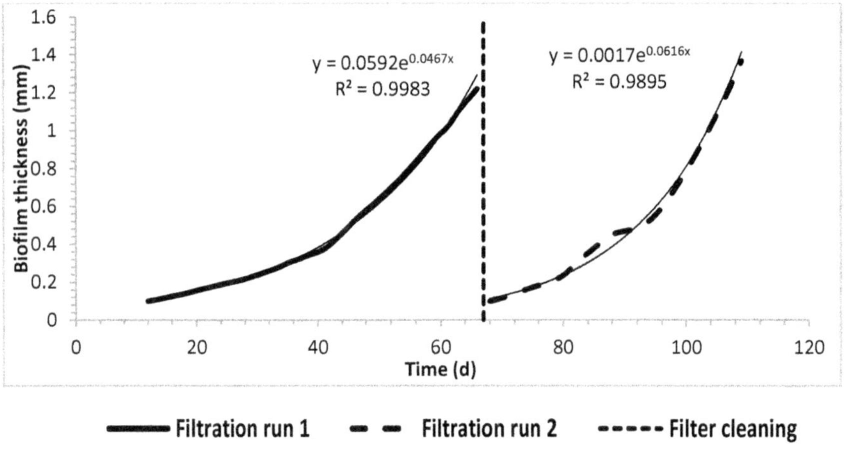

Figure 63.3 Biofilm thickness development over time.

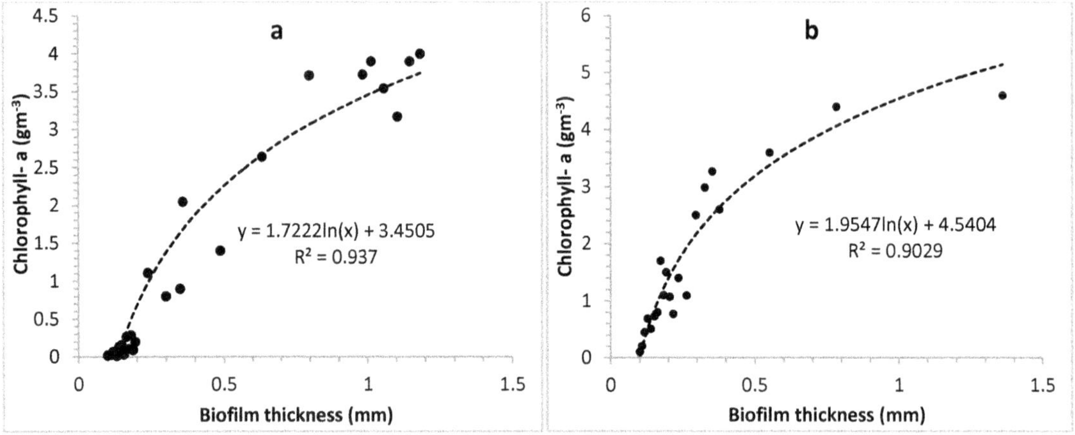

Figure 63.4 Correlation between chlorophyll-a and biofilm thickness: a) filtration run 1; b) filtration run 2.

Head loss behavior and biofilm thickness: Biofilm development contributes to the solids retention and can cause an increase of head loss in the filter fabric (Mälzer and Gimbel, 2006). For this reason head loss behavior was analyzed versus biofilm thickness and chlorophyll-a concentration. Head loss behavior for different layers in the UGF, showed a larger head loss in the gravel size of 19.1 mm (lower layer) due to the solids retention. Head loss over the filter fabric represented 26.5% of the total head loss. Head loss behavior in the filter fabric and its relation with progressive biofilm thickness is shown in Fig. 63.5a; whereas, Fig. 63.5b shows head loss behavior and its relation with chlorophyll-a concentration. The UGF with filter fabrics operating with a filtration velocity of 0.5 mh^{-1} had a filtration run between 43 and 66 d for a maximum *hf* of 15 cm and reaching a total head loss between 1.2 and 1.3 cm in filter fabric. The progressive head loss increase coincided with the biofilm thickness increase. A direct relationship was found between the behavior of the head loss in the fabric cover and Chlorophyll-a concentration that represents biofilm growth.

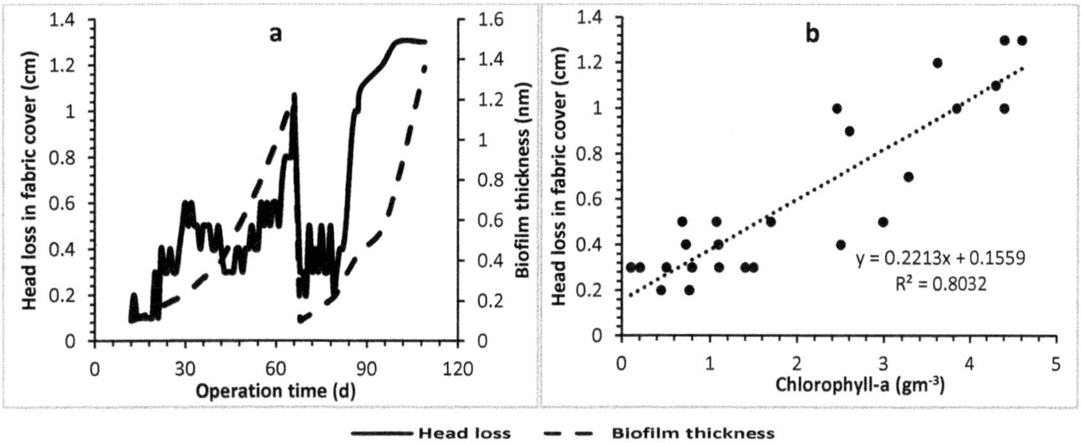

Figure 63.5 Relation between: a) head loss and biofilm thickness; b) head loss and chlorophyll-a

Solids and microorganism removal: Results indicate that all gravel layers contributed to TSS and turbidity removal but performance was better for the top layer (gravel size of 3.2 mm) and the filter fabric (Figure 63.6a). Little difference in efficiency was observed for gravel sizes between 12.7 to 3.2 mm. Net removal efficiencies in the UGF after ripening were: 60% for turbidity and 58% for TSS. The median net removal efficiency for *E-coli* in the UGF was 1.1 log CFU 100 ml^{-1}. Removal efficiency of *E-coli* was highest in the bottom layer (gravel size of 19.1 mm) and reduced towards the top (Fig. 63.6b). The top gravel layer (3.2 mm) and filter fabric did not contribute to the reduction of *E-coli.* The best solids removal in the last gravel layer and filter fabric can be explained by biofilm growth, which improves the collector efficiency in uncovered filters (Hirschi and Sims, 1991). Furthermore, solids retention in the filter fabric is enhanced because the accumulated material improves straining and adsorption in the slimy biofilm matrix formed (Langenbach *et al.*, 2010). The study suggest that removal efficiency of TSS and turbidity for a layer of fabric cover of 0.56 cm was equivalent to a gravel layer of 0.20 m in the top layer (Fig. 63.6a).Whereas turbidity and TSS were removed in the top layer of the gravel bed and the filter fabric *E-coli* levels were not reduced. The fact that no reduction of *E-coli* took place seems to confirm the point made by Campos *et al.* (2006) that the biological activity in the top layers may favors regrowth of bacteria and protozoa within the pores of the biofilm developed in the filter fabric and the gravel bed.

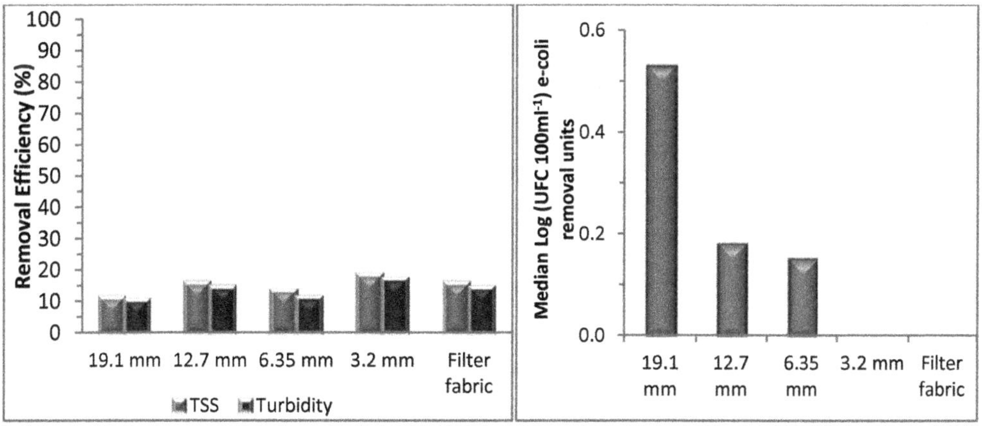

Figure 63.6 Median removal efficiency for a) TSS and Turbidity and b*) e-coli* (*vf* 0.5 mh^{-1}).

CONCLUSIONS

The behavior of algae, TSS, turbidity, *E-coli* and head loss in UGF with a filter fabric on top was studied looking at the increase in chlorophyll-a and biofilm development. The results of the study indicate that the main biological activity related to chlorophyll-a took place in the filter fabric and the top 15 cm of the gravel bed. The increase in the thickness of the biofilm on the filter fabrics was 0.0182 mmd^{-1} in the first filtration run; reaching a total thickness of 1.2 mm in 66 d. For the second filtration run growth was 0.031 mmd^{-1}, reaching 1.35 mm in 43 days. Biofilm thickness on the surface of filter fabrics simulated by AQUASIM against chlorophyll-a indicated a direct relationship, i.e., a higher concentration of chlorophyll-a coincides with a greater thickness of the biofilm (r^2 = 0.93). The condition for chlorophyll-a stabilized at a concentration of 3.8 gm^{-3} in the first filtration run and 4.2 gm^{-3} in the second run.

A direct relationship was found between the behavior of the head loss in the filter fabric and the biofilm thickness (r^2 = 0.8). Median removal efficiencies in the UGF after ripening were: for turbidity 60%, for total suspended solids 58% and for *E-coli* 1.1 log CFU 100 ml^{-1}. Removal efficiency of *E-coli* was highest in the bottom layer (gravel size of 19.1 mm) and reduced to the top of the filter. All gravel layers contributed to TSS and turbidity removal but the efficiency was higher in the top layer (gravel size of 3.2 mm) and the filter fabric. The UGF-filter fabric pilot system, operating with a filtration velocity of 0.5 mh^{-1}, showed a total head loss between 4.5-5.0 cm for a filtration run between 43 and 66 d. The study suggested that removal efficiency of TSS and turbidity for a layer of fabric cover of 0.56 cm may be equivalent to a gravel layer of 0.20 m with a gravel size of 3.2 mm at the top of the filter bed.

ACKNOWLEDGEMENTS

The authors thank the Universidad del Valle for their financial support to this research and the Cinara Institute Research Center at Puerto Mallarino for their laboratory support.

REFERENCES

APHA, AWWA, WPCF, (2005) Standard Methods for the examination of water and wastewater". U.S.A. 21ª ed.

Campos, L.C., Smith, S.R. and Graham, N.J.D. (2006) Use of a novel simulation model to define the behaviour of covered and uncovered slow sand filters in: *proceeding of recent progress in slow sand and alternative biofiltration processes*, IWA publishing, 104-112.

Di Bernardo, L (1995) Algae and Their Influences on water quality and the treatment technologies. Brazilian Association of Sanitary and Environmental Engineering, ABES, Rio de Janeiro, Brazil, September.

Di Bernardo, L., Sabogal, L.P. (2008) Selection of water treatment technologies. University Sao Carlos-Brazil. Publisher LDIBE LTDA. Vol. I, 878.

Galvis, G. (1999) Development and evaluation of multistage filtration plants; an innovative, robust and efficient water treatment technology. Ph.D. thesis, Guilford, UK, CEHE, University of Surrey, 228.

Hirschi, S.D. and Sims, R.C. (1991) Particles and microorganisms in slow rate sand filtration. *International Slow Sand Filtration Workshop*, 27-30, October, Durham, USA.

Huisman, L. and Wood, W.E. (1974) *Slow Sand Filtration*, World Health Organization, Geneva.

Ives, K. J (1990) Testing of filter media, *Water SRT Aqua,* 39, 144–151.

Langenbach, K., Kuschk, P., Horn, H., Kastner, M. (2010) Modeling of slow sand filtration for disinfection of secondary clarifier effluent. *Environmental Science and Technology* 44, 159-166.

Madigan, T.M., Martinko, M.J. y Parker, J. (2003). Brock Biology of Microorganisms, Prentice Hall, 10th edition.

Mälzer, H.-J and Gimbel, R (2006) Extension of slow sand filter running times by protection layer in: *Proceeding of recent progress in slow sand and alternative biofiltration processes*, IWA publishing, 251–259.

Reichert, P. (1998). Aquasim 2.0-User Manual. Computer program for the identification and simulation of aquatic systems. EAWAG, Dubendorf, Switzerland.

Sánchez, L.D., Latorre, J., Galvis, G., and Visscher J.T. (2006) Microbiological conditions before and after cleaning in slow sand filters under tropical conditions in: *proceeding of recent progress in slow sand and alternative biofiltration processes*, IWA publishing, 152–161.

Chapter 64

Up-flow biological contact filter for advanced water treatment system

Norihito Kurahashi, Ryota Sato, Susumu Kumano, Mizuki Fujimoto, Yutaka Ishimaru
Kobelco Eco-Solutions Co., Ltd.

ABSTRACT

Water pollution from household wastewater and eutrophication due to a system failure may cause an increase in the concentration of ammonia nitrogen or surfactant, the generation of offensive odors and tastes including earthy musty odors, and an increase in the level of dissolved manganese, consequently leading to some purification troubles. Of the many types of water treatment, the up-flow biological contact filtration using granular activated carbon as filter media is the most effective towards removing these dissolved solids. This paper reports the actual operating situation and the improvement in water quality at the Kitakyushu City Waterworks Bureau, which introduced this system in the past, as well as the results of water quality improvement in U-BCF demonstration tests carried out with low-temperature dam water in a cold area.

Keywords: Up-flow biological contact filter, Granular activated carbon, Earthy-musty odor, Ammonia nitrogen, Dissolved manganese

INTRODUCTION

In association with the deterioration in water quality of rivers and lakes that serve as the source of tap water, a water purification plant recently experienced some purification troubles caused by an increase in the concentration of ammonia nitrogen or dissolved manganese, the generation of an offensive odor and taste due to the growth of algae, and increased TOC concentration. It is impossible to eliminate these water-soluble substances by the conventional process of coagulation, sedimentation and filtration. To degrade ammonia nitrogen, a chlorine dose (as sodium hypochlorite) of approximately 10 times the ammonia concentration is required (on a weight basis). In addition, there exists a problem in that trihalomethanes are formed as chlorination byproducts.

Photo 64.1 U-BCF facility

Today, many types of advanced water treatments based on biological contact process, ozonation, granular activated carbon filtration and nanofiltration (NF) are widely utilized to remove the above dissolved solids. Given the improvement in water quality, costs of water treatment and installation area, biological treatment is the most effective means above all. In biological treatment, a higher water treatment performance can be obtained by direct filtration with raw water. However, when a general down-flow biological system is used, it is difficult to maintain stable operation because of filter clogging caused by turbidity in the raw water. Therefore, we have studied the method of passing water, particle size of the filter media, water collection and distribution devices, and washing process, to develop an up-flow biological contact filter with granular activated carbon as a biological carrier (hereinafter referred to as "U-BCF"). Since this system can provide a significant improvement in water quality through the high-efficiency direct filtration of raw water, a stable supply of safer and better tasting water has been secured. Furthermore, since the amount of sodium hypochlorite, coagulant and powdered activated carbon used by the entire water purification facility has been reduced, an improvement in system maintenance and a reduction in chemical costs can be attained. And as a result thereof, it contributes considerably to reducing the environmental load of the plant as a whole.

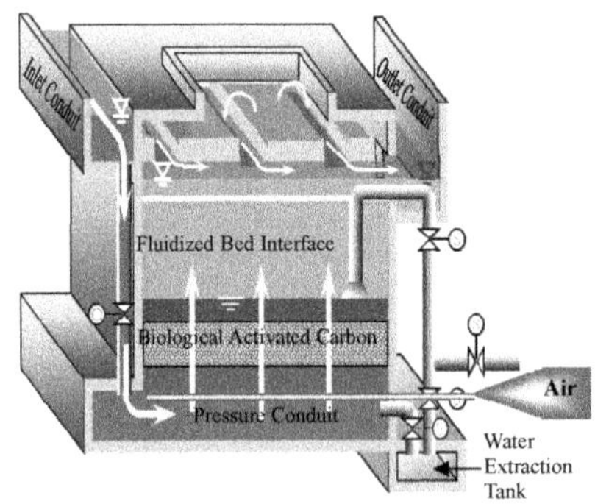

Figure 64.1 U-BCF system

In Japan, the first U-BCF system was introduced to the Honjo Water Purification Plant (Kitakyushu City Waterworks Bureau, Photo 64.1) in August 2000 and it is still running smoothly. This paper provides an overview of the U-BCF system and reports the actual operating situation.

Table 64.1 U-BCF specifications

Item	Specification
Treatment process	Up-flow biological contact filtration
Treatment flow rate	Approx. 15 m/h (360 m/d)
Contact time	Approx. 6 min
Contact filter media	Granular activated carbon, effective diameter: 0.4–0.5 mm, Filter bed depth: 1.5 m
Washing method	Air-scouring/water washing

U-BCF structure

The structure of the U-BCF system is shown in Fig. 64.1 and the typical specifications are in Table 64.1. The raw water flows from the inlet conduit into the pressure conduit through the raw water regulating valve. It is evenly distributed from the pressure conduit by the bottom water distribution device, and passes upward through the supporting gravel bed and filter bed (biological activated carbon layer) to undergo biological treatment.

The filtered water is collected in overflow troughs located in the upper part and then flows out into the outlet conduit. The use of this up-flow, biological, fluidized bed process causes only minimal head loss and, thereby, the filter can perform adequately with approximately a 1.0 m difference in water level between the inlet and outlet conduits. The U-BCF system removes mainly dissolved solids, and it is not suitable for turbidity removal. Therefore, a clarifier such as coagulation, sedimentation and sand filtration equipment or a membrane filter should be used in the subsequent treatment stage. It is best for the system not to capture turbidity, however a small amount of turbidity is in fact captured in the bottom water distribution device and supporting gravel bed. In addition, organism growth is observed in the bottom water distribution device and supporting gravel bed, as well as in the filter bed because of biological treatment. Under these undesirable conditions, head loss increases with the continuation of water passing, and thus the system must be washed periodically. Turbidity and part of adhering organisms are efficiently discharged by cleaning the system in the order of “Air washing” → “Air + Water washing” → “Water washing”.

In practical use, the system should be operated within a certain range of flow rates in order to maintain good fluidization in the filter bed. There are two control methods for achieving such operating conditions: one is to “control the number of filter beds” that keep the flow rate per filter bed constant, and the other is to “control the amount of raw water” so that a constant volume of raw water is supplied by using a circulation pump. The long-term suspension of operations may contribute to a decreased concentration in dissolved oxygen and eventually affect the biological treatment function. Therefore, in order to maintain system performance, a system operator should periodically change the water of the non-operating filter beds when “controlling the number of filter beds”, and should circulate the filtered water when “controlling the amount of raw water.”

Features of U-BCF

- An up-flow fluidized bed enables the effective utilization of the whole filter layer and thus provides high contact efficiency.
- A system built with an up-flow fluidized bed is space-saving because contact filtration of high-turbidity raw water can also be carried out at a high flow rate.
- Granular activated carbon of a small particle size has a large specific surface area and delivers high biological treatment performance.
- It is estimated that the adsorption capacity of activated carbon is maintained for a long period of time owing to the regenerative function of organisms adhered to granular activated carbon particles.
- Only a small amount of turbidity is captured and so head loss is low; the system can employ gravity flow without an intermediate pump.
- Since ammonia nitrogen and other substances are biologically oxidized through the direct biological treatment of raw water, the amount of sodium hypochlorite injected for prechlorination and intermediate chlorination can be reduced and balanced.
- Turbidity in original raw water has a better sedimentability after biological filtration.

WATER TREATMENT PERFORMANCE

The U-BCF system was introduced to the Honjo Water Purification Plant (71,000 m^3/d) and the Ano Water Purification Plant (171,000 m^3/d) under the control of the Kitakyushu City Waterworks Bureau. Since granular activated carbon is used as filter media, total organic carbon is removed by the physical adsorption effect of activated carbon at the beginning stage of operation. After that, the concentration of ammonia nitrogen and dissolved manganese begins to decrease as raw water is biologically treated. It takes about one month to obtain the desired effects of biological filtration although the quality and temperature of raw water has a significant impact on the said interval. Tables 64.2 and 64.3 show the actual water quality and removal rate of dissolved solids that were obtained using this system at both water purification plants.

Table 64.2 Water treatment performance

Water quality item	Unit	Raw water*1	U-BCF-treated water*1	Removal rate
Ammonia nitrogen	(mg/L)	0.046	0.008	90.5 %
Dissolved manganese (D-Mn)	(mg/L)	0.013	0.001	91.4 %
Threshold odor number	(-)	4.7	2	59.7 %
Turbidity	(degree)	5.2	3.9	19.6 %
Potassium permanganate consumption (KMnO4)	(mg/L)	7.2	5.4	23.0 %
Ultraviolet absorbance (UV260)	(-)	0.043	0.031	22.2 %
Trihalomethane formation potential (THM-FP)	(mg/L)	0.040	0.030	19.9 %
Anionic surface active agent	(mg/L)	0.035	0.016	49.4 %
2-MIB	(ng/L)	50	14	72.0 %
		<50	*2 <10	100 %
pH	(-)	7.98	7.76	-

*1 Honjo water purification plant, Kitakyushu City Waterworks Bureau (71,000 m^3/d), August 2000–March 2002

*2 Limit of determination (at that time): 10 ng/L

The removal of 2-methylisoborneol (2-MIB), a causative agent of earthy musty odors, is shown in Table 64.4. A removal rate of at least 90% was achieved for both ammonium nitrogen and dissolved manganese, and that for threshold odor number was about 60%. 2-MIB, the removal rate is 72% at a concentration of 50 ng/L 2-MIB in raw water, and 100% at concentrations below 50 ng/L; a superior 2-MIB removal performance is demonstrated. The total organic carbon represented by potassium permanganate consumption, ultraviolet absorbance, and trihalomethane formation potential, is reduced by approximately reduced by 20–30%, and anionic surface active agents are also reduced remarkably.

Figures 64.2 and 64.3 show the daily variations in ammonia nitrogen and dissolved manganese at the Ano Water Purification Plant. The maximum level of ammonia nitrogen in raw water was approximately 0.5 mg/L during the winter months, but it was completely removed after treatment. The dissolved manganese concentration in raw water greatly varied according to the water intake timing. Nevertheless, this dissolved solid was stably and significantly removed even at the highest level of 0.1 mg/L (raw water).

Table 64.3 Water treatment performance Average concentration and average removal rate

Water quality item	Unit	Raw water*1	U-BCF-treated water*1	Removal rate
Ammonia nitrogen	(mg/L)	0.1	Not detected	100 %
Dissolved manganese (D-Mn)	(mg/L)	0.017	0.001	94 %
Threshold odor number	(−)	1.7	0.3	80 %
Turbidity	(degree)	7.8	6.7	14 %
Potassium permanganate consumption (KMnO4)	(mg/L)	6.9	5.1	26 %
TOC	(mg/L)	1.7	1.3	24 %
DOC	(mg/L)	1.4	1.0	29 %
Ultraviolet absorbance (UV-260)	(−)	0.037	0.025	32 %
Anionic surface active agent	(mg/L)	0.06	0.01	79 %
pH	(−)	7.5	7.4	—

*1 Ano water purification plant, Kitakyushu City Waterworks Bureau (171,000 m^3/day), April 2003–March 2004

*2 Limit of determination (at that time): 10 ng/L

Table 64.4 2-MIB removal performance

Date of water sampling*1		5/15	5/25	5/28	6/1	6/4	6/6	6/8
Raw water	(ng/L)	50	48	39	31	43	28	35
U-BCF	(ng/L)	14	ND*2	ND	ND	ND	ND	ND
Removal rate	(%)	72	100	100	100	100	100	100

*1 Honjo water purification plant, Kitakyushu City Waterworks Bureau (71,000 m^3/day), 2001

*2 ND: 10 ng/L or less

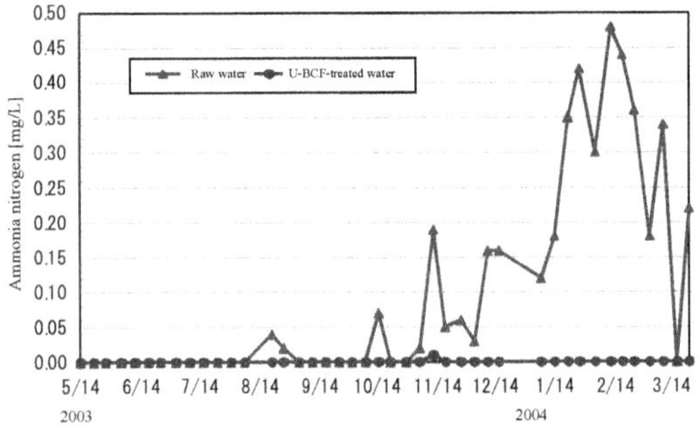

Figure 64.2 Daily variation in ammonia nitrogen

Note: Ano Water Purification Plant, Kitakyushu City Waterworks Bureau May 2003 to March 2004

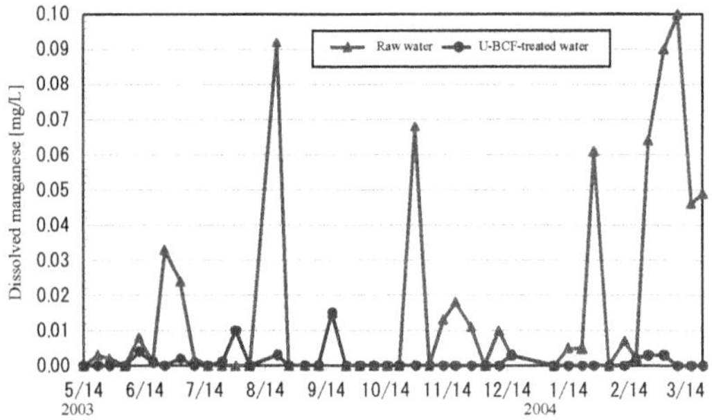

Figure 64.3 Daily variation in dissolved managanese

Note: Ano Water Purification Plant, Kitakyushu City Waterworks Bureau May 2003 to March 2004

The effects of raw water turbidity on the head loss are shown in Figure 64.4. Due to heavy rains, the highest raw water turbidity, which was above 350°, continued for approximately 24 hours, and then turbidities above 100° for some 3 weeks at the Ano Water Purification Plant. But, no significant fluctuation of head loss values was observed during a corresponding period. As shown above, when the system experienced an inflow of highly turbid raw water due to heavy rains, head loss did not rapidly increase and could be recovered sufficiently by means of cleaning. Thus, the system can still run stably.

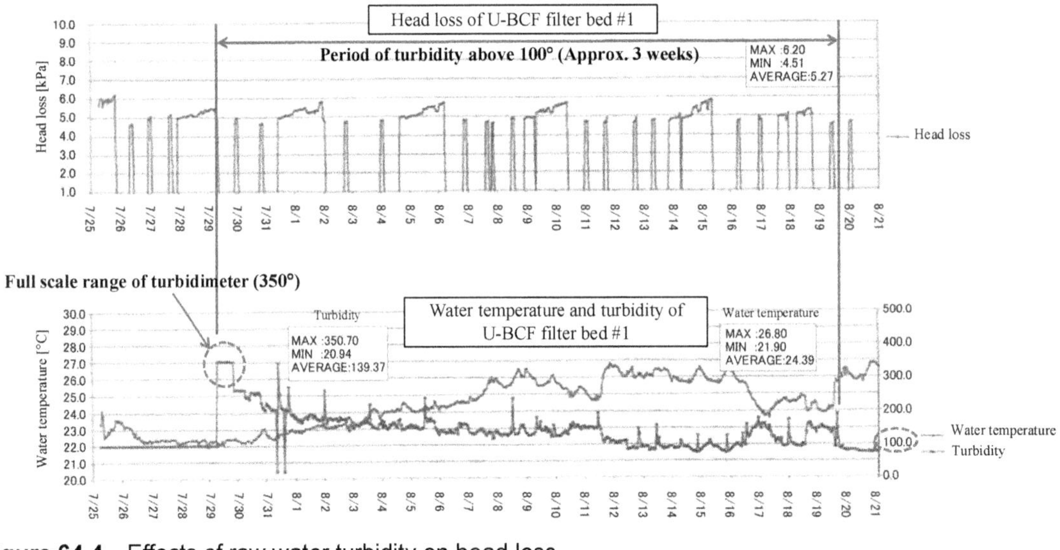

Figure 64.4 Effects of raw water turbidity on head loss

Note: Ano Water Purification Plant, Kitakyushu City Waterworks Bureau July 2009 to August 2009

It is concerning that biological water treatment is susceptible to the effects of water temperature, particularly, decreased water temperature. Accordingly, we carried out U-BCF demonstration tests with dam water in a cold area of the Tohoku region. A flow chart of the field experimental apparatus is shown in Fig. 64.5 and its specifications in Table 64.5.

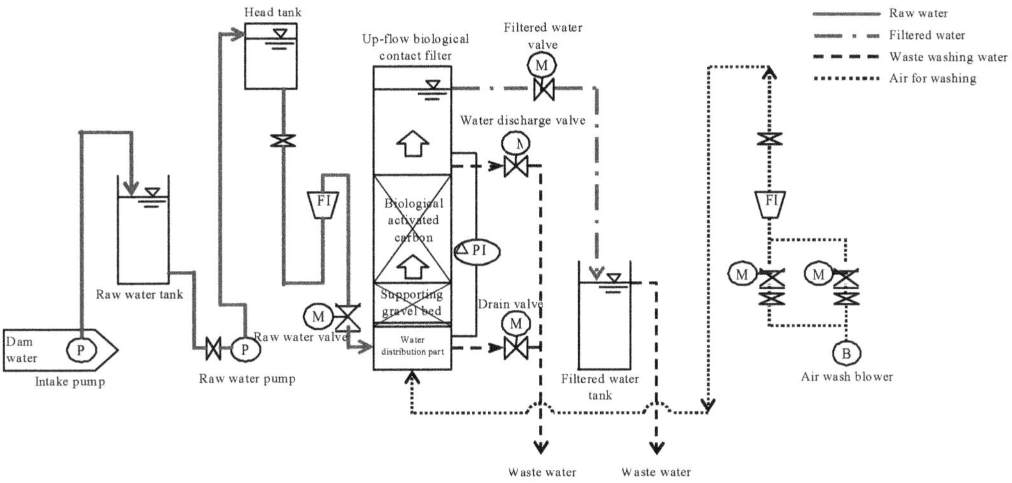

Figure 64.5 Flow chart of field experimental apparatus

Table 64.5 Specifications of field experimental apparatus

Biological contact filter tower	
Model	PVC vertical cylinder type
Number of towers	One tower
Size	200 mmφ × 4,500 mmH
Water passing method	Up-flow
Linear velocity (LV)	15 m/h
Space velocity (SV)	10 h^{-1}
Filter media	Granular activated carbon: 1,500 mmH (Effective diameter: 0.4–0.5 mm)
Supporting gravel bed	300 mmH
Water distribution device	Perforated block type

The experimental results of removing geosmin or 2-MIB (a causative agent of earthy musty odor) and dissolved manganese are shown in Figs. 64.6 to 64.8. As for geosmin, the concentration in filtered water was roughly reduced to below 1 ng/L within a raw water concentration range between less than 1 ng/L and 4 ng/L over the whole experimental period. Desirable water treatment results were also obtained during the low water temperature period. As for 2-MIB, filtered water recorded a maximum concentration of 3 ng/L during the low water temperature period and the concentration during the other period was roughly reduced to below 1 ng/L within a raw water concentration range between less than 1 ng/L and 17 ng/L. Therefore, 2-MIB was effectively removed like with geosmin. As for dissolved manganese, the concentration in filtered water was roughly reduced to below 0.001 mg/L within a raw water concentration range between less than 0.001 mg/L and 0.048 mg/L over the whole experimental period including the low water temperature period. Therefore, a high removal performance was also confirmed for this dissolved solid.

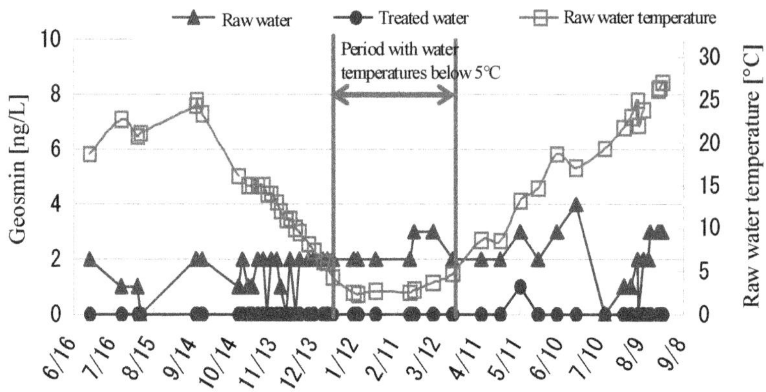

Figure 64.6 Geosmin removal performance

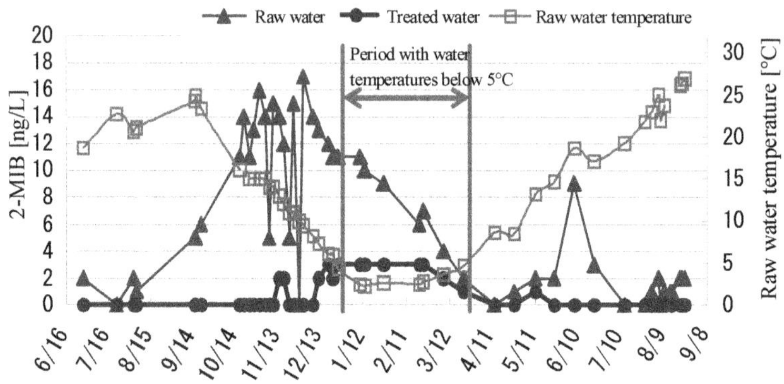

Figure 64.7 2-MIB removal performance

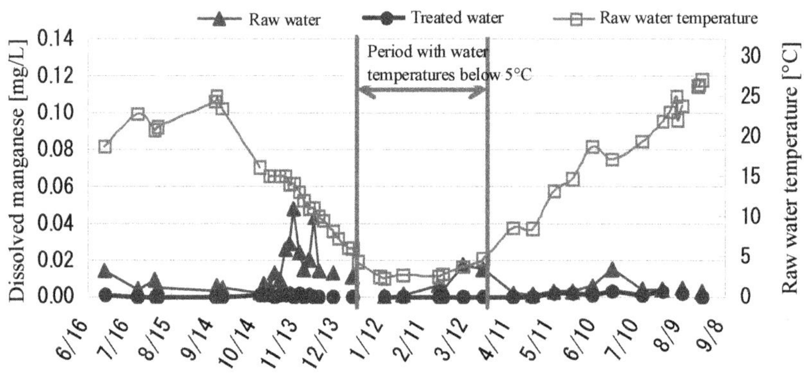

Figure 64.8 Dissolved manganese removal performance

Based on the experiments carried out in a cold area of the Tohoku region, the desirable removal of odor causing substances and dissolved manganese was demonstrated. Afterwards, in order to adequately evaluate the water treatment performance of the system at a higher dissolved solid concentration of raw water during the low water temperature period, additional laboratory experiments were conducted with the raw water in the Tohoku region and the filter media (granular activated carbon) used in the experiments. A flow chart of the laboratory experimental setup is shown in Fig. 64.9 and the specifications in Table 64.6.

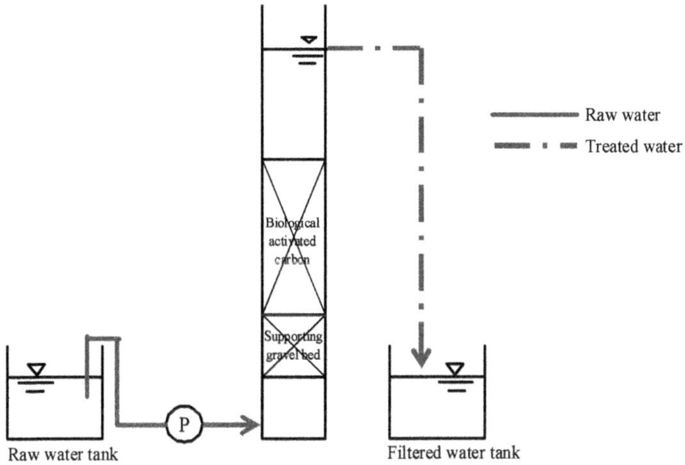

Figure 64.9 Flow chart of laboratory experimental setup

Table 64.6 Specifications of laboratory-scale experimental setup

Biological contact filter tower	
Model	PVC vertical cylinder type
Number of towers	One tower
Size	13 mmφ × 2,050 mmH
Water passing method	Up-flow
Linear velocity (LV)	10 m/h
Space velocity (SV)	10 h^{-1}
Filter media	Granular activated carbon: 1,000 mmH (Effective diameter: 0.4–0.5 mm)
Supporting gravel bed	150 mmH

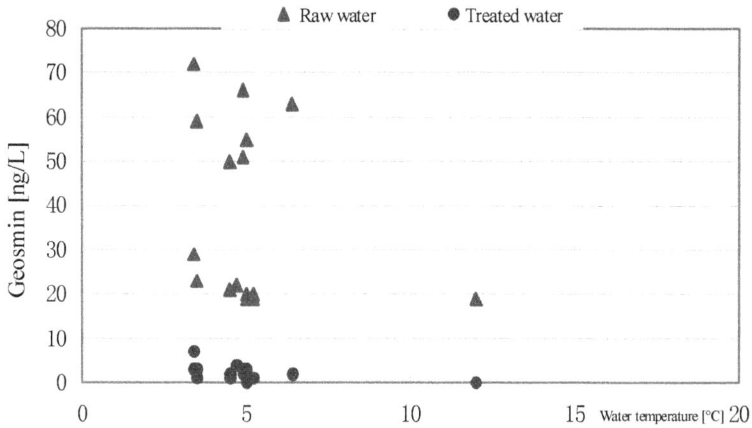

Figure 64.10 Geosmin removal performance (Laboratory experiment at low temperatures)

The laboratory results are shown in Figs. 64.10 to 64.12. As for geosmin, when passing the raw water at a concentration of 72 ng/L under the condition of a low raw water temperature below 5°C, the concentration of filtered water was 7 ng/L. Therefore, a high removal performance was obtained even under low water temperature. As for 2-MIB, when likewise passing the raw water at 67 ng/L below 5°C, the concentration of filtered water was 8 ng/L, therefore a high removal performance was obtained at low water temperature as with geosmin. As for dissolved manganese, when passing the raw water at 0.5 mg/L below 5°C, the concentration of filtered water was less than 0.001 mg/L. Therefore, similarly, a high removal performance was also obtained for this dissolved solid.

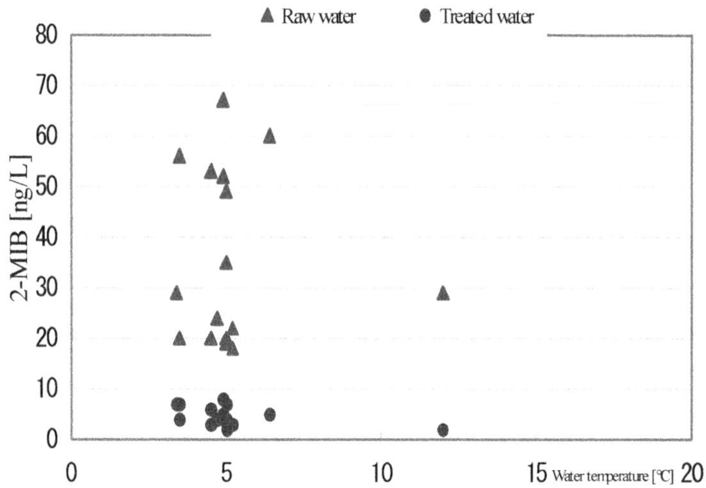

Figure 64.11 2-MIB removal performance (Laboratory experiment at low temperatures)

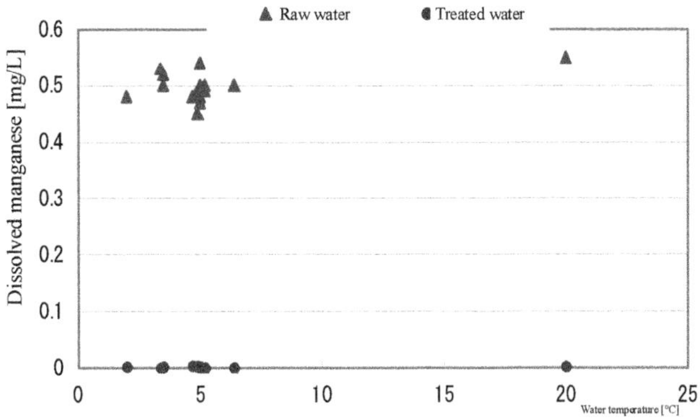

Figure 64.12 Geosmin removal performance (Laboratory experiment at low temperatures)

ECONOMIC ASPECTS

Table 64.7 shows an estimated comparison between advanced water treatments by the U-BCF system and by the combined use of ozone and activated carbon in a treatment capacity of 100,000 m^3/d, and Tables 64.7 and 64.8 show the chemical reduction effect at the actual facility of Kitakyushu City, which is compared between before and after the U-BCF introduction.

Table 64.7 Comparison between two water treatment systems

Item	Ozone - activated carbon treatment	U-BCF treatment
Construction cost	100	60
Running costs	100	55
Installation area	100	35

Note: Calculated from each cost graph in JWRC (e-water II) "Establishment of methods for selecting water purification system by water quality".

Table 64.8 Chemical reduction effect

Name of chemicals	Injection efficiency before U-BCF introduction: January-June 2000 (mg/L)	Injection efficiency after U-BCF introduction: January-June 2001 (mg/L)	Reducing rate (%)
Polyaluminum chloride (PAC)	29.2	27.2	7
Aluminum sulfate	41.5	21.7	48
Sodium hypochlorite	3.3	2.6	21

Note: Honjo Water Purification Plant, Kitakyushu City Waterworks Bureau (Proceedings of the 54th National Waterworks Conference and Symposium)

PAC combined with aluminum sulfate was used as a coagulant.

Its injection amount was adjusted by only the aluminium sulfate volume.

CONCLUSION

Recently, the surface water of dam reservoirs and lakes has increasingly been used as a source of tap water in comparison with that of rivers. As a result, measures against increased concentrations of musty odor substances or TOC have been taken frequently. Although the U-BCF system delivers slightly lower water treatment performance than advanced water treatment by the combined use of ozone and activated carbon, it is safer due to no bromate production from ozone injection and enables easy maintenance. This system has great advantages in the costs and installation area, and it thus seems that the system will become widely used alongside the replacement and/enhancement of aging facilities in the future.

In addition, we expect that the system will be constructively introduced to waterworks facilities that are tackling the deteriorating quality of drinking water sources as an important issue both at home and abroad.

REFERENCES

T. Koezuka et al. (1997) Proc. of the 48th National Waterworks Conf & Symposium, p. 172
T. Koezuka et al. (1998) Proc. of the 49th National Waterworks Conf & Symposium, p. 160
M. Ishibashi et al. (2002) Proc. of the 53rd National Waterworks Conf & Symp, p. 278
M. Nojiri et al. (2002) Proc. of the 53rd National Waterworks Conf & Symposium, p. 280
C. Yamazaki et al. (2003) Proc. of the 54th National Waterworks Conf and Symp, p. 230
A. Aoki et al. (2007) Journal of Japan Water Works Association, Vol. 76, No. 8.
Y. Ishimaru et al. (2008) The Japan-China Water Supply Technology Symposium.

Chapter 65

pH conditioning using limestone with upflow gravel filters

Andrés Araya, Johnny Valverde **, Luis D. Sánchez****
**Escuela de Ingeniería en Construcción, Instituto Tecnológico de Costa Rica, P.O. Box 159-7050, Cartago, Costa Rica*
***Escuela de Química, Instituto Tecnológico de Costa Rica, P.O. Box 159-7050, Cartago, Costa Rica*
****Instituto Cinara, Facultad de Ingeniería, Universidad del Valle, P.O. Box 25157 Cali, Colombia*

ABSTRACT
Neutral pH values from pH lower raw water in a Costa Rican village are reached by using upflow gravel filter in composition of 25 and 50% with limestone. Presence of alkaline composition in gravel and size particle is taking into account. Results of the study indicate that upflow gravel filter with limestone and gravel media to condition the pH will improve the biological processes of the slow sand filtration (photosynthesis, respiration, microbial growth). Another benefit is that it will increase the buffer capacity to neutralize acids from the disinfection process and it will protect the infrastructure of water supply systems.

Keywords limestone; pH conditioning; upflow gravel filter; drinking water

INTRODUCTION

Although pH usually has no direct impact on water consumers, plays the most important operational role in water-quality parameters. The degree of corrosion of water mains and pipes in household water systems can increase by lower pH values, affording health issues in human populations. Careful attention to pH control must be taken at all stages of water treatment to ensure satisfactory water clarification and disinfection (WHO, 2011). In biological filtration treatment processes such as upflow gravel filters and slow sand filtration, the relationship between nutrient concentrations in raw water, and environmental conditions (temperature, pH, dissolved oxygen, and sun radiation) can affect considerably the filters performance, ripening and treatment efficiency. A expected solution can be tackled with a built and integrated design, along with good operational practices (Sánchez et al., 2005).

Water sources with lower pH values are regular in rural areas of Costa Rica, due to the composition of adjacent soils, superficial deposits and bedrocks. These factors affect the low cost water treatment process such as slow sand filtration, conferring an impaired infrastructure of water supply systems (valves, pipes, meters, and concrete durability). The latter is exemplify by raw water aqueducts of Costa Rican villages of Santa Cecilia de Tarrazú, Jesús María de San Mateo and Zagala de Puntarenas, with pH values between 3.4 and 4.2.(Historic records provided to Authors) pH control is not an easy task in Costa Rican rural areas since difficulties as high cost of operation and control of the dosage of alkaline substances are the main obstacles.

Optimal growing medium for some microorganism such as acidophiles (Baker & Dopson, 2007), *Cryptococcus humicola*, *Rhodotorula glutinis*, *Aspergillus flavus* Link, *Penicillium sp.*, *Penicillium janthinellum* Biourge, and *Trichoderma asperellum* (Kawai et al., 2000), are found by pH values around 2 and 3, usually those values are less favorable for many microorganisms rather than the neutral pH values of its environments (Ledin & Pedersen, 1996). Furthermore, the optimum pH condition required for a suitable process in slow sand filtration is considered to be close to neutral pH. Surprisingly, drinking water supplies do not reach neutral pH values due to both the composition and the origin of the construction materials used in water distribution system, with ranges usually in 6.5–8.5 range (WHO, 2011). For this reason, to reach water pH values up to neutral data is urgently needed.

In recent decades, the use of limestone to neutralize acidic mine in passive-treatment systems has become popular because of an inexpensive technique to construct and to maintain. Dissolution of calcite ($CaCO_3$), as the main component of limestone in water, can neutralize its acid conditions, raising both the pH values and ionic concentrations like HCO_3^{1-}, OH^{1-} and Ca^{2+} (Cravotta & Trahan, 1999) up to the safest values.

Further investigation is needed to determine the limestone potential for operating upflow gravel filters as a good pretreatment for slow sand filters in tropical countries (Galvis, 1999). The aim of this paper is to demonstrate the use of upflow gravel filter for water pH conditioning using layers of limestone to ensure a better process of slow sand filtration, as well as conditioning the microorganism disinfection, and to prevent the damage of water distribution systems in Costa Rican areas.

METHODOLOGY

The study was developed in pilot scale and three columns of upflow gravel filters in layers were constructed in parallel (see Table 65.1) to treat raw water with pH < 4,25 in rural Costa Rican village of San Marcos de Tarrazú (Solís, 2012). The assessed filtration velocity was 0,6 m/h. Each column consists of four layers of different grain sizes with several compositions of quarry aggregate and limestone (Column 01 had 100% of limestone, Column 02 had 75% and Column 03 had 50%). The conditions of the experiment are summarized in Table 65.1.

Table 65.1 Conditions of the experiment

Layers	Grain size (mm)	Thickness (m)	Type of aggregate*		
			Column 1	Column 2	Column 3
C1	3-6	0,3	L	L	L
C2	6-12	0,2	L	L	L
C3	12-18	0,2	L	L	Q
C4	18-25	0,3	L	Q	Q

**Limestone (L) and quarry aggregate (Q)*

RESULTS AND DISCUSSION

A total of 150 pH values were obtained for each column and average data are shown in Figure 65.1.

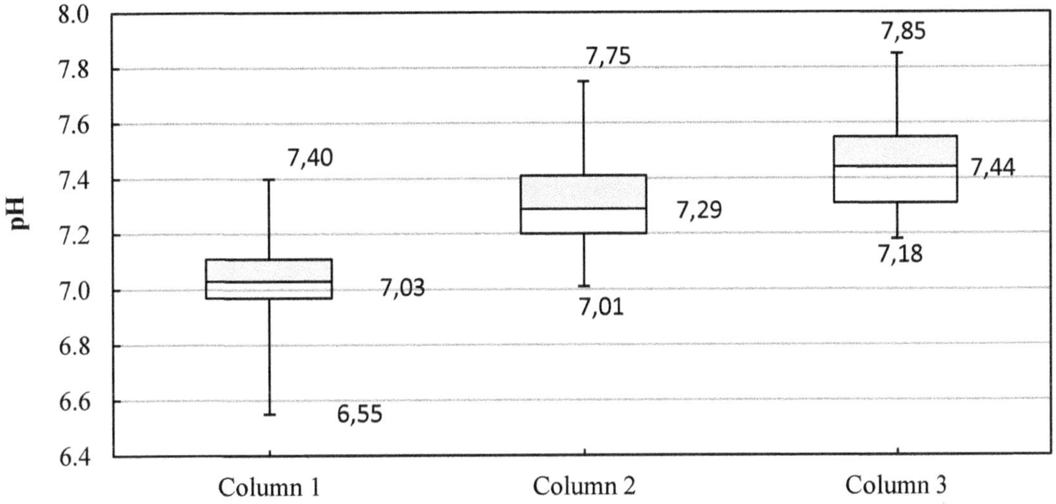

Figure 65.1 pH values for each column using upflow gravel filtration.

Starting the pH value of 4,5 for raw water, raised pH values for every tested column were cleared achieved (column 01 = 7,03 ± 0,13, column 02 = 7,30 ± 0,14, and column 03 = 7,44 ± 0,14). The latter has the best combination of limestone and quarry aggregate. According to this, the first contact of raw water occurs with the small particles of limestone. In order to study the loss of weight of the analyzed aggregates by dissolution, the three columns were weighed both at the beginning and at the end of the test over one month period. The results showed that column 03 lost 4,03% of its original weight, whereas both columns 01 and 02 reported losses of 2,11% and 2,13%, respectively due to high dissolution rate (per unit surface area) of limestone particles when the average particle size is small (Sun et al., 2010).

Gravel materials in columns 02 and 03 help reaching up to neutral pH values since dissociation of its species takes place. Important role of acid-base reactions is taking into account since alkaline species from Gravel like aluminum and calcium oxide plus calcium carbonate react with acid species from raw water. Taking into consideration the high surface area from gravel, plays a important role as the driving force to achieved the neutral pH.

CONCLUSIONS

(1) Conditioning pH using upflow gravel filter made by limestone and by gravel media improves the biological processes of slow sand filtration (photosynthesis, respiration, microbial growth). Increased buffer capacity to neutralized acids from either a remarkable benefit as a increased buffer capacity to neutralize acids from either disinfection processes and protection of the infrastructure of water supply systems are established.

(2) Small size particle materials from upflow gravel filter contributes to conditioning pH and mainly its alkaline composition contributes to reach pH lower values up to neutral pH values.

ACKNOWLEDGEMENTS

The authors wish to acknowledge the important contribution of Instituto Tecnológico de Costa Rica.

REFERENCES

Baker-Austin, C., and Dopson, M. (2007). Life in acid: pH homeostasis in acidophiles. Trends Microbiol, 15(4), 165-171.

Cravotta, C. A. III, and Trahan, M. K. (1999). Limestone drains to increase pH and remove dissolved metals from acidic mine drainage. *Appl Geochem*, 14(5), 581-606.

Kawai, F., Zhang, D., and Sugimoto, M. (2000). Isolation and characterization of acid and Al-tolerant microorganisms. FEMS Microbiol Lett, 189(2), 143-147.

Ledin, M., and Pedersen, K. (1996). The environmental impact of mine wastes - Roles of microorganisms and their significance in treatment of mine wastes. Earth - Sci Rev, 41(1), 67-108.

Sánchez, L. D., Sánchez, A., Galvis, G., Latorre, J., and Visscher, J. T. (2006). Multi-Stage Filtration. Thematic overview Paper, 15.

Solis, G. (2012) "Acondicionamiento del medio filtrante para regulación del pH utilizando agregado calcáreo con un Filtro Grueso Ascendente en Capas (FGAC) del sistema de potabilización del acueducto de Santa Cecilia de San Marcos de Tarrazú". Tesis de Licenciatura, Escuela de Ingeniería en Construcción, ITCR, Costa Rica. 34 pp.

Sun, B., Zhou, Q., Chen, X., Xu, T., and Hui, S. (2010). Effect of particle size in a limestone-hydrochloric acid reaction system. *J. Hazard Mater.*, 179(1), 400-408.

World Health Organization, (2011). Guidelines for drinking - water quality - Fourth Edition. WHO Library Cataloguing-in-Publication Data, Switzerland. 564 pp.

Chapter 66
Upward biological contact filtration (U-BCF) in Vietnam

*M. Yayama**

**Water and Sewer Bureau, City of Kitaktyushu, Japan*

ABSTRACT

In May 2013, Haiphong Water decided to install U-BCF, which is the advanced treatment method used in the City of Kitakyushu, Japan, into an existing water treatment plant in the City of Haiphong, Vietnam. It is the first time introducing of Japanese advanced treatment technology on public supply of Vietnam. This paper reports the relationship between Haiphong water and Kitakyushu Water and Sewer Bureau, as well as the introduction of U-BCF.

Keywords U-BCF, Granular activated carbon, Ammonia nitrogen, Dissolved manganese, Vietnam, Haiphong

BACKGROUND

About Haiphong City

The City of Haiphong is located approximately 100km to the east-southeast of Hanoi City, a metropolis in Vietnam. The City of Haiphong has a population of about 2 million and is the third most populous city in Vietnam. The City of Haiphong is growing rapidly as an international and industrial city with the main international port of the northern part of Vietnam and a lot of industrial zones which are operated by major foreign companies.

About Haiphong Water

Haiphong Water Supply One Member Co., LTD (hereinafter referred to as "Haiphong Water") was established in 1905, and Haiphong Water is operating the water supply service using an adequate amount of water resources from the rivers in the city.

In recent years, Haiphong Water has faced a serious problem that the river water has been polluted by waste water affected from urbanization and industrialization.

On the report of water quality analysis, the ammonium nitrogen which is one of parameters shown degree of organic pollution, in the river water is high degree. Sometime, it is analyzed ten times higher number (1.0mg/l) in comparison with an indicator (0.1mg/l) as acceptable quality of water resources for water supply service.

Photo 66.1 River in Haiphong City

Table 66.1 Quality of Raw Water from Rivers in Haiphong City

Name of WTP	Turbidity (NTU)	pH	Ammonia Nitrogen (mg/l)	Manganese (mg/l)
An Duong	30.90	7.20	0.27	0.18
Vat Cat	48.08	7.38	1.60	0.30
Do Son	29.73	7.29	1.38	0.41

Other hand we know that Trihalomethane (THM) is generated by organic substance in raw water and chlorine disinfection at water purification plant, above all, it has came out that Chloroform which is one kind of THM is carcinogen. On water purification process, the generated amount of THM which is hazardous chemical substance for human health, increase in proportion to amount of organic substance in raw water and chlorine feeding. Therefore, it is serious problem for Haiphong Water Supply Company which is using a lot of chlorine in order to remove high degree ammonium nitrogen in raw water.

However, Haiphong Water Supply Company has not taken enough effective countermeasures for monitoring and purification methods against organic substance.

ADVANCED TREATMENT TRANSFERRED

In 1997, The Water and Sewer Bureau, City of Kitakyushu (hereinafter referred to as "Kitakyushu Water") created the Upward Biological Contact Filtration (hereinafter referred to as "U-BCF") as a new treatment method in order to produce safer water to drink, as a countermeasure against polluted water resources. It was the result of 11 years of research started in 1987.

The U-BCF belongs to the biological treatment. Advantage of biological treatment is lower operation cost compared to other treatment methods because biological decomposition happens without electric power or any chemicals. However, Disadvantage is to require a large space for its facility.

Photo 66.2 HONJO water treatment plant

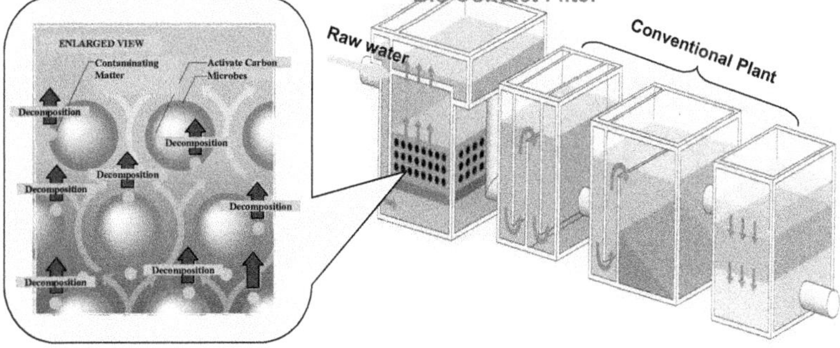

Figure 66.1 Outline of U-BCF

However, U-BCF does not require such a large space, because it is possible for raw water to efficiently come in contact with microorganisms, using upward filtration and granular activated carbon as a living bed for microorganisms. Kitakyushu Water already had installed U-BCF in the major water treatment plants, "ANOH" and "HONJO", in Kitakyushu.

BEGINNING OF TECHNICAL COOPERATION

In April 2009, the City of Kitakyushu and the City of Haiphong signed the "Agreement on Friendship and Cooperation". Then, Haiphong Water requested technical cooperation to the City of Kitakyushu to solve the serious problem of their raw water quality. In July 2010, during the periods from 2010 to 2012, Kitakyushu Water carried out on-site survey at An Duong Water Treatment Plant (100,000m3/d), which is biggest plant of Haiphong city as part of the JICA Partnership Program. In order to confirm the effectiveness of U-BCF in Vietnam, Kitakyushu Water placed the pilot plant of U-BCF in An Duong Water Treatment Plant and collected and analyzed data. An outline and a specification of the pilot plant are shown in Figure 66.2 and Table 66.2.

Photo 66.3 Agreement on Friendship and Cooperation

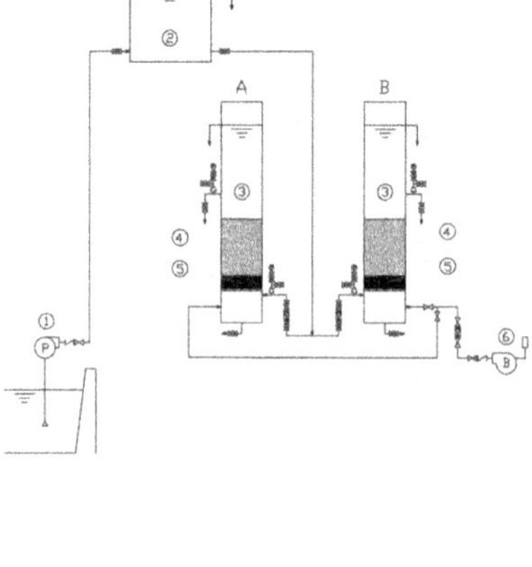

Figure 66.2 Outline of pilot plant

Table 66.2 Specification of pilot plant

Num	Specification
1	Pump for sampling
2	Distribution tank
3	Column (PVC)
	A: Space Velocity: 10 (h^{-1})
	Liner Velocity: 15 (m/h)
	B: Space Velocity: 15 (h^{-1})
	Liner Velocity: 15 (m/h)
4	Granular Activated Carbon
	Effective Size: 0.4-0.5 (mm)
	Uniformity Coefficient: 1.7
5	Gravel
6	Pump for washing

The pilot plant had two (2) columns to be able to conduct comparative experiment. The thickness of granular activated carbon in the A column was 1.5m and the thickness of granular activated carbon in the B column was 1.0m. ANO and HONJO water treatment plant in Kitakyushu is adopted the same specification as that of the A column. The result of annual pilot test is shown in Figure 66.3 to 66.5.

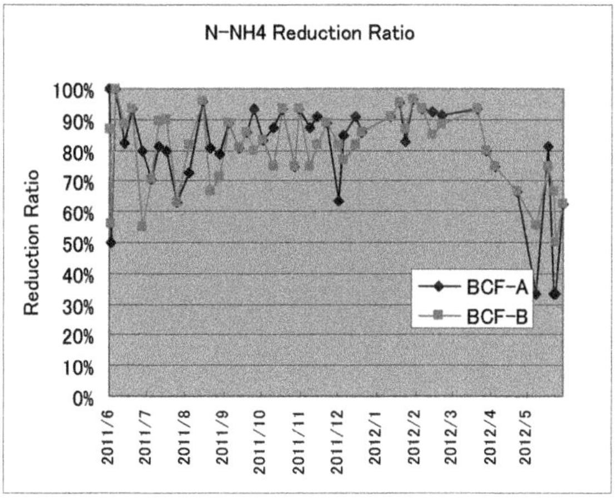

Figure 66.3 N-NH_4 Reduction Ratio

As for ammonium nitrogen, approximately 80% of reduction ratio was obtained at both of columns. As for dissolved manganese, approximately 70% of reduction ratio was obtained at both of columns. As for E260 which is an indicator of organic substances, approximately 30% of reduction ratio was obtained at the A column, and 20% of reduction ratio was obtained at the B column.

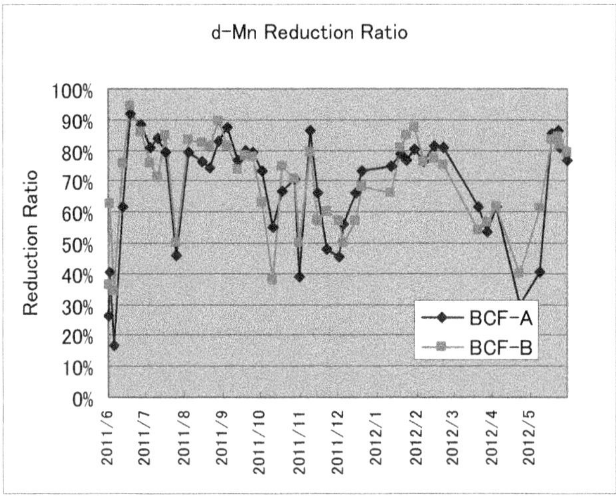

Figure 66.4 d-Mn Reduction Ratio

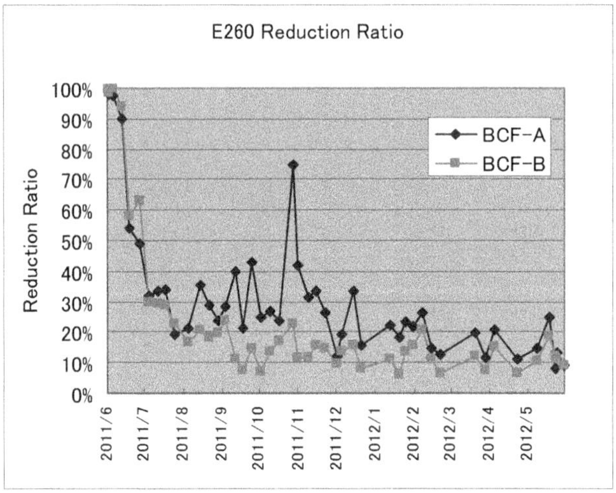

Figure 66.5 E260 Reduction Ratio

Photo 66.4 U-BCF in Vinh Bao

As a result of the survey, U-BCF was confirmed the most effective and advanced water treatment method in Vietnam and will contribute to supply safe water for the residents in Vietnam to use. Therefore, Haiphong Water decided to install U-BCF to remove the ammonia nitrogen, manganese and organic substances in Vinh Bao water treatment plant (5,000m3/d) with own budget. Additionally, Haiphong Water and Kitakyushu Water engaged in mutual cooperation to encourage broad use of U-BCF in the whole of Vietnam in that same time.

CONCLUSIONS

In Japan, we use several advanced treatment methods at water treatment plants in order to supply safer and higher quality of water for consumers, such as ozone, activated carbon, membrane, ultra violet, biological treatment etc. Also, every year, many water supply engineers from developing countries come to Japan to learn advanced treatment methods.

However, it is rare that the advanced treatment methods they learned in Japan are implemented at water treatment plants in their home countries, because of the too expensive construction and operation cost compared to the water tariffs of developing countries.

We believe that bio treatment can be a countermeasure to remove the pollutants in raw water in developing countries.

ACKNOWLEDGEDGMENTS

In this paper, we would like to express our gratitude to all the concerned persons of the Haiphong Water for their great deal of cooperation and useful data.

REFFERENCES

T. Koezuka et al.: Proceedings of the 48th National Waterworks Conference and Symposium (1997), p.172

T. Koezuka et al.: Proceedings of the 49th National Waterworks Conference and Symposium (1998), p.160

M. Ishibashi et al.: Proceedings of the 53rd National Waterworks Conference and Symposium (2002), p.278

M. Nojiri et al.: Proceedings of the 53rd National Waterworks Conference and Symposium (2002), p.280

C. Yamazaki et al: Proceedings of the 54th National Waterworks Conference and Symposium (2003), p.230

A. Aoki et al.: Journal of Japan Water Works Association, Vol. 76, No. 8

Y. Ishimaru et al.: The 2008 Japan-China Water Supply Technology Symposium

Chapter 67

Biofiltration of pig manure: behavior of nitrogen compounds

P. Lessard, G. Aubry*, Y. Gilbert*, Y. LeBihan**, G. Buelna***

**Département de génie civil et de génie des eaux, Pavillon Pouliot, Université Laval, Québec, Canada, G1V 0A6*

***Centre de recherche industrielle du Québec, 333, rue Franquet, Québec, Canada, G1P 4C7*

ABSTRACT

Biofiltration using organic media has long been studied. Even so, no thorough studies looking at inside the process have been made. Three pilot-scale organic media biofilters were operated for 450 days to treat pig manure having different COD/N ratios (approximately 5, 9, 15). Air and liquid sampling was done both on influent and effluent but also at different heights along the biofilter. Satisfactory carbon oxidation and nitrification performances were observed, with chemical oxygen demand (COD) and ammonia (NH_4^+) removals consistently reaching 95%. From 78 to 88% of total nitrogen removal was achieved, meaning denitrification also occurred although the systems were aerated. Nitrate concentration and N_2O emission were related to the nitrogen content in influent more than with the COD/N ratio. All treatment processes occurred mainly in the top 30 cm of the 190 cm biofilters.

Keywords Carbon and nitrogen removal; C/N ratio; denitrification; nitrification; pig manure; trickling biofilters.

INTRODUCTION

In Quebec (Canada), pig production has increased from one million to more than 7 million pigs per year over the last 30 years. Consequently, pig manure handling has become a problem as land available for spreading is insufficient. In areas where pig production is very important, on site manure treatment can be a solution in order to reduce water pollution problems linked with pig manure. Different processes have been proposed and biofilters using organic media have been proven to be efficient for organic matter and nitrogen removal (Buelna *et al.*, 2008).

In fact, biological filters (BFs) using organic material as media have been found efficient for the treatment of domestic and municipal effluents (Lens et al. 1994, Talbot et al. 1996, Roberge et al. 1999) and different types of leachates (Lafrance et al. 1996, Chian and DeWalle, 1996). However, even though many theoretical and practical studies have been done (Couillard, 1994), knowledge on the nitrogen removal pathways is still not well defined.

Garzón-Zúñiga *et al.* (2005) studied nitrogen removal pathways in such a system, at laboratory-scale, by measuring inlet and outlet effluents, both liquid and gaseous. They found almost complete carbon and ammonia removal, while nitrate produced by nitrification was partly denitrified by the system. To optimize denitrification, a better understanding of the nitrogen behavior inside the process was needed. The aim of this study was thus to investigate the nitrogen removal pathways at different depths inside a trickling biofilter treating pig manure. The influence of COD/N ratio on denitrification was also studied.

METHODOLOGY

Biofilters

Three pilot-scale biofilters were built, each of these having a volume of 0.24 m^3, a height of 190 cm and a diameter of 40 cm (Figure 67.1). From top to bottom, the filtering media consisted of 4 layers: Layer 1) 30 cm of pozzolana, Layer 2) 20 cm of a mix of peat and wood chips, Layer 3) 20 cm of wood chips, where a gas evacuation chimney was located, and Layer 4) 120 cm of a mix of peat, wood chips and calcite. The gas evacuation chimney was added in the system in order to avoid clogging by reducing water build-up in Layer 1. It was kept closed during the starting period and was opened when clogging started to occur in biofilters, which corresponded to day 110. To enhance nitrification, calcite was added in Layer 4 as a source of alkalinity for autotrophic biomass. Six holes for gas measurements and media sampling were present all along the depth of the columns (A to F). Pig manure was percolating downflow through the biofilters at a hydraulic rate of 0.017 m^3/m^2 of biofilter area/d. Air was applied upward at a flow rate of 4.5 m^3/m^2 of biofilter area/h.

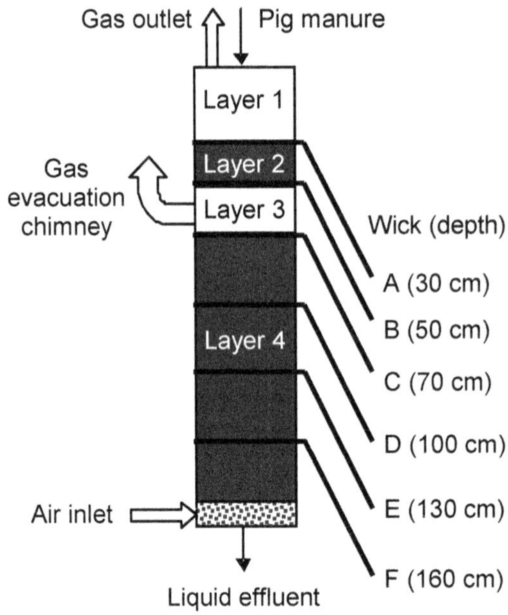

Figure 67.1 Schematic view of the pilot-scale biofilters

Pig manure

The pig manure used in this study was obtained from a local farm in Quebec (Canada). It was collected at the outlet of a physico-chemical pretreatment for solid/liquid separation, consisting in addition of ferric chloride. One biofilter was directly fed with the pretreated pig manure having a COD/N ratio of 9. In order to confirm the optimal conditions for simultaneous nitrification and denitrification, COD/N ratios of 5 and 15 were also tested. These ratios were obtained by diluting 1.5 times the pretreated pig manure, and adding NH_4HCO_3 or a synthetic carbon solution (containing lactic acid, sodium propionate, sodium acetate, ethanol and butyric acid).

Sampling

The biofilters were operated during 425 consecutive days. Liquid influents and effluents were generally sampled every two weeks. In order to draw a vertical profile of nitrogen in the biofilters, 1-meter long fiberglass wicks (Holder *et al.*, 1991) were inserted in between each of the first three layers, and at every 30 cm in Layer 4 (Figure 67.1). Gas measurements (air inlet and gas outlet, Figure 67.1) were done once a week for days 1 to 180, and then occasionally until day 425. Biomass samples were frequently collected from sampling ports located between each wick, at different heights in the biofilters for microorganisms quantification and activity measurements.

Analytical methods

Ammonia (NH_4^+), chemical oxygen demand (COD), biological oxygen demand (BOD_5), pH and alkalinity were analyzed in accordance to *Standard Methods* (APHA et al., 1995). Nitrite and nitrate were determined with an ion chromatograph (Waters Breeze HPLC System). Free ammonia (NH_3), nitrous oxide (N_2O) and carbon dioxide (CO_2) were measured weekly with a gas analyzer using photo-acoustic spectroscopy (Brüel & Kjaer model 1301). Ammonia and nitrite oxidizing bacteria as well as denitrifiers fixed on the media were quantified using real-time quantitative PCR (polymerase chain reaction). Nitrifying and denitrifying activities were evaluated *ex-situ*, respectively by respirometry and acetylene blockage technique (Gilbert *et al.*, 2007, 2008).

RESULT AND DISCUSSION

Global performances

The average characteristics of the biofilters influents and effluents are shown in Table 67.1. Satisfactory purification performances were obtained for organic matter removal, with more than 95% of COD removal and an effluent BOD_5 under 15 mg O_2/L. Since particulate organic nitrogen was removed by pretreatment, all nitrogen was in the form of ammonia in influent, of which more than 95% was consistently removed. Once nitrification was implanted, total nitrogen removal was 79%, 78% and 88% for Biofilter 1, 2 and 3, respectively. Nitrite was not detected in the effluents throughout the experiment.

Organic matter removal and, surprisingly ammonia also, were observed in a predominant way in the 30 first centimeters of biofilter 2 (Figure 67.2). Nitrifiers were able to grow in the same physical space as heterotrophic microrganisms, even when high strength wastewater was applied to the process. This is probably because the latter produces inorganic carbon essential for autotrophic bacteria and creates a biofilm where protective conditions can be found. It was expected that COD oxidation and nitrification would be observed in Layer 1. However, considering the organic load, it is a bit surprising that the level is that high. For Layer 1 of Biofilter 2, volumetric organic loading was 0.8 kg $BOD_5/m^3 \cdot d$, while nitrogen loading was 1.3 g NH_4^+-$N/m^2 \cdot d$. The latter is based on surface area of pozzolana, supposing a specific area of 100 m^2/m^3. Such loadings were higher than typical trickling filter loadings for combined BOD removal and nitrification, which are 0.1 to 0.3 kg $BOD/m^3 \cdot d$ and 0.2 to 1.0 g $TKN/m^2 \cdot d$ for trickling filter without forced aeration (Metcalf and Eddy, 2003). Thus the lower part of the biofilter can be considered as a buffer zone, which will be useful in case of a load increase and variation.

Table 67.1 Main characteristic of biofilters influents and effluents

Biofilter		CODN	COD load ($kg/m^2 \cdot d$)	COD (mg/L)	NH_4^+ load ($kg\ N/m^2 \cdot d$)	NH_4^+ (mg N/L)	NO_3^- (mg N/L)
1	Influent	4	0.22	12700 ±3100[1]	0.050	2870 ± 380	n.d.[2]
	Effluent			335 ± 93		14 ± 18	450 ± 290
2	Influent	9	0.32	18500 ± 4400	0.050	2050 ± 500	n.d.
	Effluent			440 ± 180		20 ± 31	350 ± 230
3	Influent	17	0.40	23000 ± 4500	0.024	1340 ± 280	n.d.
	Effluent			3000 ± 1800		10 ± 6	130 ± 95

1. Average ± standard deviation where n = 14 (COD), 31 ($N\text{-}NH_4^+$), 2 ($N\text{-}NO_3^-$)
2. Non-detectable

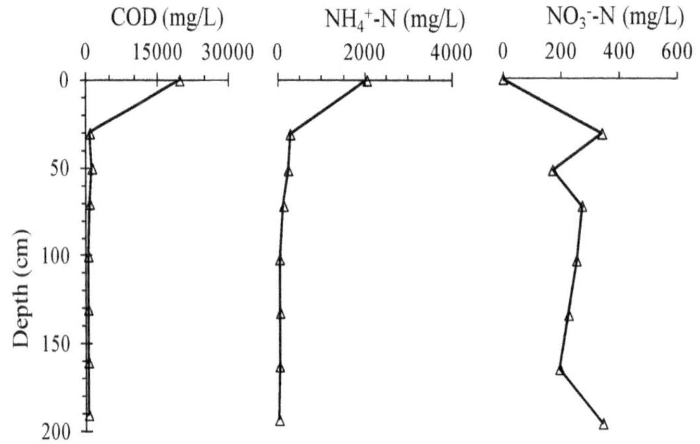

Figure 67.2 Example of COD, NH_4^+ and NO_3^- profile for biofilter 2 (average over 425 days)

Gas behavior

During the starting period (days 1 to 75), ammonia stripping was the main nitrogen removal mechanism observed in the three biofilters, as it can be seen for biofilter 2 in Figure 67.3. It was shown in a previous study that adsorption and filtration on the filtering media also play a role during the starting period (Garzón-Zúñiga *et al.*, 2005), thus explaining the low NH_4^+ concentration in effluent from the beginning of experiment, despite the high nitrogen loads. Ammonia stripping is usually negligible in the trickling biofilter once nitrification has started (Garzón-Zúñiga *et al.*, 2005).

Nitrous oxide concentrations started to be measured in the gaseous effluents as soon as nitrification began (around day 60). The N_2O reductase enzyme that is necessary to convert N_2O to N_2 in the denitrification process can be inhibited when high oxygen concentration prevails in the gas-phase (Berge *et al.*, 2006). Oxygen concentrations in the gas-phase of the biofilters varied between 17 and 21%, which suggests some nitrous oxide production was due to partial denitrification. Nevertheless, N_2O can also be a by-product of nitrification, and the exact cause of its production cannot be determined without conducting a study with labelled nitrogen species (Berge *et al.*, 2006). Other parameters can affect the production of N_2O, such as hydraulic retention time, dissolved oxygen, pH, water temperature and C/N ratio (Kishida *et al.*, 2004).

a.

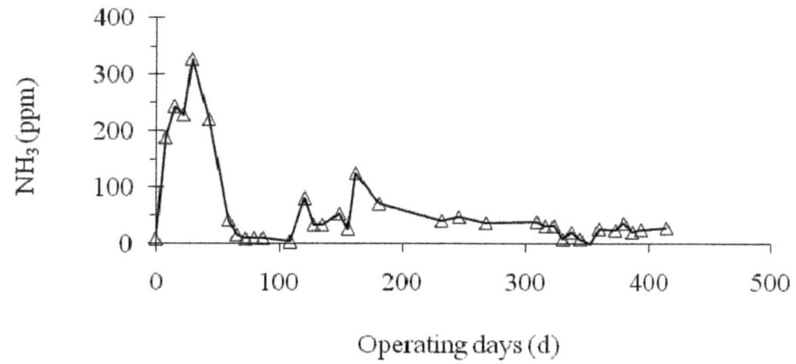

b.

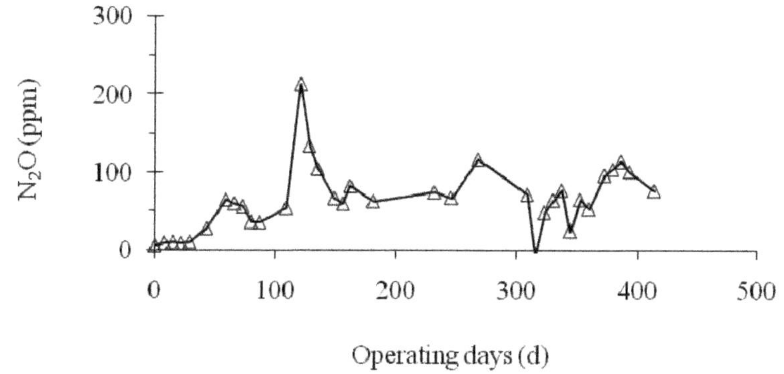

Figure 67.3 Free ammonia and nitrous oxide concentrations in gas outlet of biofilter 2

Nitrification and denitrification

In Biofilter 1 and 2, nitrate started to be detected in effluents around day 60, increased until day 93 and then a cycling trend was observed (Figure 67.4). Despite the fact that nitrate was first detected on day 86 in effluent of Biofilter 3, results from inside the column confirmed that nitrification also started around day 60 in this biofilter. The cycling trend of NO_3^- did not seem to be linked to nitrification, since no ammonia accumulation was noted in effluents after day 90. Consequently, temporary decreases in denitrification could be related to sloughing, usually known as the dominant process of biofilm detachment in highly loaded trickling filters (Morgenroth and Wilderer, 2000). When sloughing occurs, the biofilm might suddenly lose some denitrifiers, which would explain nitrate increase. Losing biofilm would also expose the deep layers to the gas phase (oxygen), which would lead to a decrease in denitrification. Further experiment should confirm if this behavior (sloughing) stabilizes with time, suggesting equilibrium is attained.

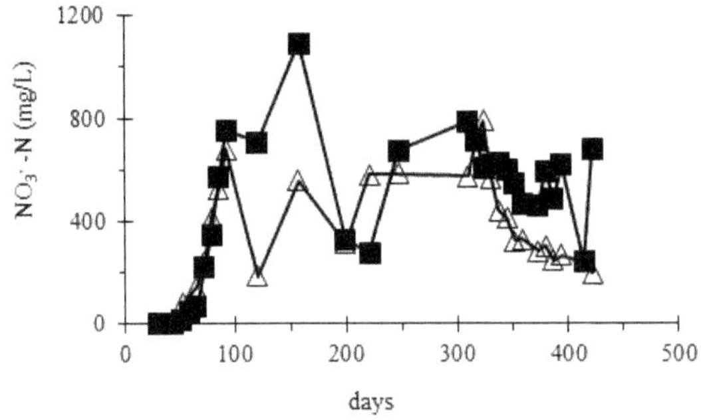

Figure 67.4 Nitrate concentration in liquid effluents (■ Biofilter 1, △ Biofilter 2)

Nitrogen mass balances

As shown on Figure 67.5, the two major N forms found in biofilters liquid and gaseous effluents are nitrate and nitrous oxide. Berge *et al.* (2006) found that N_2O corresponded to 16% of nitrogen formed under low COD/N conditions, while Kishida *et al.* (2004) found 1.71% for a BOD_5/TN of 4.5 and 17.7% for a BOD_5/TN of 2.6. N_2O is a concern because it is a potent greenhouse gas. Airflow reduction might help to reduce N_2O emission, as long as sufficient aeration is still provided for nitrification.

From our calculations, more than half of the entering nitrogen was not recovered, which is however quite common in literature (Berge *et al.*, 2006, Gupta and Sharma, 1996). Firstly, part of this missing nitrogen should be attributed to biomass assimilation, which is difficult to calculate in such systems. N_2 production is also likely to explain another part of the missing nitrogen, as N_2 was found in the gaseous effluent for a previous study (Garzón-Zúñiga *et al.*, 2005). Known as the final product of complete denitrification, N_2 production measurement was impossible during this experiment because of its large presence in the system caused by aeration.

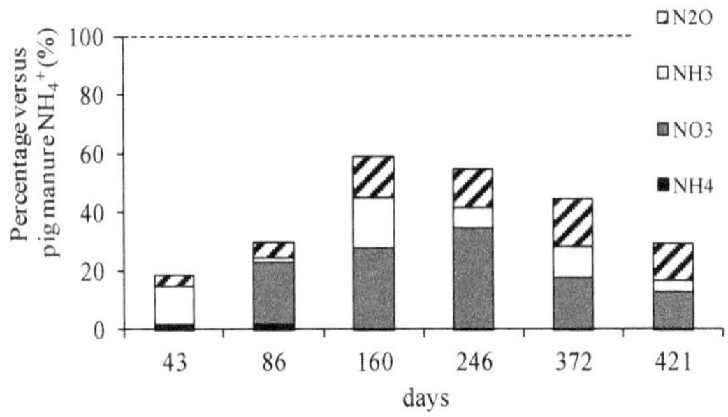

Figure 67.5 Mass balances at different time period over the sampling period (biofilter 2)

Influence of COD/N on nitrogen removal

Choi *et al.* (2004) stated that the COD/N ratio should be higher than 6 to achieve high nitrogen removal efficiency via nitrification and denitrification pathway. A pretty high total nitrogen removal was reached in the trickling biofilter, even under low COD/N ratio, but no relation could be found between this removal and the COD/N ratio (Figure 67.6).

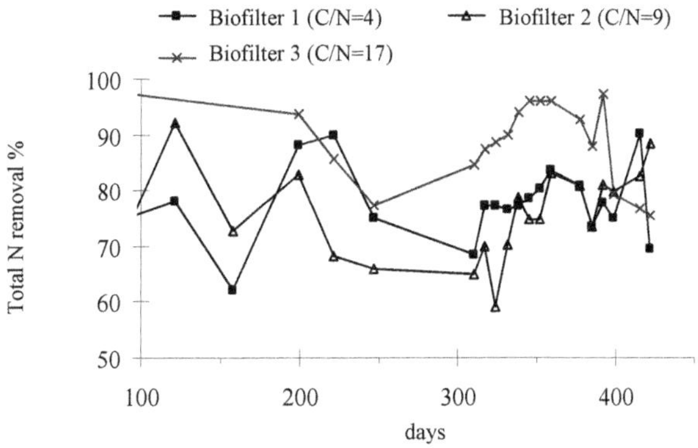

Figure 67.6 Nitrogen removal efficiency for different C/N ratios

Generally, N_2O emissions increase at lower COD/N conditions (Choi *et al.*, 2004; Kishida *et al.*, 2004). This was not observed in the trickling biofilter, which led to similar N_2O proportions for all the COD/N ratios. N_2O concentration in the gas outlet seems to be related with the influent nitrogen content more than with COD/N conditions (data not shown). The same trend was observed for nitrate concentration.

Nitrogen behavior inside biofilter

Figure 67.2 already presented results obtained inside Biofilter 2. Vertical profiles of COD and NH_4^+ did not vary much in time, while nitrate behavior offered more disparities. Similar trends were found inside Biofilter 1, but some different behaviors were observed inside Biofilter 3 (high C/N). In the latter, COD increased along the biofilter depth, probably due to wood chips and/or peat degradation caused by pH increase.

Figure 67.2 shows that oxidation of organic matter and nitrification occurred simultaneously in Layer 1 (0 to 30 cm), where the highest COD and NH_4^+ removal rates were observed. It is probable that nitrification was inhibited in Layer 3 (50 to 70 cm) due to the sharp pH drop (not shown). The pH decreased to a minimal value of 4.0, a level known as being inhibitory to nitrification activity (Metcalf and Eddy, 2003). Local inhibition of nitrification did not affect ammonia concentration in the liquid effluent, but might be partly responsible for incomplete denitrification. Wang *et al.* (2005) stated that a pH of 7.5 is optimal for simultaneous nitrification and denitrification. The pH returns to neutrality after entering Layer 4, because of the presence of calcite. This pH behavior suggests that calcite should be added in the top layers, to prevent pH drop that can cause nitrification and denitrification inhibition. Calcite is used as a source of alkalinity for nitrifying biomass. Nitrification consumes 7.14 g $CaCO_3$/g NH_4^+-N converted, while denitrification produces 3.57 g $CaCO_3$/g NO_3^--N reduced. Alkalinity profiles (data not shown) were very similar to COD and ammonia nitrogen removal, with most of it being consumed in Layer 1.

Nitrate evolution inside the biofilters varied from one sampling campaign to the other. The highest concentrations of NO_3^- were usually measured under Layers 1 or 2 for Biofilters 2 and 3. Because of the high NH_4^+ load applied to Biofilter 1, the remaining ammonia was nitrified in Layer 4. In this biofilter, the highest nitrate concentrations were measured in the final effluent, because not much denitrification could occur in Layer 4 due to the lack of organic carbon. Nitrate concentration in the effluent tended to be a little higher than its concentration in wicks F (160 cm). It has been observed by other researchers that the use of fiberglass wicks might overestimate denitrification, because of a more anoxic environment caused by water accumulation around the wicks (Holder *et al.*, 1991).

Although no nitrite was detected at any time in the effluents, nitrite accumulations were temporarily observed in wicks A (30 cm), B (50 cm) and C (70 cm) samples of the three biofilters during the nitrification start-up (days 40 to 80, data not shown). Nitrite values reached up to 600 mg NO_2^--N/L, before decreasing to negligible values.

The highest N_2O production was observed in Layer 1, which was also true for CO_2 production (Figure 67.7). These observations suggest that the biomass mainly colonized mainly the top of the biofilter, which was confirmed by microbiological monitoring. Results showed that the nitrifying and denitrifying biomasses mainly colonized Layer 1 and that they can be found together, having colonized the same ecological niche (Gilbert *et al.*, 2007). When investigating the biofilm thickness in a trickling filter, Persson *et al.* (2002) effectively found that it decreases with depth. Rodgers *et al.* (2005) also found that biomass growth occurred mainly at the top of the first sand layer in a stratified sand filter. Moreover, potential nitrifying and denitrifying activities measured *ex-situ* were much greater than activities observed inside the trickling biofilters, suggesting the possibility to enhance nitrogen removal (Gilbert *et al.*, 2007, 2008).

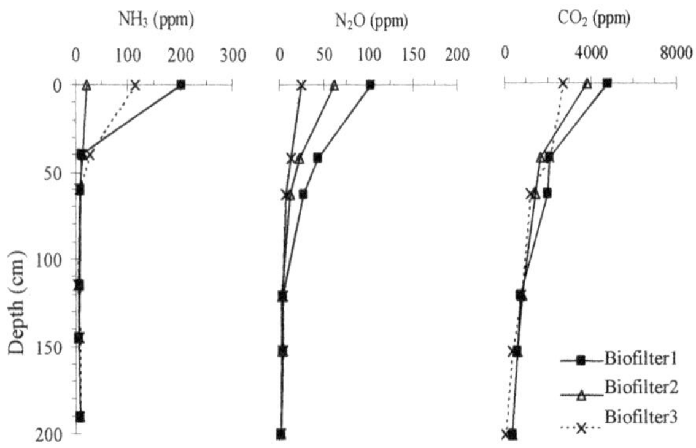

Figure 67.7 Gas profiles within the biofilters at day 65

CONCLUSIONS

The trickling biofilter using organic media has already proved its efficiency at removing carbon and ammonia from pig manure. In this study, it achieved 78% of total nitrogen removal under a COD/N ratio of 9. By investigating the nitrogen behavior inside the biofilter, it has been shown that nitrification and denitrification were primarily occurring in the top of the filter, along with organic matter oxidation. The bottom layer of the filtering media is thus not really used and simply acts as a buffer zone in case of load increases. It would eventually be possible to evaluate the possibility of using a shorter biofilter depending on the loading variations anticipated. Another improvement for total nitrogen removal would be to add calcite in the top layers of the biofilter, in order to prevent pH drops that might eventually be responsible for nitrification and denitrification inhibition.

ACKNOWLEDGEMENTS

The authors would like to thank Natural Sciences and Engineering Research Council of Canada (NSERC), Fonds Québécois de la Recherche sur la Nature et les Technologies (FQRNT), Centre de Recherche Industrielle du Québec (CRIQ) and Canada Foundation for Innovation (CFI), for their financial support

REFERENCES

APHA, AWWA, WPCF. *Standard Methods for the Examination of Water and Wastewater* (1995). 19th ed., American Public Health Association/AWWA/Water Environment Federation, Washington DC, USA.

Berge N.D., Reinhart D.R., Dietz J., Townsend T. (2006). In situ ammonia removal in bioreactor landfill leachate. *Waste Management*, 26, 334-343.

Buelna G., Dubé R., Turgeon N. (2008). Pig manure treatment by organic biofiltration. *Desalination*, 231: 297-304.

Chian E.S.K., DeWalle F.B. (1996). Sanitary landfill leachates and their treatment. *J. Environ. Eng.*, **105**, 411-431.

Choi E., Yun Z., Chung T.H. (2004). High strength nitrogen removal from nightsoil and piggery wastes. *Water Sci. Technol.*, **49**(5-6), 97-104.

Couillard D. (1994). The use of peat in wastewater treatment. *Water Res.*, **28**, 1261- 1274.

Garzón-Zúñiga M., Lessard P., Aubry G., Buelna G. (2005). Nitrogen elimination mechanisms in an aerated peat biofilter treating pig manure. *Environmental Technol.*, **26**(4), 361-371.

Gilbert, Y., Le Bihan, Y., Aubry, G., Veillette, M., Duchaine, C. and Lessard P. 2008. Microbiological and molecular characterization of denitrification in an organic biofilter treating pig manure. *Biores. Technol.*, 99: 4495-4502.

Gilbert, Y., Lessard P., Le Bihan, Y., Aubry, G., Duchaine, C. and Veillette, M. 2007. Nitrification monitoring in a biofilter treating pig manure. *J. Chem. Technol. Biotechnol.*, 82: 905-912.

Gupta S.K., Sharma R. (1996). Biological oxidation of high strength nitrogenous wastewater. *Water Res.*, **30**(3), 593-600.

Holder M., Brown K.W., Thomas J.C., Zabcik D., Murray H.E. (1991). Capillary-wick unsaturated zone soil pore water sampler. *SSSAJ*, Division S1-Soil physics. **55**(5), 1195-1202.

Kishida N., Kim J.H., Kimochi Y., Nishimura O., Sasaki H., Sudo R. (2004). Effect of C/N ratio on nitrous oxide emission from swine wastewater treatment process. *Water Sci. Technol.*, **49**(5-6), 359-365.

Lafrance C., Lessard P., Buelna G. (1996). Évaluation de la filtration sur tourbe et compost pour le traitement d'une usine de compostage des résidus verts. *Can. J. Civ. Eng.*, **23**, 1041-1050.

Lens P.N., Vochten P.M., Speleers L., Verstraete W.H. (1994). Direct treatment of domestic wastewater by percolation over peat, bark and woodchips. *Water Res.*, **28**, 17-26.

Metcalf and Eddy (2003). *Wastewater engineering: Treatment and reuse.* 4th edn, McGraw-Hill, New-York.

Morgenroth E., Wilderer P.A. (2000). Influence of detachment mechanisms on competition in biofilms. *Water Res.*, **34**(2), 417-426.

Persson F., Wik T., SÖrenson F., Hermansson M. (2002). Distribution and activity of ammonia oxidizing bacteria in a large full-scale trickling filter. *Water Res.*, **36**, 1439-1448.

Roberge G., Blais J. F., Mercier G. (1999). Enlèvement du phosphore des eaux usées par traitement à base de tourbe dopée aux boues rouges. *Can. J. Chem. Eng.*, **77**, 1185-1194.

Rodgers M., Healy M.G., Mulqueen J. (2005). Organic carbon removal and nitrificationof high strength wastewaters using stratified sand filters. *Water Res.*, **39**, 3279-3286.

Talbot P., Bélanger G., Pelletier M., Laliberté G., Arcand Y. (1996). Development of a biofilter using an organic medium for on-site wastewater treatment. *Water Sci. Technol.*, **34**, 435-441.

Wang B., He S., Wang L., Shuo L. (2005). Simultaneous nitrification and de-nitrification in MBR. *Water Sci. and Technol.*, **52**(10-11), 435-442.

Chapter 68

Exploring sand and bentonite - enhanced sand as filter media for nitrate removal

R. Malini, M. R. Sudhakar**
**Centre for Sustainable Technologies, Indian Institute of Science, Bangalore-560012, India.*
***Department of Civil Engineering, Indian Institute of Science, Bangalore-560012, India.*

ABSTRACT

Increasing nitrate concentrations in ground water is deleterious to human health as ingestion of such water can cause methemoglobinemia in infants and even cancer in adults (desirable limit for nitrate as NO_3 – 45 mg/L, IS code 10500-1991). Excess nitrate concentrations in ground water is contributed by reason being disposal of sewage and excessive use of fertilizers. Though numerous technologies such as reverse osmosis, ion exchange, electro-dialysis, permeable reactive barriers using zerovalent iron etc exists, nitrate removal continues to be one of challenging issue as nitrate ion is highly mobile within the soil strata. The tapping the denitrification potential of soil denitrifiers which are inherently available in the soil matrix is the most sustainable approach to mitigate accumulation of nitrate in ground water. The *insitu* denitrification of sand and bentonite enhanced sand (bentonite content = 5%) in presence of easily assimilable organic carbon such as ethanol was studied. Batch studies showed that nitrate reduction by sand follows first order kinetics with a rate constant $5.3\text{x}10^{-2}$ hr^{-1} and rate constant 4.3 $\text{x}10^{-2}$ hr^{-1} was obtained for bentonite-enhanced sand (BS) at 25°C. Filter columns (height = 5 cm and diameter = 8.2 cm) were constructed using sand and bentonite-enhanced sand as filter media. The filtration rate through both the filter columns was maintained at average value of 2.60 cm/h. The nitrate removal rates through both the filter media was assessed for solution containing 22.6 mg NO_3–N/L concentrations while keeping C/N mass ratio as 3. For sand filter column, the nitrate removal efficiency reached the average value of 97.6% after passing 50 pore volumes of the nitrate solution. For bentonite-enhanced sand filter column, the average nitrate removal efficiency was 83.5%.The time required for effective operation for sand filter bed was 100 hours, while bentonite-enhanced sand filter bed did not require any maturation period as that of sand filter bed for effective performance because the presence of micropores in bentonite increases the hydraulic retention time of the solution inside the filter bed.

Keywords Biodenitrification; Bentonite; Slow sand filtration; Drinking water

INTRODUCTION

Nitrate concentration has dramatically increased in groundwater, mainly due to anthropogenic activities such as unabated use of fertilizers and improper disposal of sewage. Although NO_3^-itself is relatively non-toxic, NO_2^-, derived by reduction of NO_3^- in the digestive system can cause methemoglobinemia in infants, liver damage, and even cancer (Walton, 1951; Mirvish, 1985). As per IS 10500-1991, the maximum allowable limit for NO_3-N is 10mg/L. Though numerous technologies such as reverse osmosis, ion exchange, electro-dialysis, etc exists, nitrate removal continues to be one of the challenging problems. These processes have been proven effective in NO_3–N removal. However, disadvantages, including poor selectivity for NO_3–N, concentrated waste disposal issues, cost, and susceptibility to fouling (in the case of reverse osmosis) have fed the search for alternative NO_3–Nremoval technologies (Aslan & Cakici, 2007). *Insitu* denitrification is the most useful process to remove NO_3–N from water and wastewater. Biological denitrification is the nitrate reduction process, where the nitrate is the terminal electron acceptor instead of oxygen, with the micro-organism as the electron donor and the generated energy is utilized for cell growth and maintenance. In this process, microorganisms first reduce nitrate to nitrite and then produce nitric oxide, nitrous oxide, and nitrogen gas. The pathway for nitrate reduction is $NO_3 \rightarrow NO_2 \rightarrow NO \rightarrow N_2O \rightarrow N_2$(Aslan & Cakici, 2007).

Slow sand filtration has traditionally been employed as one of the effective water treatment techniques for the removal of removal of carbon (C), pathogenic bacteria, protozoan parasite, and suspended solids (Aslan & Cakici, 2007). Stevik et al(2004) showed that the variability in sizes, surface texture and charge of porous media employed in the slow sand filter are considered to greatly influence the adhesion process. Smaller particle sizes expose a larger surface area compared to coarse particles and hence provide greater adhesion sites. Huysman and Verstraete (1993) and Fletcher and Loebhave(1996) observed that presence of clay particles in soil media enhances the adhesion of bacteria due to the charge and surface characteristics of the clay minerals.

Bentonite, a type of monitmorillonite clay mineral exposes very large surface area (800 m^2/g), has very small particle size (< 2 microns) and is characterized by charged surface owing to presence of isomorphous substitution within the clay lattice and presence of broken bonds at the particle edges(M. E. Essington, 2003). In this work, bentonite-sand mix was used as porous media for the slow sand filter and its performance was compared against the conventional filter where the sand is used as the porous media.

METHODOLOGY

The physical properties of sand and bentonite enhanced sand (BS) such as pH, electric conductivity, cation exchange capacity, specific gravity, grain size distribution, compaction characteristics and hydraulic conductivity were determined according to Indian Standards IS 2720 and are listed in Table 68.1.

Table 68.1 Properties of sand and bentonite enhanced sand

Parameter	Sand	BS
Physico-chemical properties		
pH	7.825	8.765
Electrical conductivity (mS/cm)	0.1375	0.6715
Porewater salinity (mg/L)	89.3	436.0
Specific gravity	2.6163	2.6256
Cation Exchange Capacity (meq/100g)	2.878	5.334
Index and Engineering properties		
Grain size distribution		
Percent Coarse Sand (4.75-2.00mm)	1.42	0.58
Percent Medium Sand (2.00-0.425mm)	63.66	59.32
Percent Fine Sand (0.425-0.075mm	32.78	22.28
Percent Silt/Clay (<0.075mm)	2.14	7.82
Effective Size, D_{10} (mm)	0.17	0.104
Coefficient of Curvature, C_c	1.15	1.49
Uniformity Coefficient, C_u	4.35	4.98
Coefficient of Permeability, k (cm/s)	2.36×10^{-3}	4.3×10^{-6}

To assess the denitrification potential of sand and BS, batch experiments was carried out where sand and bentonite enhanced sand was amended with nitrate solution and nutrient solution at a water content of 50% and were incubated at 25°C for different intervals of time (6, 12, 18, 24, 36 and 48h) in a bacteriological incubator (Servell Instruments). At the end of each incubation period, the soil specimens are centrifuged at 6800 rpm for 12 min to separate the soil and solution phase. The residual nitrate, nitrite and ammonium ion concentrations in the extracted soil solutions were determined.

The concentration of nitrate and nitrite in the filtrate was determined by ion chromatography system configured with hydroxyl based anion retention column (Dionex ICS 2000). The carbonate and bicarbonate concentrations were measured using Metrohm 877 TitrinoPlus Automatic Titrator. Dissolved oxygen (DO) in water samples was measured using DO probe connected to handheld portable meter (Cyberscan PD300). The ammonium ions in water samples and soil extracts were measured by phenate method using UV visible spectrometer (Thermo Scientific Evolution 201) at a wavelength of 640nm (APHA, 2005). The chemical oxygen demand (COD) was also measured by dichromate oxidation method (APHA, 2005).

Both batch and column experiments were carried out with distilled water spiked with nitrate and nutrient solution containing nitrate and carbon. The nitrate in nutrient solution was prepared from AR grade potassium nitrate salt. Ethanol was used as source of carbon. The carbon concentration in the nutrient solution was varied as per the desired C/N ratio and the volume of ethanol to be added to 1L of solution was calculated using the following formula:

$$Volume of ethanol\ added\ (mL) = 5.49 \times 10^{-4} \times C_0 \times \frac{C}{N} ratio$$

where, C_0 is the initial concentration of nitrate in the nutrient solution and C/N ratio is the ratio of carbon to nitrogen on mass basis. It was found from earlier work that when ethanol is used as carbon source, the optimum C/N ratio for denitrification is 3 (data not shown). In this work, the nitrate-nitrogen concentration used was 22.6 mg/L and C/N used is 3.

RESULTS AND DISCUSSION

The properties of filter columns constructed using sand and bentonite-enhanced sand as filter media are given in Table 68.2. The sand and bentonite enhanced sand was molded and equilibrated to initial water content of 5 and 17% respectively. The dry density used for packing was 1.5 $Mg/m^3$and the height of the compacted sand and BS soils was 4cm.The flow rate was controlled at 120 mL/ h through both filter columns. The filtration rate through both the filter columns was maintained at average value of 2.60 cm/h.

Table 68.2 Sand and BS filter bed properties

Parameter	Sand	BS
Height of the filter bed (cm)	4	4
Diameter of the filter bed (cm)	8.2	8.2
Mass of dry soil in filter bed(g)	317.0	308.6
Filter bed Volume(cm^3)	211.3	211.3
Packing density (g/cm^3)	1.5	1.5
Specific gravity of sand(g/cm^3)	2.62	2.625
Void ratio	0.747	0.745
Porosity	0.427	0.4279
Pore Volume(cm^3)	90.33	90.23

The permeating fluid was distilled water spiked with nitrate (initial concentration of 100mg/L) and ethanol at a C/N mass ratio of 3. The filter bed was allowed to mature until the colony of denitrifying bacteria was well established and the performance of the filter bed improved. This time is called maturation period. The maturation period varied depending on the permeability of filter bed and availability of the nutrients to denitrifying bacteria. The changes in nitrate and nitrite concentrations of the filtrate as function of number of pore volumes passed through the filter are shown in Figure 68.1 and 68.2 respectively for sand and BS filter columns respectively. The concentration of nitrate concentration continues to decrease until there was significant build up of head loss formed due to microbial growth on soil particles and accumulation of gas molecules. When the filtration rate reduced, the backwashing was performed by nitrate solution against gravity through the alternate backwash flow system. After backwash, there is distinct rise in the concentration of nitrate than the average nitrate concentration in the filtrate. This may be due to washing away of bacterial cells attached to soil during backwashing. Once the filter bed is again matured, the performance of the filter improves and higher nitrate removal efficiency was observed.

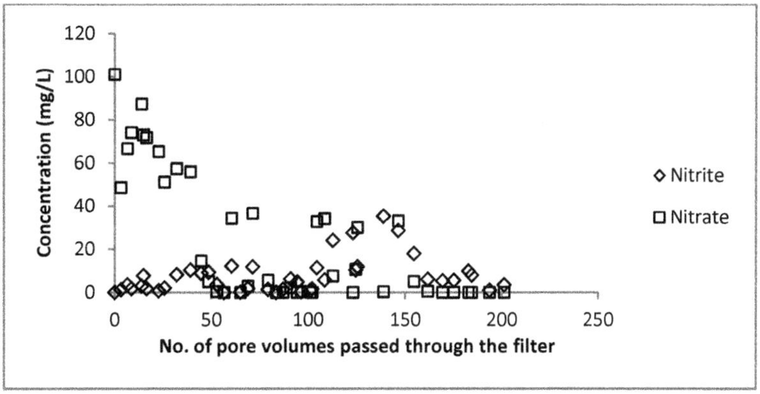

Figure 68.1 Variations in concentration of nitrate and nitrite as a function of number of pore volumes of nitrate solution through the sand filter bed

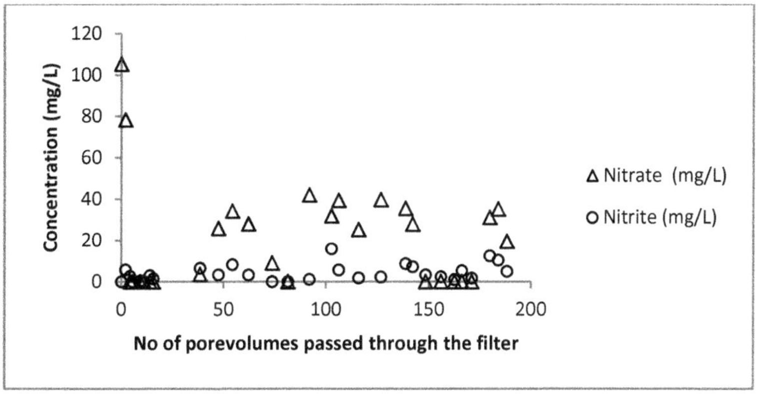

Figure 68.2 Variations in concentration of nitrate and nitrite as a function of number of pore volumes of nitrate solution through the BS filter bed

The nitrate-nitrogen removal efficiency was calculated using the following relation: $\eta = \left(\frac{C_0 - C_t}{C_0}\right) \times 100$

where η is the removal efficiency at the end of the incubation period t, C0 is the initial concentration of the reactant (mg/L) and C_t is the concentration of the reactant (mg/L) after incubation time t. The efficiency of nitrate removal process through both the filter media was assessed for solution containing 22.6 mg NO_3–N/L concentrations while keeping C/N mass ratio as 3. For sand filter column, the nitrate removal efficiency reached the average value of 97.6% after passing 50 pore volumes of the nitrate solution. For bentonite-enhanced sand filter column, the average nitrate removal efficiency was 83.5%.The time required for effective operation for sand filter bed was 100 hours, while bentonite-enhanced sand filter bed did not require any maturation period as that of sand filter bed for effective performance because the presence of micropores in bentonite increases the hydraulic retention time of the solution inside the filter bed. In both these filters where external carbon source is used, there was a constant build of bacterial cells and accumulation of gas bubbles and frequent backwashing was performed.

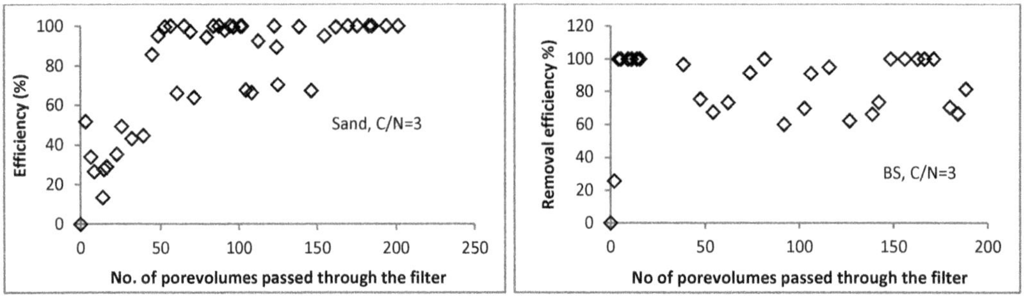

Figure 68.3 Variations in filter bed performance as a function of number of pore volumes of nitrate solution through the Sand and BS filter bed at C/N=3

In another set of column experiments, the nitrate solution was used as permeating fluid without any carbon addition. It was found that the average nitrate removal efficiency was 5.2% for sand filter bed. This may be due to low mean residence time of 4.88h due to high filtration rate of 0.82cm/h. In the bentonite enhanced sand filter, when bentonite content in sand was as low as 5%, the clay present in the filter was washed away. This may be due to higher filtration rate, hence the nitrate removal study could not be performed. To promote the inclusion of bentonite in the BS filter bed, bentonite content in sand was increased to 10%. This filter bed yielded a high mean residence time of 22.22h and filtration rate of 0.18cm/h. The nitrate removal efficiency increased with the volume of filtrate passed through the BS filter bed and attained an average value of 85.1% after the passage of 39 pore volumes of filtrate. This filter bed was operated until 50 pore volumes of nitrate solution was passed and the backwashing was not performed (Figure 68.5).When the filter was operated without external carbon source, the maturation period required was 105h.

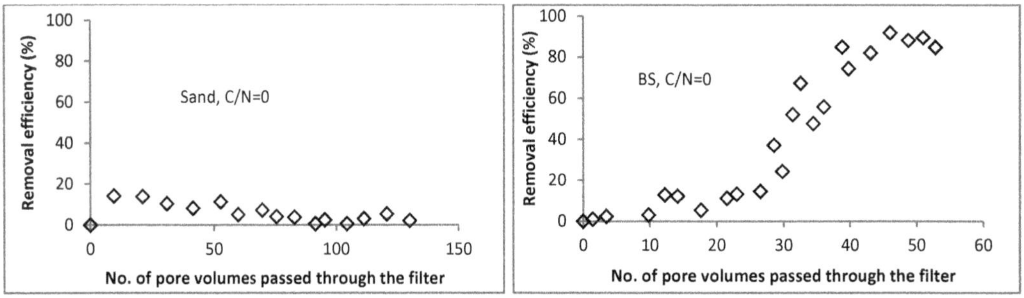

Figure 68.4 Variations in filter bed performance as a function of number of pore volumes of nitrate solution through the sand and BS filter bed at C/N=0

In both batch and column experiments, the external inoculum rich in denitrifying bacteria was not employed; instead, indigenous micro-organisms present in the soil was utilized for the process of nitrate removal. When nitrate solution (without external carbon) was used as permeating fluid, the filtration rate was low and the mean residence time was 4.88 h for sand filter bed and 22.22h for BS filter bed. When nutrient solution (with nitrate and carbon) was used as permeating fluid, the filtration rate was high and mean residence time was 1.76h. Since the external carbon was added to promote the nitrate removal process,excess unconsumed carbon was present in the filtrate with the COD of 52.3 and 47.3 mg/L for sand and BS filter bed respectively. To counteract this, the carbon filter with activated charcoal as porous media was attached and after the passage through the carbon filter, the carbon was adsorded onto the filter and the final mean COD was found be less than 5 mg/L. The final mean COD in the filtrate of the carbon filter was 4.92 and 4.39mg /L for sand and BS filter bed respectively. The COD removal efficiency of the carbon filter was found to be 90.59 and 90.72% for sand and BS filter bed respectively. The residual nitrate concentration data obtained from batch studies were fitted into rate equations and the process of nitrate removal followed first-order kinetics. The batch experiments conducted for nitrate removal for sand and BS soils show that the biochemical process of nitrate removal follows the first order kinetics with rate constant of 5.3×10^{-2} h^{-1} for sand and 4.3×10^{-2} h^{-1} for BS soil at 25°C when ethanol was used as carbon source at the carbon-nitrogen mass ratio of 3. The time needed for the reactant concentration to decrease by half of its initial concentration was 16.1 hours for sand and 13.1 hours for bentonite enhanced sand. Similarly, rate constant was found to 2.9 and 3.4×10^{-2} $h^{-1}$for sand and bentonite enhanced sand respectively at 25°Cwhen there was no external carbon addition. The time needed for the reactant concentration to decrease by half of its initial concentration was 23.9 hours for sand and 20.4 hours for bentonite enhanced sand.

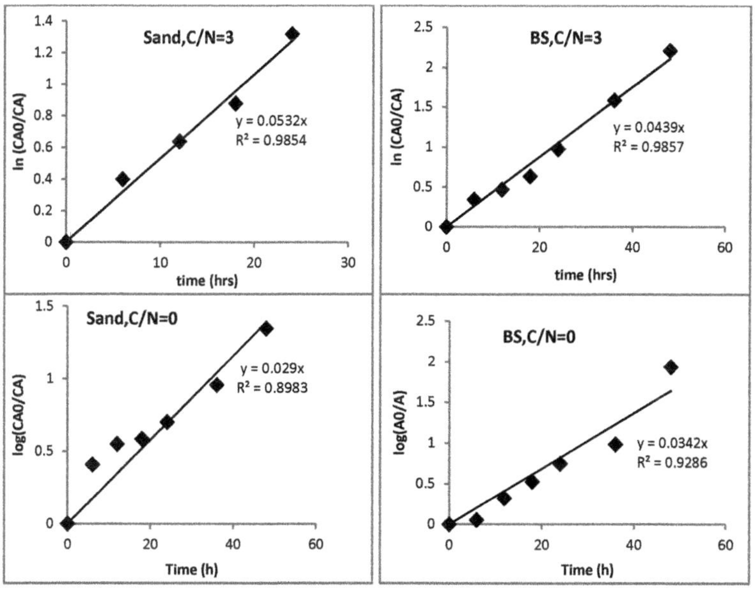

Figure 68.5 First order kinetics plot for nitrate removal for sand and BS soil

CONCLUSIONS

The main conclusions of this work are:

(1) The experimental results presented in this work show that the SSF with sand as porous media can be used as used for nitrate removal process only when external carbon source was added.

(2) The bentonite enhanced sand filters can be used for nitrate removal process with/without external carbon. When the filter was operated without external carbon source, the high maturation period was observed.

(3) High filtration rate was observed for both sand and bentonite enhanced sand filters, when external carbon was added; Without external carbon, sand filter does not impact the nitrate removal process; the bentonite enhanced sand filter require extremely low filtration rate and high maturation period for nitrate removal process.

(4) The batch experiments showed that rate constant for the biochemical process of denitrification obtained for bentonite enhanced sand was higher than that of sand; hence the bentonite enhanced sand has potential advantage over sand as porous media for the slow sand filter.

REFERENCES

1. APHA, (2005) *Standard methods for the examination of water and wastewater*, 21st Edition, A. D. Eaton, L. S. Clesceri, E. W. Rice, A. E. Greenberg, M. H. Franson (Editors) , American Public Health Association (2005) Washington D.C.
2. D. E. Fontes, A. L. Mills, G. M Hornberger. and J. S. Herman, (1991) Physical and chemical factors influencing the transport of micro-organisms through porous media, *Appl. Environ. Microbiol.* 57,2473-2481.
3. F. Huysman and W. Verstraete, (1993) Water facilitated transport of bacteria in unsaturated soil columns: influence of cell surface hydrophobicity and soil properties, *Soil Biol. Biochem.* 25, 83-90.
4. G. Walton, (1951) Survey of literature relating to infant methemoglobinemia due to nitrate-contaminated water, *Am J Public Health*, 41, 986–995.
5. IS 2720, Bureau of Indian Standards (2002) New Delhi
6. M. E. Essington (2003), *Soil and water chemistry; an integrative approach*, CRC Press, New York.
7. M. Fletcher and G. I. Loeb, (1996)Influence of sub-stratum characteristics on the attachment of a marine pseudomnad to soil surfaces, *App. Environ. Microbiol.* 37, 67-72.
8. S.Aslan, H.Cakici,(2007) Biological denitrification of drinking water in a slow sand filter, *J Haz Mat* 148, 253–258
9. S.S. Mirvish, Gastric cancer and salivary nitrate and nitrite, *Nature*, 315 (1985), pp. 461–462.
10. T. K. Stevik, K. Aa, G. Ausland and J. F. Hanssen, (2004) Retention and removal of pathogenic bacteria in wastewater percolating through porous media: a review, *Water Research*, 38 1355-1367.
11. Y. Tan, W. J. Bond and D. M. Griffen, (1992) Transport of bacteria during unsteady unsaturated soil water flow, *Soil Sci. Soc. Am. J.* 56 1331-1340.

Part IX

River Bank Filtration and Groundwater Recharge

Chapter 69

Reduction of the formation potentials of trihalomethanes (THMs) and haloacetic acids (HAAs) in reclaimed water by soil aquifer treatment

S. Rahmawati, X. Wang*, S. Phattarapattamawong**, R. Suzuki*, S. Echigo*, Y. Takabe*, S. Itoh**

**Department of Environmental Engineering, Kyoto University, Japan*

***Department of Environmental Engineering, Faculty of Engineering, King Mongkut's University of Technology Thonburi, Thailand*

ABSTRACT

Soil aquifer treatment (SAT) has been known as one option for water reclamation. However disinfectants can react with organic matter to form disinfection byproducts (DBPs) which maybe a threat to human health. Lab-scale column experiments with different soils and retention time were performed to evaluate the feasibility of soil aquifer treatment (SAT) on the reduction of formation potentials (FPs) of trihalomethanes (THMs) and haloacetic acids (HAAs). The sand column with a hydraulic retention time (HRT) of 7 days showed the highest reduction level of THM and HAA FPs. A positive correlation between the FPs and UV254 also found. Therefore UV254 could be used as an index to monitor the THM and HAA FPs in SAT effluent.

Keywords Trihalomethanes (THMs); haloacetic acids (HAAs); soil aquifer treatment (SAT)

INTRODUCTION

Increasing fresh water demand caused by population growth, industrial development, limitation of fresh water sources, and climate change are causing water scarcity around the world. To deal with this situation, effective methods of water resource management are required such as water recycling and reclamation for direct and indirect potable uses. One of the options for water reclamation is soil aquifer treatment (SAT). SAT has been recognized as a useful treatment technology because of its ability to improve water quality, low cost, and compatibility with other wastewater treatment options.

In groundwater recovery for potable uses, disinfectants (eg. chlorine) are almost always added to inactivate pathogens. Chlorination so far is still the predominant disinfection method due to low cost and broad-spectrum disinfectant capability (Sun *et al.*, 2009). However, chlorine or other disinfectants reacts with organic and inorganic compounds in water and form disinfection byproducts (DBPs). Some of DBPs are considered to be carcinogenic, mutagenic or teratogenic (Kanitz *et al.*, 1996 and Black *et al.*, 1996). Due to the effects of DBPs, the World Health Organization (WHO, 1996) published drinking water guidelines of several DBPs including THMs and HAAs. The US EPA also established the maximum allowable contaminant level for total THMs (0.08 mg/L) and the sum of five HAA (HAA5) 0.06 mg/L (USEPA, 2001). Therefore recycling wastewater into potable water by soil aquifer treatment sould be considered in term of the form of DBPs.

Several studies have been reported on the formation potential of THMs of reclaimed water by SAT (Amy *et al.*, 1993; Xue *et al.*, 2008). These previous studies focusing on DOC removal as DBPs precursors during SAT using column or basin in single condition. However, studies on the reduction levels of THMs and HAAs with different soil type and retention time are limited. Therefore, further investigation of THMs and HAAs fate in different condition such as soil type and retention time should is necessary to obtain a better understanding on SAT performance.

UV absorbance at 254 nm (UV254) has been known as an indicator of the type and characteristicts of dissolved organic matter (DOM) and DBP precursor level. Quanrud *et al.* (1996a) used UV254 to measure DBPs precursors during SAT with different soil type. Furthermore, they also used the same parameter (UV254) to investigate the fate of organics during column studies of soil aquifer system (Quanrud *et al.*, 1996b). Edzwald *et al.* (1985) and Gallard and Gunten (2002) also found that UV254 was well correlated to DBP formation potential (DBPFP) especially for total THM formation potential (THMFP).

There were two spesific objectives in this study: (1) to evaluate the reduction levels of the formation potential of THMs and HAAs during SAT with different soils and retention time, and (2) to assess their relationships with UV254.

METHODOLOGY

To investigate the reduction levels of THMFP and HAAFP, two type of soil with different water content, TOC, cation and anion exchange were employed (Table 69.1) with the operating conditions shown in Table 69.2. Four laboratory-scale soil columns were prepared to stimulate SAT. The columns were made of acrylic with 150 cm height and an internal diameter of 15 cm (Fig. 69.1).

An effluent from an actual wastewater treatment plant with anaerobic-anoxic-oxic (A2O) process was used as the influent for each column. A peristaltic pump was installed to feed the influent to the columns and the water passed through the columns by controlling the flowrate to obtain the desired HRT. The influent and the effluents after SAT were collected and analyzed periodically from January to Desember 2013. Water samples were preserved in a refrigerator at 4 °C until analysis.

All chemical solutions such as methanol methyl-tert-buthyl ether (MTBE) for extraction were purchased from Wako and of reagent grade or better unless otherwise noted. Four THMs (chloroform (TCM), bromoform (TBM), dichlorobromomethane (DCBM), and dibromochloromethane (DBCM)) and nine HAAs (monochloroacetic acid (MCAA), dichloroacetic acid (DCAA), trichloroacetic acid (TCAA), monobromoacetic acid (MBAA), dibromoacetic acid (DBAA), tribromoacetic acid (TBAA), dichlorobromoacetic acid (DCBAA), chlorodibromoacetic acid (CDBAA), and tribromoacetic acid (TBAA) were analyzed in this study. Both THMs and HAAs standards were of analytical grade, and obtained from Wako and Kanto Chemical. Stock solutions were prepared by diluting 10% of standard THMs (2 mg/mL) and HAAs (0.1 mg/mL) in methanol. All the stock solutions were stored at 4 °C. Working standard solutions were prepared by diluting stock standard solution into several different concentration levels in every analysis. Fresh stock solution of chlorine was prepared before each analysis by diluting sodium hypochlorite (Chemical grade, NaOCl) with milli-Q water and was standardized with the DPD/FAS titration method.

Table 69.1 Characteristics of soils.

Parameter	Unit	Soil Type	
		Common sand	Toyoura sand
Water content	(%)	0.567	0.0654
pH	(-)	6.64	6.87
Cation exchange capacity	(meq/100mg-dry)	2.39	0.192
Anion exchange capacity	(meq/100mg-dry)	1.49	2.3
TOC	(%)	0.0094	0.0144
Porosity	(%)	43.5	42.5
Size distribution (>0.05mm)	(%)	>99	100

Table 69.2 Operating conditions of SAT columns.

Code	Soil type	Hydraulic retention time (HRT)-days	Sand height (cm)	Vadozc zone (20 cm)
SAND 3.5	Common sand	3.5	70	With
SAND 7	Common sand	7	120	With
SAND SAT	Common sand	7	120	Without
TOYOURA	Toyoura sand	7	120	With

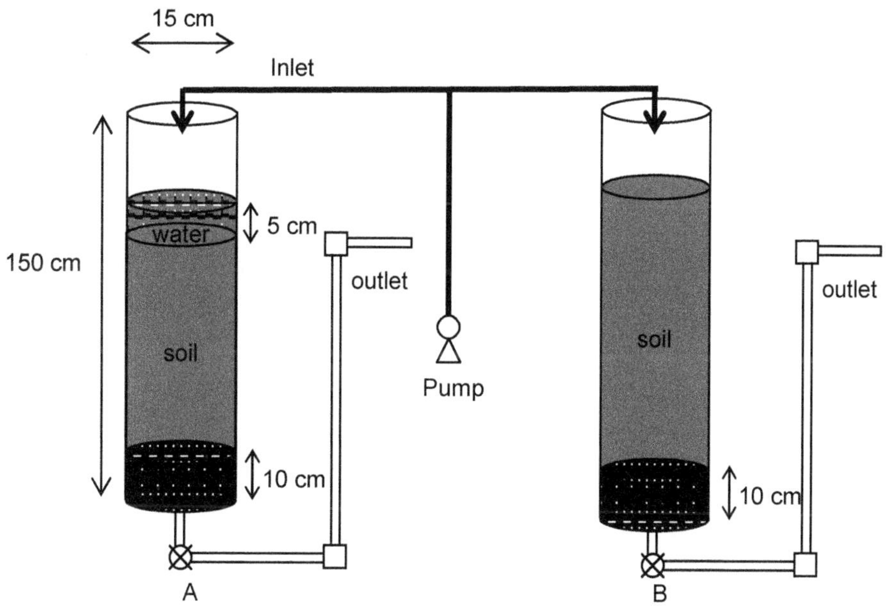

Figure 69.1 Schematic of the laboratory-scale SAT column: (A) SAND SAT (B) SAND 3.5, SAND 7 and TOYOURA

The analysis was started with the chlorine demand estimation for each sample. The second step was chlorinating water samples for 1 day with a residual chlorine concentration of 1 mg/L at the end of reaction. Then the reaction was stopped by dechlorinating agent mixture (Na_2HPO_4, KH_2PO_4, and Na_2SO_3) for THMs and ammonium chloride (NH_4Cl) for HAAs, and all the samples were individually analyzed. THMs and HAAs were analyzed using the USEPA 551.1 (USEPA, 1995) and 552.3 (USEPA, 2003) method with minor modification respectively. Water samples were extracted with MTBE. Decafluorobiphenyl and 2,3-dibromopropanac acid were added as surrogate standards for THMs and HAAs, respectively. Finally samples were analyzed using a gas chromatograph (Agilent 6890N) with electron capture detector (ECD) equipped with DB-1701 column (0.25 mm x 30m x 0.25μm) from Agilent, USA. The GC-ECD operating conditions for THMs and HAAs analysis were as follows: inlet 160 °C, detector temperature 300 °C, split mode, injection volume 1 μL, temperature program 37 °C for 21 min, ramped to 136 °C at 11 °C/min held for 3 min, then ramped to 236 °C at 20 °C/min held for 3 min. Helium was used as carrier gas while N_2 as make up gas with velocity 23 mL/min.

HAA concentrations were calculated using 1,2,3-Thrichloropropane as internal standard. While THM concentrations were determined using calibration curve of THMs standard. Undetected THMFP and HAAFP concentrations in samples were calculated as a half of its minimum detection limit. UV absorbance at 254 nm (UV254) was measured using a spectral photometer Multi-Spec-1500 (Shimadzu, Japan) using 1 cm cuvette cell. Dissolved organic carbon (DOC) was measured by TOC Analyzer TOC-L (Shimadzu, Japan).

RESULTS AND DISCUSSION

Table 69.3 shows the average concentration of THMFPs in each SAT column. TCM was the dominant THMFP followed by DCBM, DBCM, and TBM in influent and the TOYOURA column effluent. SAND 3.5, SAND 7, and SAND SAT columns had a similar pattern of THMFP: DCBM > DBCM > TCM > TBM. In addition, THMFP concentration from all columns were below standard limit according to the drinking water standard of Japan except for average concentration of TCM (73.21 μg/L) and DCBM (33.73 μg/L) of the influent.

Table 69.3 THMFP (average ± standard deviation) before and after SAT (n=16).

THMFP	Influent (μg/L)	MCL* (μg/L)	Effluent (μg/L)			
			TOYOURA	SAND 3.5	SAND 7	SAND SAT
TCM	73.21 ± 12.87	60	26.83 ± 3.54	9.46 ± 3.98	6.03 ± 1.71	7.33 ± 2.60
DCBM	33.73 ± 14.13	30	21.75 ± 9.96	12.52 ± 7.75	10.26 ± 5.97	10.87 ± 4.91
DBCM	8.19 ± 4.31	100	9.53 ± 4.81	10.14 ± 5.53	9.43 ± 4.29	9.48 ± 4.66
TBM	2.70 ± 3.49	90	3.61 ± 3.70	5.34 ± 4.80	5.86 ± 3.99	5.13 ± 4.25

*Maximum concentration limit according to the drinking water quality standards of Japan

The reduction level of the total THMFP is shown in Figure 69.2. Average concentration of THMs from influent (Fig. 69.2) was above the standard limit according to the drinking water standard of Japan (100 μg/L). While effluent from all the columns were below the standard. With the same HRT (7 days), SAND 7 has the highest reduction level of THMs (average: 73%) followed by SAND SAT (72%), SAND 3.5 (68%) and TOYOURA (48%). The reduction level of THMs at the same HRT with saturated condition (SAND SAT) was only slightly different from that of the unsaturated condition (SAND 7).

Cutting down the HRT into half week (SAND 3.5) showed reduction level of THMs slightly lower comparing to reduction level in SAND 7.

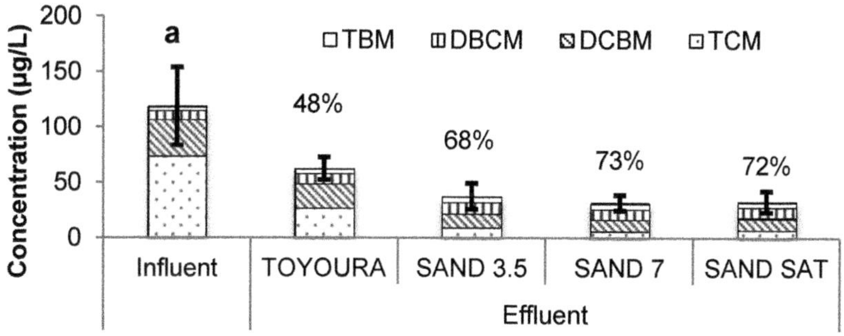

Figure 69.2 THMFPs before and after SAT (n=16). Error bars indicate standard deviations of total concentration of THMFPs. The percentages are the average reduction level of THMFP.

HAAFPs in influent and effluent are summerized in Table 69.4. The average concentrations in the influent were similar to the TOYOURA column effluent with the following order: TCAA > DCAA > BDCAA > BCAA > CDBAA > MCAA > DBAA except for MBAA and TBAA. The highest reduction level of HAAFP after SAT process in all columns was found for TCAA following by DCAA and BDCAA and the others HAAFPs only slightly decreased after SAT process. The average concentration of MCAA, DCAA, and TCAA of influent and effluent were below the maximum contaminant limits of the drinking water standards of Japan (Table 69.4).

Table 69.4 HAAFP (average ± standard deviation) before and after SAT (n=16).

HAAFP	Influent (µg/L)	MCL* (µg/L)	Effluent (µg/L)			
			TOYOURA	SAND 3.5	SAND 7	SAND SAT
MCAA	3.04 ± 2.75	20	2.68 ± 1.71	2.18 ± 1.39	1.85 ± 0.71	2.01 ± 1.69
MBAA	1.07 ± 0.63	-	0.98 ± 0.65	0.85 ± 0.53	0.85 ± 0.60	0.89 ± 0.59
DCAA	32.76 ± 7.51	40	12.22 ± 2.10	4.68 ± 2.05	2.71 ± 1.45	3.84 ± 1.71
TCAA	42.79 ± 9.65	200	10.62 ± 2.33	3.14 ± 0.95	2.28 ± 0.82	3.02 ± 1.01
BCAA	8.41 ± 2.31	-	7.26 ± 1.87	4.66 ± 1.00	3.79 ± 0.82	4.41 ± 0.98
BDCAA	11.43 ± 5.71	-	7.27 ± 2.32	6.58 ± 5.38	6.35 ± 4.80	6.07 ± 3.63
DBAA	2.87 ± 1.17	-	3.77 ± 1.23	4.37 ± 1.49	4.20 ± 1.31	4.22 ± 1.22
CDBAA	3.95 ± 2.46	-	4.13 ± 3.13	3.37 ± 2.26	2.80 ± 1.62	2.95 ± 1.83
TBAA	1.14 ± 1.68	-	0.59 ± 1.00	0.85 ± 0.95	0.96 ± 1.20	0.88 ± 1.09

*Maximum concentration limit according to the drinking water quality standards of Japan

Figure 69.3 showed reduction levels of HAAFPs. Similarly with THMs, the highest reduction levels was found in SAND 7 (average: 76%) followed by SAND SAT (74%) and SAND 3.5 (71%). While the TOYOURA column has the lowest reduction level (54%). The saturated condition in SAND SAT column did not show so much difference of HAAFP reduction level compared to SAND 7. This result also showed that SAND 3.5 was less effective by only 5% than SAND 7 and 3% with SAND SAT reduction level.

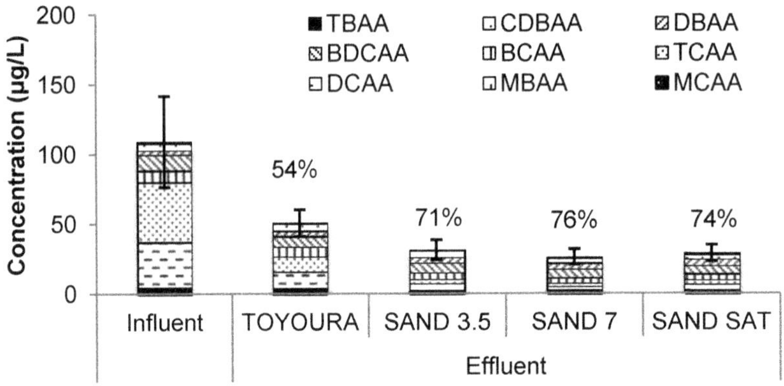

Figure 69.3 Total HAAFP in average before and after SAT (n=16). Errorbars indicate standard deviations of total concentration of HAAFP. The percentages showed the average reduction level of total HAAFP.

Base on this study, the highest reduction levels of THMFP and HAAFP were found in SAND 7 (73% and 76%) followed by SAND SAT, SAND 3.5 and TOYOURA column. However in term of hydraulic retention time, SAND 3.5 column already gave sufficient reduction level (68% and 71%) of THMFP and HAAFP. Doubling HRT 3.5 to 7 days, did not improve the reduction level significantly. The reduction level increased only by 5% compared to column with HRT 3.5 days.

DOC concentration for dissolved organic matter (DOM) from all the columns during SAT were also monitored (Fig. 69.4). The results show that DOC reduction levels followed a similar pattern to the reduction levels of THMs and HAAs (Fig. 69.4a). In addition, DOC values had a strong correlations with THMFPs ($R^2 = 0.88$) and HAAFPs ($R^2 = 0.87$) (Fig. 69.4b).

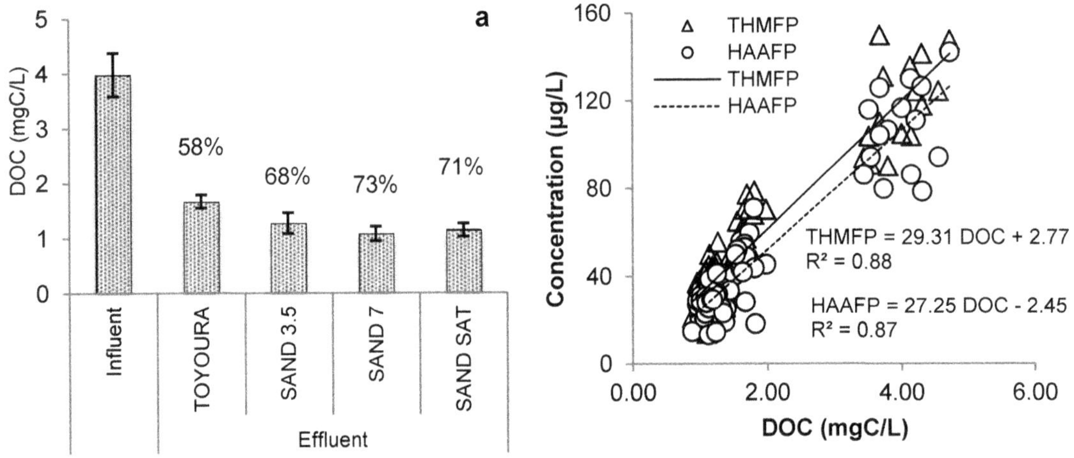

Figure 69.4 **(a)** Average DOC concentrations (n=16) before and after SAT. The percentages showed the removal of DOC **(b)** DOC values correlation with THMFP and HAAFP.

The lower reduction level of DOC after SAT is likely due to biodegradation and sorption process during soil filtration (Quanrud *et al.*, 2003 and Fox *et al.*, 2005). Moreover, the form of DBPFP especially THMFP was dependent on DOC concentration and its reactivity with chlorine. However DOC may also increase during SAT through desoption of soil or/and biomas (Xue *et al.*, 2008). Higher DOC concentration in Toyoura sand compared with the common sand may have contributed to the FPs of THMs and HAAs. Therefore, Toyoura sand showed a lower reduction level (<60%) than the common sand (>70%).

Average value of UV254 from all column effluents were shown in Figure 69.5a. UV254 values of all column effluents decreased from the influent in the following order: SAND 7 > SAND SAT > SAND 3.5 > TOYOURA column. Correlation of between UV254 values and THMFP and HAAFP was determined using a linear regression method which was shown in Figure 69.5b.

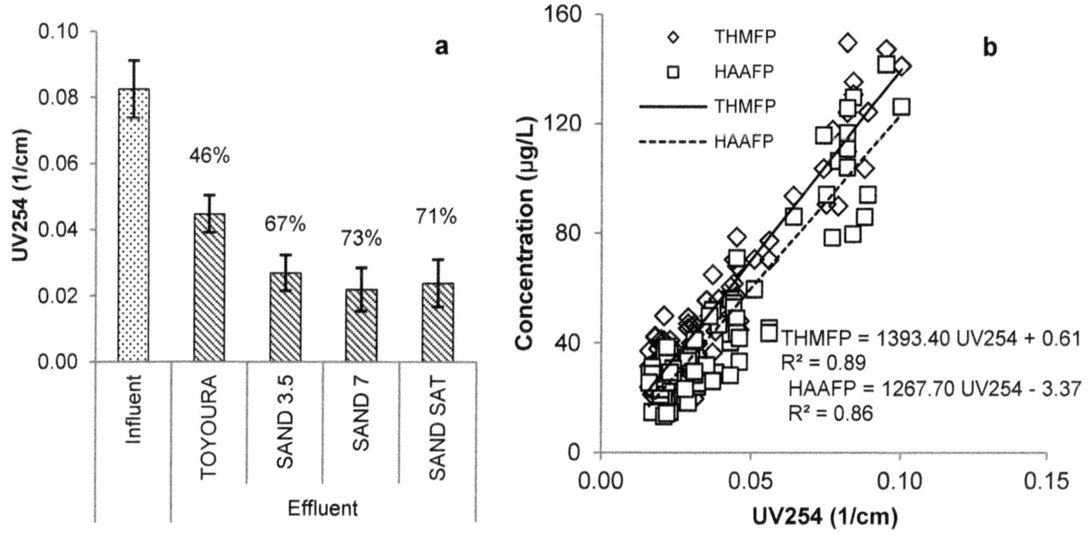

Figure 69.5 **(a)** Average UV254 values (n=16) before and after SAT. The percentages shown here are the reduction level of UV254 **(b)** UV254 values (n=16) correlation with THMFP and HAAFP

According to Figure 69.5b, UV254 values have strong relationships with THMFP and HAAFP (R^2 = 0.89 and 0.86, respectively). Thus, UV254 measurement could be used as one of parameter to monitor the THMFP and HAAFP levels for SAT effluent.

CONCLUSIONS

The main conclusions of this study are:

(1) Similar reduction level of THMFPs (68%, 73% and 72%) and HAAFPs (71%, 76%, and 74%) were observed in common sand column with HRT 3.5 and 7 days also SAT column. Therefore, saturated condition and doubling HRT from 3.5 to 7days, did not give a significant result on reduction level THMFPs and HAAFPs.

(2) According to drinking water standard of Japan, THMFP and HAAFP concentration of influent and all columns effluent was below the standard except for TCM and DCBM in influent.

(3) UV254 has a strong relationship with THMFP and HAAFP therefore UV254 could be used as one parameter to monitor THMFP and HAAFP level

ACKNOWLEDGEMENTS

The authors gratefully acknowledge funding from the Core Research for Evolutionary Science and Technology, Japan Science and Technology Agency (CREST/JST) and Kyoto City Waterworks Bureau for providing the feed water and technical supports for this project.

REFERENCES

Amy, G.L., Wilson, G., Conroy, A., Chahbandour, J., Zhai, W., and Siddiqui, M. (1993) Fate of chlorination byproducts and nitrogen species during effluent recharge and soil aquifer treatment (SAT). *Water envi. Research.*, **65**(6), 726-734.

Black, B.D., Harrington, G.W., Singer, P.C. (1996) Reducing cancer risks by improving organic carbon removal. *J. Am. Water Works Assoc.*, **88**, 40-52

Chowdhury, S., Champagne, P., and McLelan, P.J. (2009) Models for predicting disinfectant byproduct (DBP) formation in drinking waters: A chronological review. *Sci Total Environ.*, **407**, 4189-4206.

Edzwald, J.K., Becker, W.C., and Wattier, K.L. (1985) Surrogate measures for monitoring organic matter and THM precursors. *J. Am. Water Works Assoc.*, **77,** 122-132

Fox, P., Aboshanp, W., and Alsamadi, B. (2005) Analysis of soil to demonstrate sustained organic carbon removal during soil aquifer treatment. *J. Environ. Qual.*, **34**(1), 156-163.

Gallard, H., and Gunten, U.V. (2002) Chlorination of natural organic matter: kinetics of chlorination and THM formation. *Water Res.*, **36**, 65-74.

Kanitz, S., Franco, Y., Patrone, V., Caltabellotta, M., Raffo, E., Riggi, C.,Timitilli, D. and Ravera, G. (1996) Association between drinking water disinfection and somatic parameters at birth. *Eniron. Health Perspect.*, **104**, 516-520.

Quanrud, D.M., Arnold, R.G., Wilson, L.G., and Conklin, M.H. (1996a) Effect of soil type on water quality improvement during soil aquifer treatment. *Water Sci. Technol.*, **33**, 419-431

Quanrud, D.M., Arnold, R.G., Wilson, L.G., Gordon, H.J., Graham, D.W., and Amy, G.L. (1996b) Fate of organics during column studies of soil-aquifer treatment. *J. Environ. Eng.*, **122**(4), 314-321.

Quanrud, D.M., Hafer, J., Karpiscak, M.M., Zhang, J., Lansey, K.E., and Arnold, R.G. (2003) Fate of organics during soil aquifer treatment: sustainability of removals in the field. *Water Res.*, **37**(14), 3401-3411.

Sun, Y.X., Wu, Q.Y., Hu, H.Y., and Tian, J. (2009) Effect on the formation of disinfection by-products during wastewater chlorination. *Water Res.*, **43**, 2391-2398.

USEPA (1995) Methods for the determination of organic compounds in drinking water – Supplement III, EPA-600/R-95/131, Environmental Monitoring and System Laboratory. United States Environmental Protection Agency, Cincinnati, OH, USA.

USEPA (2001) National primary drinking water standards, United States Environmental Protection Agency, EPA; 816-F-01-007.

USEPA (2003) Method 552.3: Determination of haloacetic acids and dalapon in drinking water by liquid-liquid microextraction, derivatization and gas chromatography with electron capture detection. Environmental Monitoring and System Laboratory. United States Environmental Protection Agency. Cincinnati, OH. USA.

World Health Organization WHO (1993) Guidelines for drinking-water quality, 2th ed, Volume 1: Recommendations. World Health Organization, Geneva.

Xue, S., Zhao, Q.L., Wei, L.L., and Wang, L.N. (2008) Reduction of dissolved organic matter and trihalomethane formation potential during laboratory-scale soil-aquifer treatment. *Water Environ J.*, **22**,148-154.

Chapter 70

Removals of pharmaceuticals and personal care products in reclaimed water during soil aquifer treatment with different soil types, hydraulic retention time, and saturated condition

K. He, T. Yonetani*, Y. Takabe*, S. Rahmawati*, S. Echigo*, S. Itoh**

**Department of Environmental Engineering, Graduate School of Engineering, Kyoto University, Nishikyo, Kyoto 615-8540, Japan*

ABSTRACT

Pharmaceuticals and personal care products (PPCPs) are ubiquitous in aquatic environments. For controlling PPCPs in water reclamation, soil aquifer treatment (SAT) is an attractive way because of its low operational cost and stable performance. In this study, 42 PPCPs in A2O (anaerobic-anoxic-oxic process) water were measured, and the removals of 30 PPCPs in SAT with Masa soil, sand and Toyoura sand with a HRT (Hydraulic retention time) of 7 days were compared. The three SAT columns could remove antibiotic PPCPs, and Masa soil performed best on the removal of antibiotic PPCPs. Also, most non-steroidal anti-inflammatory drugs (NSAIDs, i.e., mefenamic acid, diclofenac, triclosan and ketoprofen) were removed in the three kinds of columns probably through biodegradation. Increasing HRT enhanced the removal of acetaminophen, theophylline, crotamiton, and caffeine. In addition, better removals were observed in unsaturated condition than in saturated condition for diclofenac, aspirin, sulfamethoxazole, primidone, furosemide, crotamiton, and benzophenone. Overall, most NSAIDs and most antibiotic could be removed under any conditions tested in this study.

Keywords PPCPs; SAT; Soil types; HRT; Saturated condition

INTRODUCTION

Pharmaceuticals and personal care products (PPCPs) a group of chemicals which include not only drugs and antiseptics, but also fragrances, and sunscreen (Daughton and Ternes, 1999), and was detected in aquatic environments worldwide (Focazio *et al.*, 2008). Especially PPCPs have been detected frequently in the effluents from wastewater treatment plants (WWTP) (Lin *et al.*, 2009; Okuda *et al.*, 2009; Narumiya *et al.*, 2009; Yonetani *et al.*, 2012). Though PPCPs were detected in the ng/L to low μg/L levels, toxicologists suggested a potential health risk associated with the existence of mixtures of PPCPs at low concentration (Parrott and Bennie, 2009).

To reduce the PPCPs discharged to the water environment, some advanced treatments (e.g., soil aquifer treatment (SAT), membrane technologies, and advanced oxidation processes (AOPs)) need to be employed in addition to standard wastewater treatment technologies. In this study, SAT was considered as a feasible technology for PPCPs removal. Removal mechanisms of trace organic compounds in SAT include biological, chemical, and physical processes as the effluent infiltrates and percolates through the soil (Kathryn and Edward, 2012).

Though there are some benefits such as low operation cost and stable performance, in SAT system, not all PPCPs could be degraded completely in SAT system. Therefore, their removals in SAT attracted a lot of attention recently, and were assessed in both field- and laboratory-scale studies. Drewes *et al.* (2001) mentioned that for X-ray contrast media, six to ten years of saturated anoxic flow led to 60% removal in a field-scale groundwater recharge. In another experiment, sulfamethoxazole removal was 95% in a bank filtration system, but only 53% in an artificial recharge system (Grunheid *et al.*, 2005). Removal efficiencies of trace organic compounds varied from research to research. This may be attributed to chemical structures, soil properties, and operation conditions. Several researchers suggested that degradation of EDCs and some PPCPs in soil were affected by environmental conditions such as temperature, pH, moisture content, organic carbon, the presence of specific microorganisms, and DO conditions (Colucci *et al.*, 2001; Boxall, 2008; Monteiro and Boxall, 2009). Moreover, Ying and Kookana (2005) found that the degradation efficiencies of E2 and EE2 in non-sterile aerobic soil are different from in sterile soil. However, to the best of our knowledge, no systematic study on PPCPs removal with respect to soil types, hydraulic retention time (HRT), and saturated conditions was made, especially depend on PPCPs use/category. In this study, column experiments with three different soil types (one soil and two sands) were performed to compare PPCP removals. In addition, we evaluated the effects of HRT and DO conditions on PPCPs removals in SAT system, with a special attention on the relationship between the removals and PPCPs structures.

MATERIALS AND METHODS

Reagents

As the target PPPCPs, 42 compounds were selected based on the frequency of their detection in effluent of WWTPs (Okuda *et al.*, 2009; Narumiya *et al.*, 2009; Yonetani *et al.*, 2012) (Table 70.1). These compounds consist of antibiotics such as lincomycin, analgesic drugs such as crotamiton, antiepilepsy drugs such as carbamazepine, and the others. DEET was the abbreviation of *N*, *N*-Diethyl-meta-toluamide. Clarithromycin and sulfadimethoxine were dissolved into acetone (Okuda *et al.*, 2009), ciprofloxacin, levofloxacin, norfloxacin, and enfloxacin were dissolved into a water/acetonitrile solution, and all the other compounds were dissolved into acetonitrile to prepare the stock solutions. For the analysis of PPCPs with an LC-MS/MS system, formic acid, methanol, and acetonitrile (LC-MS grade, Wako Pure Chemical Co., Ltd) were used for the preparation of mobile phase. Caffeine and chlortetracycline were purchased from Cosmo Bio Co., Ltd; isopropylantipyrine and levofloxacin were purchased from Tokyo Chemical Industry Co., Ltd; antipyrine, limcomycin, atenolol, azithromycin, ciprofloxacin, diltiazem, enfloxacin, fenoprofen, furosemide, gemfibrozil, mefenamic acid, oxytetracycline, sulfadimethoxine, tetracycline, ibuprofen, sulfamethoxazole, and trimethoprim were purchased from Funakoshi Co., Ltd, and all other compounds were purchased from Wako Pure Chemical Co., Ltd. All the aqueous solutions were prepared with ultra-pure water produced with Millipore Academic-A10 system (hereinafter call MQW).

Table 70.1 List of the target PPCPs.

Name	LOQ	Use/category	Abbr.	Name	LOQ	Use/category	Abbr.
Ampicillin	0.30	Antibiotic	AMP	Ketoprofen	1.29	Analgesic	KTP
Lincomycin	0.17	Antibiotic	LCM	Isopropylantipyrine	0.06	Analgesic	IPP
Oxytetracycline	0.15	Antibiotic	OTC	Mefenamic acid	0.21	Analgesic	MFA
Tetracycline	0.09	Antibiotic	TC	Ibuprofen	0.91	Analgesic	IBP
Azithromycin	0.31	Antibiotic	AZM	Aspirin	0.85	Analgesic	ASP
Clarithromycin	0.11	Antibiotic	CAM	Acetaminophen	0.91	Analgesic	ACEAM
Ciprofloxacin	0.39	Antibiotic	CPFX	Crotamiton	0.13	Analgesic	CRT
Levofloxacin	0.28	Antibiotic	LVFX	Atenolol	0.34	Antiarrhythmic	ATL
Norfloxacin	0.28	Antibiotic	NRFX	Disopyramide	0.12	Antiarrhythmic	DSP
Enfloxacin	0.34	Antibiotic	ERFX	Bezafibrate	0.11	Antilipidemic	BZF
Trimethoprim	0.08	Antibiotic	TRM	Clofibric acid	0.13	Antilipidemic	CFB
Sulfadimethoxine	0.17	Antibiotic	SDIME	Gemfibrozil	0.06	Antilipidemic	GFB
Sulfamethazine	0.08	Antibiotic	SMT	Benzophenone	0.04	Ultraviolet absorber	BP
Sulfamethoxazole	0.08	Antibiotic	SMETH	DEET	0.16	Rejectant	DEET
Triclosan	1.16	Antibiotic	TRC	Caffeine	0.16	Cardiac	CAF
Erythromycin	1.03	Antibiotic	ERY	Carbamazepine	0.07	Antipilepsy	CBM
Chlortetracycline	1.64	Antibiotic	CTC	Primidone	0.67	Antipilepsy	PRM
Antipyrine	0.08	Analgesic	ATP	Sulpiride	0.10	Peptic ulcer	SLP
Diclofenac	0.07	Analgesic	DCF	Theophylline	0.23	Bronchodilator	TEP
Fenoprofen	0.89	Analgesic	FNP	Diltiazem	0.09	Blood-vessel dilator	DTZ
Naproxen	0.72	Analgesic	NPX	Furosemide	0.16	Blood-vessel dilator	FSM

Unit of LOQ (limit of quantitation): μg/L; Abbr. means abbreviation.

Feed water

An effluent from an actual wastewater treatment plant (anaerobic-anoxic-oxic process, A2O) in Kyoto City (Japan) was used as the feed water for experiment.

Soil aquifer treatment (SAT)

Four large laboratory-scale acrylic columns (i.d., 15 cm; H, 150 cm), one small laboratory-scale acrylic column (i.d., 8 cm; H, 100 cm), and four pilot plant-scale stainless steel columns (L, 1.5 m; W, 1.5 m; H, 3.0 m) were used. The columns operation conditions were summarized in Table 70.2. The feed water for large laboratory-scale acrylic columns was fed by peristaltic pumps (7554-80, Thermo Fisher Scientific) and the feed water for small laboratory-scale acrylic columns was fed by peristaltic pumps (SJ1220, ATTA). Feed water for pilot plant-scale stainless steel columns was pumped into columns by centrifugal pumps. Laboratory-scale columns consisted of the bottom layer (gravel level) and the soil/sand layer; sand pilot plant-scale column consisted of sand layer and soil pilot plant-scale columns consisted of the bottom layer (gravel level) and the soil layer. In the unsaturated condition, the water head was maintained at 20 cm below the top of the soil/sand layer. In the saturated condition, the water head was maintained at 20 cm above the top of the soil/sand layer. The operation of the large laboratory-scale acrylic columns started on January 2012; sand plant in Toba started on October 2011; and three Masa soil plants in Toba started on April 2012. From August to December in 2013, A2O water and laboratory scale columns effluents were sampled every two weeks while pilot plant-scale columns effluents from columns were sampled monthly.

Table 70.2 General properties and operating conditions of the columns.

Columns	Scale	Soil types	Hydraulic retention time (HRT) (days)	Saturation condition
TS	Lab	Toyoura sand	7	Unsaturated
Sand-3.5	Lab	Common sand	3.5	Unsaturated
Sand-7	Lab	Common sand	7	Unsaturated
Sand-sat	Lab	Common sand	7	Saturated
Masa-7	Lab	Masa soil	7	Unsaturated
Sand-30	Pilot plant	Common sand	30	Unsaturated
Masa-30	Pilot plant	Masa soil	30	Unsaturated
Masa-180	Pilot plant	Masa soil	180	Unsaturated
Masa-sat	Pilot plant	Masa soil	30	Saturated

Analytical methods

200 mL of water samples was filtered through glass fiber membrane filter (ADVANTEC GF-75, 90 mm) prior to analysis. All the water samples were concentrated with solid-phase extraction (SPE). The SPE cartridge (Oasis HLB cartridges (Waters), 250mg bed) was conditioned with 5 mL of methanol and 5 mL of MQW at pH 3 (hydrochloric acid). Samples were passed through the cartridge at flow rate of 10 mL/min for 20 min. After drying the SPE cartridge with nitrogen gas, the retained compounds were eluted with 5 mL of methanol. The eluted solution was evaporated to dryness under a gentle stream of nitrogen gas (Okuda *et al.*, 2009; Yonetani *et al.*, 2012). The residue of A2O water samples and SAT effluent samples were re-dissolved in 2 mL and 1 mL of 0.1% formic acid-acetonitrile (90/10, v/v), respectively. The solution was analyzed by LC-MS/MS. In this study, prominence 20A (Shimadzu system) interfaced to 4000 QTRAP (AB SCIEX) using a C-18 column (Kinetex, 2.150mm, 2.6μm) was used in negative mode analysis, and VP-ODS column (Shimadzu, 4.6mm, 4.6 μm) was used in positive mode analysis. The way to calculate the recovery is as follows: 2 samples were conducted without any operation before extraction (BL) and the other 2 samples were added with each compound before extraction to increase 100 μg/L (ADD) in samples. After filtration, concentration was conducted with SPE and measurement was conducted. Then, the content of each PPCPs in samples between BL and ADD was compared, and the recovery was calculated as following Eq. (1) (Okuda *et al.*, 2009). In addition, recovery experiments were operated monthly to calculate the PPCPs concentration.

$$\text{Recovery} = (\text{ADD} - \text{BL}) / \text{the amount of addition} \qquad (1)$$

RESULTS AND DISCUSSION

Occurrence of PPCPs in A2O effluent

The highest, lowest and average concentrations of the selected PPCPs (n=10) A2O effluent samples were presented in Figure 70.1. Among the selected PPCPs, crotamiton, an analgesic agent, was the most abundant in the A2O water (806 ng/L). This compound is a very common analgesic agent and is known for its low removal efficiency in WWTP (Narumiya *et al.*, 2009). For antibiotics, clarithromycin, azithromycin, levofloxacin and ciprofloxacin were observed at relatively high concentrations (769 ng/L, 175 ng/L, 28 ng/L, and 23 ng/L, respectively), while enrofloxacin was not detected in the A2O effluent. For analgesics, ketoprofen, diclofenac, mefenamic acid, and acetaminophen were found at relatively high concentrations (639 ng/L, 302 ng/L, 80 ng/L, 75 ng/L, and 17 ng/L, respectively) while ibuprofen was not detected A2O effluent. For other groups of selected PPCPs, bezafibrate, sulpiride, theophylline, DEET, caffeine, disopyramide and furosemide were observed at high concentrations (193 ng/L, 397 ng/L, 108 ng/L, 455 ng/L, 94 ng/L, 237 ng/L and 203 ng/L, respectively).

Since PPCP concentrations fluctuated both for A2O effluent (Adcharee *et al.*, 2011) and SAT effluents during the monitoring period, the monthly average values were used to calculate the removal efficiencies in the following subsections

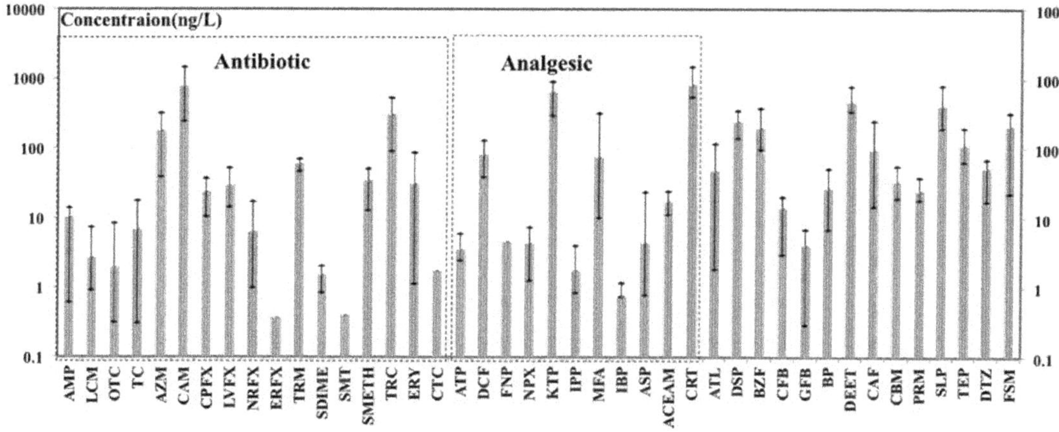

Figure 70.1 PPCPs Concentration in the A2O effluent (n=10).

Behaviors of target compounds in SAT columns with different materials

Considering the occurrence of PPCPs in A2O water, some PPCPs were chosen as target compounds to study PPCPs removal mechanism in SAT (Figure 70.2). The removal efficiencies of target PPCPs in Masa soil, Toyoura sand and sand column are compared in Figure 70.2. The removal efficiencies could be divided into high level (>80%), mean level (40-80%) and low level (<40%). Among target PPCPs, most PPCPs belonged to high or mean levels, which suggested that SAT systems had a potential to control PPCPs. For the antibiotic group, all the three materials appeared equally effective, in antibiotic group, the structure of azithromycin and clarithromycin confer a relatively low degree of unsaturation, and this results in low biodegradability, and most of antibiotic compounds contain tertiary amines, which inhibit bio-activity strongly. In addition, compared to Toyoura sand, Masa soil performed better for antibiotic removal, the organic contents in Masa soil and Toyoura sand were 0.0872(%) and 0.0144(%), and the removal of antibiotic PPCPs could be attributed to soil properties such as organic contents present on soil solids (Garmon *et al.*, 2011). However, for the analgesic group, the removal efficiencies appeared different each other: the removal efficiencies of NSAIDs except aspirin in the three columns were above 90%. Most NSAIDs contain carboxyl acid and simple carbon chain which are biodegradable function groups (Richardson and Bowron 1985; Ternes 1998). On the other hand, the removal efficiencies of antipyrine, isopropylantipyrine, and acetaminophen were mostly less than 50% probably due to their low biodegradability, and increasing HRT should be considered for better removal. This will be discussed in the next subsection. Overall, Masa soil column had better performance for most PPCPs removals, especially crotamiton and caffeine, than Toyoura sand except antipyrine.

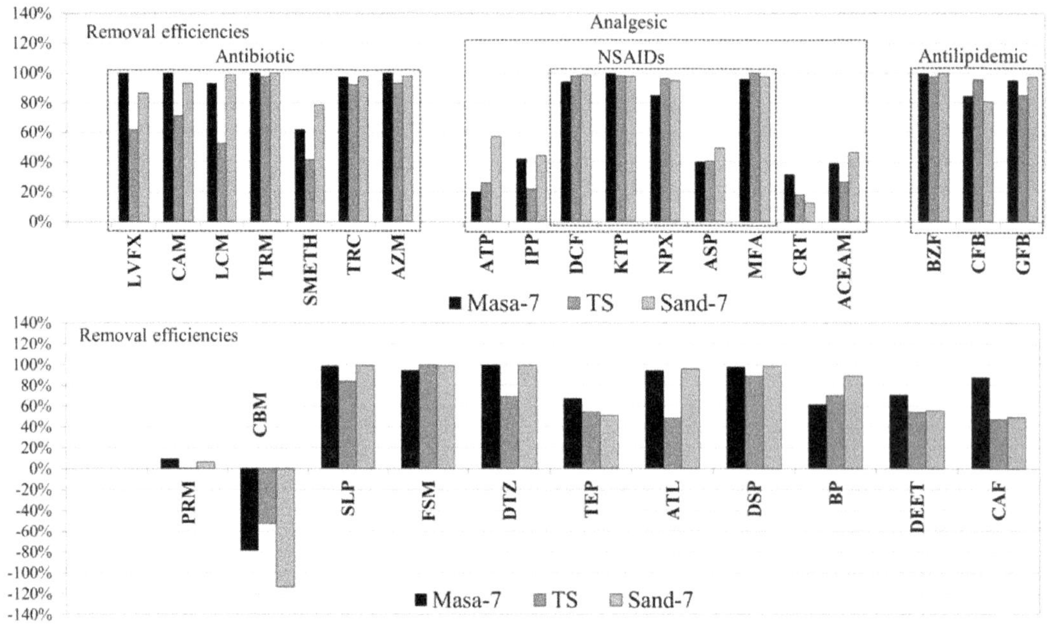

Figure 70.2 Comparison of PPCPs removals in SAT with different soil types.

Behaviors of target compounds in SAT columns with different HRT

The removal efficiencies of target PPCPs in sand columns and soil columns with different HRT were shown in Figure 70.3. Even with increasing HRT, the removals of antibiotics remained similar. This result may suggest that the main removal mechanism of these compounds be sorption (Karnjanapiboonwong *et al.*, 2011). For acetaminophen, theophylline, crotamiton and caffeine, however, longer HRT enhanced the removals for both Masa soil and sand columns. Especially, the removal efficiency of acetaminophen increased from 43.38% to 79.37% in sand columns, and from 39.03% to 77.19% in soil columns. This could be attributed to the biodegradation reinforcement due to increasing HRT (Lin *et al.*, 2010). Overall, the extension of HRT was effective for some PPCPs, and soil columns could have better performance than sand columns with increasing HRT.

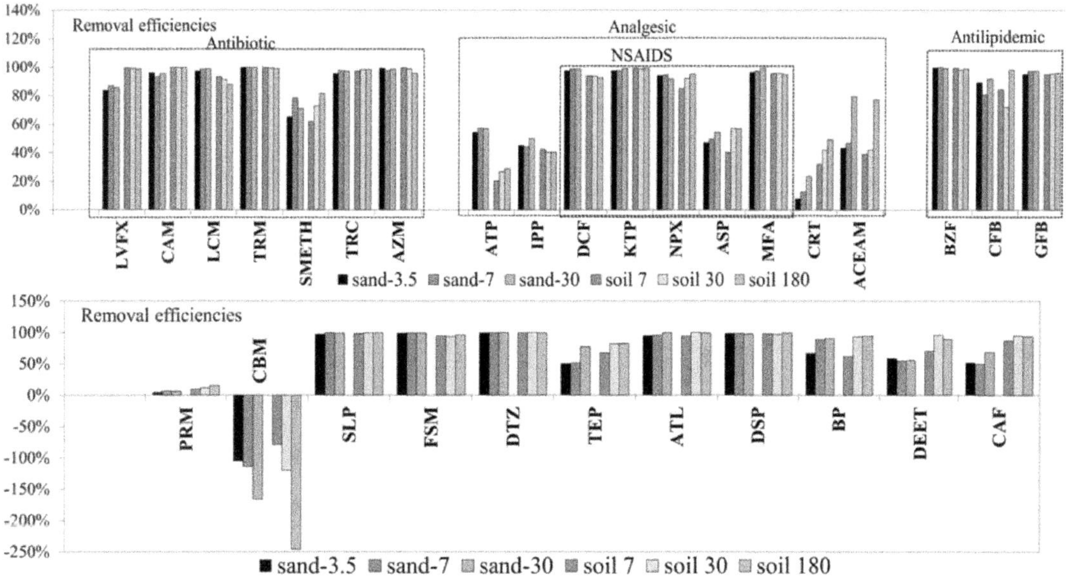

Figure 70.3 PPCPs removal efficiencies with different HRT in sand and soil columns.

Behaviors of target compounds in saturated and unsaturated SAT columns

In actual operation, the aerobic and anaerobic conditions were considered as key operation conditions in SAT. To study the influence of saturated/unsaturated condition on PPCPs removal, saturated columns and unsaturated columns experiments were performed.

The removal efficiencies of target PPCPs in sand columns and soil columns in saturated and unsaturated condition were shown in Figure 70.4. For sulfamethoxazole, diclofenac, aspirin, primidone, furosemide, crotamiton, and benzophenone, the removal efficiencies soil columns in unsaturated condition were 72.70%, 93.83%, 92.10%, 11.33%, 93.63%, 41.94% and 93.15%, respectively, which were better than those in the saturated condition. Because unsaturated columns contained unsaturated layer and saturated layer and oxygen went through unsaturated layer easily, these results suggested that the unsaturated level was useful for these PPCPs. Degradation, which could be attributed the present of oxygen and micro-community in soil (Monteiro and Boxall, 2009). For most antibiotic PPCPs, the removal efficiencies were above 80% in both saturated and unsaturated conditions. This may suggest that sorption be the main mechanism of removal of these compounds as mentioned earlier. In addition, for ketoprofen, naproxen, and mefenamic acid, the removal efficiencies were high both under saturated and unsaturated conditions. Overall, unsaturated condition was better for PPCPs removal.

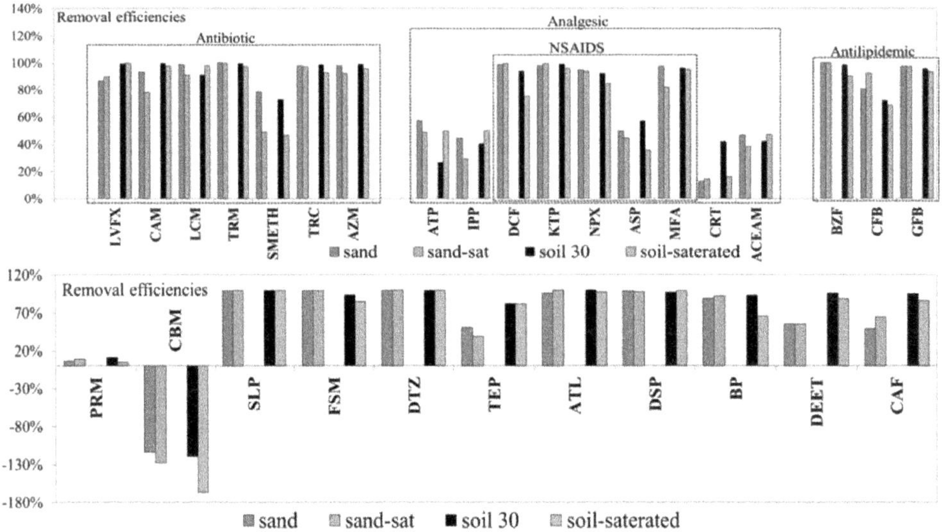

Figure 70.4 PPCPs removal efficiencies in saturated and unsaturated sand columns and soil columns.

CONCLUSIONS

In A2O water, antibiotic and analgesic PPCPs were observed at relatively high concentration. Beside these, bezafibrate, sulpiride, theophylline, DEET, caffeine, disopyramide and furosemide were detected at high concentration in A2O water. Masa soil performed better removals on antibiotic PPCPs due to sorption; Also, NSAIDs in analgesic group were removed effectively in Masa soil, sand and Toyoura sand columns due to biodegradability. In addition, Masa soil could remove crotamiton and caffeine more effectively than Toyoura sand and common sand. Increasing HRT was helpful for the removal of acetaminophen, theophylline, crotamiton and caffeine, For sulfamethoxazole, diclofenac, aspirin, primidone, furosemide, crotamiton, and benzophenone, the removal efficiencies soil columns in unsaturated were better than in saturated condition; For antibiotic PPCPs except sulfamethoxazole, ketoprofen, naproxen and mefenamic acid, SAT columns under both saturated condition and unsaturated condition could effectively remove these compounds.

ACKNOWLEDGEMENTS

This study was supported by JST, CREST. We also appreciate Kyoto City Waterworks Bureau for providing the feed water and technical supports for this project.

REFERENCES

Lin, A.Y., Yu, T., and Lateef, S.K. (2009) Removal of pharmaceuticals in secondary wastewater treatment processes in Taiwan. Journal of Hazardous Materials, 167(2009), 1163-1169.

Okuda, T., Naoyuki Yamashita, N., Tanaka, H., Matsukawa, H., and Tanabe, K. (2009) Development of extraction method of pharmaceuticals and their occurrences found in Japanese wastewater treatment plants. Environment International, 35(5), 815-820.

Narumiya, M., Okuda, T., Nakada, N., Yamashita, N., Tanaka, H., Sato, K., Sueoka, M., and Oiwa, T. (2009) Occurrence and Fate of Pharmaceuticals and Personai Care Products during Wastewater Treatments. Environmental Engineering Research, 46, 175-186.

Colucci, M.S., Bork, H., and Topp, E. (2001) Persistence of estrogenic hormones in agricultural soils: I. 17β-estradiol and estrone. Journal of Environmental Quality, 30(6), 2070–2076.

Yonetani, T., Phattarapattamawong, S., Echigo, S., Morita, Y., Ohkouchi, Y., and Itoh, S. (2012) Fate of pharmaceutical and personal care productes in treated sewage. Journal of JSCE, G68(7), III_419-III_428.

Parrott, J.L., and Bennie, D.T. (2009) Life-Cycle Exposure of Fathead Minnows to a Mixture of Six Common Pharmaceuticals and Triclosan. Journal of Toxicology and Environmental Health, Part A: Current Issues, 72(10), 633-641.

Onesios, K.M., and Bouwer, E.J. (2012) Biological removal of pharmaceuticals and personal care products during laboratory soil aquifer treatment simulation with different primary substrate concentrations. Water Research, 46(7), 2365-2375.

Drewes, J.E., Fox, P., and Jekel, M. (2001) Occurrence of iodinated x-ray contrast media in domestic effluents and their fate during indirect potable reuse. Journal of Environmental Science and Health, Part A, 36(9), 1633-1645.

Grunheid, S., Amy, G., and Jekel, M. (2005) Removal of bulk dissolved organic carbon (DOC) and trace organic compounds by bank filtration and artificial recharge. Water Research, 39(14), 3219-3228.

Boxall, A. B. A. (2008). Fate and transport of veterinary medicines in the soil environment. In Fate of Pharmaceuticals in the Environment and in Water Treatment Systems, D. S. Aga (Ed.), CRC Press, USA, 123-137.

Monteiro, S. C., and Boxall, A. B. (2009) Factors affecting the degradation of pharmaceuticals in agricultural soils. Environmental Toxicology and Chemistry, 28(12), 2546–2554.

Ying, G. G., & Kookana, R. S. (2005) Sorption and degradation of estrogen-like-endocrine disrupting chemicals in soil. Environmental Toxicology and Chemistry, 24(10), 2640–2645.

Karnjanapiboonwong, A., Suski, J.G., Shah, A.A., Cai, Q., Morse, A.N., and Anderson, T.A. (2011) Occurrence of PPCPs at a Wastewater Treatment Plant and in Soil and Groundwater at a Land Application Site, 216(1-4), 257-273.

Garmon, J.S., Drew, J.E., Khan, S.J., Mcdonald, J.A., and Dickenson, E.R.V. (2011) Sorption of emerging trace organic compounds onto wastewater sludge solids. Water Research, 45(2011), 3417-3426.

Richardson, M.L., and Bowron, J.M. (1985) The fate of pharmaceutical chemicals in the aquatic environment. The Journal of Pharmacy and Pharmacology, 37, 1–12.

Ternes, T. A. (1998) Occurrence of drugs in German sewage treatment plants and rivers. Water Research, 32(11), 3245–3260.

Lin, A.Y., Lin, C., Tung, H., and Chary, N.S. (2010) Potential for biodegradation and sorption of acetaminophen, caffeine, propranolol and acebutolol in lab-scale aqueous environments. Journal of Hazardous Materials, 183(1-3), 242-250.

Chapter 71

Behavior of trace organic contaminants in soil aquifer treatment (SAT) process for reuse of secondary sewage effluent

F. Nishimura, N. Kurita*, Y. Takabe*, T. Mizuno*, T. Hidaka**, S. Ito**

**Department of Environmental Engineering, Graduate School of Engineering, Kyoto University, Nishikyo, Kyoto 615-8540, Japan*

***Institute of Public Works, 1-6 Minamihara, Tsukuba 305-8516, Japan*

ABSTRACT

In this study, fate of endocrine disrupting chemicals, Ethylenediaminetetraacetic acid (EDTA) and 1,4-dioxane, whose risks were concerned in the case when treated wastewater is reused for potable purpose, in soil aquifer treatment (SAT) was evaluated with various experiments. Using SAT columns and pilot scale reactors, which continuously received treated sewage effluent, it was cleared that bisphenol-A(BPA) was effectively removed; meanwhile, the removal of 1,4-dioxane by SAT was not so expected. Removal of EDTA was expected; however, the removal was limited (66% of removal ratio with 30 days of Hydraulic retention time). In order to made clear the mechanism of treatment and removal efficiencies of these trace contaminants in SAT, adsorption and biodegradation characteristics were also investigated and evaluated quantitatively from the points of reaction kinetics. It was understood that BPA, E1, and E2 can be removed easily by biological reaction, whereas EDTA and 1,4-dioxane cannot be degraded biologically in the system. On the other hand EDTA can be adsorbed to the soil. These findings corroborate the results of SAT columns experiment, and can be used for mathematical models which express and predict behaviors of the chemicals in SAT.

Keywords Indirect potable reuse; Soil aquifer treatment; Endocrine disrupting chemicals; Ethylenediaminetetraacetic acid; 1,4-dioxane; Adsorption; Biodegradation

INTRODUCTION

Water demand rises concomitantly with population growth, and water shortage is being experienced in many parts of the world. Securement of clean drinking water sources might be more difficult in the future. Water reuse can help solving this problem, and reuse system with soil aquifer treatment (SAT) is as one of the major methods. The Groundwater Replenishment System in South California has provided drinking water with wastewater effluent by SAT (Jansen *et al.*, 2007).

In the case when wastewater is reused with SAT for portable purpose, fate of chemicals, whose concentrations in treated wastewater are comparable to the drinking water quality standards and risks were concerned, in SAT should be specially paid to be attention.

Endocrine disrupting chemicals (EDCs), such as 17β-estradiol (E2) and bipshnol-A (BPA), are listed in 'Items for Further Study' in the Drinking water quality standards in Japan (80 ng/L and 0.1 mg/L, respectively) (Ministry of Health, Labour and Welfare). Meanwhile, the maximum E2 concentration detected in wastewater effluent in Japan was 66 ng/L (Ministry of Land, Infrastructure, Transport and Tourism, 2001), which was close to the standard, and the risks were concerned.

Ethylenediaminetetraacetic acid (EDTA) is also listed in 'Items for Further Study' in the Drinking water quality standards in Japan (0.5 mg/L). It was reported that EDTA was detected with the concentration of 0.15 mg/L in treated effluent (Suzuki *et al.*, 2009); therefore, the fate of EDTA in SAT also should be paid attentions. Drewes *et al.* (2003) and Yoo *et al.* (2006) reported the removal of EDTA in SAT; meanwhile, there were still few knowledge of its removal, including removal mechanism, in SAT.

1,4-dioxane is listed in the Drinking water quality standards in Japan (0.05 mg/L), and the reported concentrations in treated effluent (0.046 mg/L) is also comparable to the standard (Suzuki *et al*, 2005). Therefore, its fate in SAT should be well understood; however, there were few researches which focused on the fate.

In this study, SAT columns and SAT pilot-scale reactor were operated in order to analyze the fate and removal performance of estrogenic compounds, EDTA and 1,4-dioxane in SAT. In addition, batch and continuous experiments was conducted to comprehend the removal mechanism of the chemicals in SAT.

MATERIALS AND METHODS

SAT columns

Three Lab-scale SAT columns were made from acrylic. The height and inner diameter were 150 cm and 15 cm, respectively. Fine sand, which was collected in Shiga prefecture, was packed and used as a kind of soil for SAT. An effluent from an actual wastewater treatment plant (anaerobic-anoxic-oxic process, A2O) in Kyoto City (Japan) which treats mainly domestic wastewater, was used as the feed water. EDCs was spiked because the concentrations of some EDCs were comparatively low. Columns were operated with different conditions (Table 71.1). The reactor operations started on 3 January 2012. The water samples, including A2O effluent and SAT effluent, were collected from 25 October, 2012 to 29 January, 2013 for EDC measurement and from 23 October to 24 December, 2013 for EDTA and 1,4-dioxane measurement.

Table 71.1 Operating conditions of the SAT columns

Column name	HRT (day)	With or without vadose zone
Sand 3.5	3.5	With
Sand 7	7.0	With
Sand-sat	7.0	Without

Removal mechanism and behaviors of these trace organic contaminants in SAT system are represented by differential equations. The examples of the equations are summarized in Table 71.2. Transformation paths and rate equation, and constants and coefficients related to the mathematical model are summarized in Table 71.3 and 71.4, respectively.

The behaviors of targeted trace compounds are characterized and evaluated concretely by the reaction constants in the equations, i.e. both adsorption isotherm and adsorption rate constant can characterize the adsorption of the trace organic contaminants to soil in SAT, and biological removal rate of the contaminants can characterize the biodegradation of these contaminants by microorganisms. Therefore, these reaction rates are determined by conducting experiments and used for the evaluation of behaviors of these trace organic contaminants.

Table 71.2 Mathematical models for behaviors of chemicals in SAT system

State variables	Symbol and unit	Changing rate
Trace contaminants in water phase	C_t [μg/L]	$\alpha V \frac{dC_t}{dt} = Q(C_{t-in} - C_t) - V(1-\alpha)\beta R_1 - \alpha V R_2$
Trace contaminants in soil phase	q_t [μg/g-soil]	$V(1-\alpha)\beta \frac{dq_t}{dt} = V(1-\alpha)\beta R_1$
Organic compounds in water phase	C_s [mgC/L]	$\alpha V \frac{dC_s}{dt} = Q(C_{s-in} - C_s) - V(1-\alpha)\beta R_3 - \alpha V R_4 - \alpha V R_5$
Organic compounds in soil phase	q_s [mgC/g- soil]	$V(1-\alpha)\beta \frac{dq_s}{dt} = V(1-\alpha)\beta R_3$
Heterotrophic microbes	X_H [mgC/g- soil]	$V(1-\alpha)\beta \frac{dX_H}{dt} = \alpha V Y_1 R_2 + \alpha V Y_2 R_4 + \alpha V Y_3 R_5 - V(1-\alpha)\beta R_{11}$
Autotrophic microbes	X_A [mgC/g- soil]	$V(1-\alpha)\beta \frac{dX_A}{dt} = \alpha V Y_4 R_9 - V(1-\alpha)\beta R_{12}$

Table 71.3 Transformation paths and rate equations

Transformation paths	rate equations
Adsorption of trace organic contaminants to soil [μg/(g-soil·hr)]	$R_1 = \omega_t(a_t C_t^{bt} - q_t)$
Biological degradation of trace organic contaminants by heterotrophic microbes [μg/(L·hr)]	$R_2 = i_t \cdot C_t \cdot \frac{(1-\alpha)\beta}{\alpha} \cdot X_H$
Adsorption of organic compounds to soil [mgC/(g-soil·hr)]	$R_3 = \omega_s(a_s C_s^{bs} - q_s)$
Biological degradation of trace organic compounds by heterotrophic microbes [μg/(L·hr)]	$R_4 = k_4 \cdot \frac{C_s}{K_{Cs}+C_s} \cdot \frac{C_p}{K_{Cp}+C_p} \cdot \frac{C_{NH}}{K_{CNH1}+C_{NH}} \cdot \frac{(1-\alpha)\beta}{\alpha} \cdot X_H$
Denitrification [mgC/(L·hr)]	$R_5 = k_5 \cdot \frac{C_s}{K_{Cs}+C_s} \cdot \frac{C_p}{K_{Cp}+C_p} \cdot \frac{C_{NO}}{K_{CNO1}+C_{NO}} \cdot \frac{(1-\alpha)\beta}{\alpha} \cdot X_H$
Self-degradation of heterotrophic microbes [mgC/(g-soil·hr)]	$R_{11} = k_{11} \cdot X_H$
Self-degradation of autotrophic microbes [mgC/(g-soil·hr)]	$R_{12} = k_{12} \cdot X_A$

Table 71.4 Constants and coefficients

Constants and coefficients	Symbol	unit
Maximum removal rate constant of organic compounds	k_4	hr^{-1}
Maximum removal rate constant of organic compounds at denitrification	k_5	hr^{-1}
Self-degradation rate constant of heterotrophic microbes	k_{11}	hr^{-1}
Self-degradation rate constant of heterotrophic microbes	k_{12}	hr^{-1}
Half-saturation constant of organic compounds	K_{Cs}	mgC/L
Half-saturation constant of ammonium nitrogen	K_{CNH}	mgN/L
Half-saturation constant of oxidized nitrogen	K_{CNO}	mgN/L
Half-saturation constant of organic nitrogen	K_{CN}	mgN/L
Yield	Y_2, Y_3	—
Flow rate	Q	L/hr
Volume of a section of SAT	V	L
Biological removal rate constant of C_t	i_t	L/(mgC · hr)
Adsorption rate constant	ω	hr^{-1}
■voidage	α	—
specific gravity of soil	β	g-soil/L
Adsorption coefficients of Freundlich isotherm	a	$(\mu g/g\text{-soil})(L/\mu g)^b$
Adsorption coefficients of Freundlich isotherm	b	—

SAT pilot-scale reactor

A pilot-scale reactors were set in a wastewater treatment plant. It was cuboidal, and the width, depth and height of the reactors were 150 cm, 150 cm and 300 cm, respectively. The sand was packed in the reactors. The sand height of 237 cm, and the reactor had 17 cm vadose zone and 220 cm saturated zone. The influent water to the reactors was the A2O effluent. The hydraulic retention time (HRT) was set at 30 days, and the reactor operations started on 20 October 2011. The reactor is called "Sand 30" in this study. The water samples, including A2O effluent and SAT effluent, were collected from 22 October, 2012 to 28 January, 2013 for EDC measurement and from 22 November 2012 to 23 January, 2013 for EDTA and 1,4-dioxane measurement.

Batch experiments for biodegradation

7 sets of glass vial, which included 5 g-wet of the sand, which was collected from the surface of the pilot-scale reactor, and 40 mL of sterile A2O effluent with the spiked target chemicals were mixed for 48 h, and the vials were serially collected during the period. After the centrifuged separation with 1500 rpm for 20 min, the water sample was obtained by the filtration with GF/B, and the concentrations of the taget chemicals were measured. The same experimental sets in which the sand was also sterilized were used as control. Only adsorption was expected to occur, and biodegradation can be evaluated by the comparison of removal efficiencies between the experimental cases.

Batch experiments and continuous experiments for adsorption

5 sets of glass vial, which included 5 g-wet of the sand, which was collected from the surface of the pilot-scale reactor and sterilized by autoclave treatment, and 40 mL of sterile A2O effluent with the spiked target chemical were mixed. The vial was shaken for 48 h, and the vials were serially collected during the period. After the centrifuged separation with 1500 rpm for 20 min, the water sample was obtained by the filtration with GF/B, and the concentrations of the target chemicals were measured.

A glass column, whose inner diameter and height were 4.6 cm and 50 cm, respectively, was set, and the sand, which was not received the A2O effluent, was packed with the height of 25 cm. The A2O treated effluent with the spiked target chemicals and 2 mM NaN_3 was continuously feed to the column with the water height was 30 cm. The effluent from the column was continuously collected. After the centrifuged separation and filtration by GF/B, the concentration in the water was measured.

Analytical methods

All water samples were filtrated with GF/B (Whatman, U.S.). EDCs was measured by GC (6890series, Agilent, U.S.)/MS (5973 network, Agilent, U.S.) after solid-phase extraction and trimethylsilylation (Takabe *et al.*, 2010). EDTA was measured by HPLC (Shimadzu LC-10ATvp, Japan) after derivatization (Kemmei *et al.*, 2007). 1,4-dioxane was measured by the GC/MS after solid-phase extraction (Nishimura et al., 2010).

RESULTS AND DISCUSSION

Removal of the target chemicals in SAT

The concentrations of the chemicals in the A2O effluent and SAT effluent were shown in Fig. 71.1. BPA, represented as EDCs, was effectively removed by SAT, and over 89% removal ratio was obtained.

The median concentrations of EDTA in A2O effluent for the columns and reactor were 180 and 180 μg/L, respectively. Comparing the Drinking water quality standards in Japan, the concentrations exceed 0.1 times of the standard; therefore, this study also suggested the need of EDTA monitoring when the effluent is reused. The median value in Sand 3.5, Sand 7, Sand-sat and Sand 30 was 140, 75, 83 and 59, respectively. The values were decreased with the increase of HRT. There were little differences in the concentrations between Sand 7 and Sand-sat, and it was found that there were little influences on the EDTA removal by the existence of the vadoze zone.

The median concentrations of 1,4-dioxane in A2O effluent for the column, Sand 3.5, Sand 7 and Sand-sat were 770, 570, 590 and 530 ng/L, respectively. The concentrations in the SAT effluent was slightly lower than the A2O effluent; however increase of the HRT had no effects on the removal. In addition, There were little differences in the concentrations between Sand 7 and Sand-sat. The median concentrations of A2O effluent for the reactor and Sand 30 were 830 and 840 ng/L, respectively, and the removal was not observed. Based on the results, the removal of 1,4-dioxane by SAT was not so expected.

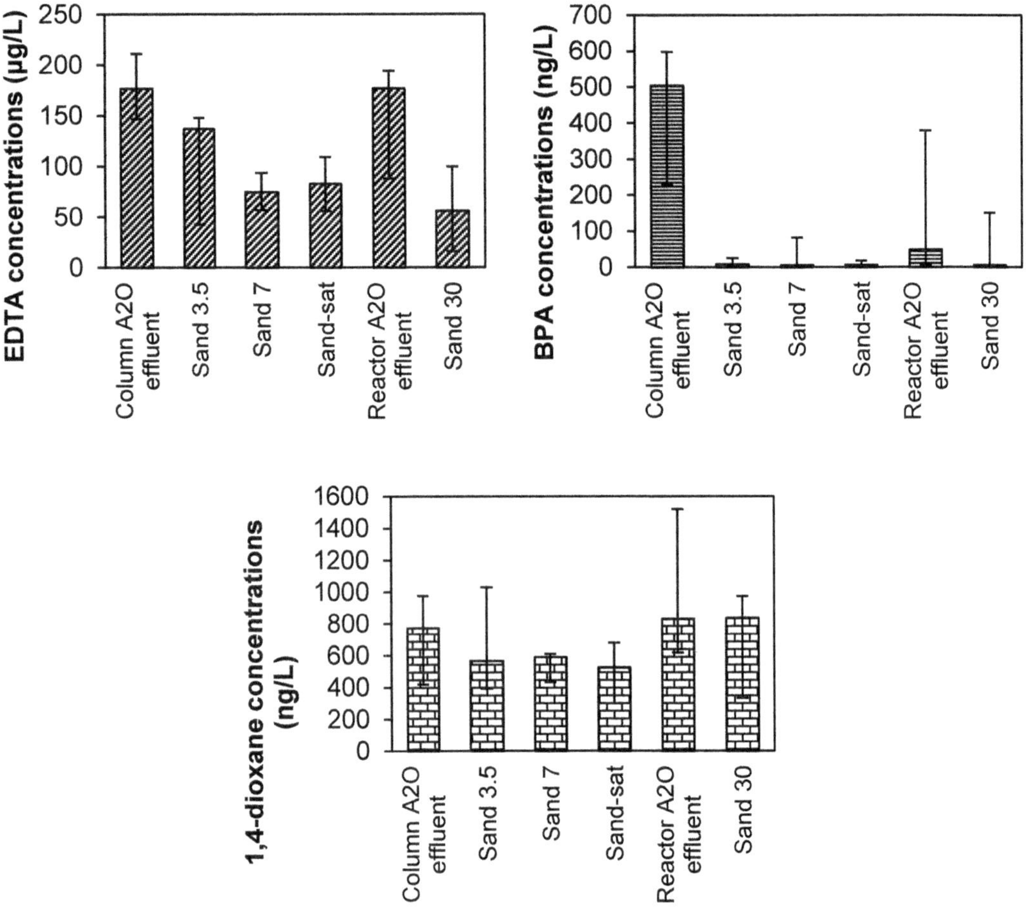

Figure 71.1 Concentrations of the chemicals in the A2O effluent and SAT effluent.

Experiments for investigation of adsorption

Adsorption isotherms of E1 and EDTA were shown in Fig. 71.2 and 71.3, respectively. Adsorptions obey the Freundlich equation of isotherm. Amount of adsorption of E2 was relatively large, whereas that of EDTA was relatively small. Adsorption coefficients of Freundlich isotherm of the trace organic compounds are summarized in Table 71.5. SAT can do good adsorption performance for E1 and BPA removal potentially. On the other hand, the adsorption rate is also important for the removal of these contaminants. The adsorption rate can be expressed with the equation listed in Table 71.3, and adsorption rate constant is one of the evaluation indexes for adsorption rate which is also important for treatment. The calculated adsorption rate constant from the experimental results are listed in Table 71.6. BPA had the largest value of the rate constant of the 5 contaminants, whereas that of EDTA was the smallest. EDCs can be adsorbed to the sand relatively well, and SAT can be useful for the treatment of these trace organic contaminants.

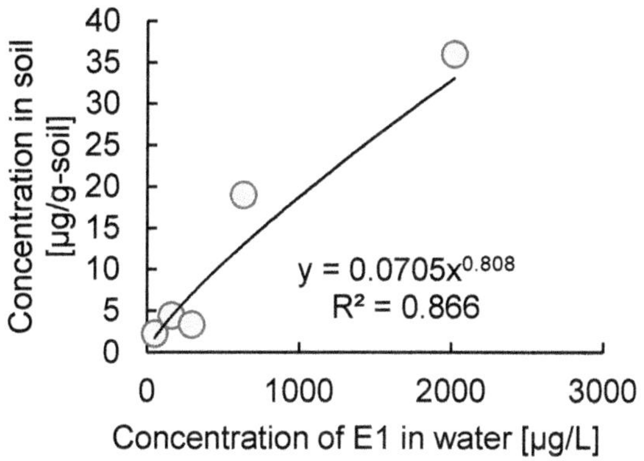

Figure 71.2 Adsorption isotherm of E1

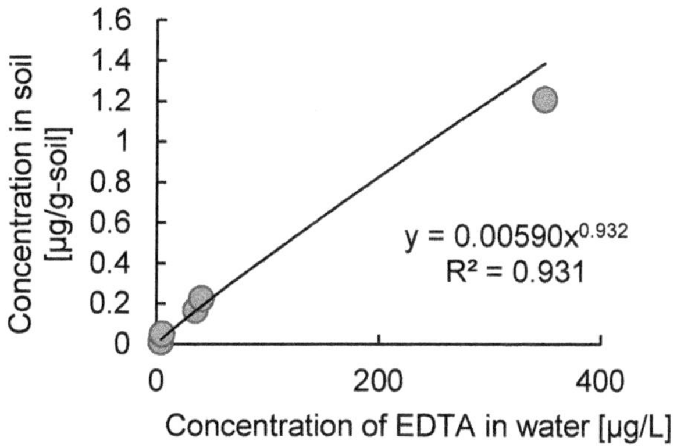

Figure 71.3 Adsorption isotherm of EDTA.

Table 71.5 Adsorption coefficients of Freundlich isotherm

Trace organic compounds	a [(μg/g-soil)(L/μg)^b]	b [-]
BPA	0.0105	0.705
E1	0.0705	0.808
E2	0.00247	0.625
EDTA	0.00590	0.932
1,4-dioxane	0.00290	0.924

Table 71.6 Adsorption rate constant

Trace organic compounds	ω [hr^{-1}]
BPA	0.238
E1	0.0261
E2	0.226
EDTA	0.00244
1,4-dioxane	0.201

Experiments for investigation of biological degradation

One of the main, and the ultimate removal mechanism of the trace contaminants is biological degradation in SAT. The results of biodegradation of E2 and EDTA are shown in Fig. 71.4 and 71.5, respectively. E2 was removed in the case without sand sterilization effectively, whereas the removal was not clearly observed in the case with sand sterilization. It is understood that E2 is easily degraded biologically in SAT. Similar results were obtained in BPA and E1 removal. On the other hand, EDTA removal patterns in both cases with and without sand sterilization were almost the same, which means biological degradation did not occur. Similar results to EDTA case were obtained in 1,4-dioxane. The concentration of microorganisms estimated by OUR measurement was 0.712 mgC-microorganisms/g-sand. According to the material balance equations shown in Table 71.2, the biological degradation constant of these trace contaminants were calculated. The results were summarized in Table 71.7. E2 is the easiest biodegradable contaminants in SAT, and E1 and BPA follow in the order under aerobic condition. It is clear that biological degradations of both EDTA and 1,4-dioxane were not expected in SAT. Considering the performance of adsorption and biodegradation, EDCs can be treated effectively by SAT, whereas 1,4-dioxane cannot be treated from the results of both treatment principle. Other treatment process such as advanced oxidation process(AOP) will be required for effective removal of 1,4-dioxane.

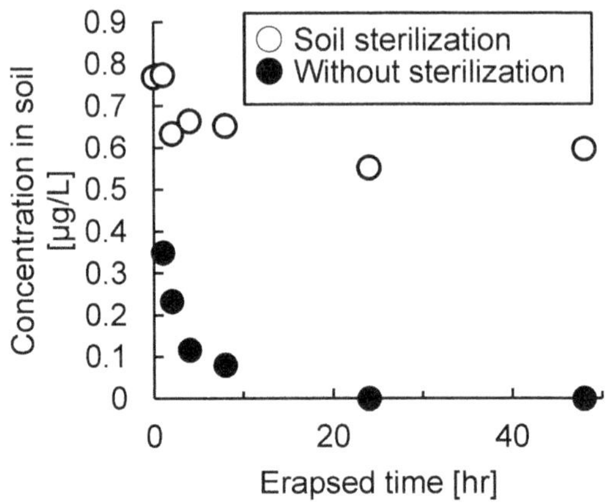

Figure 71.4 Biological degradation of E2

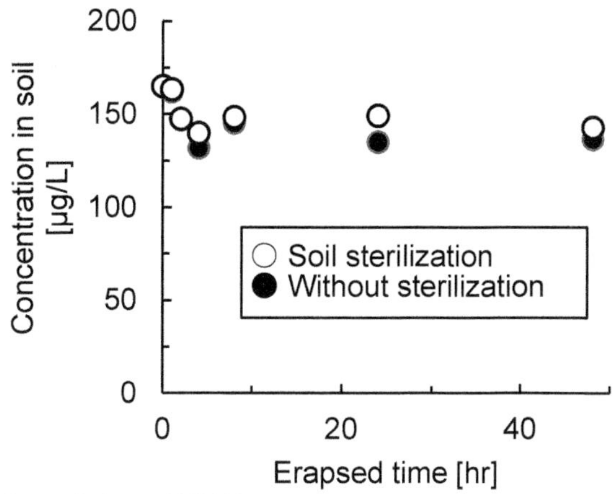

Figure 71.5 Biological degradation of EDTA

Table 71.7 Biological removal rate constant

Trace organic compounds	Biological removal rate constant [L/(mgC · hr)]
BPA	1.98×10^{-4}
E1	6.52×10^{-4}
E2	1.97×10^{-3}

CONCLUSIONS

In this study, fate of endocrine disrupting chemicals, Ethylenediaminetetraacetic acid (EDTA) and 1,4-dioxane, whose risks were concerned in the case when treated wastewater is reused with for potable purpose, in soil aquifer treatment (SAT) was evaluated with various experiments. Using SAT columns and pilot scale reactors, which continuously received treated effluent, it was cleared that bisphenol-A was effectively removed; meanwhile, the removal of 1,4-dioxane by SAT was not so expected. Removal of EDTA was expected; however, the removal was limited (66% of removal ratio with 30 days of Hydraulic retention time)

According to the mathematical model which expresses major reactions of contaminants in SAT, i.e. adsorption to soil and biodegradation, kinetic constants and coefficients of the contaminants were determined and evaluated for behaviors of the contaminants. In order to made clear the mechanism of treatment and removal of these trace contaminants in SAT, adsorption and biodegradation characteristics were also investigated and evaluated quantitatively from the points of reaction kinetics. It is understood that BPA, E1, and E2 can be removed easily by biological reaction, whereas EDTA and 1,4 dioxane can not be degraded biologically. On the other hand EDTA can be adsorbed to the soil. These findings corroborate the results of SAT columns experiment, and can be used for mathematical models which express and predict behaviors of the chemicals in SAT.

ACKNOWLEDGEMENTS

We thank the Core Research for Evolutionary Science and Technology, Japan Science and Technology Agency (CREST/JST) for financial support of this work and Kyoto City Waterworks Bureau for support of the experiments.

REFERENCES

F. Nishimura, K. Okuda, E. Hasegawa, H. Tsuno (2010) Removal of 1,4-dioxane from landfill leachate by O_3 and O_3/H_2O_2 process. Environmental Engineering Research, Vol.47, pp.497-505.

J.E. Drewes, M. Reinhard, P. Fox (2003) Comparing microfiltration-reverse osmosis and soil-aquifer treatment for indirect potable reuse of water, Water Research, Vol.37 (15), pp.3612-3621.

H.H. Yoo, J.H. Miller, K. Lansey, M. Reinhard (2006) EDTA, NTA, Alkylphenol Ethoxylate and DOC attenuation during soil aquifer treatment, Journal of environmental engineering, Vol.132 (6), pp.674-682.

H.P. Jansen, M.K. Stenstrom, J. de Koning (2007) Development of indirect potable reuse in impacted areas of the United States, Water Science and Technology, Vol.55, pp.357-366.,

Ministry of Health, Labour and Welfare, Japan: http://www.mhlw.go.jp/english/policy/health/water_supply/4.html

Ministry of Land, Infrastructre, Transport and Tourism, Japan (2001) Survey of endocrine dispurupting chemicals.

T. Kemmei, S. Kodama, A. Yamamoto, Y. Inoue, K. Hayakawa (2007) Determination of sequestering agents in cosmetics and synthetic detergents by high-performance liquid chromatography with ultraviolet detection, Journal of Chromatography A, Vol.1171 (1-2), pp.63-68.

Y. Suzuki, T. Igarashi, M. Usami, K. Yasuda, K. Yaguchi (2005) Monitoring of 1,4-Dioxane in groundwater and river wate rat the Tama District in Tokyo, Vol.28 (2), pp.139-143.

Y. Suzuki, Y. Komori, Y. Okayasu, T. Kitamura (2009) Fate of chemicals in sewage system, FY 2008 Aunnual Report of Wastewater Management and Water Quality Control, pp.101-114.

Y. Takabe, H. Tsuno, F. Nishimura, M. Sakamoto, Y. Terasoma (2010) Removal characteristics of natural estrogens in biological treatment process, Environmental Engineering Research, Vol.47, pp.707-717.

Author Index

Y

Z

Keyword Index